Diesel-Engine Management

Robert Bosch GmbH

Robert Bosch GmbH

Diesel-Engine Management

4th
Edition,
completely revised and extended

Published by:
© Robert Bosch GmbH, 2005
Postfach 1129,
D-73201 Plochingen
Automotive Aftermarket
Business Unit Diagnostics
Marketing – Test Equipment
(AA-DG/MKT5)

2nd edition updated and expanded, 1998
3rd edition, completely revised and extended,
January 2004
4th edition, completely revised and extended,
October 2005

All rights reserved

John Wiley & Sons Ltd,
The Atrium,
Southern Gate, Chichester,
West Sussex PO19 8SQ, England
Telephone (+44) 1243 779777
Internet: www.wiley.com or www.wileyeurope.com
Email: cs-books@wiley.co.uk

ISBN 0-470-02689-8

"My engine is still making enormous progress..."
This quotation from Rudolf Diesel in 1895 still holds true today. Ever since the launch of the first in-line fuel-injection pumps in 1927, this progress has been significantly shaped by Bosch. Above all, improvements in diesel fuel-injection technology have made possible favorable torque curves and high performance figures with simultaneously reduced fuel consumption. The introduction of the unit-injector and common-rail high-pressure fuel-injection systems represented another milestone in diesel fuel-injection technology and has made a significant contribution to the diesel boom in passenger cars in Western Europe. Bosch's latest development, which makes diesel engines even quieter, more economical, cleaner and even more powerful, is the piezo-inline injector for the common-rail system.

Modern diesel engines, in conjunction with Bosch diesel fuel-injection systems, create the conditions for improved fuel combustion. This results in fewer emissions: however, future exhaust-gas emissions standards – at least in larger-sized passenger cars – can no longer be met without additional exhaust-gas treatment systems. The particulate filter, with which it is possible to drastically reduce particulate emissions, is gaining in importance.

These developments in diesel technology are now described in detail in this completely revised and updated 4th Edition of the "Diesel-Engine Management" reference book.
The major part of this book concentrates on detailed descriptions of fuel-injection systems and their components. The mechanical governors and control systems for in-line fuel-injection pumps have been omitted to make way for new subjects. The descriptions of these governors can now be found in the "Diesel In-line Fuel-Injection Pump" publication from the Bosch "Expert Know-How on Automotive Technology" Series.

Compared with the 3rd Edition, the following subjects are new or have been updated and extended:
● History of the diesel engine
● Common-rail system
● Minimizing emissions inside the engine
● Exhaust-gas treatment systems
● Electronic Diesel Control (EDC)
● Start-assist systems
● Diagnostics (On-Board Diagnosis)

With these extensions and revisions, the 4th Edition of the "Diesel-Engine Management" reference book gives the reader a comprehensive insight into today's diesel fuel-injection technology.

The Editorial Team

Authors

Areas of use for diesel engines
Dipl.-Ing. Joachim Lackner,
Dr.-Ing. Herbert Schumacher,
Dipl.-Ing. (FH) Hermann Gries-
haber.

Basic principles of the diesel engine
Dr.-Ing. Thorsten Raatz,
Dipl.-Ing. (FH) Hermann Gries-
haber.

Fuels
Dr. rer. nat. Jörg Ullmann.

Cylinder-charge control systems
Dr.-Ing. Thomas Wintrich,
Dipl.-Betriebsw. Meike Keller.

Basic principles of diesel fuel injection
Dipl.-Ing. Jens Olaf Stein,
Dipl.-Ing. (FH) Hermann Gries-
haber.

Fuel supply
Dipl.-Ing. (FH) Rolf Ebert,
Dipl.-Betriebsw. Meike Keller,
Ing. grad. Peter Schelhas,
Dipl.-Ing. Klaus Ortner,
Dr.-Ing. Ulrich Projahn.

In-line fuel-injection pumps and their governors
Henri Bruognolo,
Dr.-Ing. Ernst Ritter.

Helix- and port-controlled distributor injection pumps and their add-on modules
Dipl.-Ing. (FH) Helmut Simon.

Solenoid-valve controlled distributor injection pumps
Dipl.-Ing. Johannes Feger,
Dr. rer. nat. Dietmar Ottenbacher.

Single-plunger fuel-injection pumps
Dr. techn. Theodor Stipek.

Unit injector systems and unit pump systems
Dipl.-Ing. Roger Potschin,
Dipl.-Ing. (HU) Carlos Alvarez-
Avila,
Dr.-Ing. Ulrich Projahn,
Dipl.-Ing. Nestor Rodriguez-
Amaya.

Common-rail system
Dipl.-Ing. Felix Landhäußer,
Dr.-Ing. Ulrich Projahn,
Dipl.-Ing. Thilo Klam,
Dipl.-Ing. (FH) Andreas Rettich,
Dr. techn. David Holzer,
Dipl.-Ing. (FH) Andreas Koch,
Dr.-Ing. Patrick Mattes,
Dipl.-Ing. Werner Brühmann,
Dipl.-Ing. Sandro Soccol,
Ing. Herbert Strahberger,
Ing. Helmut Sattmann.

Injection nozzles and nozzle holders
Dipl.-Ing. Thomas Kügler.

High-pressure lines
Kurt Sprenger.

Start-assist systems
Dr. rer. nat. Wolfgang Dreßler.

Electronic Diesel Control
Dipl.-Ing. Felix Landhäußer,
Dr.-Ing. Andreas Michalske,
Dipl.-Ing. (FH) Mikel Lorente
Susaeta,
Dipl.-Ing. Martin Grosser,
Dipl.-Inform. Michael Heinzel-
mann,
Dipl.-Ing. Johannes Feger,
Dipl.-Ing. Lutz-Martin Fink,
Dipl.-Ing. Wolfram Gerwing,
Dipl.-Ing. (BA) Klaus Grabmaier,
Dipl.-Math. techn. Bernd Illg,
Dipl.-Ing. (FH) Joachim Kurz,
Dipl.-Ing. Rainer Mayer,
Dr. rer. nat. Dietmar Ottenbacher,
Dipl.-Ing. (FH) Andreas Werner,
Dipl.-Ing. Jens Wiesner,
Dr. Ing. Michael Walther.

Sensors
Dipl.-Ing. Joachim Berger,

Diagnosis
Dr.-Ing. Günter Driedger,
Dr. rer. nat. Walter Lehle,
Dipl.-Ing. Wolfgang Schauer.

Service technology
Dipl.-Wirtsch.-Ing. Stephan Sohnle,
Dipl.-Ing. Rainer Rehage,
Rainer Heinzmann,
Rolf Wörner,
Günter Mauderer,
Hans Binder.

Minimizing emissions inside the engine
Dipl.-Ing. Jens Olaf Stein.

Exhaust-gas treatment systems
Dr. rer. nat. Norbert Breuer,
Priv.-Doz. Dr.-Ing. Johannes
K. Schaller,
Dr. rer. nat. Thomas Hauber,
Dr.-Ing. Ralf Wirth,
Dipl.-Ing. Stefan Stein.

Emissions-control legislation
Dr.-Ing. Stefan Becher,
Dr.-Ing. Torsten Eggert.

Exhaust-gas measuring techniques
Dipl.-Ing. Andreas Kreh,
Dipl.-Ing. Bernd Hinner,
Dipl.-Ing. Rainer Pelka.

and the editorial team in
cooperation with the responsible
technical departments of
Robert Bosch GmbH.

Unless otherwise stated,
the authors are employees of
Robert Bosch GmbH, Stuttgart.

History of the diesel engine

As early as 1863, the Frenchman Etienne Lenoir had test-driven a vehicle which was powered by a gas engine which he had developed. However, this drive plant proved to be unsuitable for installing in and driving vehicles. It was not until Nikolaus August Otto's four-stroke engine with magneto ignition that operation with liquid fuel and thereby mobile application were made possible. But the efficiency of these engines was low. Rudolf Diesel's achievement was to theoretically develop an engine with comparatively much higher efficiency and to pursue his idea through to readiness for series production.

In 1897, in cooperation with Maschinenfabrik Augsburg-Nürnberg (MAN), Rudolf Diesel built the first working prototype of a combustion engine to be run on inexpensive heavy fuel oil. However, this first diesel engine weighed approximately 4.5 tonnes and was three meters high. For this reason, this engine was not yet considered for use in land vehicles.

"It is my firm conviction that the automobile engine will come, and then I will consider my life's work complete." (Quotation by Rudolf Diesel shortly before his death)

However, with further improvements in fuel injection and mixture formation, Diesel's invention soon caught on and there were no longer any viable alternatives for marine and fixed-installation engines.

2   Rudolf Diesel

1   Patent document for the diesel engine and its first design from 1894

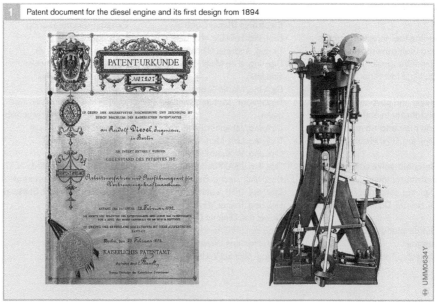

Rudolf Diesel

Rudolf Diesel (1858–1913), born in Paris, decided at 14 that he wanted to become an engineer. He passed his final examinations at Munich Polytechnic with the best grades achieved up to that point.

Idea for a new engine

Diesel's idea was to design an engine with significantly greater efficiency than the steam engine, which was popular at the time. An engine based on the isothermal cycle should, according to the theory of the French physicist Sadi Carnot, be able to be operated with a high level of efficiency of over 90%.

Diesel developed his engine initially on paper, based on Carnot's models. His aim was to design a powerful engine with comparatively small dimensions. Diesel was absolutely convinced by the function and power of his engine.

Diesel's patent

Diesel completed his theoretical studies in 1890 and on 27 February 1892 applied to the Imperial Patent Office in Berlin for a patent on "New rational thermal engines". On 23 February 1893, he received patent document DRP 67207 entitled "Operating Process and Type of Construction for Combustion Engines", dated 28 February 1892.

This new engine initially only existed on paper. The accuracy of Diesel's calculations had been verified repeatedly, but the engine manufacturers remained skeptical about the engine's technical feasibility.

Realizing the engine

The companies experienced in engine building, such as Gasmotoren-Fabrik Deutz AG, shied away from the Diesel project. The required compression pressures of 250 bar were beyond what appeared to be technically feasible. In 1893, after many months of endeavor, Diesel finally succeeded in reaching an agreement to work with Maschinenfabrik Augsburg-Nürnberg (MAN). However, the agreement contained concessions by Diesel in respect of the ideal engine. The maximum pressure was reduced from 250 to 90 bar, and then later to 30 bar. This lowering of the pressure, required for mechanical reasons, naturally had a disadvantageous effect on combustibility. Diesel's initial plans to use coal dust as the fuel were rejected.

Finally, in Spring 1893, MAN began to build the first, uncooled test engine. Kerosene was initially envisaged as the fuel, but what came to be used was gasoline, because it was thought (erroneously) that this fuel would auto-ignite more easily. The principle of auto-ignition – i.e. injection of the fuel into the highly compressed and heated combustion air during compression – was confirmed in this engine.

In the second test engine, the fuel was not injected and atomized directly, but with the aid of compressed air. The engine was also provided with a water-cooling system.

It was not until the third test engine – a new design with a single-stage air pump for compressed-air injection – that the breakthrough made. On 17 February 1897, Professor Moritz Schröder of Munich Technical University carried out the acceptance tests. The test results confirmed what was then for a combustion engine a high level of efficiency of 26.2%.

Patent disputes and arguments with the Diesel consortium with regard to development strategy and failures took their toll, both mentally and physically, on the brilliant inventor. He is thought to have fallen overboard on a Channel crossing to England on 29 September 1913.

Mixture formation in the first diesel engines

Compressed-air injection

Rudolf Diesel did not have the opportunity to compress the fuel to the pressures required for spray dispersion, spray disintegration and droplet formation. The first diesel engine from 1897 therefore worked with compressed-air injection, whereby the fuel was introduced into the cylinder with the aid of compressed air. This process was later used by Daimler in its diesel engines for trucks.

The fuel injector had a port for the compressed-air feed (Fig. 1, 1) and a port for the fuel feed (2). A compressor generated the compressed air, which flowed into the valve. When the nozzle (3) was open, the air blasting into the combustion chamber also swept the fuel in and in this two-phase flow generated the fine droplets required for fast droplet vaporization and thus for auto-ignition.

A cam ensured that the nozzle was actuated in synchronization with the crankshaft. The amount of fuel to be injected as controlled by the fuel pressure. Since the injection pressure was generated by the compressed air, a low fuel pressure was sufficient to ensure the efficacy of the process.

The problem with this process was – on account of the low pressure at the nozzle – the low penetration depth of the air/fuel mixture into the combustion chamber. This type of mixture formation was therefore not suitable for higher injected fuel quantities (higher engine loads) and engine speeds. The limited spray dispersion prevented the amount of air utilization required to increase power and, with increasing injected fuel quantity, resulted in local over-enrichment with a drastic increase in the levels of smoke. Furthermore, the vaporization time of the relatively large fuel droplets did not permit any significant increase in engine speed. Another disadvantage of this engine was the enormous amount of space taken up by the compressor. Nevertheless, this principle was used in trucks at that time.

Precombustion-chamber engine

The Benz diesel was a precombustion-chamber engine. Prosper L'Orange had already applied for a patent on this process in 1909. Thanks to the precombustion-chamber principle, it was possible to dispense with the complicated and expensive system of air

Fig. 1
1   Compressed-air feed
2   Fuel feed
3   Nozzle

Fig. 2
(Picture source: DaimlerChrysler)
1   Fuel valve
2   Glow filament for heating precombustion chamber
3   Precombustion chamber
4   Ignition insert

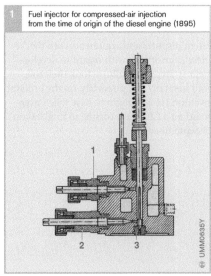

1   Fuel injector for compressed-air injection from the time of origin of the diesel engine (1895)

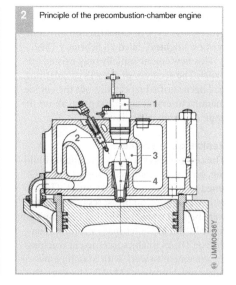

2   Principle of the precombustion-chamber engine

injection. Mixture formation in the main combustion chamber of this process, which is still used to this day, is ensured by partial combustion in the precombustion chamber. The precombustion-chamber engine has a specially shaped combustion chamber with a hemispherical head. The precombustion chamber and combustion chamber are interconnected by small bores. The volume of the precombustion chamber is roughly one fifth of the compression chamber.

The entire quantity of fuel is injected at approximately 230 to 250 bar into the precombustion chamber. Because of the limited amount of air in the precombustion chamber, only a small amount of the fuel is able to combust. As a result of the pressure increase in the precombustion chamber caused by the partial combustion, the unburned or partially cracked fuel is forced into the main combustion chamber, where it mixes with the air in the main combustion chamber, ignites and burns.

The function of the precombustion chamber here is to form the mixture. This process – also known as indirect injection – finally caught on and remained the predominant process until developments in fuel injection were able to deliver the injection pressures required to form the mixture in the main combustion chamber.

Direct injection

The first MAN diesel engine operated with direct injection, whereby the fuel was forced directly into the combustion chamber via a nozzle. This engine used as its fuel a very light oil, which was injected by a compressor into the combustion chamber. The compressor determined the huge dimensions of the engine.

In the commercial-vehicle sector, direct-injection engines resurfaced in the 1960s and gradually superseded precombustion-chamber engines. Passenger cars continued to use precombustion-chamber engines because of their lower combustion-noise levels until the 1990s, when they were swiftly superseded by direct-injection engines.

Use of the first vehicle diesel engines

Diesel engines in commercial vehicles

Because of their high cylinder pressures, the first diesel engines were large and heavy and therefore wholly unsuitable for mobile applications in vehicles. It was not until the beginning of the 1920s that the first diesel engines were able to be deployed in commercial vehicles.

Uninterrupted by the First World War, Prosper L'Orange – a member of the executive board of Benz & Cie – continued his development work on the diesel engine. In 1923 the first diesel engines for road vehicles were installed in five-tonne trucks. These four-cylinder precombustion-chamber engines with a piston displacement of 8.8 l delivered 45...50 bhp. The first test drive of the Benz truck took place on 10 September with brown-coal tar oil serving as the fuel. Fuel consumption was 25% lower than benzene engines. Furthermore, operating fluids such as brown-coal tar oil cost much less than benzene, which was highly taxed.

The company Daimler was already involved in the development of the diesel engine prior

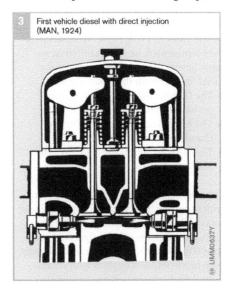

3 First vehicle diesel with direct injection (MAN, 1924)

4 The most powerful diesel truck in the world from 1926 from MAN with 150 bhp (110 kW) for a payload of 10 t

to the First World War. After the end of the war, the company was working on diesel engines for commercial vehicles. The first test drive was conducted on 23 August 1923 – at virtually the same time as the Benz truck. At the end of September 1923, a further test drive was conducted from the Daimler plant in Berlin to Stuttgart and back.

The first truck production models with diesel engines were exhibited at the Berlin Motor Show in 1924. Three manufacturers were represented, each with different systems, having driven development of the diesel forward with their own ideas:

- The Daimler diesel engine with compressed-air injection
- The Benz diesel with precombustion chamber
- The MAN diesel engine with direct injection

Diesel engines became increasingly powerful with time. The first types were four-cylinder units with a power output of 40 bhp. By 1928, engine power-output figures of more than 60 bhp were no longer unusual. Finally, even more powerful engines with six and eight cylinders were being produced for heavy commercial vehicles. By 1932, the power range stretched up to 140 bhp.

The diesel engine's breakthrough came in 1932 with a range of trucks offered by the company Daimler-Benz, which came into being in 1926 with the merger of the automobile manufacturers Daimler and Benz. This range was led by the Lo2000 model with a payload of 2 t and a permissible total weight of almost 5 t. It housed the OM59 four-cylinder engine with a displacement of 3.8 l and 55 bhp. The range extended up to the L5000 (payload 5 t, permissible total weight 10.8 t). All the vehicles were also available with gasoline engines of identical power output, but these engines proved unsuccessful when up against the economical diesel engines.

To this day, the diesel engine has maintained its dominant position in the commercial-vehicle sector on account of its economic efficiency. Virtually all heavy goods vehicles are driven by diesel engines. In Japan, large-displacement conventionally aspirated engines are used almost exclusively. In the USA and Europe, however, turbocharged engines with charge-air cooling are favored.

Diesel engines in passenger cars

A few more years were to pass before the diesel engine made its debut in a passenger car. 1936 was the year, when the Mercedes 260D appeared with a four-cylinder diesel engine and a power output of 45 bhp.

The diesel engine as the power plant for passenger cars was long relegated to a fringe existence. It was too sluggish when compared with the gasoline engine. Its image was to change only in the 1990s. With exhaust-gas turbocharging and new high-pressure fuel-injection systems, the diesel engine is now on an equal footing with its gasoline counterpart. Power output and environmental performance are comparable. Because the diesel engine, unlike its gasoline counterpart, does not knock, it can also be turbocharged in the lower speed range, which results in high torque and very good driving performance. Another advantage of the diesel engine is, naturally, its excellent efficiency. This has led to it becoming increasingly accepted among car drivers – in Europe, roughly every second newly registered car is a diesel.

Further areas of application

When the era of steam and sailing ships crossing the oceans came to an end at the beginning of the 20th century, the diesel engine also emerged as the drive source for this mode of transport. The first ship to be fitted with a 25-bhp diesel engine was launched in 1903. The first locomotive to be driven by a diesel engine started service in 1913. The engine power output in this case was 1,000 bhp. Even the pioneers of aviation showed interest in the diesel engine. Diesel engines provided the propulsion on board the Graf Zeppelin airship.

6 Bosch in-line fuel-injection pump on the engine of the Mercedes 260D

UMM0640Y

5 First diesel car: Mercedes-Benz 260D from 1936 with an engine power output of 45 bhp (33 kW) and a fuel consumption of 9.5 l/100 km

UMM0639Y

Bosch diesel fuel injection

1 Robert Bosch

Bosch's emergence onto the diesel-technology stage

In 1886, Robert Bosch (1861–1942) opened a "workshop for light and electrical engineering" in Stuttgart. He employed one other mechanic and an apprentice. At the beginning, his field of work lay in installing and repairing telephones, telegraphs, lightning conductors, and other light-engineering jobs.

The low-voltage magneto-ignition system developed by Bosch had provided reliable ignition in gasoline engines since 1897. This product was the launching board for the rapid expansion of Robert Bosch's business. The high-voltage magneto ignition system with spark plug followed in 1902. The armature of this ignition system is still to this day incorporated in the logo of Robert Bosch GmbH.

In 1922, Robert Bosch turned his attention to the diesel engine. He believed that certain accessory parts for these engines could similarly make suitable objects for Bosch high-volume precision production

like magnetos and spark plugs. The accessory parts in question for diesel engines were fuel-injection pumps and nozzles.

Even Rudolf Diesel had wanted to inject the fuel directly, but was unable to do this because the fuel-injection pumps and nozzles needed to achieve this were not available. These pumps, in contrast to the fuel pumps used in compressed-air injection, had to be suitable for back-pressure reactions of up to several hundred atmospheres. The nozzles had to have quite fine outlet openings because now the task fell upon the pump and the nozzle alone to meter and atomize the fuel.

The injection pumps which Bosch wanted to develop should match not only the requirements of all the heavy-oil low-power engines with direct fuel injection which existed at the time but also future motor-vehicle diesel engines. On 28 December 1922, the decision was taken to embark on this development.

Demands on the fuel-injection pumps

The fuel-injection pump to be developed should be capable of injecting even small amounts of fuel with only quite small differences in the individual pump elements. This would facilitate smoother and more uniform engine operation even at low idle speeds. For full-load requirements, the delivery quantity would have to be increased by a factor of four or five. The required injection pressures were at that time already over 100 bar. Bosch demanded that these pump properties be guaranteed over 2,000 operating hours.

These were exacting demands for the then state-of-the-art technology. Not only did some feats of fluid engineering have to be achieved, but also this requirement represented a challenge in terms of production engineering and materials application technology.

Development of the fuel-injection pump
Firstly, different pump designs were tried out. Some pumps were spool-controlled, while others were valve-controlled. The injected fuel quantity was regulated by altering the plunger lift. By the end of 1924, a pump design was available which, in terms of its delivery rate, its durability and its low space requirement, satisfied the demands both of the Benz precombustion-chamber engine presented at the Berlin Motor Show and of the MAN direct-injection engine.

In March 1925, Bosch concluded contracts with Acro AG to utilize the Acro patents on a diesel-engine system with air chamber and the associated injection pump and nozzle. The Acro pump, developed by Franz Lang in Munich, was a unique fuel-injection pump. It had a special valve spool with helix, which was rotated to regulate the delivery quantity. Lang later moved this helix to the pump plunger.

The delivery properties of the Acro injection pump did not match what Bosch's own test pumps had offered. However, with the Acro engine, Bosch wanted to come into contact with a diesel engine which was particularly suitable for small cylinder units and high speeds and in this way gain a firm foothold for developing injection pumps and nozzles. At the same time, Bosch was led by the idea of granting licenses in the Acro patents to engine factories to promote the spread of the vehicle diesel engine and thereby contribute to the motorization of traffic.

After Lang's departure from the company in October 1926, the focus of activity at Bosch was again directed toward pump development. The first Bosch diesel fuel-injection pump ready for series production appeared soon afterwards.

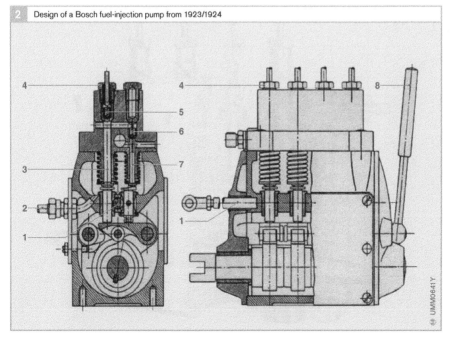

2   Design of a Bosch fuel-injection pump from 1923/1924

Fig. 2
1   Control rack
2   Inlet port
3   Pump plunger
4   Pressure-line port
5   Delivery valve
6   Suction valve
7   Valve tappet
8   Shutdown and
    pumping lever

Bosch diesel fuel-injection pump ready for series production

In accordance with the design engineer's plans of 1925 and like the modified Acro pump, the Bosch fuel-injection pump featured a diagonal helix on the pump plunger. Apart from this, however, it differed significantly from all its predecessors.

The external lever apparatus of the Acro pump for rotating the pump plunger was replaced by the toothed control rack, which engaged in pinions on control sleeves of the pump elements.

In order to relieve the load on the pressure line at the end of the injection process and to prevent fuel dribble, the delivery valve was provided with a suction plunger adjusted to fit in the valve guide. In contrast to the load-relieving techniques previously used, this new approach achieved increased steadiness of delivery at different speeds and quantity settings and significantly simplified and shortened the adjustment of multicylinder pumps to identical delivery by all elements.

The pump's simple and clear design made it easier to assemble and test. It also significantly simplified the replacement of parts compared with earlier designs. The housing conformed first and foremost to the requirements of the foundry and other manufacturing processes. The first specimens of this Bosch fuel-injection pump really suitable for volume production were manufactured in April 1927. Release for production in greater batch quantities and in versions for two-, four- and six-cylinder engines was granted on 30 November 1927 after the specimens had passed stringent tests at Bosch and in practical operation with flying colors.

3   First series-production diesel fuel-injection pump from Bosch (1927)

Fig. 3

1   Camshaft
2   Roller tappet
3   Control-sleeve gear
4   Control rack
5   Inlet port
6   Pump cylinder
7   Control sleeve
8   Pressure-line port
9   Delivery valve with plunger
10 Oil level
11 Pump plunger

Nozzles and nozzle holders

The development of nozzles and nozzle holders was conducted in parallel to pump development. Initially, pintle nozzles were used for precombustion-chamber engines. Hole-type nozzles were added at the start of 1929 with the introduction of the Bosch pump in the direct-injection diesel engine.

The nozzles and nozzle holders were always adapted in terms of their size to the new pump sizes. It was not long before the engine manufacturers also wanted a nozzle holder and nozzle which could be screwed into the cylinder head in the same way as the spark plug on a gasoline engine. Bosch adapted to this request and started to produce screw-in nozzle holders.

Governor for the fuel-injection pump

Because a diesel engine is not self-governing like a gasoline engine, but needs a governor to maintain a specific speed and to provide protection against overspeed accompanied by self-destruction, vehicle diesel engines had to be equipped with such devices right from the start. The engine factories initially manufactured these governors themselves. However, the request soon came to dispense with the drive for the governor, which was without exception a mechanical governor, and to combine it with the injection pump. Bosch complied with this request in 1931 with the introduction of the Bosch governor.

Spread of Bosch diesel fuel-injection technology

By August 1928, one thousand Bosch fuel-injection pumps had already been supplied. When the upturn in the fortunes of the vehicle diesel engine began, Bosch was well prepared and fully able to serve the engine factories with a full range of fuel-injection equipment. When the Bosch pumps and nozzles proved their worth, most companies saw no further need to continue manufacturing their own accessories in this field.

Bosch's expertise in light engineering (e.g., in the manufacture of lubricating pumps) stood it in good stead in the development of diesel fuel-injection pumps. Its products could not be manufactured "in accordance with the pure principles of mechanical engineering". This helped Bosch to obtain a market advantage. Bosch had thus made a significant contribution towards enabling the diesel engine to develop into what it is today.

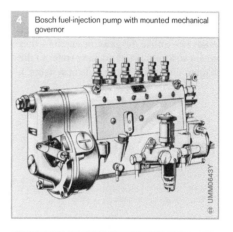

4 Bosch fuel-injection pump with mounted mechanical governor

5 Billboard advertisement for Bosch diesel fuel injection

Meisterwerke der Technik:
Diesel-Einspritzpumpen
und Einspritzdüsen
von
BOSCH

Areas of use for diesel engines

No other internal-combustion engine is as widely used as the diesel engine[1]. This is due primarily to its high degree of efficiency and the resulting fuel economy.

The chief areas of use for diesel engines are
- Fixed-installation engines
- Cars and light commercial vehicles
- Heavy goods vehicles
- Construction and agricultural machinery
- Railway locomotives and
- Ships

Diesel engines are produced as inline or V-configuration units. They are ideally suited to turbocharger or supercharger aspiration as – unlike the gasoline engine – they are not susceptible to knocking (refer to the chapter "Cylinder-charge control systems").

Suitability criteria

The following features and characteristics are significant for diesel-engine applications (examples):
- Engine power
- Specific power output
- Operational safety
- Production costs
- Economy of operation
- Reliability
- Environmental compatibility
- User-friendliness
- Convenience (e.g. engine-compartment design)

The relative importance of these characteristics affect engine design and vary according to the type of application.

Applications

Fixed-installation engines

Fixed-installation engines (e.g. for driving power generators) are often run at a fixed speed. Consequently, the engine and fuel-injection system can be optimized specifically

---

[1] Named after Rudolf Diesel (1858 to 1913) who first applied for a patent for his "New rational thermal engines" in 1892. A lot more development work was required, however, before the first functional diesel engine was produced at MAN in Augsburg in 1897.

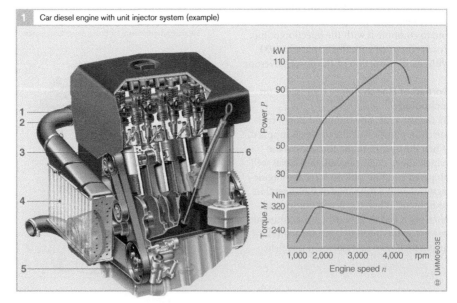

1   Car diesel engine with unit injector system (example)

Fig. 1
1   Valve gear
2   Injector
3   Piston with gudgeon
     pin and conrod
4   Intercooler
5   Coolant pump
6   Cylinder

for operation at that speed. An engine governor adjusts the quantity of fuel injected dependent on engine load. For this type of application, mechanically governed fuel-injection systems are still used.

Car and commercial-vehicle engines can also be used as fixed-installation engines. However, the engine-control system may have to be modified to suit the different conditions.

Cars and light commercial vehicles
Car engines (Fig. 1) in particular are expected to produce high torque and run smoothly. Great progress has been made in these areas by refinements in engine design and the development of new fuel-injection with Electronic Diesel Control (EDC). These advances have paved the way for substantial improvements in the power output and torque characteristics of diesel engines since the early 1990s. And as a result, the diesel engine has forced its way into the executive and luxury-car markets.

Cars use fast-running diesel engines capable of speeds up to 5,500 rpm. The range of sizes extends from 10-cylinder 5-liter units used in large saloons to 3-cylinder 800-cc models for small subcompacts.

In Europe, all new diesel engines are now Direct-Injection (DI) designs as they offer fuel consumption reductions of 15 to 20% in comparison with indirect-injection engines. Such engines, now almost exclusively fitted with turbochargers, offer considerably better torque characteristics than comparable gasoline engines. The maximum torque available to a vehicle is generally determined not by the engine but by the power-transmission system.

The ever more stringent emission limits imposed and continually increasing power demands require fuel-injection systems with extremely high injection pressures. Improving emission characteristics will continue to be a major challenge for diesel-engine developers in the future. Consequently, further innovations can be expected in the area of exhaust-gas treatment in years to come.

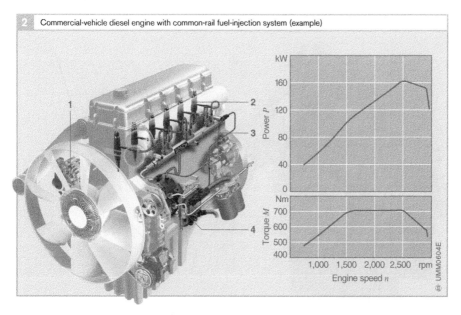

2   Commercial-vehicle diesel engine with common-rail fuel-injection system (example)

Fig. 2
1   Alternator
2   Injector
3   Fuel rail
4   High-pressure pump

UMM0604E

Heavy goods vehicles

The prime requirement for engines for heavy goods vehicles (Fig. 2) is economy. That is why diesel engines for this type of application are exclusively direct-injection (DI) designs. They are generally medium-fast engines that run at speeds of up to 3,500 rpm.

For large commercial vehicles too, the emission limits are continually being lowered. That means exacting demands on the fuel-injection system used and a need to develop new emission-control systems.

Construction and agricultural machinery

Construction and agricultural machinery is the traditional domain of the diesel engine. The design of engines for such applications places particular emphasis not only on economy but also on durability, reliability and ease of maintenance. Maximizing power utilization and minimizing noise output are less important considerations than they would be for car engines, for example. For this type of use, power outputs can range from around 3 kW to the equivalent of HGV engines.

Many engines used in construction-industry and agricultural machines still have mechanically governed fuel-injection systems. In contrast with all other areas of application, where water-cooled engines are the norm, the ruggedness and simplicity of the air-cooled engine remain important factors in the building and farming industries.

Railway locomotives

Locomotive engines, like heavy-duty marine diesel engines, are designed primarily with continuous-duty considerations in mind. In addition, they often have to cope with poorer quality diesel fuel. In terms of size, they range from the equivalent of a large truck engine to that of a medium-sized marine engine.

Ships

The demands placed on marine engines vary considerably according to the particular type of application. There are out-and-out high-performance engines for fast naval vessels or speedboats, for example. These tend to be 4-stroke medium-fast engines that run at speeds of 400...1,500 rpm and have up to 24 cylinders (Fig. 3). At the other end of

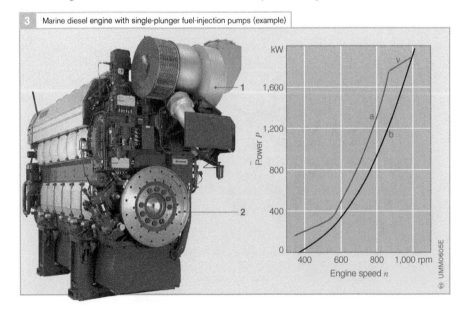

3   Marine diesel engine with single-plunger fuel-injection pumps (example)

Fig. 3
1   Turbocharger
2   Flywheel

a   Engine power
     output
b   Running-resistance
     curve
v   Full-load limitation
     zone

the scale there are 2-stroke heavy-duty engines designed for maximum economy in continuous duty. Such slow-running engines (< 300 rpm) achieve effective levels of efficiency of up to 55%, which represent the highest attainable with piston engines.

Large-scale engines are generally run on cheap heavy oil. This requires pretreatment of the fuel on board. Depending on quality, it has to be heated to temperatures as high as 160°C. Only then is its viscosity reduced to a level at which it can be filtered and pumped.

Smaller vessels often use engines originally intended for large commercial vehicles. In that way, an economical propulsion unit with low development costs can be produced. Once again, however, the engine management system has to be adapted to the different service profile.

Multi-fuel engines

For specialized applications (such as operation in regions with undeveloped infrastructures or for military use), diesel engines capable of running on a variety of different fuels including diesel, gasoline and others have been developed. At present they are of virtually no significance whatsoever within the overall picture, as they are incapable of meeting the current demands in respect of emissions and performance characteristics.

Engine characteristic data

Table 1 shows the most important comparison data for various types of diesel and gasoline engine.

The average pressure in petrol engines with direct fuel injection is around 10% higher than for the engines listed in the table with inlet-manifold injection. At the same time, the specific fuel consumption is up to 25% lower. The compression ratio of such engines can be as much as 13:1.

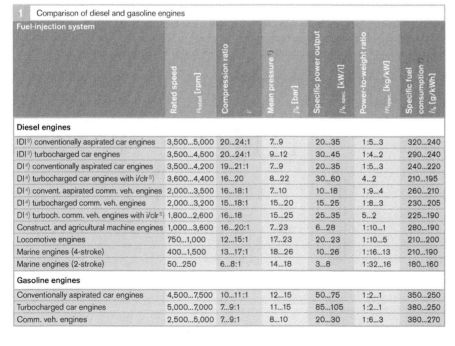

1   Comparison of diesel and gasoline engines

Fuel-injection system	Rated speed n_{rated} [rpm]	Compression ratio	Mean pressure [1] p_e [bar]	Specific power output $P_{e,spec}$ [kW/l]	Power-to-weight ratio m_{spec} [kg/kW]	Specific fuel consumption b_e [g/kWh]
Diesel engines						
IDI[2] conventionally aspirated car engines	3,500...5,000	20...24:1	7...9	20...35	1:5...3	320...240
IDI[3] turbocharged car engines	3,500...4,500	20...24:1	9...12	30...45	1:4...2	290...240
DI[4] conventionally aspirated car engines	3,500...4,200	19...21:1	7...9	20...35	1:5...3	240...220
DI[4] turbocharged car engines with i/clr[5]	3,600...4,400	16...20	8...22	30...60	4...2	210...195
DI[4] convent. aspirated comm. veh. engines	2,000...3,500	16...18:1	7...10	10...18	1:9...4	260...210
DI[4] turbocharged comm. veh. engines	2,000...3,200	15...18:1	15...20	15...25	1:8...3	230...205
DI[4] turboch. comm. veh. engines with i/clr[5]	1,800...2,600	16...18	15...25	25...35	5...2	225...190
Construct. and agricultural machine engines	1,000...3,600	16...20:1	7...23	6...28	1:10...1	280...190
Locomotive engines	750...1,000	12...15:1	17...23	20...23	1:10...5	210...200
Marine engines (4-stroke)	400...1,500	13...17:1	18...26	10...26	1:16...13	210...190
Marine engines (2-stroke)	50...250	6...8:1	14...18	3...8	1:32...16	180...160
Gasoline engines						
Conventionally aspirated car engines	4,500...7,500	10...11:1	12...15	50...75	1:2...1	350...250
Turbocharged car engines	5,000...7,000	7...9:1	11...15	85...105	1:2...1	380...250
Comm. veh. engines	2,500...5,000	7...9:1	8...10	20...30	1:6...3	380...270

Table 1

[1] The average pressure p_e can be used to calculate the specific torque M_{spec} [Nm]:

$$M_{spec.} = \frac{25}{\pi \cdot p_e}$$

[2] Best consumption
[3] Indirect Injection
[4] Direct Injection
[5] Intercooler

Basic principles of the diesel engine

The diesel engine is a compression-ignition engine in which the fuel and air are mixed inside the engine. The air required for combustion is highly compressed inside the combustion chamber. This generates high temperatures which are sufficient for the diesel fuel to spontaneously ignite when it is injected into the cylinder. The diesel engine thus uses heat to release the chemical energy contained within the diesel fuel and convert it into mechanical force.

The diesel engine is the internal-combustion engine that offers the greatest overall efficiency (more than 50% in the case of large, slow-running types). The associated low fuel consumption, its low-emission exhaust and quieter running characteristics assisted, for example, by pre-injection have combined to give the diesel engine its present significance.

Diesel engines are particularly suited to aspiration by means of a turbocharger or supercharger. This not only improves the engine's power yield and efficiency, it also reduces pollutant emissions and combustion noise.

In order to reduce NO_x emissions on cars and commercial vehicles, a proportion of the exhaust gas is fed back into the engine's intake manifold (exhaust-gas recirculation). An even greater reduction of NO_x emissions can be achieved by cooling the recirculated exhaust gas.

Diesel engines may operate either as two-stroke or four-stroke engines. The types used in motor vehicles are generally four-stroke designs.

Method of operation

A diesel engine contains one or more cylinders. Driven by the combustion of the air/fuel mixture, the piston (Fig. 1, 3) in each cylinder (5) performs up-and-down movements. This method of operation is why it was named the "reciprocating-piston engine".

The connecting rod, or conrod (11), converts the linear reciprocating action of the piston into rotational movement on the part of the crankshaft (14). A flywheel (15) connected to the end of the crankshaft helps to maintain continuous crankshaft rotation and reduce unevenness of rotation caused by the periodic nature of fuel combustion in the individual cylinders. The speed of rotation of the crankshaft is also referred to as engine speed.

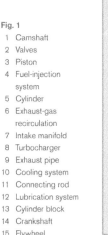

1   Four-cylinder diesel engine without auxiliary units (schematic)

Fig. 1
1 Camshaft
2 Valves
3 Piston
4 Fuel-injection system
5 Cylinder
6 Exhaust-gas recirculation
7 Intake manifold
8 Turbocharger
9 Exhaust pipe
10 Cooling system
11 Connecting rod
12 Lubrication system
13 Cylinder block
14 Crankshaft
15 Flywheel

SMM0660BY

2 Operating cycle of a four-stroke diesel engine

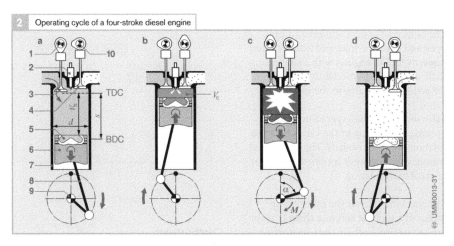

Fig. 2

a Induction stroke
b Compression stroke
c Ignition stroke
d Exhaust stroke

1 Inlet-valve camshaft
2 Fuel injector
3 Inlet valve
4 Exhaust valve
5 Combustion
   chamber
6 Piston
7 Cylinder wall
8 Connecting rod
9 Crankshaft
10 Exhaust-valve
   camshaft

α Crankshaft angle
   of rotation
d Bore
M Turning force
s Piston stroke
V_c Compression
   volume
V_h Swept volume

TDC Top dead center
BDC Bottom dead
   center

Four-stroke cycle

On a four-stroke diesel engine (Fig. 2), inlet and exhaust valves control the intake of air and expulsion of burned gases after combustion. They open and close the cylinder's inlet and exhaust ports. Each inlet and exhaust port may have one or two valves.

1. Induction stroke (a)

Starting from Top Dead Center (TDC), the piston (6) moves downwards increasing the capacity of the cylinder. At the same time the inlet valve (3) is opened and air is drawn into the cylinder without restriction by a throttle valve. When the piston reaches Bottom Dead Center (BDC), the cylinder capacity is at its greatest ($V_h + V_c$).

2. Compression stroke (b)

The inlet and exhaust valves are now closed. The piston moves upwards and compresses the air trapped inside the cylinder to the degree determined by the engine's compression ratio (this can vary from 6:1 in large-scale engines to 24:1 in car engines). In the process, the air heats up to temperatures as high as 900°C. When the compression stroke is almost complete, the fuel-injection system injects fuel at high pressure (as much as 2,000 bar in modern engines) into the hot, compressed air. When the piston reaches top dead center, the cylinder capacity is at its smallest (compression volume, V_c).

3. Ignition stroke (c)

After the ignition lag (a few degrees of crankshaft rotation) has elapsed, the ignition stroke (working cycle) begins. The finely atomized and easily combustible diesel fuel spontaneously ignites and burns due to the heat of the compressed air in the combustion chamber (5). As a result, the cylinder charge heats up even more and the pressure in the cylinder rises further as well. The amount of energy released by combustion is essentially determined by the mass of fuel injected (quality-based control). The pressure forces the piston downwards. The chemical energy released by combustion is thus converted into kinetic energy. The crankshaft drive translates the piston's kinetic energy into a turning force (torque) available at the crankshaft.

4. Exhaust stroke (d)

Fractionally before the piston reaches bottom dead center, the exhaust valve (4) opens. The hot, pressurized gases flow out of the cylinder. As the piston moves upwards again, it forces the remaining exhaust gases out.

On completion of the exhaust stroke, the crankshaft has completed two revolutions and the four-stroke operating cycle starts again with the induction stroke.

Valve timing

The cams on the inlet and exhaust camshafts open and close the inlet and exhaust valves respectively. On engines with a single camshaft, a rocker-arm mechanism transmits the action of the cams to the valves.

Valve timing involves synchronizing the opening and closing of the valves with the rotation of the crankshaft (Fig. 4). For that reason, valve timing is specified in degrees of crankshaft rotation.

The crankshaft drives the camshaft by means of a toothed belt or a chain (the timing belt or timing chain) or sometimes by a series of gears. On a four-stroke engine, a complete operating cycle takes two revolutions of the crankshaft. Therefore, the speed of rotation of the camshaft is only half that of the crankshaft. The transmission ratio between the crankshaft and the camshaft is thus $2:1$.

At the changeover from exhaust to induction stroke, the inlet and exhaust valves are open simultaneously for a certain period of time. This "valve overlap" helps to "flush out" the remaining exhaust and cool the cylinders.

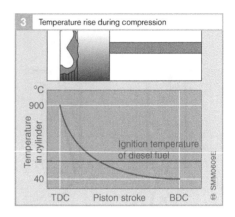

3 Temperature rise during compression

Compression

The compression ratio, ε, of a cylinder results from its swept volume, V_h, and its compression volume, V_c, thus:

$$\varepsilon = \frac{V_h + V_c}{V_c}$$

The compression ratio of an engine has a decisive effect on the following:
● The engine's cold-starting characteristics
● The torque generated
● Its fuel consumption
● How noisy it is and
● The pollutant emissions

The compression ratio, ε, is generally between 16:1 and 24:1 in engines for cars and commercial vehicles, depending on the engine design and the fuel-injection method. It is therefore higher than in gasoline engines ($\varepsilon = 7:1...13:1$). Due to the susceptibility of gasoline to knocking, higher compression ratios and the resulting higher combustion-chamber temperatures would cause the air/fuel mixture to spontaneously combust in an uncontrolled manner.

The air inside a diesel engine is compressed to a pressure of 30...50 bar (conventionally aspirated engine) or 70...150 bar (turbocharged/supercharged engine). This generates temperatures ranging from 700 to 900°C (Fig. 3). The ignition temperature of the most easily combustible components of diesel fuel is around 250°C.

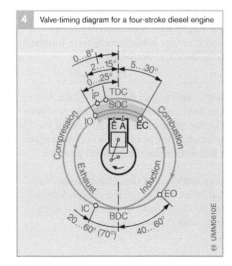

4 Valve-timing diagram for a four-stroke diesel engine

Torque and power output

Torque

The conrod converts the linear motion of the piston into rotational motion of the crankshaft. The force with which the expanding air/fuel mixture forces the piston downwards is thus translated into rotational force or torque by the leverage of the crankshaft.

The output torque M of the engine is, therefore, dependent on mean pressure p_e (mean piston or operating pressure). It is expressed by the equation:

$$M = p_e \cdot V_H / (4 \cdot \pi)$$

where
V_H is the cubic capacity of the engine and $\pi \approx 3.14$.

The mean pressure can reach levels of 8...22 bar in small turbocharged diesel engines for cars. By comparison, gasoline engines achieve levels of 7...11 bar.

The maximum achievable torque, M_{max}, that the engine can deliver is determined by its design (cubic capacity, method of aspiration, etc.). The torque output is adjusted to the requirements of the driving situation essentially by altering the fuel and air mass and the mixing ratio.

Torque increases in relation to engine speed, n, until maximum torque, M_{max}, is reached (Fig. 1). As the engine speed increases beyond that point, the torque begins to fall again (maximum permissible engine load, desired performance, gearbox design).

Engine design efforts are aimed at generating maximum torque at low engine speeds (under 2,000 rpm) because at those speeds fuel consumption is at its most economical and the engine's response characteristics are perceived as positive (good "pulling power").

Power output

The power P (work per unit of time) generated by the engine depends on torque M and engine speed n. Engine power output increases with engine speed until it reaches its maximum level, or rated power P_{rated} at the engine's rated speed, n_{rated}. The following equation applies:

$$P = 2 \cdot \pi \cdot n \cdot M$$

Figure 1a shows a comparison between the power curves of diesel engines made in 1968 and in 1998 in relation to engine speed.

Due to their lower maximum engine speeds, diesel engines have a lower displacement-related power output than gasoline engines. Modern diesel engines for cars have rated speeds of between 3,500 and 5,000 rpm.

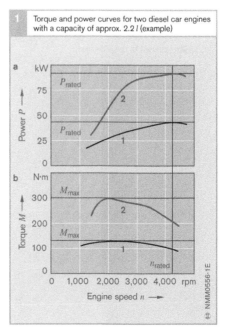

1 Torque and power curves for two diesel car engines with a capacity of approx. 2.2 l (example)

Fig. 1
a   Power curve
b   Torque curve

1   1968 engine
2   1998 engine

M_{max} Maximum torque
P_{rated} Rated power
n_{rated} Rated speed

Engine efficiency

The internal-combustion engine does work by changing the pressure and volume of a working gas (cylinder charge).

Effective efficiency of the engine is the ratio between input energy (fuel) and useful work. This results from the thermal efficiency of an ideal work process (Seiliger process) and the percentage losses of a real process.

Seiliger process

Reference can be made to the Seiliger process as a thermodynamic comparison process for the reciprocating-piston engine. It describes the theoretically useful work under ideal conditions. This ideal process assumes the following simplifications:
- Ideal gas as working medium
- Gas with constant specific heat
- No flow losses during gas exchange

The state of the working gas can be described by specifying pressure (p) and volume (V). Changes in state are presented in the p-V chart (Fig. 1), where the enclosed area corresponds to work that is carried out in an operating cycle.

In the Seiliger process, the following process steps take place:

Isentropic compression (1-2)
With isentropic compression (compression at constant entropy, i.e. without transfer of heat), pressure in the cylinder increases while the volume of the gas decreases.

Isochoric heat propagation (2-3)
The air/fuel mixture starts to burn. Heat propagation (q_{BV}) takes place at a constant volume (isochoric). Gas pressure also increases.

Isobaric heat propagation (3-3')
Further heat propagation (q_{Bp}) takes place at constant pressure (isobaric) as the piston moves downwards and gas volume increases.

Isentropic expansion (3'-4)
The piston continues to move downwards to bottom dead center. No further heat transfer takes place. Pressure drops as volume increases.

Isochoric heat dissipation (4-1)
During the gas-exchange phase, the remaining heat is removed (q_A). This takes place at a constant gas volume (completely and at infinite speed). The initial situation is thus restored and a new operating cycle begins.

p-V chart of the real process
To determine the work done in the real process, the pressure curve in the cylinder is measured and presented in the p-V chart (Fig. 2). The area of the upper curve corresponds to the work present at the piston.

Fig. 1

1-2   Isentropic compression
2-3   Isochoric heat propagation
3-3'   Isobaric heat propagation
3'-4   Isentropic expansion
4-1   Isochoric heat dissipation

TDC Top dead center
BDC Bottom dead center

q_A Quantity of heat dissipated during gas exchange
q_{Bp} Combustion heat at constant pressure
q_{BV} Combustion heat at constant volume
W Theoretical work

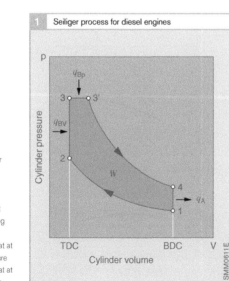

1   Seiliger process for diesel engines

2 Real process in a turbocharged/supercharged diesel engine represented by p-V indicator diagram (recorded using a pressure sensor)

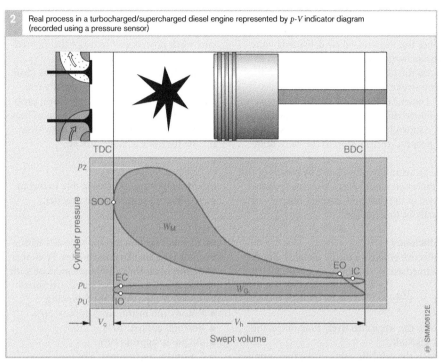

Fig. 2

EO Exhaust opens
EC Exhaust closes
SOC Start of combustion
IO Inlet opens
IC Inlet closes
TDC Top dead center
BDC Bottom dead center

p_U Ambient pressure
p_L Charge-air pressure
p_Z Maximum cylinder pressure
V_c Compression volume
V_h Swept volume
W_M Indexed work
W_G Work during gas exchange (turbocharger/supercharger)

SMM0612E

3 Pressure vs. crankshaft rotation curve (p-α diagram) for a turbocharged/supercharged diesel engine

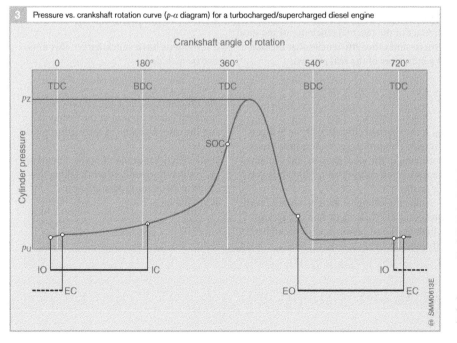

Fig. 3
EO Exhaust opens
EC Exhaust closes
SOC Start of combustion
IO Inlet opens
IC Inlet closes
TDC Top dead center
BDC Bottom dead center

p_U Ambient pressure
p_L Charge-air pressure
p_Z Maximum cylinder pressure

SMM0613E

For assisted-aspiration engines, the gas-exchange area (W_G) has to be added to this since the compressed air delivered by the turbocharger/supercharger also helps to press the piston downwards on the induction stroke.

Losses caused by gas exchange are overcompensated at many operating points by the supercharger/turbocharger, resulting in a positive contribution to the work done.

Representation of pressure by means of the crankshaft angle (Fig. 3, previous page) is used in the thermodynamic pressure-curve analysis, for example.

Efficiency
Effective efficiency of the diesel engine is defined as:

$$\eta_e = \frac{W_e}{W_B}$$

W_e is the work effectively available at the crankshaft.
W_B is the calorific value of the fuel supplied.

Effective efficiency η_e is representable as the product of the thermal efficiency of the ideal process and other efficiencies that include the influences of the real process:

$$\eta_e = \eta_{th} \cdot \eta_g \cdot \eta_b \cdot \eta_m = \eta_i \cdot \eta_m$$

η_{th}: thermal efficiency
η_{th} is the thermal efficiency of the Seiliger process. This process considers heat losses occurring in the ideal process and is mainly dependent on compression ratio and excess-air factor.

As the diesel engine runs at a higher compression ratio than a gasoline engine and a high excess-air factor, it achieves higher efficiency.

η_g: efficiency of cycle factor
η_g specifies work done in the real high-pressure work process as a factor of the theoretical work of the Seiliger process.

Deviations between the real and the ideal processes mainly result from use of a real working gas, the finite velocity of heat propagation and dissipation, the position of heat propagation, wall heat loss, and flow losses during the gas-exchange process.

η_b: fuel conversion factor
η_b considers losses occurring due to incomplete fuel combustion in the cylinder.

η_m: mechanical efficiency
η_m includes friction losses and losses arising from driving ancillary assemblies. Frictional and power-transmission losses increase with engine speed. At nominal speed, frictional losses are composed of the following:
- Pistons and piston rings approx. 50%
- Bearings approx. 20%
- Oil pump approx. 10%
- Coolant pump approx. 5%
- Valve-gear approx. 10%
- Fuel-injection pump approx. 5%

If the engine has a supercharger, this must also be included.

η_i: efficiency index
The efficiency index is the ratio between "indexed" work present at the piston W_i and the calorific value of the fuel supplied.

Work effectively available at the crankshaft W_e results from indexed work taking mechanical losses into consideration:
$W_e = W_i \cdot \eta_m$.

Operating statuses

Starting
Starting an engine involves the following stages: cranking, ignition and running up to self-sustained operation.

The hot, compressed air produced by the compression stroke has to ignite the injected fuel (combustion start). The minimum ignition temperature required for diesel fuel is approx. 250°C.

This temperature must also be reached in poor conditions. Low engine speeds, low outside temperatures, and a cold engine lead to relatively low final compression temperatures due to the fact that:

- The lower the engine speed, the lower the ultimate pressure at the end of the compression stroke and, accordingly, the ultimate temperature (Fig. 1). The reasons for this phenomenon are leakage losses through the piston ring gaps between the piston and the cylinder wall and the fact that when the engine is first started, there is no thermal expansion and an oil film has not formed. Due to heat loss during com-

pression, maximum compression temperature is reached a few degrees before TDC (thermodynamic loss angle, Fig. 2).

- When the engine is cold, heat loss occurs across the combustion-chamber surface area during the compression stroke. On indirect-injection (IDI) engines, this heat loss is particularly high due to the larger surface area.
- Internal engine friction is higher at low temperatures than at normal operating temperature because of the higher viscosity of the engine oil. For this reason, and also due to low battery voltage, the starter-motor speed is only relatively low.
- The speed of the starter motor is particularly low when it is cold because the battery voltage drops at low temperatures.

The following measures are taken to raise temperature in the cylinder during the starting phase:

Fuel heating
A filter heater or direct fuel heater (Fig. 3 on next page) can prevent the precipitation of paraffin crystals that generally occurs at low

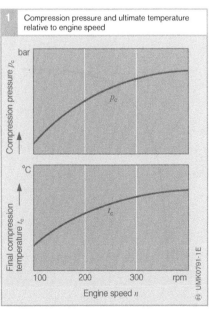

1   Compression pressure and ultimate temperature relative to engine speed

Compression pressure p_c (bar)
Final compression temperature t_c (°C)

p_c
t_c

100   200   300   rpm
Engine speed n

UMK0791-1E

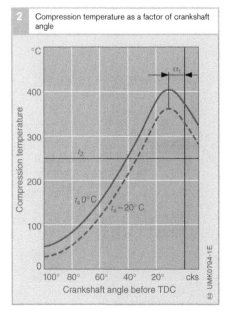

2   Compression temperature as a factor of crankshaft angle

Compression temperature (°C)

400

300

200

t_z

100

t_a 0°C
t_a −20°C

α_T

0

100°   80°   60°   40°   20°   cks
Crankshaft angle before TDC

UMK0794-1E

Fig. 2
t_a Outside temperature
t_z Ignition temperature of diesel fuel
α_T Thermodynamic loss angle

n = 200 rpm

temperatures (during the starting phase and at low outside temperatures).

Start-assist systems

The air/fuel mixture in the combustion chamber (or in the prechamber or whirl chamber) is normally heated by sheathed-element glow plugs in the starting phase on direct-injection (DI) engines for passenger cars, or indirect-injection engines (IDI). On direct-injection (DI) engines for commercial vehicles, the intake air is preheated. Both the above methods assist fuel vaporization and air/fuel mixing and therefore facilitate reliable combustion of the air/fuel mixture.

Glow plugs of the latest generation require a preheating time of only a few seconds (Fig. 4), thus allowing a rapid start. The lower post-glow temperature also permits longer post-glow times. This reduces not only harmful pollutant emissions but also noise levels during the engine's warm-up period.

Injection adaptation

Another means of assisted starting is to inject an excess amount of fuel for starting to compensate for condensation and leakage losses in the cold engine, and to increase engine torque in the running-up phase.

Advancing the start of injection during the warming-up phase helps to offset longer ignition lag at low temperatures and to ensure reliable ignition at top dead center, i.e. at maximum final compression temperature.

The optimum start of injection must be achieved within tight tolerance limits. As the internal cylinder pressure (compression pressure) is still too low, fuel injected too early has a greater penetration depth and precipitates on the cold cylinder walls. There, only a small proportion of it vaporizes since then the temperature of the air charge is too low.

If the fuel is injected too late, ignition occurs during the downward stroke (expansion phase), and the piston is not fully accelerated, or combustion misses occur.

Fig. 3

1   Fuel tank
2   Fuel heater
3   Fuel filter
4   Fuel-injection pump

Fig. 4

Filament material:
1   Nickel (conventional glow plug type S-RSK)
2   CoFe alloy (2nd-generation glow plug type GSK2)

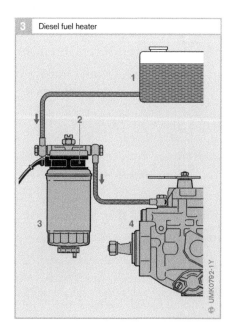

3   Diesel fuel heater

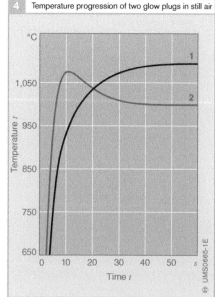

4   Temperature progression of two glow plugs in still air

No load

No load refers to all engine operating statuses in which the engine is overcoming only its own internal friction. It is not producing any torque output. The accelerator pedal may be in any position. All speed ranges up to and including breakaway speed are possible.

Idle

The engine is said to be idling when it is running at the lowest no-load speed. The accelerator pedal is not depressed. The engine does not produce any torque. It only overcomes its internal friction. Some sources refer to the entire no-load range as idling. The upper no-load speed (breakaway speed) is then called the upper idle speed.

Full load

At full load (or Wide-Open Throttle (WOT)), the accelerator pedal is fully depressed, or the full-load delivery limit is controlled by the engine management dependent on the operating point. The maximum possible fuel volume is injected and the engine generates its maximum possible torque output under steady-state conditions. Under non steady-state conditions (limited by turbocharger/supercharger pressure) the engine develops the maximum possible (lower) full-load torque with the quantity of air available. All engine speeds from idle speed to nominal speed are possible.

Part load

Part load covers the range between no load and full load. The engine is generating an output between zero and the maximum possible torque.

Lower part-load range

This is the operating range in which the diesel engine's fuel consumption is particularly economical in comparison with the gasoline engine. "Diesel knock" that was a problem on earlier diesel engines – particularly when cold – has virtually been eliminated on diesels with pre-injection.

As explained in the "Starting" section, the final compression temperature is lower at lower engine speeds and at lower loads. In comparison with full load, the combustion chamber is relatively cold (even when the engine is running at operating temperature) because the energy input and, therefore, the temperatures, are lower. After a cold start, the combustion chamber heats up very slowly in the lower part-load range. This is particularly true for engines with prechamber or whirl chambers because the larger surface area means that heat loss is particularly high.

At low loads and with pre-injection, only a few mm^3 of fuel are delivered in each injection cycle. In this situation, particularly high demands are placed on the accuracy of the start of injection and injected fuel quantity. As during the starting phase, the required combustion temperature is reached also at idle speed only within a small range of piston travel near TDC. Start of injection is controlled very precisely to coincide with that point.

During the ignition-lag period, only a small amount of fuel may be injected since, at the point of ignition, the quantity of fuel in the combustion chamber determines the sudden increase in pressure in the cylinder.

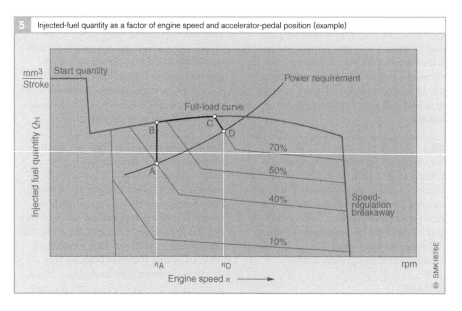

5   Injected-fuel quantity as a factor of engine speed and accelerator-pedal position (example)

The greater the increase in pressure, the louder the combustion noise. Pre-injection of approx. 1 mm³ (for cars) of fuel virtually cancels out ignition lag at the main injection point, and thus substantially reduces combustion noise.

Overrun
The engine is said to be overrunning when it is driven by an external force acting through the drivetrain (e.g. when descending an incline). No fuel is injected (overrun fuel cutoff).

Steady-state operation
Torque delivered by the engine corresponds to the torque required by the accelerator-pedal position. Engine speed remains constant.

Non-steady-state operation
The engine's torque output does not equal the required torque. The engine speed is not constant.

Transition between operating statuses
If the load, the engine speed, or the accelerator-pedal position change, the engine's operating state changes (e.g. its speed or torque output).

The response characteristics of an engine can be defined by means of characteristic data diagrams or maps. The map in Figure 5 shows an example of how the engine speed changes when the accelerator-pedal position changes from 40% to 70% depressed. Starting from operating point A, the new part-load operating point D is reached via the full-load curve (B-C). There, power demand and engine power output are equal. The engine speed increases from n_A to n_D.

Operating conditions

In a diesel engine, the fuel is injected directly into the highly compressed hot air which causes it to ignite spontaneously. Therefore, and because of the heterogeneous air/fuel mixture, the diesel engine – in contrast with the gasoline engine – is not restricted by ignition limits (i.e. specific air-fuel ratios λ). For this reason, at a constant air volume in the cylinder, only the fuel quantity is controlled.

The fuel-injection system must assume the functions of metering the fuel and distributing it evenly over the entire charge. It must accomplish this at all engine speeds and loads, dependent on the pressure and temperature of the intake air.

Thus, for any combination of engine operating parameters, the fuel-injection system must deliver:
- The correct amount of fuel
- At the correct time
- At the correct pressure
- With the correct timing pattern and at the correct point in the combustion chamber

In addition to optimum air/fuel mixture considerations, metering the fuel quantity also requires taking account of operating limits such as:
- Emission restrictions (e.g. smoke emission limits)
- Combustion-peak pressure limits
- Exhaust temperature limits
- Engine speed and full-load limits
- Vehicle or engine-specific load limits, and
- Altitude and turbocharger/supercharger pressure limits

Smoke limit
There are statutory limits for particulate emissions and exhaust-gas turbidity. As a large part of the air/fuel mixing process only takes place during combustion, localized over-enrichment occurs, and, in some cases, this leads to an increase in soot-particle emissions, even at moderate levels of excess air. The air-fuel ratio usable at the statutory full-load smoke limit is a measure of the efficiency of air utilization.

Combustion pressure limits
During the ignition process, the partially vaporized fuel mixed with air burns at high compression, at a rapid rate, and at a high

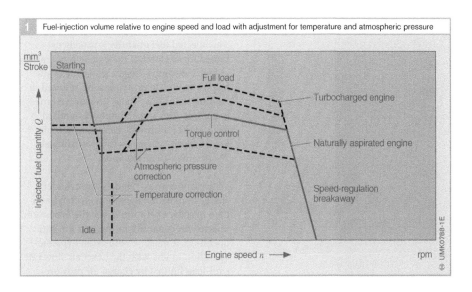

1 Fuel-injection volume relative to engine speed and load with adjustment for temperature and atmospheric pressure

initial thermal-release peak. This is referred to as "hard" combustion. High final compression peak pressures occur during this phenomenon, and the resulting forces exert stresses on engine components and are subject to periodic changes. The dimensioning and durability of the engine and drivetrain components, therefore, limit the permissible combustion pressure and, consequently, the injected fuel quantity. The sudden rise in combustion pressure is mostly counteracted by pre-injection.

Exhaust-gas temperature limits

The high thermal stresses placed on the engine components surrounding the hot combustion chamber, the heat resistance of the exhaust valves and of the exhaust system and cylinder head determine the maximum exhaust temperature of a diesel engine.

Engine speed limits

Due to the existing excess air in the diesel engine, power at constant engine speed mainly depends on injected fuel quantity. If the amount of fuel supplied to a diesel engine is increased without a corresponding increase in the load that it is working against, then the engine speed will rise. If the fuel supply is not reduced before the engine reaches a criti-

cal speed, the engine may exceed its maximum permitted engine speed, i.e. it could self-destruct. Consequently, an engine speed limiter or governor is absolutely essential on a diesel engine.

On diesel engines used to drive road-going vehicles, the engine speed must be infinitely variable by the driver using the accelerator pedal. In addition, when the engine is under load or when the accelerator pedal is released, the engine speed must not be allowed to drop below the idling speed to a standstill. This is why a minimum-maximum-speed governor is fitted. The speed range between these two points is controlled using the accelerator-pedal position. If the diesel engine is used to drive a machine, it is expected to keep to a specific speed constant, or remain within permitted limits, irrespective of load. A variable-speed governor is then fitted to control speed across the entire range.

A program map is definable for the engine operating range. This map (Fig. 1, previous page) shows the fuel quantity in relation to engine speed and load, and the necessary adjustments for temperature and air-pressure variations.

Altitude and turbocharger/supercharger pressure limits

The injected fuel quantity is usually designed for sea level. If the engine is operated at high elevations (height above mean sea level), the fuel quantity must be adjusted in relation to the drop in air pressure in order to comply with smoke limits. A standard value is the barometric elevation formula, i.e. air density decreases by approximately 7% per 1,000 m of elevation.

With turbocharged engines, the cylinder charge in dynamic operation is often lower than in static operation. Since the maximum injected fuel quantity is designed for static operation, it must be reduced in dynamic operation in line with the lower air-flow rate (full-load limited by charge-air pressure).

2   Development of diesel engines for mid-range cars

Engine variants
- Torque of largest engine [Nm]
- Torque of smallest engine [Nm]
- Rated output of largest engine [kW]
- Rated output of smallest engine [kW]

Year of manufacture	1953	1961	1968	1976	1984	1995	2000
Torque of largest engine [Nm]	101	118	126	172	185	150	470
Torque of smallest engine [Nm]		113	113	123	100	145	250
Rated output of largest engine [kW]	30	40	44	59	80	75	210
Rated output of smallest engine [kW]		40	40	53	70		

NMM0616E

Fuel-injection system

The low-pressure fuel supply conveys fuel from the fuel tank and delivers it to the fuel-injection system at a specific supply pressure. The fuel-injection pump generates the fuel pressure required for injection. In most systems, fuel runs through high-pressure delivery lines to the injection nozzle and is injected into the combustion chamber at a pressure of 200...2,200 bar on the nozzle side.

Engine power output, combustion noise, and exhaust-gas composition are mainly influenced by the injected fuel mass, the injection point, the rate of discharge, and the combustion process.

Up to the 1980s, fuel injection, i.e. the injected fuel quantity and the start of injection on vehicle engines, was mostly controlled mechanically. The injected fuel quantity is then varied by a piston timing edge or via slide valves, depending on load and engine speed. Start of injection is adjusted by mechanical control using flyweight governors, or hydraulically by pressure control (see section entitled "Overview of diesel fuel-injection systems").

Now electronic control has fully replaced mechanical control – not only in the automotive sector. Electronic Diesel Control (EDC) manages the fuel-injection process by involving various parameters, such as engine speed, load, temperature, geographic elevation, etc. in the calculation. Start of injection and fuel injection quantity are controlled by solenoid valves, a process that is much more precise than mechanical control.

▶ **Size of injection**

An engine developing 75 kW (102 HP) and a specific fuel consumption of 200 g/kWh (full load) consumes 15 kg fuel per hour. On a 4-stroke 4-cylinder engine, the fuel is distributed by 288,000 injections at 2,400 revs per minute. This results in a fuel volume of approx. 60 mm³ per injection. By comparison, a raindrop has a volume of approximately 30 mm³.

Even greater precision in metering requires an idle with approx. 5 mm³ fuel per injection and a pre-injection of only 1 mm³. Even the minutest variations have a negative effect on the smooth running of the engine, noise, and pollutant emissions.

The fuel-injection system not only has to deliver precisely the right amount of fuel for each individual, it also has to distribute the fuel evenly to the individual cylinder of an engine. Electronic Diesel Control (EDC) adapts the injected fuel quantity for each cylinder in order to achieve a particularly smooth-running engine.

Combustion chambers

The shape of the combustion chamber is one of the decisive factors in determining the quality of combustion and therefore the performance and exhaust characteristics of a diesel engine. Appropriate design of combustion-chamber geometry combined with the action of the piston can produce whirl, squish, and turbulence effects that are used to improve distribution of the liquid fuel or air/fuel vapor spray inside of the combustion chamber.

The following technologies are used:
- Undivided combustion chamber (Direct Injection (DI) engines) and
- Divided combustion chamber (Indirect Injection (IDI) engines)

The proportion of direct-injection engines is increasing due to their more economical fuel consumption (up to 20% savings). The harsher combustion noise (particularly under acceleration) can be reduced to the level of indirect-injection engines by pre-injection. Engines with divided combustion chambers now hardly figure at all among new developments.

Undivided combustion chamber (direct-injection engines)

Direct-injection engines (Fig. 1) have a higher level of efficiency and operate more economically than indirect-injection engines. Accordingly, they are used in all types of commercial vehicles and most modern diesel cars.

As the name suggests, the direct-injection process involves injecting the fuel directly into the combustion chamber, part of which is formed by the shape of the piston crown (piston crown recess, 2). Fuel atomization, heating, vaporization and mixing with the air must therefore take place in rapid succession. This places exacting demands on fuel and air delivery. During the induction and compression strokes, the special shape of the intake port in the cylinder head creates an air vortex inside of the cylinder. The shape of the combustion chamber also contributes to the air flow pattern at the end of the compression stroke (i.e. at the moment of fuel injection). Of the combustion chamber designs used over the history of the diesel engine, the most widely used at present is the ω piston crown recess.

In addition to creating effective air turbulence, the technology must also ensure that fuel is delivered in such a way that it is evenly distributed throughout the combustion chamber to achieve rapid mixing. A multihole nozzle is used in the direct-injection process and its nozzle-jet position is optimized as a factor of combustion-chamber design. Direct fuel injection requires very high injection pressures (up to 2,200 bar).
  In practice, there are two types of direct fuel injection:
- Systems in which mixture formation is assisted by specifically created air-flow effects and
- Systems which control mixture formation virtually exclusively by means of fuel injection and largely dispense with any air-flow effects

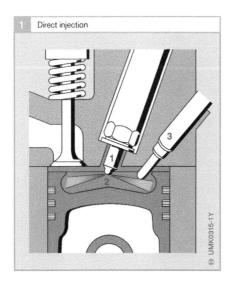

Direct injection

Fig. 1
1   Multihole injector
2   ω piston recess
3   Glow plug

In the latter case, no effort is expended in creating air-turbulence effects and this is evident in smaller gas replacement losses and more effective cylinder charging. At the same time, however, far more demanding requirements are placed on the fuel-injection system with regard to injection-nozzle positioning, the number of nozzle jets, the degree of atomization (dependent on spray-hole diameter), and the intensity of injection pressure in order to obtain the required short injection times and quality of the air/fuel mixture.

Divided combustion chamber (indirect injection)

For a long time diesel engines with divided combustion chambers (indirect-injection engines) held an advantage over direct-injection engines in terms of noise and exhaust-gas emissions. That was the reason why they were used in cars and light commercial vehicles. Now direct-injection engines are more economical than IDI engines, with comparable noise emissions as a result of their high injection pressures, electronic diesel control, and pre-injection. As a result, indirect-injection engines are no longer used in new vehicles.

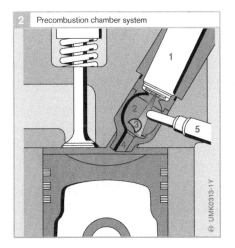

There are two types of processes with divided combustion chamber:
- The precombustion chamber system and
- The whirl-chamber system

Precombustion chamber system

In the prechamber (or precombustion chamber) system, fuel is injected into a hot prechamber recessed into the cylinder head (Fig. 2, 2). The fuel is injected through a pintle nozzle (1) at a relatively low pressure (up to 450 bar). A specially shaped baffle (3) in the center of the chamber diffuses the jet of fuel that strikes it and mixes it thoroughly with the air.

Combustion starting in the prechamber drives the partly combusted air/fuel mixture through the connecting channel (4) into the main combustion chamber. Here and further down the combustion process, the injected fuel is mixed intensively with the existing air. The ratio of precombustion chamber volume to main combustion chamber volume is approx. 1:2.

The short ignition lag[1]) and the gradual release of energy produce a soft combustion effect with low levels of noise and engine load.

A differently shaped prechamber with an evaporation recess and a different shape and position of the baffle (spherical pin) apply a specific degree of whirl to the air that passes from the cylinder into the prechamber during the compression stroke. The fuel is injected at an angle of 5 degrees in relation to the prechamber axis.

So as not to disrupt the progression of combustion, the glow plug (5) is positioned on the "lee side" of the air flow. A controlled post-glow period of up to 1 minute after a cold start (dependent on coolant temperature) helps to improve exhaust-gas characteristics and reduce engine noise during the warm-up period.

[1]) Time from start of injection to start of ignition

Fig. 2
1   Nozzle
2   Precombustion chamber
3   Baffle surface
4   Connecting channel
5   Glow plug

Swirl-chamber system

With this process, combustion is also initiated in a separate chamber (swirl chamber) that has approx. 60% of the compression volume. The spherical and disk-shaped swirl chamber is linked by a connecting channel that discharges at a tangent into the cylinder chamber (Fig. 3, 2).

During the compression cycle, air entering via the connecting channel is set into a swirling motion. The fuel is injected so that the swirl penetrates perpendicular to its axis and meets a hot section of the chamber wall on the opposite side of the chamber.

As soon as combustion starts, the air/fuel mixture is forced under pressure through the connecting channel into the cylinder chamber where it is turbulently mixed with the remaining air. With the swirl-chamber system, the losses due to gas flow between the main combustion chamber and the swirl chamber are less than with the precombustion chamber system because the connecting channel has a larger cross-section. This results in smaller throttle-effect losses and consequent benefits for internal efficiency and fuel consumption. However, combustion noise is louder than with the precombustion chamber system.

It is important that mixture formation takes place as completely as possible inside the swirl chamber. The shape of the swirl chamber, the alignment and shape of the fuel jet and the position of the glow plug must be carefully matched to the engine in order to obtain optimum mixture formation at all engine speeds and under all operating conditions.

Another demand is for rapid heating of the swirl chamber after a cold start. This reduces ignition lag and combustion noise as well as preventing unburned hydrocarbons (blue smoke) during the warm-up period.

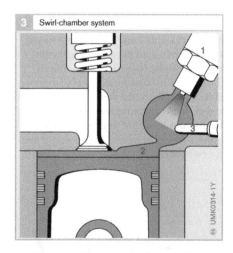

3 Swirl-chamber system

Fig. 3
1 Fuel injector
2 Tangential
   connecting
   channel
3 Glow plug

▶ **M System**

In the direct-injection system with recess-wall deposition (M system) for commercial-vehicle and fixed-installation diesel engines and multi-fuel engines, a single-jet nozzle sprays the fuel at a low injection pressure against the wall of the piston crown recess. There, it vaporizes and is absorbed by the air. This system thus uses the heat of the piston recess wall to vaporize the fuel. If the air flow inside of the combustion chamber is properly adapted, an extremely homogeneous air/fuel mixture with a long combustion period, low pressure increase and, therefore, quiet combustion can be achieved. Due to its consumption disadvantages compared with the air-distributing direct injection process, the M system is no longer used in modern applications.

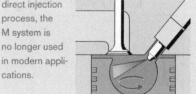

Automotive manufacturers are obliged by law to specify the fuel consumption of their vehicles. This figure is determined from the exhaust-gas emissions during the exhaust-gas test when the vehicle travels a specific route profile (test cycle). The fuel consumption figures are therefore comparable for all vehicles.

Every driver makes a significant contribution to reducing fuel consumption by his or her driving style. Reducing the fuel consumption that the driver can achieve with a vehicle depends on several factors.

Applying the measures listed below, an "economical" driver can reduce fuel consumption in everyday traffic by 20 to 30% compared to an average driver. The reduction in fuel consumption achievable by applying the individual measures depends on a number of factors, mainly the route profile (city streets, overland roads), and on traffic conditions. For this reason, it is not always practical to specify figures for fuel-consumption savings.

Positive influences on fuel consumption
- Tire pressure: Remember to increase tire pressure when the vehicle is carrying a full payload (saving: approx. 5%).
- When accelerating at high load and low engine speed, shift up at 2,000 rpm.
- Drive in the highest possible gear. You can even drive at full-load at engine speeds below 2,000 rpm.
- Avoid braking and re-accelerating by adopting a forward-looking style of driving.
- Use overrun fuel cutoff to the full.
- Switch off the engine when the vehicle is stopped for an extended period of time, e.g. at traffic lights with a long red phase, or at closed railroad crossings (3 minutes at idle consumes as much fuel as driving 1 km).
- Use high-lubricity engine oils (saving: approx. 2% according to manufacturer specifications).

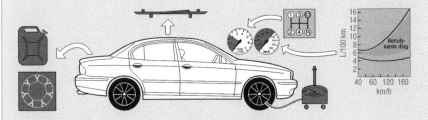

Negative influences on fuel consumption
- Greater vehicle weight due to ballast, e.g. in the trunk (additional approx. 0.3 l/100 km).
- High-speed driving.
- Greater aerodynamic drag from carrying objects on the roof.
- Additional electrical equipment, e.g. rear-window heating, foglamps (approx. 1 l/1 kW).
- Dirty air filter.

SMK1827E

Fuels

Diesel fuels are the product of graduated distillation of crude oil. They contain a whole range of individual hydrocarbons with boiling points ranging from roughly 180°C to 370°C. Diesel fuel ignites on average at approximately 350°C (lower limit 220°C), which is very early in comparison with gasoline (on average 500°C).

Diesel fuel

In order to cover the growing demand for diesel fuels, refineries are increasingly adding conversion products, i.e. thermal and catalytic-cracking products. They are obtained by cracking large heavy-oil molecules.

Quality and grading criteria

In Europe, the standard for diesel fuels is EN 590. The key parameters are listed in Table 1. Defining limits is intended to secure troublefree vehicle operation and restrict pollutants.

In many other countries around the world, fuel standards are less strict. The U.S. standard for diesel fuels, ASTM D975, for example, specifies fewer criteria and applies less stringent limits to these quality criteria. The requirements for marine and fixed-installation engines are also much less demanding.

High-quality diesel fuels are characterized by the following features:
● High cetane number
● Relatively low final boiling point
● Narrow density and viscosity spread
● Low aromatic compounds (particularly polyaromatic compounds) content
● Low sulfur content

1	European Standard EN 590: Selected requirements for diesel fuels (figures specified for moderate climate where requirements are climate-dependent)	
Criterion	**Parameter**	**Unit**
Cetane number	≥ 51	–
Cetane index	≥ 46	–
CFPP[1]) in six seasonal categories, max.	+5...−20[2])	°C
Flash point	≥ 55	°C
Density at 15°C	820...845	kg/m^3
Viscosity at 40°C	2.00...4.50	mm^2/s
Lubricity	≤ 460	µm (wear scar diameter)
Sulfur content[3])	≤ 350 (until 12-31-2004); ≤ 50 (low sulfur, starting 2005 – 2008); ≤ 10 (sulfur-free, starting 2009)[4])	mg/kg
Moisture content	≤ 200	mg/kg
Total contamination	≤ 24	mg/kg
FAME content	≤ 5	% by volume

[1]) Filtration limit
[2]) Defined by national law, for Germany 0...−20°C
[3]) In Germany, sulfur-free fuel has been on sale nationwide since 2003, throughout the EU starting 2005.
[4]) EU proposal

Table 1

In addition, the following characteristics are particularly important for the service life and constant function of fuel-injection systems:
- Good lubricity
- Absence of free water
- Limited pollution with particulate

The most important criteria are explained in detail below.

Cetane number, cetane index

The Cetane Number (CN) expresses the ignition quality of the diesel fuel. The higher the cetane number, the greater the fuel's tendency to ignite. As the diesel engine dispenses with an externally supplied ignition spark, the fuel must ignite spontaneously (auto-ignition) and with minimum delay (ignition lag) when injected into the hot, compressed air in the combustion chamber.

Cetane number 100 is assigned to n-hexadecane (cetane), which ignites very easily, while slow-igniting methyl naphthalene is allocated cetane number 0. The cetane number of a diesel fuel is defined in a standard CFR[1] single-cylinder test engine with variable compression pistons. The compression ratio is measured at constant ignition lag. The engine is run on reference fuels comprising cetane and α-methyl naphthalene (Fig. 1) at the measured compression ratio. The proportion of cetane in the mixture is altered until the same ignition lag is obtained. According to the definition, the cetane proportion specifies the cetane number. Example: A mixture comprising 52% cetane and 48% α-methyl naphthalene has the cetane number 52.

A cetane number in excess of 50 is desirable for optimized operation in modern engines (smooth running, low exhaust-gas emissions). High-quality diesel fuels contain a high proportion of paraffins with high CN ratings. Conversely, aromatic compounds reduce ignition quality.

Yet another parameter of ignition quality is provided by the cetane index, which is calculated on the basis of fuel density and various points on the boiling curve. This purely mathematical parameter does not take into account the influence of cetane improvers on ignition quality. In order to limit the adjustment of the cetane number by means of cetane improvers, both the cetane number and the cetane index have been included in the list of requirements in EN 590. Fuels whose cetane number has been enhanced by cetane improvers respond differently during engine combustion than fuels with the same natural cetane number.

Boiling range

The boiling range of a fuel, i.e. the temperature range at which the fuel vaporizes, depends on its composition.

A low initial boiling point makes a fuel suitable for use in cold weather, but also means a lower cetane number and poor lubricant properties. This raises the wear risk for central injection units.

[1] Cooperative Fuel Research

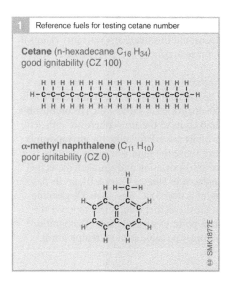

1   Reference fuels for testing cetane number

Cetane (n-hexadecane $C_{16}H_{34}$)
good ignitability (CZ 100)

α-**methyl naphthalene** ($C_{11}H_{10}$)
poor ignitability (CZ 0)

SMK1877E

Fig. 1
C   Carbon
H   Hydrogen
—   Chemical bond

On the other hand, if the final boiling point is situated at high temperatures, this can result in increased soot production and nozzle coking (deposit caused by chemical decomposition of not easily volatized fuel constituents on the nozzle cone, and deposits of combustion residues). For this reason, the final boiling point should not be too high. The requirement of the Association des Constructeurs Européens d'Automobiles (ACEA: Association of European Automobile Manufacturers) is 350°C.

Filtration limit (cold-flow properties)

Precipitation of paraffin crystals at low temperatures can result in fuel-filter blockage, ultimately leading to interruption of fuel flow. In worst-case scenarios, paraffin particles can start to form at temperatures of 0°C or even higher. The cold-flow properties of a fuel are assessed by means of the "filtration limit" (Cold Filter Plugging Point (CFPP)).

European Standard EN 590 defines the CFPP for various classes, and can be defined by individual member states depending on the prevailing geographical and climatic conditions.

Formerly, owners sometimes added regular gasoline to their vehicle fuel tanks to improve the cold response of diesel fuel. This practice is no longer necessary now that fuels conform to standards, and, in any case, this would invalidate any warranty claims if damage occurs.

Flash point

The flash point is the temperature at which the quantities of vapor which a combustible fluid emits to the atmosphere are sufficient to allow a spark to ignite the air/vapor mixture above the fluid. For safety reasons, (e.g. for transportation and storage), diesel fuel is placed in Hazard Class A III, i.e. its flash point is over 55°C. Less than 3% gasoline in the diesel fuel is sufficient to lower the flash point to such an extent that ignition becomes possible at room temperature.

Density

The energy content of diesel fuel per unit of volume increases with density. Assuming constant fuel-injection-pump settings (i.e. constant injected fuel quantity), the use of fuels with widely different densities causes variations in mixture ratios due to fluctuations in calorific value.

When an engine runs on fuel that has a high type-dependent density, engine performance and soot emissions increase; as fuel density decreases, these parameters drop. As a result, the requirements call for a diesel fuel that has a low type-dependent density spread.

Viscosity

Viscosity is a measure of a fuel's resistance to flow due to internal friction. Leakage losses in the fuel-injection pump result if diesel-fuel viscosity is too low, and this in turn results in performance loss.

Much higher viscosity – e.g. Fatty Acid Methyl Ester (biodiesel) – causes a higher peak injection pressure at high temperatures in non-pressure-regulated systems (e.g. unit injector systems). For this reason, mineral-oil diesel may not be applied at the maximum permitted primary pressure. High viscosity also changes the spray pattern due to the formation of larger droplets.

Lubricity

In order to reduce the sulfur content of diesel fuel, it is hydrogenated. In addition to removing sulfur, the hydrogenation process also removes the ionic fuel components that aid lubrication. After the introduction of desulfurized diesel fuels, wear-related problems started to occur on distributor fuel-injection pumps due to the lack of lubricity. As a result, they were replaced by diesel fuels containing lubricity enhancers.

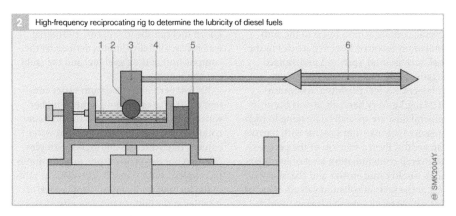

2 High-frequency reciprocating rig to determine the lubricity of diesel fuels

Fig. 2
1 Fuel bath
2 Test ball
3 Stress introduced
4 Test disk
5 Heating device
6 Oscillating
  movement

Lubricity is measured in a High-Frequency Reciprocating Rig (HFRR method). A fixed, clamped steel ball is ground on a plate by fuel at high frequency. The magnitude of the resulting flattening, i.e. the Wear Scar Diameter (WSD) measured in µm, specifies the amount of wear, and is thus a measure of fuel lubricity.

Diesel fuels complying with EN 590 must have a WSD of ≤ 460 µm.

Sulfur content
Diesel fuels contain chemically bonded sulfur, and the actual quantities depend on the quality of the crude petroleum and the components added at the refinery. In particular, crack components mostly have high sulfur contents.

To desulfurize fuel, sulfur is removed from the middle distillate by hydrogenation at high pressure and temperature in the presence of a catalyst. The initial byproduct of this process is hydrogen sulfide (H_2S) which is subsequently converted into pure sulfur.

Since the beginning of 2000 the EN 590 maximum limit for the sulfur content of diesel fuel has been 350 mg/kg. Starting 2005 all regular gasolines and diesel fuels will be subject to a minimum low-sulfur requirement (sulfur content < 50 mg/kg) throughout Europe. Starting 2009 only sulfur-free fuels (sulfur content < 10 mg/kg) will be allowed.

In Germany, a penalty tax has been levied on fuels containing sulfur since 2003. As a result, the German market only offers sulfur-free diesel fuel. This has dropped direct SO_2 emissions (sulfur dioxide), as well as emitted particle mass (sulfur adhering to soot).

Exhaust-gas treatment systems for NO_x and particulate filters use catalysts. They must run on sulfur-free fuel since sulfur poisons the active catalyst surface.

Carbon-deposit index
The carbon-deposit index describes a fuel's tendency to form carbon residue on injection nozzles. The processes of carbon depositing are highly complex. Above all, components which the diesel fuel contains at the final boiling point (particularly cracking constituents) influence carbon-deposit formation (coking).

Overall contamination

Overall contamination refers to the sum total of undissolved foreign particles in the fuel, such as sand, rust, and undissolved organic components, including aging polymers. EN 590 permits a maximum of 24 mg/kg. Very hard silicates as occur in mineral dust are specially damaging to high-pressure fuel-injection systems with narrow gap widths. Even a fraction of the permissible overall contamination level of hard particles would cause erosive and abrasive wear (e.g. at the seats of solenoid valves). Wear of this nature results in valve leakage, which lowers fuel-injection pressure and engine performance, and increases particulate emissions from the engine.

Typical European diesel fuels contain about 100,000 particles per 100 ml. Particle sizes of 6 to 7 μm are particularly critical. High-performance fuel filters with high filtration efficiency help to prevent damage caused by particles.

Water in diesel fuel

Diesel fuel can absorb approx. 100 mg/kg water. The solubility limit is defined by the composition of the diesel fuel and the ambient temperature.

EN 590 permits a maximum water content of 200 mg/kg. Although much higher water contents occur in diesel fuel in many countries, market surveys show that water content rarely exceeds 200 mg/kg. Samples often do not detect any water, or detection is incomplete, since water is deposited on walls in the form of undissolved, "free" water, or it settles at the bottom in a separate phase. Whereas dissolved water does not damage the fuel-injection system, even very small quantities of free water can cause major damage to fuel-injection pumps within a short period of time.

It is not possible to prevent the entrainment of water into the fuel tank as a result of condensation from the air. For this reason, water separators are specified as obligatory equipment in certain regions of the world. In addition, the vehicle manufacturer must design the tank ventilation system and the fuel-filler neck so as to prevent additional water from entering.

▶ **Fuel parameters**

Net and gross calorific values

Specific calorific value H_u (formerly: *lower calorific value*) is usually specified to express the energy content of fuels. The specific gross calorific value H_O (formerly: *upper calorific value* or *combustion heat*) for fuels that have water vapor in their combustion products is higher than the calorific value since the gross calorific value also includes the heat trapped in the water vapor (latent heat). This component is not used in the vehicle. The specific calorific value of diesel fuel is 42.5 MJ/kg.

Oxygenates, i.e. fuel constituents containing oxygen, such as alcohol fuels, ether, or fatty-acid methyl ester, have a lower calorific value than pure hydrocarbons because the oxygen bonded in them does not contribute to the combustion process. Performance comparable to that achievable with oxygenate-free fuels can only be attained at the cost of higher fuel-consumption rates.

Calorific value of air/fuel mixture

The calorific value of the combustible air/fuel mixture determines engine output. Assuming a constant stoichiometric ratio, this figure remains roughly the same for all liquid fuels and liquefied gases (approx. 3.5...3.7 MJ/m³).

Additives

Additives, a long-standard feature in gasolines, have become commonplace as quality improvers in diesel fuels. The various agents are generally combined in additive packages to achieve a variety of objectives. The total concentration of additives is normally about < 0.1%. This does not change the physical parameters of fuels, such as density, viscosity, or the boiling curve.

Lubricity enhancers

It is possible to improve the lubricity of diesel fuels which have poor lubrication properties by adding fatty acids, fatty-acid esters, or glycerins. Biodiesel is also a fatty-acid ester. In this case, if diesel fuel already contains a proportion of biodiesel, no further lubricity enhancers are added.

Cetane improvers

Cetane improvers are nitric acid esters of alcohols added to shorten ignition lag. They reduce emissions and noise (combustion noise).

Flow improvers

Flow improvers consist of polymer substances that lower the filtration limit. They are added in winter to ensure troublefree operation at low temperatures.

Although flow improvers cannot prevent the precipitation of paraffin crystals from diesel fuel, it can severely limit their growth. The size of the crystals produced is so small that they can still pass through the filter pores.

Detergent additives

Detergent additives are used to keep the intake system clean. They can also inhibit the formation of deposits and reduce the buildup of carbon deposits on the injection nozzles.

Corrosion inhibitors

Corrosion inhibitors are deposited on the surfaces of metal parts and protect them against corrosion if water is entrained.

Antifoaming agents (defoamants)

Adding defoamants helps to avoid excessive foaming when the vehicle is refueled quickly.

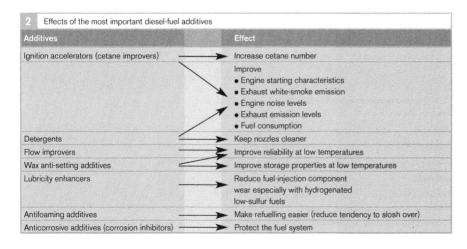

2 Effects of the most important diesel-fuel additives	
Additives	**Effect**
Ignition accelerators (cetane improvers)	Increase cetane number
	Improve • Engine starting characteristics • Exhaust white-smoke emission • Engine noise levels • Exhaust emission levels • Fuel consumption
Detergents	Keep nozzles cleaner
Flow improvers	Improve reliability at low temperatures
Wax anti-setting additives	Improve storage properties at low temperatures
Lubricity enhancers	Reduce fuel-injection component wear especially with hydrogenated low-sulfur fuels
Antifoaming additives	Make refuelling easier (reduce tendency to slosh over)
Anticorrosive additives (corrosion inhibitors)	Protect the fuel system

Table 2

Alternative fuels

Alternative fuels for diesel engine include biogenic fuels, and, in a wider sense, fossil fuels that are not produced on the basis of crude oil. This includes mainly esters that are derived from organic oils.

Alcohol fuels (methanol and ethanol) are only used in diesel engines to a minor extent, and only as an emulsion together with diesel fuel.

Fatty-Acid Methyl Ester (FAME)

Fatty-Acid Methyl Ester (FAME) – commonly known as biodiesel – is the generic term applied to vegetable or animal oils and greases which have been transesterified with methanol. FAME is produced from various raw materials, mainly from rape seed oil (Rape Seed Oil Methyl Ester (RME) Europe), or soya (Soya Methyl Ester (SME), U.S.A.). There are also sunflower and palm esters (Used Frying Oil Methyl Esters (UFOME)), and beef tallow esters (Tallow Methyl Esters (TME)), but these are mostly used in conjunction with other FAMEs. Ethanol can also be transesterified instead of methanol, as in Brazil to produce soya ethyl ester.

FAME is either used in pure form (B 100, i.e. 100% biodiesel), or it is mixed with diesel fuel to a maximum FAME proportion of 5% to form Blend B 5. B 5 is permitted as a diesel fuel in compliance with EN 590.

Since the use of low-quality FAME may lead to malfunctions or damage to the engine and fuel-injection system, FAME specifications are controlled at European level (EN 14 214). It is essential, in particular, to ensure good aging stability (oxidation stability) and to eliminate contamination caused by the process. FAME must satisfy European Standard EN 14 214, regardless of whether it is used directly as B 100, or as an additive in diesel fuel. The B 5 blend created by FAME additives must also comply with the requirements for pure diesel fuel (EN 590).

The production of FAME is uneconomical in comparison with mineral-oil-based diesel fuels and must be heavily subsidized (exemption from mineral-oil tax) in Germany.

Pure, unesterified vegetable oils are no longer used in direct-injection diesel engines since they cause considerable problems, mainly due to their high viscosity and extreme nozzle coking.

1 European Standard EN 14 214: Selected requirements for FAME		
Criterion	Parameter	Unit
CFPP[1]) in six seasonal categories, max.	+5...–20[2])	°C
Flash point	≥ 120	°C
Density at 15°C	860...900	kg/m³
Viscosity at 40°C	3.5...5.0	mm²/s
Sulfur content	10	mg/kg
Moisture content	≤ 500	mg/kg
[1]) Filtration limit, here for moderate climate		
[2]) Defined by national law, for Germany 0...-20°C		

Table 1

Synfuels® and Sunfuels®

The terms Synfuel and Sunfuel refer to fuels which are produced from synthesis gas (H_2 and CO) using the Fischer-Tropsch process.

The end product is known as Synfuel when coal, coke, or natural gas is used to produce the synthesis gas. When biomass is used, it is known as Sunfuel.

In the Fischer-Tropsch process, synthesis gas is catalytically converted to produce hydrocarbons. This results in high-quality, sulfur- and aromatic-free diesel fuels which are mainly used to improve the quality of conventional diesel fuels. Depending on the catalysts used, it is also possible to produce gasoline. The byproducts are liquefied gas and paraffins.

Due to the high costs involved, the production of synthetic fuels has been, and is, restricted to special markets (oil embargo in South Africa in the 1970s, surplus natural gas in Malaysia, research laboratories).

Dimethylether (DME)

Dimethylether (DME) is a synthetically generated fuel that is only produced from methanol in small quantities. DME has a cetane number of CN $\cong$ 55. When used in diesel engines, it produces low soot and nitrogen-oxide emissions. Its calorific value is low on account of its low density and high oxygen content. It is a gas-phase fuel, which means that the fuel-injection equipment requires modification.

Other ethers (e.g. dimethoxymethane, di-n-pentylether) are under investigation to determine their suitability as fuels.

Emulsions

Emulsions of water or ethanol in diesel fuels are undergoing trials at a number of different institutes. Water and alcohols are difficult to dissolve in diesel. Emulsifiers are required to keep the mixture stable and prevent it from demulsification. Wear- and corrosion-inhibiting measures are also necessary. The use of emulsifiers reduces soot- and nitrogen-oxide emissions since the combustion mixture is cooler due to the water content.

Their use to date has been restricted to vehicle fleets which, for the most part, are equipped with in-line fuel-injection pumps. Other fuel-injection systems are either unsuitable for operation with emulsifiers, or no trials have been made using them.

1   Damage to a fuel-injection pump caused by poor fuel quality

a

b

SMK1878Y

Fig. 1
a   Deposits on actuator mechanism caused by contaminated FAME
b   Bearing damage caused by free water (vehicle mileage approx. 5,600 km)

Cylinder-charge control systems

1) The cylinder charge
is a mixture of gases
trapped in the cylin-
der when the intake
valves are closed.
It consists of the
intake air and the
residual burned
gases from the
preceding com-
bustion cycle.

In diesel engines, both the fuel mass injected and the air mass with which it is
mixed are decisive factors in determining
torque output and, therefore, engine performance, and exhaust-gas composition. For
this reason, the systems that control the
cylinder-air charge[1]) have an important
role to play as well as the fuel-injection system. Cylinder-charge control systems clean
the intake air and affect the flow, density,
and composition (e.g. oxygen content) of
the cylinder charge.

Overview

In order to burn the fuel, the engine requires
oxygen which it extracts from the intake air.
In principle, the more oxygen there is available for combustion in the combustion
chamber, the greater the amount of fuel
that can be injected for full-load delivery.
There is, therefore, a direct relationship
between the amount of air with which
the cylinder is charged and the maximum
possible engine power output.

Air-intake systems have the function of conditioning the intake air and ensuring that
the cylinders are properly charged. Cylinder-
charge control systems are made up of the
following components (Fig. 1):
- Air filter (1)
- Turbocharger/supercharger (2)
- Exhaust-gas recirculation system (4)
- Swirl flaps (5)

Supercharging/turbocharging systems
(i.e. to precompress air before injection
in the cylinder) are fitted to most diesel
engines to raise performance.

Exhaust-gas recirculation systems are
fitted on all modern diesel cars and some
commercial vehicles for the purpose of minimizing pollutants in the exhaust gas. They
reduce the amount of oxygen in the cylinder.
Less nitrogen oxides (NO_x) caused by the
resulting drop in combustion temperature
are then formed on combustion.

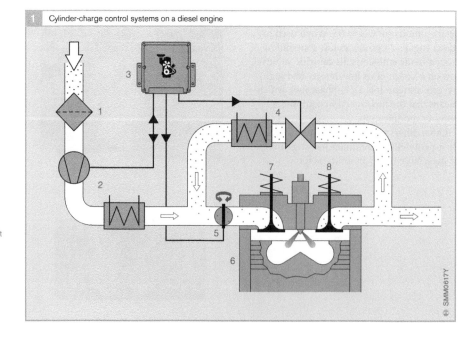

1   Cylinder-charge control systems on a diesel engine

Fig. 1
1   Air filter
2   Turbocharger/
    supercharger
    with intercooler
3   Engine control unit
4   Exhaust-gas
    recirculation
    and cooler
5   Swirl flap
6   Engine cylinder
7   Intake valve
8   Exhaust valve

SMM0617Y

Turbochargers and superchargers

Assisted aspiration by means of turbochargers or superchargers has been in existence for many years[1] on large diesel engines for fixed installations, marine propulsion systems, and commercial vehicles. It has now been adopted for fast-running diesel engines in cars[2]. In contrast to a conventionally aspirated engine, air in a turbocharged or supercharged engine is forced under pressure into the cylinders. This increases the air mass in the cylinder charge and, in combination with a greater fuel mass, this results in a greater power yield from the same engine capacity, or the same power yield from a smaller engine capacity. Lower fuel consumption is achievable by reducing engine swept volume (downsizing).

At the same time it improves exhaust-gas emission rates.

The diesel engine is particularly suited to assisted aspiration as its compressed cylinder charge consists only of air rather than a mixture of fuel and air, and it can be economically combined with a supercharger/turbocharger because of its quality-based method of control. On larger commercial-vehicle engines, a further increase in mean pressure (and, therefore, torque) is achieved by higher turbocharger pressures and lower compression, but is offset by poorer cold-starting characteristics.

A distinction is made between two types of supercharger/turbocharger:
- On the *exhaust-gas turbocharger*, compression power is won from the exhaust gas (flow of exhaust gas between engine and turbocharger).
- On the *supercharger*, compression power is tapped from the engine crankshaft (mechanical coupling between engine and supercharger).

Volumetric efficiency
Volumetric efficiency refers to the relationship between the actual air charge trapped inside the cylinder and the theoretical air charge governed by the cylinder capacity under standard conditions (air pressure $p_0 = 1,013$ hPa, temperature $T_0 = 273$ K) without supercharging/turbocharging. On supercharged/turbocharged diesel engines, volumetric efficiency is within the range of 0.85...3.0.

Intercooling
In the process of being compressed by the turbocharger, air also heats up (to as much as 180°C). Since hot air is less dense than cold air, a higher air temperature has a negative effect on cylinder charge. A charge-air cooler (intercooler) downstream of the supercharger/turbocharger (cooled by ambient air or with a separate coolant circuit) cools the compressed air, thus increasing the cylinder charge further. It means that more oxygen is available for combustion, with the result that a higher maximum torque and, therefore, greater power output is available at a given engine speed.

The lower temperature of the air entering the cylinder also reduces the temperatures generated during the compression stroke. This has a number of advantages:
- Greater thermal efficiency and, therefore, lower fuel consumption and soot emission from diesel engines
- Reduced knock tendency in gasoline engines
- Lower thermal stresses on the cylinder block/head
- Slight reduction in NO_x emissions as a result of the lower combustion temperature

Turbocharging
Of the methods of assisted aspiration, the exhaust-gas-driven turbocharger is by far the most widely used. Turbochargers are used on engines for cars and commercial vehicles as well as on large, heavy-duty marine and locomotive engines.

The exhaust-gas turbocharger is used as a means of improving the power-to-weight ratio, and improving maximum torque at low to medium engine speeds, when it is normally fitted with electronic boost-pressure control. In addition, the aspects of minimizing pollutants also play a growing role.

[1] Even the pioneers of automotive engineering, Gottlieb Daimler (1885) and Rudolf Diesel (1896), considered the possibility of precompressing intake air in order to improve performance. But it was the Swiss Alfred Büchi who first successfully produced a turbocharger in 1925 – it boosted power output by 40% (application for the patent was made in 1905). The first turbocharged commercial-vehicle engines were built in 1938. They became widespread by the early 1950s.

[2] They became more widespread from the 1970s onwards.

Design and operating concept

The hot exhaust gas expelled under pressure from an internal-combustion engine represents a substantial loss of energy. It makes sense, therefore, to utilize some of that energy to generate pressure in the intake manifold.

The turbocharger (Fig. 1) is a combination of two turbo elements:

- An exhaust-gas turbine (7) that is driven by the flow of exhaust gas.
- A centrifugal turbo-compressor (2) that is directly coupled to the turbine by means of a shaft (11) and which compresses the intake air.

The hot exhaust gas flows into the turbine and, by so doing, forces it to rotate at high speeds (in diesel engines, up to around 200,000 rpm). The inward-facing blades of the turbine divert the flow of gas into the center from where it passes out to the side (8, radial-flow turbine). The connecting shaft drives the radial compressor. This is the exact reverse of the turbine: The intake air (3) is drawn in at the center of the compressor and is driven outwards by the blades of the impeller so that it is compressed (4).

As a result of exhaust-gas pressure that builds upstream of the turbine, the engine has to work harder to expel the exhaust gas on the exhaust stroke. Besides converting the flow energy of exhaust gas into compression power, the turbine also converts the thermal energy in the exhaust gas into compression power. As a result, the increase in charge-air pressure is greater than the rise in exhaust-gas pressure upstream of the turbine (positive scavenging drop). This improves the overall efficiency of the engine across large sections of the engine map.

For fixed-installation engines running at constant speed, the turbine and turbocharger characteristics can be tuned to a high level of efficiency and turbocharger pressure. Turbocharger design becomes more complicated when it is applied to road-vehicle engines that do not run under steady-state conditions – because they are expected to produce high torque levels, particularly when accelerating from slow speeds. Low exhaust-gas temperatures, low exhaust-gas flow rates, and the mass moment of inertia of the turbocharger itself all contribute to a slow buildup of pressure in the compressor at the start of acceleration. On turbocharged car engines, this is referred to as "turbo lag". Special turbochargers have been developed for supercharging/turbocharging in passenger cars and commercial vehicles. They respond at small exhaust-gas flow rates due to their low intrinsic mass, and thus improve performance in the lower rev band to a considerable extent.

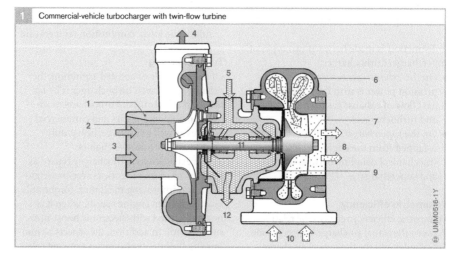

1   Commercial-vehicle turbocharger with twin-flow turbine

Fig. 1
1 Compressor housing
2 Centrifugal compressor
3 Intake air
4 Compressed intake air
5 Lubricant inlet
6 Turbine housing
7 Turbine
8 Exhaust-gas outflow
9 Bearing housing
10 Exhaust-gas inflow
11 Shaft
12 Lubricant return outlet

A distinction is made between two methods of turbocharging.

Constant-pressure turbocharging involves the use of an exhaust-gas accumulator upstream of the turbine to smooth out the pressure pulsations in the exhaust system. As a result, the turbine can accommodate a higher exhaust-gas flow rate at a lower pressure at high engine speeds. As the exhaust-gas back pressure that the engine is working against is lower under those operating conditions, fuel consumption is also lower. Constant-pressure turbocharging is used for large-scale marine, generator and fixed-installation engines.

Pulse turbocharging utilizes the kinetic energy of the pressure pulsations caused by the expulsion of the exhaust gas from the cylinders. Pulse turbocharging achieves higher torques at lower engine speeds. It is the principle used by turbochargers for cars and commercial vehicles. Separate exhaust manifolds are used for different banks of cylinders to prevent individual cylinders from interfering with each other during gas exchange, e.g. two groups of three cylinders on a six-cylinder engine. If twin-flow turbines – which have two outer channels – are used (Fig. 1), the exhaust flows are kept separate in the turbocharger as well.

In order to obtain good response characteristics, the turbocharger is positioned as close as possible to the exhaust valves in the flow of hot exhaust gas. It therefore has to be made of highly durable materials. On ships – where hot surfaces in the engine room have to be prevented because of the fire risk – turbochargers are water-cooled or enclosed in heat-insulating material. Turbochargers for gasoline engines, where the exhaust-gas temperatures can be 200...300°C higher than on diesel engines, may also be water-cooled.

Designs
Engines need to be able to generate high torque even at low speeds. For that reason, turbochargers are designed for low exhaust-gas mass flow rates (e.g. full load at an engine speed of $n \leq 1,800$ rpm). To prevent the turbocharger from overloading the engine at higher exhaust-gas mass flow rates, or being damaged itself, the turbocharger pressure has to be controlled. There are three turbocharger designs which can achieve this:
- The wastegate turbocharger
- The variable-turbine-geometry turbocharger and
- The variable-sleeve-turbine turbocharger

Wastegate turbocharger (Fig. 2)
At higher engine speeds or loads, part of the exhaust flow is diverted past the turbine by a bypass valve – the "wastegate" (5). This reduces the exhaust-gas flow passing through the turbine and lowers the exhaust-gas back pressure, thereby preventing excessive turbocharger speed.

At low engine speeds or loads, the wastegate closes and the entire exhaust flow passes through and drives the turbine.

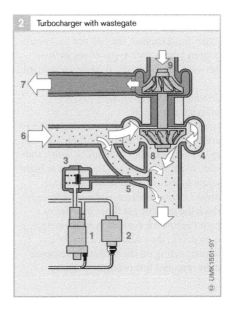

2 Turbocharger with wastegate

Fig. 2
1   Charge-pressure actuator
2   Vacuum pump
3   Pressure actuator
4   Turbocharger
5   Wastegate (bypass valve)
6   Exhaust flow
7   Intake air flow
8   Turbine
9   Centrifugal compressor

The wastegate usually takes the form of a flap integrated in the turbine housing. In the early days of turbocharger design, a poppet valve was used in a separate housing parallel to the turbine.

The wastegate is operated by an electro-pneumatic charge-pressure actuator (1). That actuator is an electrically operated 3/2-way valve that is connected to a vacuum pump (2). In its neutral position (de-energized) it allows atmospheric pressure to act on the pressure actuator (3). The spring in the pressure actuator opens the wastegate.

If a current is applied to the charge-pressure actuator by the engine control unit, it opens the connection between the pressure actuator and the vacuum pump so that the diaphragm is drawn back against the action of the spring. The wastegate closes and the turbocharger speed increases.

The turbocharger is designed in such a way that the wastegate will always open if the control system fails. This insures that, at high engine speeds, excessive turbocharger pressure which might damage the engine or the turbocharger itself cannot be produced.

On gasoline engines, sufficient vacuum is created by the intake manifold. Therefore, unlike diesel engines, they do not require a vacuum pump. Both types of engine may also use a purely electrical wastegate actuator.

Variable-turbine-geometry (VTG) turbocharger (Fig. 3)
Varying the rate of gas flow through the turbine by means of Variable Turbine Geometry (VTG) is another method by which the exhaust-gas flow rate can be limited at high engine speeds. The adjustable deflector blades (3) alter the size of the gap through which the exhaust gas flows in order to reach the turbine (variation of geometry). By so doing, they adjust the exhaust-gas pressure acting on the turbine in response to the required turbocharger pressure.

At low engine speeds or loads, they allow only a small gap for the exhaust gas to pass through so that the exhaust-gas back pressure increases. The exhaust-gas flow velocity through the turbine is then higher so that the turbine turns at a higher speed (a). In addition, the exhaust-gas flow is directed at the outer ends of the turbine blades. This generates more leverage which in turn produces greater torque.

At high engine speeds or loads, the deflector blades open up a larger gap for the exhaust gas to flow through with the result that the flow velocity is lower (b). Consequently, the turbocharger turns more slowly if the flow volume remains the same, or else its speed does not increase as much if the flow volume increases. In that way, the turbocharger pressure is limited.

3   Variable turbine geometry of VTG turbocharger

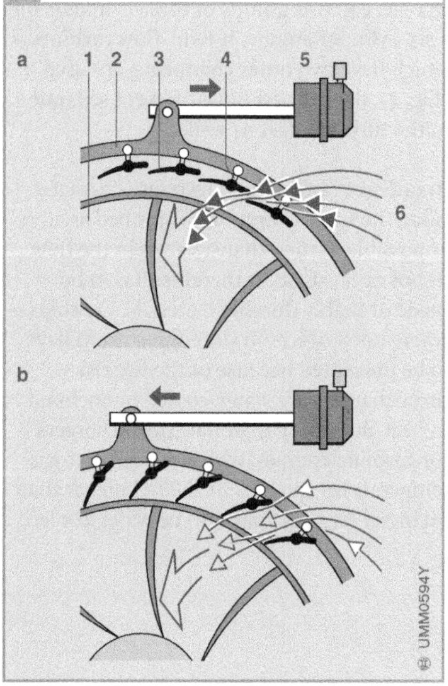

Fig. 3
a Deflector blade setting for high turbocharger pressure
b Deflector blade setting for low turbocharger pressure

1 Turbine
2 Adjusting ring
3 Deflector blade
4 Adjusting lever
5 Pneumatic actuator
6 Exhaust flow

◀— High flow rate
◁— Low flow rate

The deflector blade angle is adjusted very simply by turning an adjuster ring (2). This sets the deflector blades to the desired angle by operating them either directly using adjusting levers (4) attached to the blades or indirectly by means of adjuster cams. The adjusting ring is operated by a pneumatic actuator (5) to which positive or negative pressure is applied, or alternatively by an electric motor with position feedback (position sensor). The engine control unit controls the actuator. Thus the turbocharger pressure can be adjusted to the optimum setting in response to a range of input variables.

The VTG turbocharger is fully open in its neutral position and therefore inherently safe, i.e. if the control system fails, neither the turbocharger nor the engine suffers damage as a result. There is merely a loss of power at low engine speeds.

This is the type of turbocharger most widely used on diesel engines today. It has not been able to establish itself as the preferred choice for gasoline engines because of the high thermal stresses and the higher exhaust temperatures encountered.

Variable-sleeve-turbine (VST) turbocharger (Fig. 4)
The variable-sleeve-turbine turbocharger is used on small car engines. On this type of turbocharger, an intake slide valve (4) alters the cross-section of the inlet flow to the turbine by opening one or both of the intake ports (2, 3).

At low engine speeds or loads, only one of the intake ports is open (2). The small inlet aperture produces high exhaust-gas back pressure combined with a high exhaust-gas flow velocity, and consequently results in a high speed of rotation on the part of the turbine (1).

When the required turbocharger pressure is reached, the intake valve gradually opens the second intake port (3). The flow velocity of the exhaust gas – and therefore the turbine speed and the turbocharger pressure – then gradually reduce.
The engine control unit module controls the valve setting by means of a pneumatic actuator.

There is also a bypass channel (5) integrated in the turbine housing so that virtually the entire exhaust gas flow can be diverted past the turbine in order to obtain a very low turbocharger pressure.

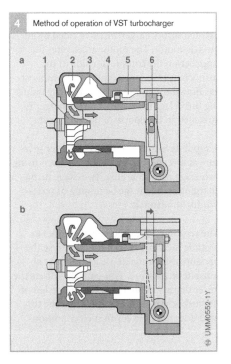

4 Method of operation of VST turbocharger

a   1   2   3   4   5   6

b

UMM0552-1Y

Fig. 4
a   Only one intake
    port open
b   Both intake
    ports open

1   Turbine
2   1st intake port
3   2nd intake port
4   Inlet slide valve
5   Bypass channel
6   Valve actuator

Advantages and disadvantages of
turbocharging

Downsizing

When compared with a conventionally
aspirated engine of equal power, the prime
advantage of a turbocharged engine is its
lighter weight and smaller dimensions. It
also has better torque characteristics within
the useful speed range (Fig. 5). Consequently, the power output at a given speed
is higher (A – B) at the same specific fuel
consumption.

The same amount of power is available at
a lower engine speed because of the superior
torque characteristics (B – C). Thus, with
a turbocharged engine, the point at which
a required amount of power is produced is
shifted to a position where frictional losses
are lower. The result of this is lower fuel
consumption (E – D).

Torque curve

At very low engine speeds, the basic torque
of a turbocharged engine is similar to that of
a conventionally aspirated engine. At that
point, the usable energy from the exhaust-
gas flow is insufficient to drive the turbine.
No turbocharger pressure is generated in
this way.

Under dynamic operating conditions,
the torque output remains similar to that
of a conventionally aspirated engine even at
medium engine speeds (c). This is because
of the delay in the build-up of the exhaust-
gas flow. On acceleration from slow speeds,
therefore, the "turbo lag" effect occurs.

On gasoline engines in particular, the turbo
lag can be minimized by utilizing the dynamic supercharging effect. This improves
the turbocharger's response characteristics.

On diesel engines, the use of turbochargers
with variable turbine geometry provides a
means of significantly reducing turbo lag.

Another design variation is the electrically
assisted turbocharger which is aided by an
electric motor. The motor accelerates the
impeller on the compressor side of the
turbocharger independently of the exhaust-
gas flow through the turbine, thereby reducing turbo lag. This type of turbocharger is
currently in the course of development.

A rapid development of turbocharger pressure at low speeds can also be achieved using
two-stage turbocharging. Two-stage turbocharging stands at the beginning of series-
production launch.

The response of turbocharged engines as
altitude increases is very good because the
pressure differential is greater at lower atmospheric pressure. This partially offsets the
lower density of air. However, the design of
the turbocharger must ensure that the turbine does not over-rev in such conditions.

Fig. 5

a Conventionally
   aspirated engine
   under steady-state
   conditions

b Turbocharged
   engine under
   steady-state
   conditions

c Turbocharged
   engine under
   dynamic conditions

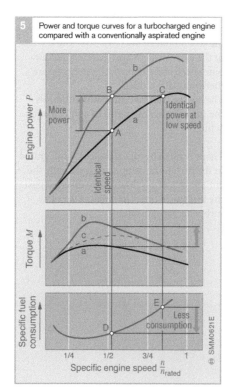

5 Power and torque curves for a turbocharged engine
compared with a conventionally aspirated engine

Multistage turbocharging
Multistage turbocharging is an improvement on single-stage turbocharging in that power limits can be significantly extended. The objective here is to improve air supply under both steady-state and dynamic operating conditions and at the same time improve the specific fuel consumption of the engine. Two methods of turbocharging have proved successful in this respect.

Sequential supercharging
Sequential supercharging involves the use of multiple turbochargers connected in parallel which successively cut in as engine load increases. Thus, in comparison with a single larger turbocharger which is geared to the engine's rated power output, two or more optimum levels of operation can be obtained. Because of the added expense of the supercharger sequencing control system, however, sequential supercharging is predominantly used on marine propulsion systems or generator engines.

Controlled two-stage turbocharging
Controlled two-stage turbocharging involves two differently dimensioned turbochargers connected in series with a controlled bypass and, ideally, two intercoolers (Fig. 6, 1 and 2). The first turbocharger is a low-pressure turbocharger (1) and the second, a high-pressure turbocharger (2). The intake air first undergoes precompression by the low-pressure turbocharger. Consequently, the relatively small high-pressure compressor in the second turbocharger is operating at a higher input pressure with a low volumetric flow rate, so that it can deliver the required air-mass flow rate. A particularly high level of compressor efficiency can be achieved with two-stage turbocharging.

At lower engine speeds, the bypass valve (5) is closed, so that both turbochargers are working. This provides for very rapid development of a high turbocharger pressure. As engine speed increases, the bypass valve gradually opens until eventually only the low-pressure turbocharger is operating. In this way, the turbocharging system adjusts evenly to the engine's requirements.

This method of turbocharging is used in automotive applications because of its straightforward control characteristics.

Electric booster
This is an additional compressor mounted upstream of the turbocharger. It is similar in design to the turbocharger's compressor but is driven by an electric motor. Under acceleration, the electric booster supplies the engine with extra air, thereby improving its response characteristics at low speeds in particular.

Supercharging
A supercharger consists of a compressor driven directly by the engine. The engine and the compressor are generally rigidly linked, e.g. by a belt drive system. Compared with turbochargers, superchargers are rarely used on diesel engines.

Positive-displacement supercharger
The most common type of supercharger is the positive-displacement supercharger. It is used mainly on small and medium-sized car engines. The following types of supercharger are used on diesel engines:

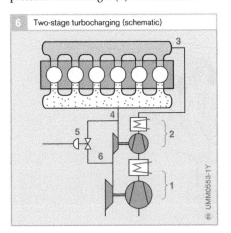

6 Two-stage turbocharging (schematic)

Fig. 6
1   Low-pressure stage
    (turbocharger with
    intercooler)
2   High-pressure stage
    (turbocharger with
    intercooler)
3   Intake manifold
4   Exhaust manifold
5   Bypass valve
6   Bypass pipe

Positive-displacement supercharger with internal compression

With this type of supercharger, the air is compressed inside the compressor. The types used on diesel engines are the reciprocating-piston supercharger and the helical-vane supercharger.

Reciprocating-piston supercharger: This type has either a rigid piston (Fig. 7) or a diaphragm (Fig. 8). A piston (similar to an engine piston) compresses the air which then passes through an outlet valve to the engine cylinder.

Helical-vane supercharger (Fig. 9): Two intermeshing helical vanes (4) compress the air.

Positive-displacement supercharger without internal compression

With this type of supercharger, the air is compressed outside of the supercharger by the action of the fluid flow generated. The only example of this type to be used on diesel engines was the Roots supercharger (Fig. 10) which was fitted to some two-stroke diesels.

Roots supercharger: Two contra-rotating rotary vanes (2) linked by gears rotate in contact with one another in similar fashion to a gear pump and in that way compress the intake air.

Fig. 7
1   Inlet valve
2   Outlet valve
3   Piston
4   Drive shaft
5   Casing

Fig. 8
1   Inlet valve
2   Outlet valve
3   Diaphragm
4   Drive shaft

Fig. 9
1   Drive pulley
2   Intake air
3   Compressed air
4   Helical vane

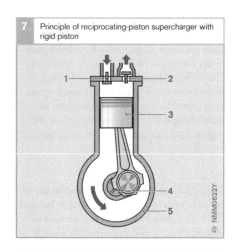

7 Principle of reciprocating-piston supercharger with rigid piston

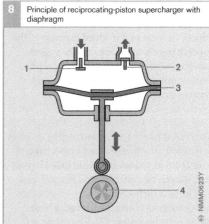

8 Principle of reciprocating-piston supercharger with diaphragm

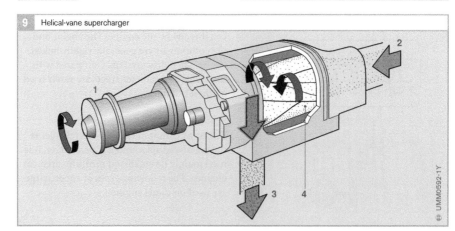

9 Helical-vane supercharger

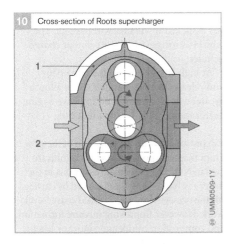

10   Cross-section of Roots supercharger

1

2

Fig. 10
1   Housing
2   Rotary vane

Centrifugal supercharger
In addition to the positive-displacement superchargers, there are also centrifugal superchargers (centrifugal-flow compressors) in which the compressor is similar to that in a turbocharger. In order to obtain the high peripheral velocity required, they are driven via a system of gears. This type of supercharger offers good volumetric efficiency over a wide range of speeds and can be seen as an alternative to the turbocharger for small engines. Mechanical centrifugal superchargers are also known as mechanical centrifugal turbo-compressors. Centrifugal turbochargers are rarely used on medium-sized or larger car engines.

Controlling supercharger pressure
The pressure generated by a supercharger can be controlled by means of a bypass. A proportion of the compressed air flow enters the cylinder and determines the cylinder charge. The remainder flows through the bypass and is returned to the intake side. The bypass valve is controlled by the engine control unit.

Advantages and disadvantages of supercharging
Because the supercharger is driven directly by the crankshaft, any increase in engine speed is instantaneously mirrored by an increase in compressor speed. This means that under dynamic operating conditions, higher engine torque and better response characteristics are obtained than with a turbocharger. If variable-speed gearing is used, the engine response to load changes can also be improved.

Since, however, the necessary power output for driving the compressor (approx. 10...15 kW for cars) is not available as effective engine output, those advantages are offset by a somewhat higher rate of fuel consumption than with a turbocharger. That disadvantage is mitigated if the compressor can be disconnected at low engine speeds and loads by means of a clutch operated by the engine control unit. This, on the other hand, makes the supercharger more expensive to produce. Another disadvantage of the supercharger is the greater amount of space it requires.

Dynamic supercharging
A degree of supercharging can be achieved simply by the utilization of dynamic effects in the intake manifold. Dynamic supercharging effects of this type are less important in diesel engines than they are for gasoline engines. In diesel engines, the main emphasis of intake-manifold design is on even distribution of the air charge between all cylinders and distribution of the recirculated exhaust gas. In addition, the creation of whirl effects inside the cylinders is also of importance. At the relatively low speeds at which diesel engines run, designing the intake manifold specifically to obtain dynamic supercharging effects would require it to be extremely long. Since virtually all modern diesel engines are equipped with turbochargers, the only benefit that could be achieved would be under non steady-state operating conditions where the turbocharger has not reached full delivery pressure.

In general, the intake manifold on a diesel engine is kept as short as possible. The advantages of this are
- Improved dynamic response characteristics and
- Better control characteristics on the part of the exhaust-gas recirculation system

Swirl flaps

The pattern of air flow inside of the cylinders of a diesel engine has a fundamental effect on mixture formation. It is mainly influenced by:
● The air flow generated by the injection jets.
● The movement of air flowing into the cylinder.
● The movement of the piston.

The whirl-assisted combustion process swirls the air during the induction and compression cycles to obtain rapid, complete mixture formation. Using special flaps and channels, the whirl can be regulated depending on engine speed and load.

The intake ducts are designed as fill channels (Fig. 1, 5) and swirl passages (2). The fill channels can be closed by a flap (swirl flap; 6). The flap is controlled by the engine control unit, depending on the program map. Besides a simple system with two positions, "Open" and "Closed", there are also position-controlled systems that allow intermediate positions.

The swirl flap is closed at low engine speeds. Air is sucked in via the swirl passage. The whirl is stronger when the cylinder charge is fuller.

At high engine speeds, the flap opens and releases the fill channels (5) to allow a greater cylinder charge and improve engine performance. Whirl is then reduced.

By controlling whirl as a function of the program map, it is possible to make significant cuts in NO_x and particulate emissions in the lower rev band. Flow losses caused by closing off the passages lead to increased charge-cycle work. However, improving mixture formation and combustion compensate more or less for any additional fuel consumption. A compromise between optimizing emission, fuel consumption, and performance is achieved dependent on engine load and speed.

Intake-duct switchoff is presently fitted to some car engines and is playing an increasingly important role in the emission-minimization concept.

However, modern truck diesel engines operate generally at very low whirl rates. Due to their smaller speed range and larger combustion chambers, the energy of the injection jets are sufficient to allow mixture formation.

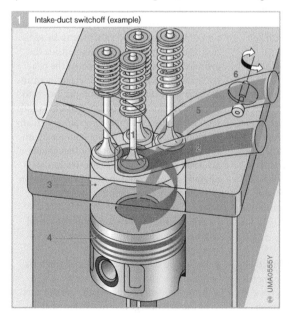

1   Intake-duct switchoff (example)

UMA0555Y

Fig. 1
1   Intake valve
2   Swirl passage
3   Engine cylinder
4   Piston
5   Fill channel
6   Flap

Intake air filters

The air filter filters the engine intake air and prevents any mineral dust or particles from entering the engine and becoming entrained in the engine oil. This reduces wear in the bearings, piston rings, cylinder walls, etc. It also protects the sensitive air-mass meter by preventing dust from depositing there. If this should happen, it could result in incorrect signals, higher fuel consumption, and higher pollutant emissions.

Typical air impurities include oil mist, aerosols, diesel soot, industrial waste gases, pollen, and dust. Dust particles drawn in together with the intake air have diameters that range from approx. 0.01 μm (soot particles) to approx. 2 mm (sand grains).

Filter medium and design

The air filters are normally deep-bed filters that retain particles in the filter-medium structure – as opposed to surface filters. Deep-bed filters with high dust retention capacities are always preferred when large volumetric flows with low particle concentrations need to be filtered efficiently.

Pressures achieve mass-related, overall separation efficiencies of up to 99.8 bar (passenger cars) and 99.95 bar (commercial vehicles). Such figures must be capable of being maintained under all prevailing conditions, including the dynamic conditions that exist in the air-intake system of an engine (pulsation). Filters of inadequate quality have greater dust passage rates under such circumstances.

The filter elements are individually designed for each engine. In this way, pressure losses can be kept to a minimum, and the high filtration rates are not dependent on the air-flow rate. The filter elements, which may be rectangular or cylindrical, consist of a filter medium that is folded so that the maximum possible filter surface area can be accommodated within the smallest possible space. Generally cellulose-fiber based, the filter medium is compressed and impreg-

nated to give it the required structural strength, wet rigidity, and chemical resistance.

The elements are changed at intervals specified by the vehicle manufacturer.

The demands for small and highly efficient filter elements (smaller space requirements) that also offer longer servicing intervals is the driving force behind the development of innovative, new air-filter media. New air-filter media made of synthetic fibers (Fig. 1), which have substantially improved performance figures in some cases, are already in production.

Better results than with purely cellulose-based media can be achieved using composite materials (e.g. paper with melt-blown layer) and special nano-fiber filter media, which consist of a relatively coarse base layer made of cellulose on which ultra-thin fibers with diameters of only 30 to 40 nm are applied. New folded structures with alternately sealed channels, similar to diesel soot filters, are soon to be launched on the market.

Conical, oval, and stepped or trapezoidal geometries add to the range of shapes available in order to optimize use of the space under the hood, which is becoming ever more confined.

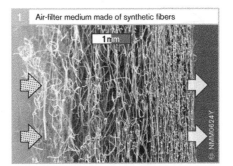

1   Air-filter medium made of synthetic fibers

1 mm

Fig. 1
Synthetic high-performance nonwoven filters, with gradually increasing density, and decreasing fiber diameter in the cross-section from the intake side to the clean-air side.
Source: Freudenberg Vliesstoffe KG

Mufflers

Previously, air-filter housings were almost exclusively designed as "muffler filters". Their large volume was designed for the supplementary function of reducing air intake noise. In the meantime, the two functions of filtration and engine-noise reduction have become increasingly separated and the different components independently optimized. This means that the filter housing can be reduced in size. And that results in very slim filters which can be integrated in the engine trim covers while the mufflers are placed in less accessible positions inside the engine compartment.

Air filters for cars

In addition to the housing (1 and 3) with the cylindrical air-filter element, the passenger-car air-intake module (Fig. 2) incorporates all the intake ducts (5 and 6) and the air-intake module (4). Arranged in-between

are Helmholtz resonators and lambda quarter pipes for acoustics. With the aid of this type of overall system optimization, the individual components can be better matched to one another. This helps to comply with the ever stricter noise-output restrictions.

Increasingly in demand are water-separating components, which are integrated in the air-intake system. They are used primarily to protect the air-mass sensor, which measures the air-mass flow. Water droplets which are drawn in through poorly situated intake fittings in the event of heavy rain, heavy splash water (e.g., on off-road vehicles) or snowfall and which reach the sensor can cause incorrect cylinder-charge readings to be taken.

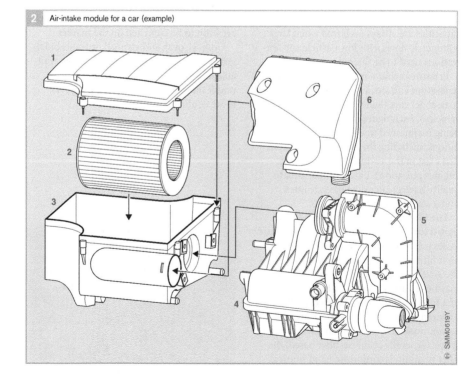

2   Air-intake module for a car (example)

Fig. 2

1   Housing lid
2   Filter element
3   Filter housing
4   Air-intake module
5   Intake duct
6   Intake duct

SMM0619Y

Splash plates or cyclone-like designs installed in the intake duct are used to separate out the water droplets. The shorter the distance from the air intake to the filter element, the more difficult it is to obtain a solution because only very low flow-pressure losses are permitted. However, it is also possible to use appropriately fitted filter elements which collect (coalesce) the water droplets and deflect the water film outwards ahead of the actual particulate-filter element. A housing specially designed for this purposes aids this process. This arrangement can also be successfully used for water separation even in the case of very short raw-air ducts.

Air filters for commercial vehicles

Figure 3 shows an easy-to-maintain and weight-optimized plastic air filter for commercial vehicles. In addition to having a very high filtration rate, the elements for this filter are dimensioned for servicing intervals of over 100,000 km. The servicing intervals are thus significantly longer than those for passenger cars.

In countries with high levels of atmospheric dust, and on construction and agricultural machines, a pre-filter is fitted upstream of the filter element. The pre-filter filters out coarse-grained, heavy dust particles, thereby substantially increasing the service life of the fine filter element.

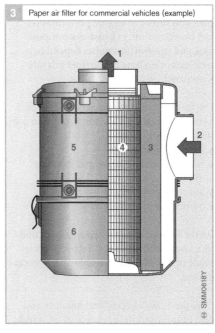

3   Paper air filter for commercial vehicles (example)

SMM0618Y

Fig. 3
1   Air outlet
2   Air inlet
3   Filter element
4   Supporting tube
5   Housing
6   Dust collector

In its most simple form, it is a ring of deflector vanes which set the air flow into a rotating motion. The resulting centrifugal force separates out the coarse dust particles. However, only mini-cyclone pre-filter batteries optimized for use in conjunction with the main filter element can properly utilize the potential of centrifugal separators in commercial-vehicle air filters.

Basic principles of diesel fuel injection

The combustion processes in the diesel engine, also linked to engine performance, fuel consumption, exhaust-gas composition, and combustion noise, depend to a great extent on how the air/fuel mixture is prepared.

The fuel-injection parameters that are decisive on the quality of the mixture formation are primarily:
- start of injection
- rate-of-discharge curve and injection duration
- injection pressure
- number of injection events

On the diesel engine, exhaust-gas and noise emissions are largely reduced by measures inside of the engine, i.e. combustion-process control.

Until the 1980s injected fuel quantity and start of injection were controlled on vehicle engines by mechanical means only. However, compliance with prevailing emission limits requires the high-precision adjustment of injection parameters, e.g. pre-injection, main injection, injected fuel quantity, injection pressure, and start of injection, adapted to the engine operating state. This is only achievable using an electronic control unit that calculates injection parameters as a factor of temperature, engine speed, load, altitude (elevation), etc. Electronic Diesel Control (EDC) has generally become widespread on diesel engines.

As exhaust-gas emission standards become more severe in future, further measures for minimizing pollutants will have to be introduced. Emissions, as well as combustion noise, can continue to be reduced by means of very high injection pressures, as achieved by the Unit Injector System, and by a rate-of-discharge curve that is adjustable independent of pressure buildup, as implemented by the common-rail system.

Mixture distribution

Excess-air factor λ

The excess-air factor λ (lambda) was introduced to indicate the degree by which the actual air/fuel mixture actually deviates from the stoichiometric[1]) mass ratio. It indicates the ratio of intake air mass to required air mass for stoichiometric combustion, thus:

$$\lambda = \frac{Air\ mass}{Fuel\ mass \cdot Stoichiometric\ ratio}$$

$\lambda = 1$: The intake air mass is equal to the air mass theoretically required to burn all of the fuel injected.

$\lambda < 1$: The intake air mass is less than the amount required and therefore the mixture is rich.

$\lambda > 1$: The intake air mass is greater than the amount required and therefore the mixture is lean.

---

[1]) The stoichiometric ratio indicates the air mass in kg required to completely burn 1 kg of fuel (m_L/m_K). For diesel fuel, this is approx. 14.5.

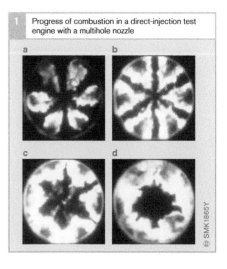

1 Progress of combustion in a direct-injection test engine with a multihole nozzle

a   b

c   d

SMK1865Y

Lambda levels in diesel engines

Rich areas of mixture are responsible for sooty combustion. In order to prevent the formation of too many rich areas of mixture, diesel engines – in contrast to gasoline engines – have to be run with an overall excess of air.

The lambda levels for turbocharged diesel engines at full load are between $\lambda = 1.15$ and $\lambda = 2.0$. When idling and under no-load conditions, those figures rise to $\lambda > 10$.

These excess-air factors represent the ratio of total masses of fuel and air in the cylinder. However, the lambda factor, which is subject to strong spatial fluctuation, is primarily responsible for auto-ignition and the production of pollutants.

Diesel engines operate with heterogeneous mixture formation and auto-ignition. It is not possible to achieve completely homogeneous mixing of the injected fuel with the air charge prior to or during combustion. Within the heterogeneous mixture encountered in a diesel engine, the localized excess-air factors can cover the entire range from $\lambda = 0$ (pure fuel) in the eye of the jet close to the injector to $\lambda = \infty$ (pure air) at the outer extremities of the spray jet. Around the outer zone of a single liquid droplet (vapor envelope), there are localized lambda levels of 0.3 to 1.5 (Figs. 2 and 3). From this, it can be deduced that optimized atomization (large numbers of very small droplets), high levels of excess air, and "metered" motion of the air charge produce large numbers of localized zones with lean, combustible lambda levels. This results in less soot occurring during combustion. EGR compatibility then rises, and NO_x emissions are reduced.

Optimized fuel atomization is achieved by high injection pressures that range up to max. 2,200 bar for UIS. Common-rail systems (CRS) operate at an injection pressure of max. 1,800 bar. This results is a high relative velocity between the jet of fuel and the air in the cylinder which has the effect of scattering the fuel jet.

With a view to reducing engine weight and cost, the aim is to obtain as much power as possible from a given engine capacity. To achieve this, the engine must run on the lowest possible excess air at high loads. On the other hand, a deficiency in excess air increases the amount of soot emissions. Therefore, soot has to be limited by precisely metering the injected fuel quantity to match the available air mass as a factor of engine speed.

Low atmospheric pressures (e.g. at high altitudes) also require the fuel volume to be adjusted to the smaller amount of available air.

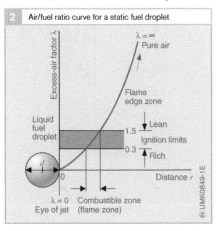

2   Air/fuel ratio curve for a static fuel droplet

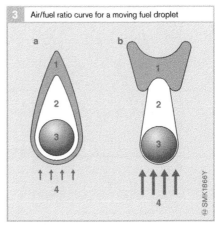

3   Air/fuel ratio curve for a moving fuel droplet

Fig. 2
d Droplet diameter (approx. 2...20 μm)

Fig. 3
a   Low relative velocity
b   High relative velocity

1   Flame zone
2   Vapor envelope
3   Fuel droplet
4   Air flow

Fuel-injection parameters

Start of injection and delivery

Start of injection

The point at which injection of fuel into the combustion chamber starts has a decisive effect on the point at which combustion of the air/fuel mixture starts, and therefore on emission levels, fuel consumption and combustion noise. For this reason, start of injection plays a major role in optimizing engine performance characteristics.

Start of injection specifies the position stated in degrees of crankshaft rotation relative to crankshaft Top Dead Center (TDC) at which the injection nozzle opens, and fuel is injected into the engine combustion chamber.

The position of the piston relative to top dead center at that moment influences the flow of air inside of the combustion chamber, as well as air density and temperature. Accordingly, the degree of mixing of air and fuel is also dependent on start of injection.

Thus, start of injection affects emissions such as soot, nitrogen oxides (NO_x), unburned hydrocarbons (HC), and carbon monoxide (CO).

The start-of-injection setpoints vary according to engine load, speed, and temperature. Optimized values are determined for each engine, taking into consideration the impacts on fuel consumption, pollutant emission, and noise. These values are then stored in a start-of-injection program map (Fig. 4). Load-dependent start-of-injection variability is controlled across the program map.

Compared with cam-controlled systems, common-rail systems offer more freedom in selecting the quantity and timing of injection events and injection pressure. As a consequence, fuel pressure is built up by a separate high-pressure pump, optimized to every operating point by the engine management system, and fuel injection is controlled by a solenoid valve or piezoelectric element.

Fig. 4
1   Cold start (< 0°C)
2   Full load
3   Medium load

Fig. 5
Example of an application:
α_N Optimum start of injection at no-load: low HC emissions while NO_x emissions at no load are low anyway.
α_V Optimum start of injection at full load: low NO_x emissions while HC emissions are low at full load anyway.

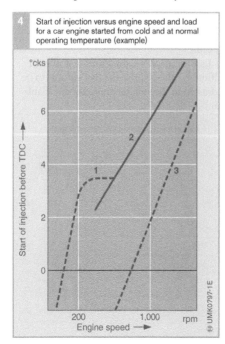

4 Start of injection versus engine speed and load for a car engine started from cold and at normal operating temperature (example)

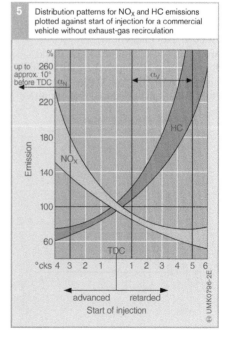

5 Distribution patterns for NO_x and HC emissions plotted against start of injection for a commercial vehicle without exhaust-gas recirculation

Standard values for start of injection
On a diesel-engine data map, the optimum points of combustion start for low fuel consumption are in the range of 0...8° crankshaft angle before TDC. As a result, and based on statutory exhaust-gas emission limits, the start of injection points are as follows:

Direct-injection car engines:
- No load: 2° crankshaft angle before TDC to 4° crankshaft angle after TDC
- Part load: 6° crankshaft angle before TDC to 4° crankshaft angle after TDC
- Full load: 6 to 15° crankshaft angle before TDC

Direct-injection commercial-vehicle engines (without exhaust-gas recirculation):
- No load: 4 to 12° crankshaft angle before TDC
- Full load: 3 to 6° crankshaft angle before TDC to 2° crankshaft angle after TDC

When the engine is cold, the start of injection for car and commercial-vehicle engines is 3 to 10° earlier. Combustion time at full load is 40 to 60° crankshaft angle.

Advanced start of injection
The highest compression temperature (final compression temperature) occurs shortly before piston Top Dead Center (TDC). If combustion starts a long way before TDC, combustion pressure rises steeply, and acts as a retarding force against the piston stroke. Heat lost in the process diminishes engine efficiency and, therefore, increases fuel consumption. The steep rise in compression pressure also makes combustion much noisier.

An advanced start of injection increases temperature in the combustion chamber. As a result, NO_x emission levels rise, but HC emissions are lower (Fig. 5).

Minimizing blue and white smoke levels requires advanced start of injection and/or pre-injection when the engine is cold.

Retarded start of injection
A retarded start of injection at low-load conditions can result in incomplete combustion and, therefore, in the emission of unburned hydrocarbons (HC) and carbon monoxide (CO) since the temperature in the combustion chamber is already dropping (Fig. 5).

The partially conflicting tradeoffs of specific fuel consumption and hydrocarbon emissions on the one hand, and soot (black smoke) and NO_x emissions on the other, demand compromises and very tight tolerances when modifying the start of injection to suit a particular engine.

Start of delivery
In addition to start of injection, start of delivery is another aspect that is often considered. It relates to the point at which the fuel-injection pump starts to deliver fuel to the injector.

On older fuel-injection systems, start of delivery plays an important role since the inline or distributor injection pump must be allocated to the engine. The relative timing between pump and engine is fixed at start of delivery, since this is easier to define than the actual start of injection. This is made possible because there is a definite relationship between start of delivery and start of injection (injection lag[1]).

Injection lag results from the time it takes the pressure wave to travel from the high-pressure pump through to the injection nozzle. Therefore, it depends on the length of the line. At different engine speeds, there is a different injection lag measured as a crankshaft angle (degrees of crankshaft rotation). At higher engine speeds, the engine has a greater ignition lag[2] related to the crankshaft position (in degrees of crankshaft angle). Both of these effects must be compensated for – which is why a fuel-injection system must be able to adjust the start of delivery/start of injection in response to engine speed, load, and temperature.

[1] Time or crankshaft angle swept from start of delivery through start of injection

[2] Time or crankshaft angle swept from start of injection through start of ignition

Injected-fuel quantity

The required fuel mass, m_e, for an engine cylinder per power stroke is calculated using the following equation:

$$m_e = \frac{P \cdot b_e \cdot 33.33}{n \cdot z} \text{ [mg/stroke]}$$

where:
P engine power in kilowatts
b_e engine specific fuel consumption in g/kWh
n engine speed in rpm
z number of engine cylinders

The corresponding fuel volume (injected fuel quantity), Q_H, in mm³/stroke or mm³/injection cycle is then:

$$Q_H = \frac{P \cdot b_e \cdot 1,000}{30 \cdot n \cdot z \cdot \rho} \text{ [mm}^3\text{/stroke]}$$

Fuel density, ρ, in g/cm³ is temperature-dependent.

Engine power output at an assumed constant level of efficiency ($\eta \sim 1/b_e$) is directly proportional to the injected fuel quantity.

The fuel mass injected by the fuel-injection system depends on the following variables:
● The fuel-metering cross-section of the injection nozzle
● The injection duration
● The variation over time of the difference between the injection pressure and the pressure in the combustion chamber
● The density of the fuel

Diesel fuel is compressible, i.e it is compressed at high pressures. This increases the injected fuel quantity. The deviation between the setpoint quantity in the program map and the actual quantity impacts on performance and pollutant emissions. In high-precision fuel-injection systems controlled by electronic diesel control, the required injected fuel quantity can be metered with a high degree of accuracy.

Injection duration

One of the main parameters of the rate-of-discharge curve is injection duration. During this period, the injection nozzle is open, and fuel flows into the combustion chamber. This parameter is specified in degrees of crankshaft or camshaft angle, or in milli-seconds. Different diesel combustion processes require different injection durations, as illustrated by the following examples (approximate figures at rated output):
● Passenger-car direct-injection (DI) engine approx. 32...38° crankshaft angle
● Indirect-injection car engines: 35...40° crankshaft angle
● Direct-injection commercial-vehicle engines: 25...36° crankshaft angle

A crankshaft angle of 30° during injection duration is equivalent to a camshaft angle of 15°. This results in an injection pump speed [1]) of 2,000 rpm, equivalent to an injection duration of 1.25 ms.

In order to minimize fuel consumption and emissions, the injection duration must be defined as a factor of the operating point and start of injection (Figs. 6 through 9).

---

[1]) Equivalent to half the engine speed on four-stroke engines

6
Specific fuel consumption b_e in g/kWh versus start of injection and injection duration

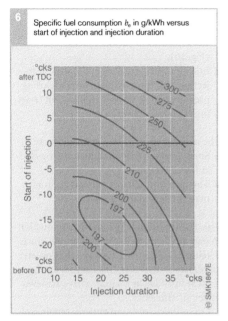

7
Specific nitrogen oxide (NO_x) emission in g/kWh versus start of injection and injection duration

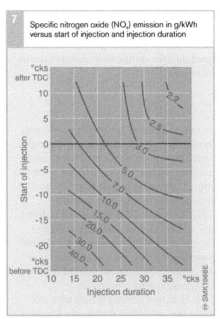

8
Specific emission of unburned hydrocarbons (HC) in g/kWh versus start of injection and injection duration

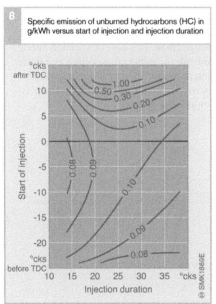

9
Specific soot emission in g/kWh versus start of injection and injection duration

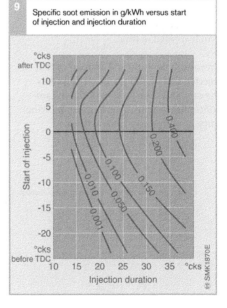

Figs. 6 to 9
Engine:
Six-cylinder commercial-vehicle diesel engine with common-rail fuel injection
Operating conditions:
$n = 1,400$ rpm,
50% full load.

The injection duration is varied in this example by changing the injection pressure to such an extent that a constant injected fuel quantity results for each injection event.

Rate-of-discharge curve

The rate-of-discharge curve describes the fuel-mass flow plotted against time when injected into the combustion chamber during the injection duration.

Rate-of-discharge curve on cam-controlled fuel-injection systems

On cam-controlled fuel-injection systems, pressure is built up continuously throughout the injection process by the fuel-injection pump. Thus, the speed of the pump has a direct impact on fuel delivery rate and, consequently, on injection pressure.

Port-controlled distributor and in-line fuel-injection pumps do not permit any pre-injection. With two-spring nozzle-and-holder assemblies, however, the injection rate can be reduced at the start of injection to improve combustion noise.

Pre-injection is also possible with solenoid-valve controlled distributor injection pumps. Unit Injector Systems (UIS) for passenger cars are equipped with hydro-mechanical pre-injection, but its control is only limited in time.

Pressure generation and delivery of the injected fuel quantity are interlinked by the cam and the injection pump in cam-controlled systems. This has the following impacts on injection characteristics:

- Injection pressure rises as engine speed and injected fuel quantity increase, and until maximum pressure is reached (Fig. 10).
- Injection pressure rises at the start of injection, but drops back to nozzle-closing pressure before the end of injection (starting at end of delivery).

The consequences of this are as follows:
- Small injected fuel quantities are injected at lower pressure.
- The rate-of-discharge curve is approximately triangular in shape.

This triangular curve promotes combustion in part-load and at low engine speeds since it achieves a shallower rise, and thus quieter combustion; however, this curve is less beneficial at full-load as a square curve achieves better air efficiency.

On indirect-injection engines (engines with prechamber or whirl chambers), throttling-pintle nozzles are used to produce a single jet of fuel and define the rate-of-discharge curve. This type of injection nozzle controls the outlet cross-section as a function of needle lift. It produces a gradual increase in pressure and, consequently, "quiet combustion".

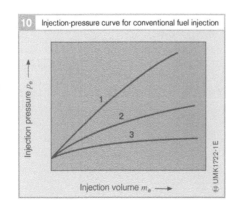

10   Injection-pressure curve for conventional fuel injection

Injection pressure p_e →

Injection volume m_e →

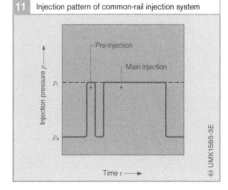

11   Injection pattern of common-rail injection system

Injection pressure p →

Pre-injection

Main injection

p_r

p_0

Time t →

Rate-of-discharge curve in the common-rail system

A high-pressure pump generates the fuel-rail pressure independently of the injection cycle. Injection pressure during the injection process is virtually constant (Fig. 11). At a given system pressure, the injected fuel quantity is proportional to the length of time the injector is open, and it is independent of engine or pump speed (time-based injection).

This results in an almost square rate-of-discharge curve which intensifies with short injection durations and the almost constant, high spray velocities at full-load, thus permitting higher specific power outputs.

However, this is not beneficial to combustion noise since a large quantity of fuel is injected during ignition lag because of the high injection rate at the start of injection. This leads to a high pressure rise during premixed combustion. As it is possible to exclude up to two pre-injection events, the combustion chamber can be preconditioned. This shortens ignition lag and achieves the lowest possible noise emissions.

Since the electronic control unit triggers the injectors, start of injection, injection duration, and injection pressure are freely definable for the various engine operating points in an engine application. They are controlled by Electronic Diesel Control (EDC). EDC balances out injected-fuel-quantity spread in individual injectors by means of injector delivery compensation (IMA).

Modern piezoelectric common-rail fuel-injection systems permit several pre-injection and secondary injection events. In fact, up to five injection events are possible during a power cycle.

Fig. 12
Adjustments aimed at low NO_x levels require starts of injection close to TDC.
The fuel delivery point is significantly in advance of the start: injection lag is dependent on the fuel-injection system

1   Pre-injection
2   Main injection
3   Steep pressure gradient (common-rail system)
4   "Boot-shaped" pressure rise (UPS with 2-stage opening solenoid-valve needle (CCRS). Dual-spring nozzle holders can achieve a boot-shaped curve of the needle lift (not pressure curve!).
5   Gradual pressure gradient (conventional fuel injection)
6   Flat pressure drop (in-line and distributor injection pumps)
7   Steep pressure drop (UIS, UPS, slightly less steep with common rail)
8   Advanced secondary injection
9   Retarded post-injection

p_s Peak pressure
p_o Nozzle-opening pressure
b   Duration of combustion for main injection phase
v   Duration of combustion for pre-injection phase
IL   Ignition lag of main injection

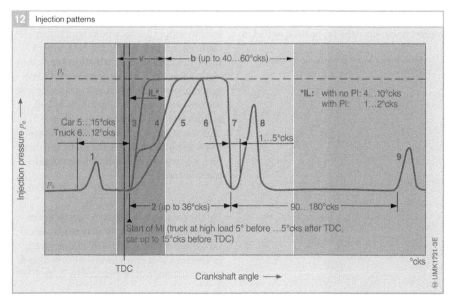

12   Injection patterns

b (up to 40...60°cks)

*IL: with no PI: 4...10°cks
       with PI: 1...2°cks

Car 5...15°cks
Truck 6...12°cks

1...5°cks

2 (up to 36°cks)   90...180°cks

Start of MI (truck at high load 5° before ...5°cks after TDC, car up to 15°cks before TDC)

TDC   Crankshaft angle ⟶   °cks

Injection pressure p_e

UMK1721-3E

Injection functions

Depending on the application for which the engine is intended, the following injection functions are required (Fig. 12):

- *Pre-injection* (1) reduces combustion noise and NO_x emissions, in particular on DI engines.
- *Positive-pressure gradient* during the main injection event (3) reduces NO_x emissions on engines without exhaust-gas recirculation.
- *Two-stage pressure gradient (4)* during the main injection event reduces NO_x and soot emissions on engines without exhaust-gas recirculation.
- *Constant high pressure* during the main injection event (3, 7) reduces soot emissions when operating the engine with exhaust-gas recirculation.
- *Advanced secondary injection* (8) reduces soot emissions.
- *Retarded secondary injection* (9).

Pre-injection

The pressure and temperature levels in the cylinder at the point of main injection rise if a small fuel quantity (approx. 1 mg) is burned during the compression phase. This shortens the ignition lag of the main injection event and has a positive impact on combustion noise, since the proportion of fuel in the premixed combustion process decreases. At the same time the quantity of diffuse fuel combusted increases. This increases soot and NO_x emissions, also due to the higher temperature prevailing in the cylinder.

On the other hand, the higher combustion-chamber temperatures are favorable mainly at cold start and in the low load range in order to stabilize combustion and reduce HC and CO emissions.

A good compromise between combustion noise and NO_x emissions is obtainable by adapting the time interval between pre-injection and main injection dependent on the operating point, and metering the pre-injected fuel quantity.

Retarded secondary injection

With retarded secondary injection, fuel is not combusted, but is evaporated by residual heat in the exhaust gas. The secondary-injection phase follows the main-injection phase during the expansion or exhaust stroke at a point up to 200° crankshaft angle after TDC. It injects a precisely metered quantity of fuel into the exhaust gas. The resulting mixture of fuel and exhaust gas is expelled through the exhaust ports into the exhaust-gas system during the exhaust stroke.

Retarded secondary injection is mainly used to supply hydrocarbons which also cause an increase in exhaust-gas temperature by oxidation in an oxidation-type catalytic converter. This measure is used to regenerate downstream exhaust-gas treatment systems, such as particulate filters or NO_x accumulator-type catalytic converters.

Since retarded secondary injection may cause thinning of the engine oil by the diesel fuel, it needs clarification with the engine manufacturer.

Advanced secondary injection

On the common-rail system, secondary injection can occur directly after main injection while combustion is still taking place. In this way, soot particles are reburned, and soot emissions can be reduced by 20 to 70%.

13   Effect of pre-injection on combustion-pressure pattern

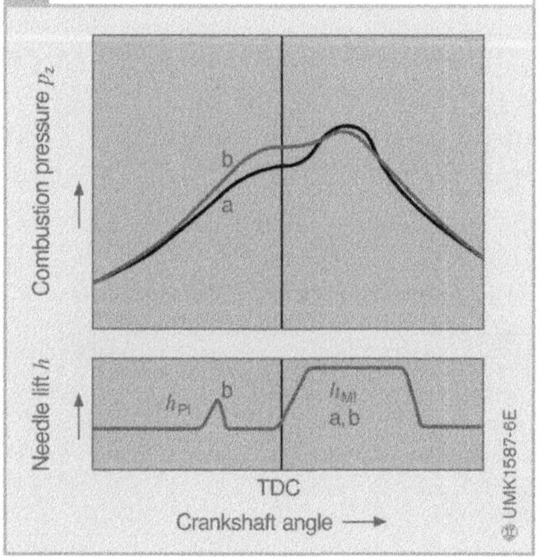

Fig. 13
a   Without pre-injection
b   With pre-injection

h_{PI} Needle lift during pre-injection
h_{MI} Needle lift during main injection

Timing characteristics of fuel-injection systems

Figure 14 presents an example of a radial-piston distributor pump (VP44). The cam on the cam ring starts delivery, and fuel then exits from the nozzle. It shows that pressure and injection patterns vary greatly between the pump and the nozzle, and are determined by the characteristics of the components that control injection (cam, pump, high-pressure valve, fuel line, and nozzle). For this reason, the fuel-injection system must be precisely matched to the engine.

The characteristics are similar for all fuel-injection systems in which pressure is generated by a pump plunger (in-line injection pumps, unit injectors, and unit pumps).

Detrimental volume in conventional injection systems

The term "detrimental volume" refers to the volume of fuel on the high-pressure side of the fuel-injection system. This is made up of the high-pressure side of the fuel-injection pump, the high-pressure fuel lines, and the volume of the nozzle-and-holder assembly. Every time fuel is injected, the detrimental volume is pressurized and depressurized. As a result, compression losses occur, thus retarding injection lag. The fuel volume inside of the pipes is compressed by the dynamic processes generated by the pressure wave.

The greater the detrimental volume, the poorer the hydraulic efficiency of the fuel-injection system. A major consideration when developing a fuel-injection system is, therefore, to minimize detrimental volume as much as possible. The unit injector system has the smallest detrimental volume.

In order to guarantee uniform control of the engine, the detrimental volume must be equal for all cylinders.

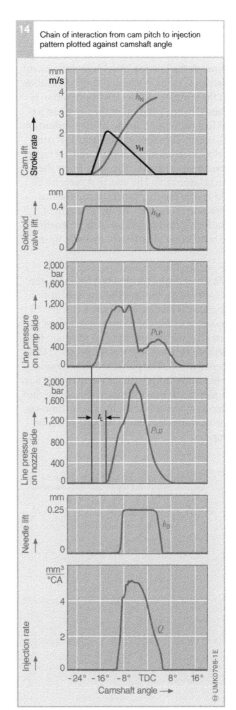

14 Chain of interaction from cam pitch to injection pattern plotted against camshaft angle

Fig. 14

Example of radial-piston distributor injection pump (VP44) at full load without pre-injection

t_L Time for fuel to pass through line

Injection pressure

The process of fuel injection uses pressure in the fuel system to induce the flow of fuel through the injector jets. A high fuel-system pressure results in a high rate of fuel outflow at the injection nozzle. Fuel atomization is caused by the collision of the turbulent jet of fuel with the air inside of the combustion chamber. Therefore, the higher the relative velocity between fuel and air, and the higher the density of the air, the more finely the fuel is atomized. The injection pressure at the nozzle may be higher than in the fuel-injection pump because of the length of the high-pressure fuel line, whose length is matched to the reflected pressure wave.

Direct-injection (DI) engines

On diesel engines with direct injection, the velocity of the air inside of the combustion chamber is relatively slow since it only moves as a result of its mass moment of inertia (i.e. the air "attempts" to maintain the velocity at which it enters the cylinder; this causes whirl). The piston stroke intensifies whirl in the cylinder since the restricted flow forces the air into the piston recess, and thus into a smaller diameter. In general, however, air motion is less and in indirect-injection engines.

The fuel must be injected at high pressure due to low air flow. Modern direct-injection systems now generate full-load peak pressures of 1,000...2,050 bar for car engines, and 1,000...2,200 bar for commercial vehicles. However, peak pressure is available only at higher engine speeds – except on the common-rail system.

A decisive factor to obtain an ideal torque curve with low-smoke operation (i.e. with low particulate emission) is a relatively high injection pressure adapted to the combustion process at low full-load engine speeds. Since the air density in the cylinder is relatively low at low engine speeds, injection pressure must be limited to avoid depositing fuel on the cylinder wall. Above about 2,000 rpm, the maximum charge-air pressure becomes available, and injection pressure can rise to maximum.

To obtain ideal engine efficiency, fuel must be injected within a specific, engine-speed-dependent angle window on either side of TDC. At high engine speeds (rated output), therefore, high injection pressures are required to shorten the injection duration.

Engines with indirect injection (IDI)

On diesel engines with divided combustion chambers, rising combustion pressure expels the charge out of the prechamber or whirl chamber into the main combustion chamber. This process runs at high air velocities in the whirl chamber, in the connecting passage between the whirl chamber, and the main combustion chamber.

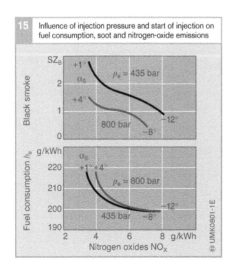

15 Influence of injection pressure and start of injection on fuel consumption, soot and nitrogen-oxide emissions

Fig. 15

Direct-injection engine, engine speed 1,200 rpm, mean pressure 16.2 bar

p_e Injection pressure
α_S Start of injection after TDC
SZ_B Black smoke number

Nozzle and nozzle holder designs

Secondary injection

Unintended secondary injection has a particularly undesirable effect on exhaust-gas quality. Secondary injection occurs when the injection nozzle shortly re-opens after closing and allows poorly conditioned fuel to be injected into the cylinder at a late stage in the combustion process. This fuel is not completely burned, or may not be burned at all, with the result that it is released into the exhaust gas as unburned hydrocarbons. This undesirable effect can be prevented by rapidly closing nozzle-and-holder assemblies, at sufficiently high closing pressure and low static pressure in the supply line.

Dead volume

Dead volume in the injection nozzle on the cylinder side of the needle-seal seats has a similar effect to secondary injection. The fuel accumulated in such a volume runs into the combustion chamber on completion of combustion, and partly escapes into the exhaust pipe. This fuel component similarly increases the level of unburned hydrocarbons in the exhaust gas (Fig. 1). Sac-less (vco) nozzles, in which the injection orifices are drilled into the needle-seal seats, have the smallest dead volume.

Injection direction

Direct-injection (DI) engines

Diesel engines with direct injection generally have hole-type nozzles with between 4 and 10 injection orifices (most commonly 6 to 8 injection orifices) arranged as centrally as possible. The injection direction is very precisely matched to the combustion chamber. Divergences of the order of only 2 degrees from the optimum injection direction lead to a detectable increase in soot emission and fuel consumption.

Engines with indirect injection (IDI)

Indirect-injection engines use pintle nozzles with only a single injection jet. The nozzle injects the fuel into the precombustion or whirl chamber in such a way that the glow plug is just within the injection jet. The injection direction is matched precisely to the combustion chamber. Any deviations in injection direction result in poorer utilization of combustion air and, therefore, to an increase in soot and hydrocarbon emissions.

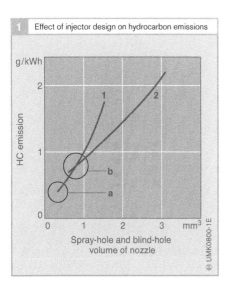

1 Effect of injector design on hydrocarbon emissions

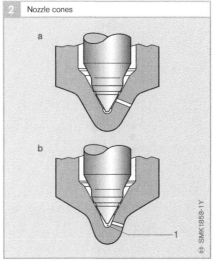

2 Nozzle cones

Fig. 1
a Sac-less (vco) nozzle
b Injector with micro-blind hole

1 Engine with 1 l/cylinder
2 Engine with 2 l/cylinder

Fig. 2
a Sac-less (vco) nozzle
b Injector with micro-blind hole

1 Dead volume

Overview of diesel fuel-injection systems

The fuel-injection system injects fuel into the combustion chamber at high pressure, at the right time, and in the right quantity. The main components of the fuel-injection system are the injection pump that generates high pressure, and the injection nozzles that are linked to the injection pump via high-pressure delivery lines – except with the Unit Injector System. The injection nozzles project into the combustion chamber of each cylinder.

In most systems, the nozzle opens when the fuel pressure reaches a specific opening pressure, and closes when it drops below this pressure. The nozzle is only controlled externally by an electronic controller in the common-rail system.

Designs

The main differences between fuel-injection systems are in the high-pressure generation system, and in the control of start of injection and injection duration. Whereas older systems still have mechanical control only, electronic control is now widespread.

In-line fuel-injection pumps
Standard in-line fuel-injection pumps
In-line fuel-injection pumps (Fig. 1) have a separate pump element consisting of a barrel (1) and plunger (4) for each engine cylinder. The pump plunger is moved in the delivery direction (in this case upwards) by the camshaft (7) integrated in the fuel-injection pump, is driven by the engine, and is returned to its starting position by the plunger spring (5). The individual pump units are generally arranged in-line (hence the name in-line fuel-injection pump).

The stroke of the plunger is invariable. The point at which the top edge of the plunger closes off the inlet port (2) on its upward stroke marks the beginning of the pressure generation phase. This point is referred to as the start of delivery. The plunger continues to move upwards. The fuel pressure therefore increases, the nozzle opens and fuel is injected into the combustion chamber.

When the helix (3) of the plunger clears the inlet port, fuel can escape and pressure is lost. The nozzle closes and fuel injection ceases.

The piston travel between opening and closing the inlet opening is the effective stroke.

Fig. 1
a   Standard in-line
    fuel-injection pump
b   In-line control-sleeve
    fuel-injection pump

1   Pump barrel
2   Inlet port
3   Helix
4   Pump plunger
5   Plunger spring
6   Adjustment range
    using control rack
    (injected-fuel
    quantity)
7   Camshaft
8   Control sleeve
9   Adjustment range
    using actuator shaft
    (start of delivery)
10 Fuel outflow to
    injection nozzle
X   Effective stroke

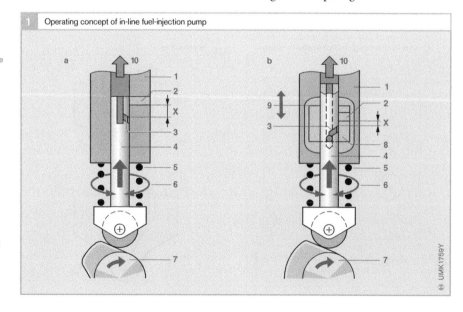

1   Operating concept of in-line fuel-injection pump

2   Method of operation of port-controlled axial-piston distributor injection pump

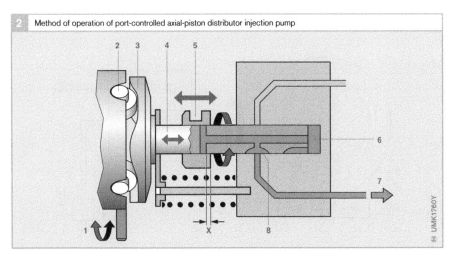

Fig. 2
1   Injection timing
    adjustment range
    on roller ring
2   Roller
3   Cam plate
4   Axial piston
5   Control sleeve
6   High-pressure
    chamber
7   Fuel outflow to
    injection nozzle
8   Spill port
X   Effective stroke

The greater the effective stroke, the greater the delivery quantity and injected fuel quantity.

The pump plunger is turned by a control rack to control the injected fuel quantity as a factor of engine speed and load. This changes the position of the helix relative to the inlet opening, and thus the effective stroke. The control rack is controlled by a mechanical governor, or an electrical actuator.

Injection pumps that work according to this principle are called "port-controlled".

Control-sleeve in-line fuel-injection pumps
The control-sleeve in-line fuel-injection pump has a control sleeve on the pump plunger (Fig. 1, 8) to change the LPC – i.e. the plunger lift to port closing – via an actuator shaft that shifts the start of delivery.

Control-sleeve in-line fuel-injection pumps are always electronically controlled. The injected fuel quantity and start of injection are adjusted according to calculated setpoint values.

With the standard in-line fuel-injection pump, however, the start of injection is dependent on engine speed.

Distributor injection pumps
Distributor injection pumps have only one pump unit that serves all cylinders (Figs. 2 and 3). A vane pump forces the fuel into the high-pressure chamber (6). High pressure is generated by an axial piston (Fig. 2, 4), or several radial pistons (Fig. 3, 4). A rotating central distributor plunger opens and closes metering slots (8) and spill ports, thereby distributing fuel to the individual engine cylinders. The injection duration is controlled by a control collar (Fig. 2, 5) or a high-pressure solenoid valve (Fig. 3, 5).

Axial-piston distributor pumps
A rotating cam plate (Fig. 2, 3) is driven by the engine. The number of cam lobes on the bottom of the cam plate is equal to the number of engine cylinders. They travel over rollers (2) on the roller ring and thus cause the distributor piston to describe a rotating as well as a lifting motion. In the course of each rotation of the drive shaft, the piston completes a number of strokes equal to the number of engine cylinders to be supplied.

3 Operating concept of solenoid-valve controlled radial-piston distributor injection pump

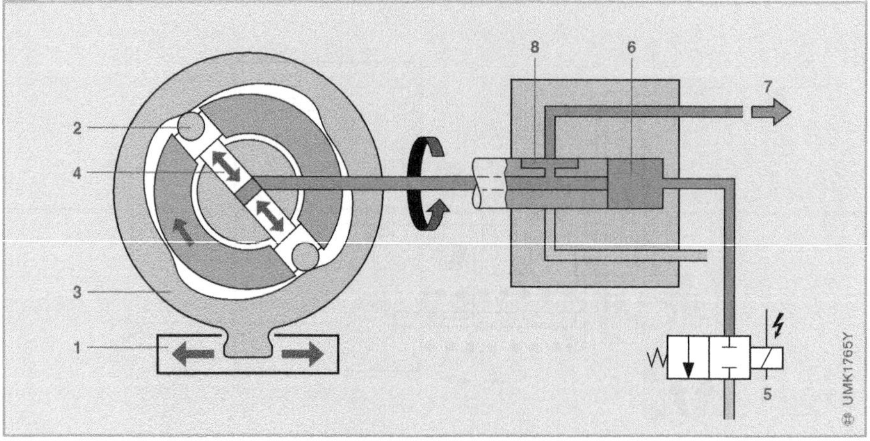

Fig. 3
1 Injection timing
   adjustment range
   on cam ring
2 Roller
3 Cam ring
4 Radial piston
5 High-pressure
   solenoid valve
6 High-pressure
   chamber
7 Fuel outflow to
   injection nozzle
8 Spill port

In a port-controlled axial-piston distributor pump with mechanical governor or electronically controlled actuator mechanism, a control collar (5) determines the effective stroke, thereby controlling the injected-fuel quantity.

The timing device can vary the pump's start of delivery by turning the roller ring.

Radial-piston distributor pumps
High pressure is generated by a radial-piston pump with cam ring (Fig. 3, 3), and from two to four radial pistons (4). Radial-piston pumps can generate higher injection pressures than axial-piston pumps. However, they have to be capable of withstanding greater mechanical stresses.

The cam ring can be rotated by the timing device (1), and this shifts the start of delivery. With radial-piston distributor pumps, the start of injection and injection duration are always controlled by solenoid valve.

Solenoid-valve-controlled distributor injection pumps
With this type of distributor injection pump, an electronically controlled high-pressure solenoid valve (5) meters the injected-fuel quantity and varies the start of injection. When the solenoid valve is closed, pressure can build up in the high-pressure chamber (6). When it is open, the fuel escapes so that no pressure buildup occurs and therefore fuel injection is not possible. One or two electronic control units (pump control unit and engine control unit) generate the control and regulation signals.

Individual injection pumps type PF
Individual injection pumps of type PF (pump with external drive) are directly driven by the engine camshaft. They are mainly fitted to marine engines, diesel locomotives, construction machinery, and low-power engines. Besides the cam for engine valve timing, the engine camshaft has drive cams for the individual injection pumps.

Otherwise, the single-plunger fuel-injection pump of type PF basically operates in the same way as in-line fuel-injection pumps.

Unit Injector System (UIS)

In a Unit Injector System, UIS, the fuel-injection pump and the injection nozzle form a single unit (Fig. 4). There is a unit injector fitted in the cylinder head for each cylinder. It is actuated either directly by a tappet, or indirectly by a rocker arm driven by the engine camshaft.

The integrated construction of the unit injector dispenses with the high-pressure line between the fuel-injection pump and the injection nozzle otherwise required on other fuel-injection systems. The unit-injector system can therefore be designed to operate at higher injection pressures. The maximum injection pressure is presently about 2,200 bar (on commercial vehicles).

The unit injector system is controlled electronically. An electronic control unit calculates start of injection and injection duration, which are controlled by a high-pressure solenoid valve.

Unit Pump System (UPS)

The modular Unit Pump System (UPS) operates on the same principle as the unit injector system (Fig. 5). In contrast to the unit injector system, however, the nozzle-and-holder assembly (2) and the fuel-injection pump are linked by a short high-pressure line (3) specifically designed for the system components. This separation of high-pressure generation and nozzle-and-holder assembly allows for a simpler attachment to the engine. There is one unit pump assembly (fuel-injection pump, fuel line, and nozzle-and-holder assembly) for each engine cylinder. The unit pump assemblies are driven by the engine camshaft (6).

As with the unit injector system, the unit pump system uses an electronically controlled fast-switching high-pressure solenoid valve (4) to regulate injection duration and start of injection.

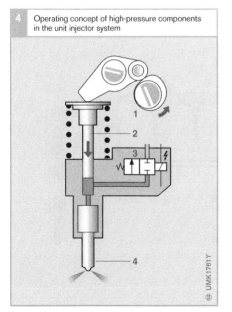

4 Operating concept of high-pressure components in the unit injector system

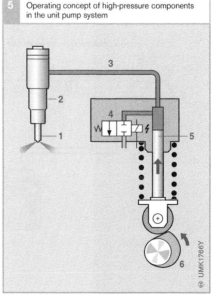

5 Operating concept of high-pressure components in the unit pump system

Fig. 4
1 Drive cam
2 Pump plunger
3 High-pressure solenoid valve
4 Injection nozzle

Fig. 5
1 Injection nozzle
2 Nozzle-and-holder assembly
3 High-pressure fuel line
4 High-pressure solenoid valve
5 Pump plunger
6 Drive cam

Common-Rail system (CR)

In the common-rail pressure-accumulator fuel-injection system, the functions of pressure generation and fuel injection are separate. This takes place by means of an accumulator volume composed of the common rail and the injectors (Fig. 6). Injection pressure is largely independent of engine speed or injected-fuel quantity, and is generated by a high-pressure pump. This system thus offers a high degree of flexibility in designing the fuel-injection process.

Presently, pressures range up to 1,600 bar (passenger cars) and 1,800 bar (commercial vehicles).

Functional description

A presupply pump feeds fuel via a filter and water separator to a high-pressure pump. The high-pressure pump ensures that the required fuel pressure in the rail is constantly high.

The Electronic Diesel Control (EDC) calculates the injection point and injected fuel quantity dependent on engine operating state, ambient conditions, and rail pressure.

Fuel is metered by controlling injection time and injection pressure. Pressure is controlled by the pressure-control valve which returns excess fuel to the fuel tank. In a more recent CR generation, metering is performed by a metering unit in the low-pressure stage to control the pump delivery rate.

The injector is connected to the fuel rail by short supply lines. Solenoid-valve injectors were used on previous CR generations. The latest system uses piezo-inline injectors. Their moves masses and inner friction has been reduced. This allows very short intervals between injection events and has a positive impact on fuel consumption.

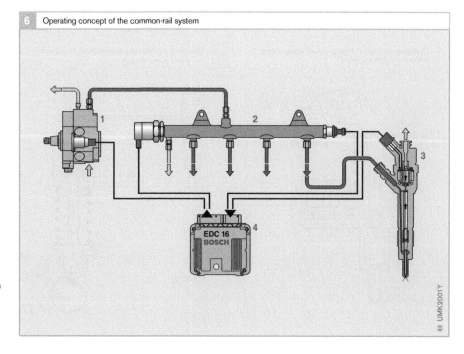

6   Operating concept of the common-rail system

UMK2001Y

Fig. 6
1   High-pressure pump
2   Fuel rail
3   Injection nozzle
4   EDC electronic
     control unit

History of diesel fuel injection

Bosch started development on a fuel-injection system for diesel engines in 1922 The technological omens were good: Bosch had experience with internal-combustion engines, its production systems were highly advanced and, above all, expertise developed in the production of lubrication pumps could be utilized. Nevertheless, this step was still a substantial risk for Bosch as there were many difficulties to be overcome.

The first volume-production fuel-injection pumps appeared in 1927. At the time, the level of precision of the product was unmatched. They were small, light, and enabled diesel engines to run at higher speeds. These in-line fuel-injection pumps were used on commercial vehicles from 1932 and in cars from 1936. Since that time, the technological advancement of the diesel engine and its fuel-injection systems has continued unabated.

In 1962, the distributor injection pump with automatic timing device developed by Bosch gave the diesel engine an additional boost. More than two decades later, many years of intensive development work at Bosch culminated in the arrival of the electronically controlled diesel fuel-injection system.

The pursuit of ever more precise metering of minute volumes of fuel delivered at exactly the right moment coupled with the aim of increasing the injection pressure is a constant challenge for developers. This has led to many more innovations in the design of fuel-injection systems (see figure).

In terms of fuel consumption and energy efficiency, the compression-ignition engine remains the benchmark.

New fuel-injection systems have helped to further exploit its potential. In addition, engine performance has been continually improved while noise and exhaust-gas emissions have been consistently lowered.

Milestones in diesel fuel injection

1927
First series-production in-line pump

1962
First axial-piston distributor pump, the EP-VM

1986
The first electronically controlled axial-piston distributor pump

1994
First Unit Injector System (UIS) for commercial vehicles

1995
First Unit Pump System (UPS)

1996
First radial-piston distributor pump

1997
First Common Rail accumulator injection system (CRS)

1998
First Unit Injector System (UIS) for passenger cars

UMK1753E

Fuel supply system to the low-pressure stage

The function of the fuel supply system is to store and filter the required fuel, and to provide the fuel-injection system with fuel at a specific supply pressure in all operating conditions. For some applications, the fuel return flow is also cooled.

Essentially, the fuel-supply system differs greatly, depending on the fuel-injection system used, as the following figures for radial-piston pump, common-rail system and passenger-car UIS show.

Overview

The fuel-supply system comprises the following main components (Figs. 1, 2, 3):
● Fuel tank
● Pre-filter
● Control unit cooler (optional)
● Presupply pump (optional, also in-tank pump on cars)
● Fuel filter
● Fuel pump (low-pressure)
● Pressure-control valve (overflow valve)
● Fuel cooler (optional)
● Low-pressure fuel lines

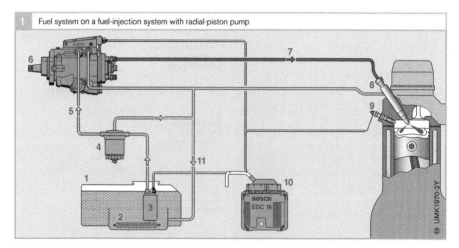

1 Fuel system on a fuel-injection system with radial-piston pump

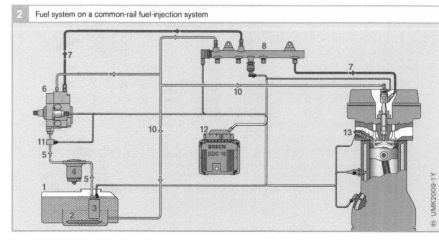

2 Fuel system on a common-rail fuel-injection system

Fuel tank

The fuel tank stores the fuel. It must be cor-rosion-resistant and leakproof at double the operating pressure, but at least at 0.3 bar. Any gauge pressure must be relieved auto-matically by suitable vents or safety valves. When the vehicle is negotiating corners, inclines or bumps, fuel must not escape past the filler cap or leak out of the pressure-relief vents or valves.

The fuel tank must be separated from the engine to prevent the fuel from igniting in case of an accident.

Fuel lines

Besides metallic tubes, flexible, flame-retar-dant tubes reinforced with braided-steel armoring can be used in the low-pressure stage. They must be routed so as to avoid contact with moving components that might damage them and in such a way that any leak fuel or evaporation cannot collect or ignite. The function of the fuel lines must not be impaired by twisting of the chassis, movement of the engine or any other similar effects.

All fuel-conveying parts must be pro-tected against heat that may affect their proper operation. On buses, fuel lines may not be routed though the passenger cabin or the driver's cab. Fuel may not be gravity-fed.

Diesel fuel filter

Fuel-injection equipment for diesel engines are manufactured with great precision and are sensitive to the slightest contamination in the fuel. The fuel filter has the following functions:

- Reduce particulate impurities to avoid particulate erosion
- Separate emulgated water from free water to avoid corrosion damage

The fuel filter must be adapted to the fuel-injection system.

Fuel-supply pump

The fuel-supply pump draws fuel from the fuel tank and conveys it continuously to the high-pressure pump. The fuel pump is inte-grated in the high-pressure pump on axial-piston and radial-piston distributor pumps, and in a few instances in common-rail sys-tems.

Alternatively, an additional fuel pump can be provided as a presupply pump.

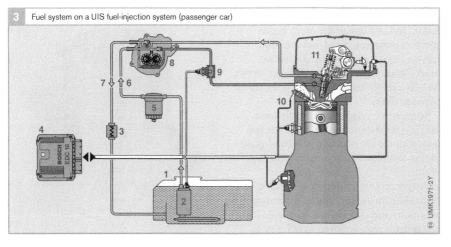

3   Fuel system on a UIS fuel-injection system (passenger car)

Fig. 3
1 Fuel tank
2 Presupply pump
3 Fuel cooler
4 ECU
5 Fuel filter
6 Fuel supply line
7 Fuel return line
8 Tandem pump
9 Fuel-temperature
   sensor
10 Glow-plug
11 Nozzle

Fuel filter

Design and requirements

Modern direct-injection (DI) systems for gasoline and diesel engines are highly sensitive to the smallest impurities in the fuel. Damage mainly occurs as a result of particulate erosion and water corrosion. The service-life design of the fuel-injection system depends on a specific minimum purity of the fuel.

Particulate filtration

Reducing particulate impurities is one of the functions of the fuel filter. In this way, it protects the wear-prone components of the fuel-injection system. In other words, the fuel-injection system prescribes the necessary filter fineness. Besides wear protection, fuel filters must also have a sufficient particulate storage capacity, otherwise they could become blocked before the end of the change interval. If they do become blocked, they would reduce fuel delivery quantity as well as engine performance. It is therefore essential to fit a fuel filter tailored to the requirements of the fuel-injection system. Fitting the incorrect filters would have unpleasant results at best; at worst, it would have very expensive consequences (from replacing components through to renewing the complete fuel-injection system).

Compared to gasoline fuel, diesel fuel contains many more impurities. For this reason, and also due to the much higher injection pressures, diesel fuel-injection systems require greater wear protection, larger filtration capacities, and longer service lives. As a result, diesel filters are designed as exchange filters.

Requirements regarding filter fineness have increased dramatically in the last few years with the introduction of second-generation common-rail systems and further advances in Unit Injector Systems for passenger cars and commercial vehicles. Depending on the application (operating conditions, fuel contamination, engine life), new systems require filtration efficiencies ranging from 65% to 98.6% (particle size 3 to 5 µm, ISO/TR 13353:1994). Longer servicing intervals in more recent vehicles require greater particulate storage capacities as well as intensive fine particulate filtration.

Water separation

The second main function of diesel fuel filters is to separate emulgated and undissolved water from the fuel in order to avoid corrosion damage. Efficient water separation greater than 93% at maximum flow (ISO 4020:2001) is a specially important factor for distributor injection pumps and common-rail systems.

Designs

The filter must be carefully selected depending on the fuel-injection system and the operating conditions.

Main filter

The diesel fuel filter is normally fitted in the low-pressure circuit between the electric fuel pump and the high-pressure pump in the engine compartment.

1 Diesel exchange filter with spiral vee-shaped filter element

UMK2020Y

The use of screw-on exchange filters, in-line filters, and metalfree filter elements is widespread. The replacement parts are inserted in filter housings made of aluminum, solid plastic, or sheet steel (to meet higher crash requirements). Only the filter element is replaced in these filters. The filter elements are mainly spiral vee-shaped (Fig. 1).

Two filters can also be fitted in parallel, resulting in greater particulate storage capacity. Connecting the filter in series produces a higher filtration efficiency. Stepped filters, or a fine filter with a matched pre-filter, are used in series connections.

Pre-filter for presupply pump

If requirements are particularly high, an additional pre-filter is fitted on the suction or pressure side with a filter fineness matched to the main filter (fine filter). Pre-filters are mainly used for commercial vehicles in countries that have poor diesel fuel quality. These filters are mainly designed as strainers with a mesh width of 300 μm.

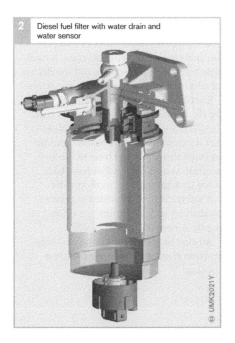

2   Diesel fuel filter with water drain and water sensor

UMK2021Y

Water separator

Water is separated by the filter medium using the *repellent effect* (droplets forming due to the different surface tensions of water and fuel). Separated water collects in the chamber at the bottom of the filter housings (Fig. 2). Conductivity sensors are used in some cases to monitor the water level. The water is drained manually by opening a water drain plug or pressing a pushbutton switch. Fully automatic water-disposal systems are currently under development.

Filter media

Increased demands with respect to fuel filters in engines of the new generation require the use of special filter media composed of several synthetic layers and cellulose. The filter media employ a preliminary fine filtering effect and guarantee maximum particulate retention capacity by separating particles within each filtering layer.

The new filter generation is also deployable with biodiesel (Fatty Acid Methyl Ester (FAME). However, the higher concentration of organic particles in FAME means taking account of a shorter filter service life in the servicing concept.

Additional functions

Modern filter modules integrate additional modular functions such as:
- Fuel preheating: Electrically, by cooling water, or by return fuel flow. Preheating prevents paraffin crystals from blocking the filter pores in winter.
- Displaying the servicing interval by measuring differential pressure.
- Filling and venting facilities: After a filter change, the fuel system is filled and vented by hand pump. The pump is usually integrated in the filter cover.

Fuel-supply pump

The fuel-supply pump in the low-pressure stage (the so-called presupply pump) is responsible for maintaining an adequate supply of fuel to the high-pressure components. This applies:

- irrespective of operating state,
- with a minimum of noise,
- at the necessary pressure, and
- throughout the vehicle's complete service life.

The fuel-supply pump draws fuel out of the fuel tank and conveys it continuously in the required quantity (injected fuel quantity and scavenging flow) to the high-pressure fuel-injection installation (60...500 l/h, 300...700 kPa or 3...7 bar). Many pumps bleed themselves automatically so that starting is possible even when the fuel tank has run dry.

There are three designs:

- Electric fuel pump (as used in passenger cars)
- Mechanically driven gear-type supply pumps, and
- Tandem fuel pumps (passenger cars, UIS)

In axial-piston and radial-piston distributor pumps, a vane-type supply pump is used as presupply pump and is integrated directly in the fuel-injection pump.

Electric fuel pump

The Electric Fuel Pump (EFP, Figs. 1 and 2) is only fitted to passenger cars and light-duty trucks. As part of the system-monitoring strategy, it is responsible, besides fuel delivery, for cutting off the fuel supply, if this is necessary in an emergency.

Electric fuel pumps are available as in-line or in-tank versions. In-line pumps are fitted to the vehicle's body platform outside of the fuel tank in the fuel line between tank and fuel filter. On the other hand, in-tank pumps are located inside of the fuel tank in a special retainer that normally includes a suction-side fuel strainer, a fuel-level sensor, a swirl plate acting as fuel reservoir, and the electrical and hydraulic connections to the exterior.

Starting with the engine cranking process, the electric fuel pump runs continuously, irrespective of engine speed. This means that it permanently delivers fuel from the fuel tank through a fuel filter to the fuel-injection system. Excess fuel flows back to the tank through an overflow valve.

A safety circuit is provided to prevent the delivery of fuel if ignition is on and the engine is stopped.

An electric fuel pump comprises three function elements inside of a common housing:

1 Single-stage electric fuel pump

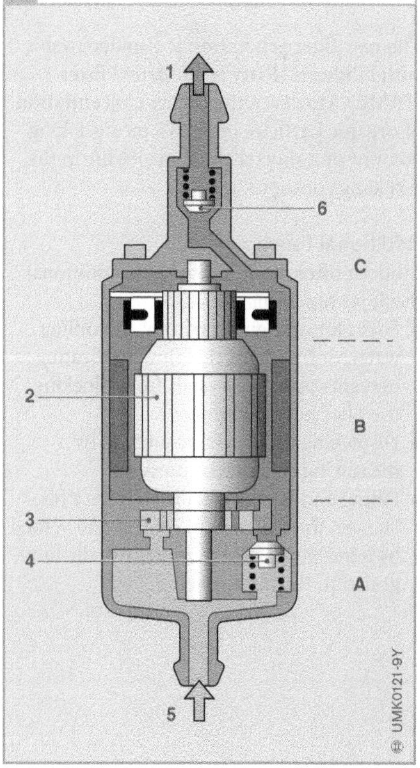

UMK0121-9Y

Fig. 1

A Pumping element
B Electric motor
C End cover

1 Pressure side
2 Motor armature
3 Pumping element
4 Pressure limiter
5 Suction side
6 Non-return valve

Pump element (Fig. 1, A)

There are a variety of different pump-element designs available, depending on the electric fuel pump's specific operating concept. Diesel applications mainly use Roller-Cell Pumps (RCP).

The roller-cell pump (Fig. 2) is a positive-displacement pump consisting of an eccentrically located base plate (4) in which a slotted rotor (2) is free to rotate. There is a movable roller in each slot (3) which, when the rotor rotates, is forced outwards against the outside roller path and against the driving flanks of the slots by centrifugal force and the pressure of the fuel. The result is that the rollers now act as rotating seals, whereby a chamber is formed between the rollers of adjacent slots and the roller path. The pumping effect is due to the fact that, once the kidney-shaped inlet opening (1) has closed, the chamber volume reduces continuously.

Electric motor (Fig. 1, B)

The electric motor comprises a permanent-magnet system and an armature (2). Design is determined by the required delivery quantity at a given system pressure. The electric motor is permanently flushed by fuel so that it remains cool. This design permits high motor performance without the necessity for complicated sealing elements between pumping element and electric motor.

End cover (Fig. 1, C)

The end cover contains the electrical connections as well as the pressure-side hydraulic connection. A non-return valve (6) is incorporated to prevent the fuel lines from emptying once the fuel pump has been switched off. Interference suppressors can also be fitted in the end cover.

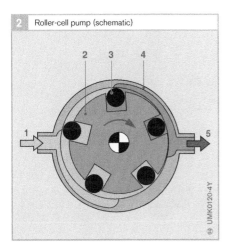

2   Roller-cell pump (schematic)

UMK0120-4Y

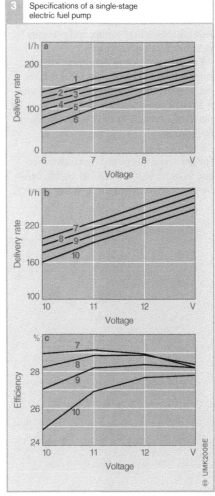

3   Specifications of a single-stage electric fuel pump

UMK2008E

Fig. 2
1   Suction (inlet) side
2   Slotted rotor
3   Roller
4   Base plate
5   Pressure side

Fig. 3
Parameter: delivery pressure
a   Delivery rate at low voltage
b   Delivery rate dependent on voltage in normal operation
c   Efficiency dependent on voltage

1   at 200 kPa
2   at 250 kPa
3   at 300 kPa
4   at 350 kPa
5   at 400 kPa
6   at 450 kPa
7   at 450 kPa
8   at 500 kPa
9   at 550 kPa
10   at 600 kPa

Gear-type fuel pump

The gear-type supply pump (Figs. 4 and 6) is used to supply the fuel-injection modules of single-cylinder pump systems (commercial vehicles) and common-rail systems (passenger cars, commercial vehicles, and off-road vehicles). It is directly attached to the engine or is integrated in the common-rail high-pressure pump. Common forms of drive are via coupling, gearwheel, or toothed belt.

The main components are two rotating, engaged gearwheels that convey the fuel in the tooth gaps from the suction side (Fig. 6, 1) to the pressure side (5). The line of contact between the rotating gearwheels provides the seal between the suction and pressure sides of the pump, and prevents fuel from flowing back again.

The delivery quantity is approximately proportional to engine speed, For this reason, fuel-delivery control takes place either by throttle control on the suction side, or by means of an overflow valve on the pressure side (Fig. 5).

The gear-type fuel pump is maintenance-free. In order to bleed the fuel system before the first start, or when the tank has run dry,

a hand pump can be installed directly on the gear pump, or in the low-pressure lines.

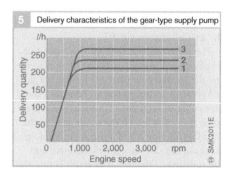

5 Delivery characteristics of the gear-type supply pump

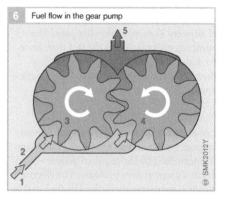

6 Fuel flow in the gear pump

Fig. 5

Pressure at pump outlet: 8 bar

Parameter: suction-side pressure at pump inlet
1   500 mbar
2   600 mbar
3   700 mbar

Fig. 6
1   Suction side (fuel inlet)
2   Suction throttle
3   Primary gearwheel (drive wheel)
4   Secondary gearwheel
5   Pressure side

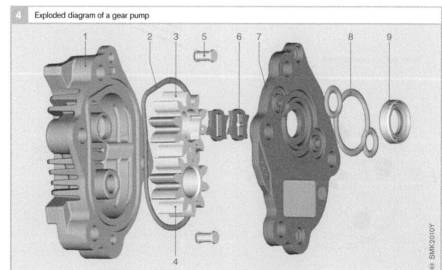

4   Exploded diagram of a gear pump

Fig. 4
1   Pump housing
2   O-ring seal
3   Primary gearwheel
4   Secondary gearwheel
5   Rivet
6   Coupling
7   Cover plate
8   Molded seal
9   Shaft seal

Vane-type pump with separating vanes

In the version of this pump used with the passenger-car UIS (Fig. 7), two separating vanes (4) are pressed by springs (3) against a rotor (1). When the rotor rotates, volume increases at the inlet (suction) end (2) and fuel is drawn into two chambers. With continued rotation, chamber volumes decrease, and fuel is forced out of the chambers at the outlet (pressure) end (5). This pump delivers fuel even at very low rotational speeds.

Tandem pump

The tandem pump used on the passenger-car UIS is a unit comprising the fuel pump (Fig. 8) and the vacuum pump for the brake booster. It is attached to the engine's cylinder head and driven by the engine's camshaft. The fuel pump itself is either a vane-type pump with separating vanes or a gear pump (3), and even at low speeds (cranking speeds) delivers enough fuel to ensure that the engine starts reliably. The pump contains a variety of valves and throttling orifices:

Suction throttling orifice (6): Essentially, the quantity of fuel delivered by the pump is proportional to the pump's speed. The pump's maximum delivery quantity is limited by the suction throttling orifice so that not too much fuel is delivered.

Overpressure valve (7): This is used to limit the maximum pressure in the high-pressure stage.

Throttling bore (4): Vapor bubbles in the fuel-pump outlet are eliminated in the fuel-return throttling bore (1).

Bypass (12): If there is air in the fuel system (for instance, if the vehicle has been driven until the fuel tank is empty), the low-pressure pressure-control valve remains closed. The air is forced out of the fuel system through the bypass by the pressure of the pumped fuel.

Thanks to the ingenious routing of the pump passages, the pump's gearwheels never run dry even when the fuel tank is empty. When restarting after filling the tank, therefore, this means that the pump draws in fuel immediately.

The fuel pump is provided with a connection (8) for measuring the fuel pressure in the pump outlet.

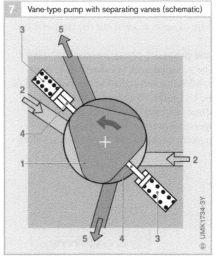

7 Vane-type pump with separating vanes (schematic)

Fig. 7
1 Rotor
2 Inlet (suction) side
3 Spring
4 Separating vane
5 Outlet (pressure) end

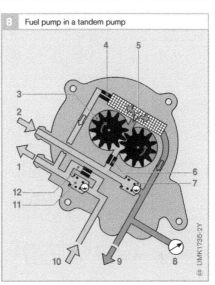

8 Fuel pump in a tandem pump

Fig. 8
1 Return to tank
2 Entry from the fuel tank
3 Pumping element (gearwheel)
4 Throttling bore
5 Filter
6 Suction throttling orifice
7 Overpressure valve
8 Connection for pressure measurement
9 Outlet to the injector
10 Return from the injector
11 Non-return valve
12 Bypass

Miscellaneous components

Distributor tube

The passenger-car UIS is provided with a distributor tube which, as its name implies, distributes the fuel to the unit injectors. This form of distribution ensures that the individual injectors all receive the same quantities of fuel at the same temperature, and smooth engine running is the result. In the distributor tube, fuel flowing to the unit injectors mixes with fuel flowing back from them in order to even out the temperature.

Low-pressure pressure-control valve

The pressure-control valve (Fig. 1) is an overflow valve installed in the fuel return of the UIS and UPS systems. Independent of operating status, it provides for adequate operating pressure in the respective low-pressure stages so that the pumps are always well filled with a consistently even charge of fuel. The accumulator plunger (5) opens at a "snap-open pressure" of 3...3.5 bar, so that the conical seat (7) releases the accumulator volume (6). Only very little leakage fuel can escape through the gap seal (4). The spring (3) is compressed as a function of the fuel pressure, so that the accumulator volume changes and compensates for minor pressure fluctuations.

When pressure has increased to 4...4.5 bar, the gap seal also opens and the flow quantity increases abruptly. The valve closes again when the pressure drops. Two threaded elements (2), each with a different spring seat, are available for preliminary adjustment of opening pressure.

ECU cooler

On commercial vehicles, ECU cooling must be provided if the ECU for the UIS or UPS systems is mounted directly on the engine. In such cases, fuel is used as the cooling medium. It flows past the ECU in special cooling channels and in the process absorbs heat from the electronics.

Fuel cooler

Due to the high pressures in the injectors for the passenger-car UIS, and some Common Rail Systems (CRS), the fuel heats up to such an extent that in order to prevent damage to fuel tank and level sensor it must be cooled down before returning. The fuel returning from the injector flows through the fuel cooler (heat exchanger, Fig. 2, 3) and gives off heat energy to the coolant in the fuel-cooling circuit. This is separated from the engine-cooling circuit (6) since at normal engine temperatures the engine coolant is too hot to absorb heat from the fuel. In order that the fuel-cooling circuit can be filled and temperature fluctuations compensated for, the fuel-cooling circuit is connected to the engine-cooling circuit near the equalizing reservoir (5). Connection is such that the fuel-cooling circuit is not adversely affected by the engine-cooling circuit which is at a higher temperature.

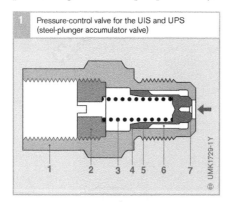

Fig. 1

1 Valve body
2 Threaded element
3 Spring
4 Gap seal
5 Accumulator plunger
6 Accumulator volume
7 Conical seat

Pressure-control valve for the UIS and UPS (steel-plunger accumulator valve)

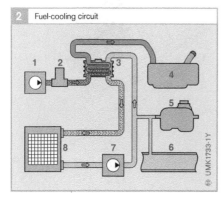

Fig. 2

1 Fuel pump
2 Fuel-temperature sensor
3 Fuel cooler
4 Fuel tank
5 Equalizing reservoir
6 Engine-cooling circuit
7 Coolant pump
8 Auxiliary cooler

Fuel-cooling circuit

Diesel aircraft engines of the 1920s and 30s

In the 1920s and 1930s numerous two and four-stroke diesel engines were developed for use as aircraft engines. Apart from their economical consumption and the lower price of diesel fuel, diesels had a number of other features in their favor such as a lower fire risk and simpler maintenance due to the absence of carburetor, spark plugs and magneto. Engineers also hoped that the compression-ignition engine would provide good performance at high altitudes. In those days, spark-ignition engines were liable to misfire because the ignition system was subject to atmospheric pressure. The main problems associated with the development of a diesel aircraft engine involved controlling the fuel/air mixture effectively and handling the higher mechanical and thermal stresses.

The most successful production aircraft diesel engine was the Jumo 205 6-cylinder two-stroke opposed-piston heavy-oil engine (see illustration). Following its introduction in 1933 it was fitted in numerous planes. It had a take-off power output of up to 645 kW (880 hp). Its strengths primarily lay in its suitability for long-distance flights at constant speeds, e.g. for transatlantic postal services. Around 900 units of this reliable engine were built.

The fuel injection system for the Jumo 205 consisted of two pumps and two injectors for each cylinder. The injection pressure was in excess of 500 bar. It was that fuel-injection system which was a major factor in the breakthrough of the Jumo 205. Based on the experience gained from that engine, development work was also started on direct fuel-injection for spark-ignition aircraft engines in the 1930s.

The Jumo 205 was followed in 1939 by the Jumo 207 high-altitude engine which also had a take-off power output of 645 kW (880 hp). Thanks to its turbocharger aspiration, aircraft with the new engine could reach altitudes of up to 14,000 metres.

The technical high point in the development of diesel aircraft engines was the experimental 24-cylinder opposed-piston Jumo 224 produced in the early 1940s which developed as much as 3,330 kW (4,400 hp) take-off power. This "square configuration" engine had its cylinders arranged in a cross formation driving four separate crankshafts.

A whole series of diesel aircraft engines were developed by other manufacturers as well. However, none of them progressed beyond the experimental stage. In later years interest in diesel aircraft engines waned because of progress made with high-performance spark-ignition engines with fuel injection.

▼ Junkers Jumo 205 two-stroke opposed-piston diesel aircraft engine

(Source: Deutsches Museum, Munich)

SMM0606Y

Supplementary valves for in-line fuel-injection pumps

In addition to the overflow valve, electronically controlled in-line fuel-injection pumps also have an electric shutoff valve (Type ELAB) or an electrohydraulic shutoff device (Type EHAB).

Overflow valve

The overflow valve is fitted to the pump's fuel-return outlet. It opens at a pressure (2...3 bar) that is set to suit the fuel-injection pump concerned and thereby maintains the pressure in the fuel gallery at a constant level A valve spring (Fig. 1, 4) acts on a spring seat (2) which presses the valve cone (5) against the valve seat (6). As the pressure, p_i in the fuel-injection pump rises, it pushes the valve seat back, thus opening the valve. When the pressure drops, the valve closes again. The valve seat has to travel a certain distance before the valve is fully open. The buffer volume thus created evens out rapid pressure variations, which has a positive effect on valve service life.

Type ELAB electric shutoff valve

The Type ELAB electric shutoff valve acts as a redundant (i.e. duplicate) back-up safety device. It is a 2/2-way solenoid valve which is screwed into the fuel inlet of the in-line fuel-injection pump (Fig. 2). When not energized, it cuts off the fuel supply to the pump's fuel gallery. As a result, the fuel-injection pump is prevented from delivering fuel to the nozzles even if the actuator mechanism is defective, and the engine cannot overrev. The engine control unit closes the electric shutoff valve if it detects a permanent governor deviation or if a fault in the control unit's fuel-quantity controller is detected.

When it is energized (i.e. when the status of Terminal 15 is "Ignition on"), the electro-magnet (Fig. 2, 3) draws in the solenoid armature (4) (12 or 24 V, stroke approx. 1.1 mm). The sealing cone seal (7) attached to the armature then opens the channel to the inlet passage (9). When the engine is switched off using the starter switch ("ignition switch"), the supply of electricity to the solenoid coil is also disconnected. This causes the magnetic field to collapse so that the compression spring (5) pushes the armature and the attached sealing cone back against the valve seat.

Fig. 1
1 Sealing ball
2 Spring seat
3 Sealing washer
4 Valve spring
5 Valve cone
6 Valve seat
7 Hollow screw housing
8 Fuel return

p_i Pump fuel gallery pressure

Fig. 2
1 Electrical connection to engine control unit
2 Solenoid valve housing
3 Solenoid coil
4 Solenoid armature
5 Compression spring
6 Fuel inlet
7 Plastic sealing cone
8 Constriction plug for venting
9 Inlet passage to pump
10 Connection for overflow valve
11 Housing (ground)
12 Mounting-bolt eyes

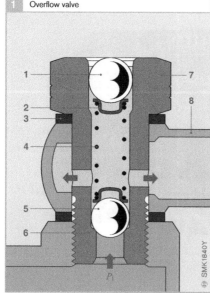

1 Overflow valve

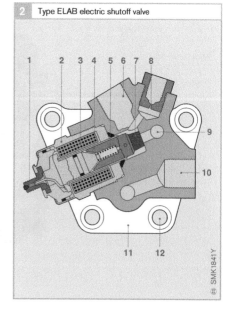

2 Type ELAB electric shutoff valve

Type EHAB electrohydraulic shutoff device

The Type EHAB electrohydraulic shutoff device is used as a safety shutoff for fuel-injection pumps with relatively high fuel gallery pressures. In such cases, the capabilities of the Type ELAB electric shutoff valve are insufficient. With high fuel-gallery pressures and in the absence of any special compensating devices, it can take up to 10 s for the pressure to drop sufficiently for fuel injection to stop. The electrohydraulic shutoff device thus ensures that fuel is drawn back out of the fuel-injection pump by the pre-supply pump. Thus, when the valve is de-energized, the fuel gallery pressure in the fuel-injection pump is dissipated much more quickly and the engine can be stopped within a period of no more than 2 s. The electrohydraulic shutoff device is mounted directly on the fuel-injection pump. The EHAB housing also incorporates an integrated fuel-temperature sensor for the electronic governing system (Fig. 3, 8).

Normal operation setting (Fig. 3a)

As soon as the engine control unit activates the electrohydraulic shutoff device ("Ignition on"), the electromagnet (6) draws in the solenoid armature (5, operating voltage 12 V). Fuel can then flow from the fuel tank (10) via the heat exchanger (11) for cold starting and the preliminary filter (3) to port A. From there, the fuel passes through the right-hand valve past the solenoid armature to port B. This is connected to the presupply pump (1) which pumps the fuel via the main fuel filter (2) to port C of the electrohydraulic shutoff device. The fuel then passes through the open left-hand valve to port D and finally from there to the fuel-injection pump (12).

Reversed-flow setting (Fig. 3b)

When the ignition is switched off, the valve spring (7) presses the solenoid armature back to its resting position. The intake side of the presupply pump is then connected directly to the fuel-injection pump's inlet passage so that fuel flows back from the fuel gallery to the fuel tank. The right hand valve opens the connection between the preliminary filter and main fuel filter, allowing fuel to return to the fuel tank.

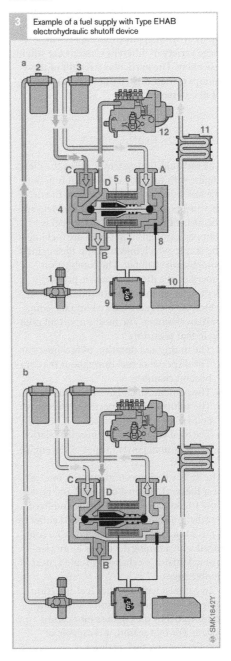

3 Example of a fuel supply with Type EHAB electrohydraulic shutoff device

Fig. 3
a Normal operation setting
b Reversed-flow/ emergency shutoff setting

1 Presupply pump
2 Main fuel filter
3 Preliminary filter
4 Type EHAB electrohydraulic shutoff device
5 Solenoid armature
6 Electromagnet
7 Valve spring
8 Fuel-temperature sensor
9 Engine control unit
10 Fuel tank
11 Heat exchanger
12 Fuel-injection pump

A...D valve ports

SMK1842Y

Overview of in-line fuel-injection pump systems

No other fuel-injection system is as widely used as the in-line fuel-injection pump – the "classic" diesel fuel-injection technology. Over the years, this system has been continually refined and adapted to suit its many areas of application. As a result, a large variety of different versions are still in use today. The particular strength of these pumps is their rugged durability and ease of maintenance.

Areas of application

The fuel-injection system supplies the diesel engine with fuel. To perform that function, the fuel-injection pump generates the necessary fuel pressure for injection and delivers the fuel at the required rate. The fuel is pumped through a high-pressure fuel line to the nozzle, which injects it into the engine's combustion chamber. The combustion processes in a diesel engine are primarily dependent on the quantity and manner in which the fuel is introduced into the combustion chamber. The most important criteria in that regard are
● The timing and duration of fuel injection
● The dispersal of fuel throughout the combustion chamber
● The point at which ignition is initiated
● The volume of fuel injected relative to crankshaft rotation and
● The total volume of fuel injected relative to the desired power output of the engine

The in-line fuel-injection pump is used all over the world in medium-sized and heavy-duty trucks as well as on marine and fixed-installation engines. It is controlled either by a mechanical governor, which may be combined with a timing device, or by an electronic actuator mechanism (Table 1, next double page).

In contrast with all other fuel-injection systems, the in-line fuel-injection pump is lubricated by the engine's lubrication system. For that reason, it is capable of handling poorer fuel qualities.

Types

Standard in-line fuel-injection pumps
The range of standard in-line fuel-injection pumps currently produced encompasses a large number of pump types (see Table 1, next double page). They are used on diesel engines with anything from 2 to 12 cylinders and ranging in power output from 10 to 200 kW per cylinder (see also Table 1 in the chapter "Overview of diesel fuel-injection systems"). They are equally suitable for use on direct-injection (DI) or indirect-injection (IDI) engines.

Depending on the required injection pressure, injected-fuel quantity and injection duration, the following versions are available:
● Type M for 4...6 cyl. up to 550 bar
● Type A for 2...12 cyl. up to 750 bar
● Type P3000 for 4...12 cyl. up to 950 bar
● Type P7100 for 4...12 cyl. up to 1,200 bar
● Type P8000 for 6...12 cyl. up to 1,300 bar
● Type P8500 for 4...12 cyl. up to 1,300 bar
● Type R for 4...12 cyl. up to 1,150 bar
● Type P10 for 6...12 cyl. up to 1,200 bar
● Type ZW(M) for 4...12 cyl. up to 950 bar
● Type P9 for 6...12 cyl. up to 1,200 bar
● Type CW for 6...10 cyl. up to 1,000 bar

The version most commonly fitted in commercial vehicles is the Type P.

Control-sleeve in-line fuel-injection pump
The range of in-line fuel-injection pumps also includes the control-sleeve version (Type H), which allows the start-of-delivery point to be varied in addition to the injection quantity. The Type H pump is controlled by a Type RE electronic controller which has two actuator mechanisms. This arrangement enables the control of the start of injection and the injected-fuel quantity with the aid of two control rods and thus makes the automatic timing device superfluous. The following versions are available:
● Type H1 for 6...8 cyl. up to 1,300 bar
● Type H1000 for 5...8 cyl. up to 1,350 bar

Design

Apart from the in-line fuel-injection pump, the complete diesel fuel-injection system (Figs. 1 and 2) comprises
- A fuel pump for pumping the fuel from the fuel tank through the fuel filter and the fuel line to the injection pump
- A mechanical governor or electronic control system for controlling the engine speed and the injected-fuel quantity
- A timing device (if required) for varying the start of delivery according to engine speed
- A set of high-pressure fuel lines corresponding to the number of cylinders in the engine and
- A corresponding number of nozzle-and-holder assemblies

In order for the diesel engine to function properly, all of those components must be matched to each other.

Control

The operating parameters are controlled by the injection pump and the governor which operates the fuel-injection pump's control rod. The engine's torque output is approximately proportional to the quantity of fuel injected per piston stroke.

Mechanical governors

Mechanical governors used with in-line fuel-injection pumps are centrifugal governors. This type of governor is linked to the accelerator pedal by means of a rod linkage and an adjusting lever. On its output side, it operates the pump's control rod. Depending on the type of use, different control characteristics are required of the governor:
- The Type RQ maximum-speed governor limits the maximum speed.
- The Type RQ and RQU minimum/maximum-speed governors also control the idle speed in addition to limiting the maximum speed.

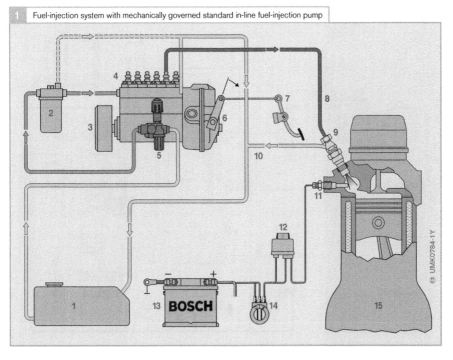

1 Fuel-injection system with mechanically governed standard in-line fuel-injection pump

Fig. 1
1 Fuel tank
2 Fuel filter with overflow valve (option)
3 Timing device
4 In-line fuel-injection pump
5 Fuel pump (mounted on injection pump)
6 Governor
7 Accelerator pedal
8 High-pressure fuel line
9 Nozzle-and-holder assembly
10 Fuel-return line
11 Type GSK glow plug
12 Type GZS glow plug control unit
13 Battery
14 Glow plug/starter switch ("ignition switch")
15 Diesel engine (IDI)

- The Type RQV, RQUV, RQV..K, RSV and RSUV variable-speed governors also control the intermediate speed range.

Timing devices

In order to control start of injection and compensate for the time taken by the pressure wave to travel along the high-pressure fuel line, standard in-line fuel-injection pumps use a timing device which "advances" the start of delivery of the fuel-injection pump as the engine speed increases. In special cases, a load-dependent control system is employed. Diesel-engine load and speed are controlled by the injected-fuel quantity without exerting any throttle action on the intake air.

Electronic control systems

If an electronic control system is used, there is an accelerator-pedal sensor which is connected to the electronic control unit. The control unit then converts the accelerator-position signal into a corresponding nominal control-rack travel while taking into account the engine speed.

An electronic control system performs significantly more extensive functions than the mechanical governor. By means of electrical measuring processes, flexible electronic data processing and closed-loop control systems with electrical actuators, it enables more comprehensive response to variable factors than is possible with the mechanical governor.

Electronic diesel control systems can also exchange data with other electronic control systems on the vehicle (e.g. Traction Control System, electronic transmission control) and can therefore be integrated in a vehicle's overall system network.

Electronic control of diesel engines improves their emission characteristics by more precise metering of fuel delivery.

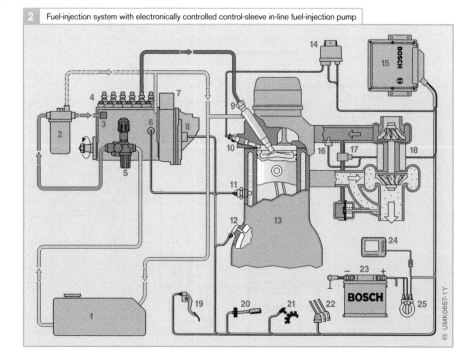

2   Fuel-injection system with electronically controlled control-sleeve in-line fuel-injection pump

1 Areas of application for the most important in-line fuel-injection pumps and their governors						
Area of application	Cars	Fixed-installation engines	Commercial vehicles	Construction and agricultural machinery	Railway locomotives	Ships
Pump type						
Standard in-line fuel-injection pump Type M	●	–	–	●	–	–
Standard in-line fuel-injection pump Type A	–	●	–	●	–	–
Standard in-line fuel-injection pump Type MW [1]	–	–	●	●	–	–
Standard in-line fuel-injection pump Type P	–	●	●	●	●	●
Standard in-line fuel-injection pump Type R [2]	–	–	●	●	●	●
Standard in-line fuel-injection pump Type P10	–	●	–	●	●	●
Standard in-line fuel-injection pump Type ZW(U)	–	–	–	–	●	●
Standard in-line fuel-injection pump Type P9	–	●	–	●	●	●
Standard in-line fuel-injection pump Type CW	–	●	–	–	●	●
Control-sleeve in-line fuel-injection pump Type O	–	–	●	–	–	–
Governor type						
Minimum/maximum speed governor Type RSF	●	–	–	●	–	–
Minimum/maximum speed governor Type RQ	–	–	●	●	–	–
Minimum/maximum speed governor Type RQU	–	–	–	–	–	●
Variable-speed governor Type RQV	–	●	●	●	–	–
Variable-speed governor Type RQUV	–	–	–	–	●	●
Variable-speed governor Type RQV..K	–	–	–	–	–	–
Variable-speed governor Type RSV	–	●	–	●	–	–
Variable-speed governor Type RSUV	–	–	–	–	–	●
Type RE (electric actuator mechanism)	●	–	●	–	–	–

Table 1
[1] This type of pump is no longer used with new systems.
[2] Same design as Type P but for heavier duty.

3 Examples of in-line fuel-injection pumps

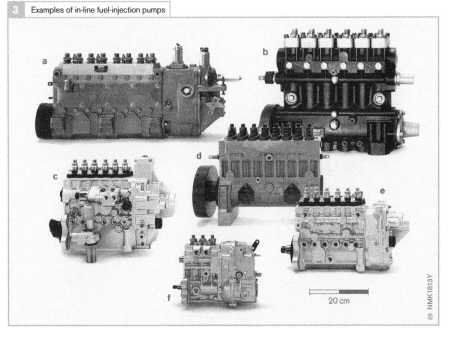

20 cm

Ⓓ NMK1813Y

Fig. 3
Pump types:
a ZWM (8 cylinders)
b CW (6 cylinders)
c H (control-sleeve type) (6 cylinders)
d P9/P10 (8 cylinders)
e P7100 (6 cylinders)
f A (3 cylinders)

Presupply pumps for in-line fuel-injection pumps

The presupply pump's job is to supply the in-line fuel-injection pump with sufficient diesel fuel under all operating conditions. In addition, it "flushes" the fuel-injection pump with fuel to cool it down by extracting heat from the fuel and returning it through the overflow valve to the fuel tank. In addition to the presupply pumps described in this section, there are also multifuel and electric presupply pumps. In certain relatively rare applications, the in-line fuel-injection pump can be operated without a presupply pump in a gravity-feed fuel-tank system.

Applications

In applications where there is an insufficient height difference or a large distance between the fuel tank and the fuel-injection pump, a presupply pump (Bosch type designation FP) is fitted. This is normally flange-mounted on the in-line fuel-injection pump. Depending on the conditions in which the engine is to be used and the specifics of the engine design, various fuel line arrange-

ments are required. Figures 1 and 2 illustrate two possible variations.

If the fuel filter is located in the immediate vicinity of the engine, the heat radiated from the engine can cause bubbles to form in the fuel lines. In order to prevent this, the fuel is made to circulate through the fuel-injection pump's fuel gallery so as to cool the pump. With this line arrangement, the excess fuel flows through the overflow valve (6) and the return line back to the fuel tank (1).

If, in addition, the ambient temperature in the engine compartment is high, the line arrangement shown in Figure 2 may also be used. With this system, there is an overflow restriction (7) on the fuel filter through which a proportion of the fuel flows back to the fuel tank during normal operation, taking any gas or vapor bubbles with it. Bubbles that form inside the fuel-injection pump's fuel gallery are removed by the excess fuel that escapes through the overflow valve (6) to the fuel tank. The presupply pump must therefore be dimensioned to be able to deliver not only the fuel volume

Fig. 1

1 Fuel tank
2 Presupply pump
3 Fuel filter
4 In-line fuel-injection pump
5 Nozzle-and-holder assembly
6 Overflow valve

— Supply line
– – Return line

Fig. 2

1 Fuel tank
2 Presupply pump
3 Fuel filter
4 In-line fuel-injection pump
5 Nozzle-and-holder assembly
6 Overflow valve
7 Overflow restriction

— Supply line
– – Return line

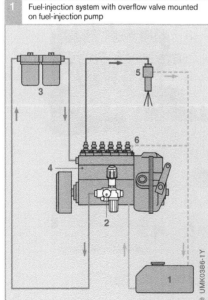

1 Fuel-injection system with overflow valve mounted on fuel-injection pump

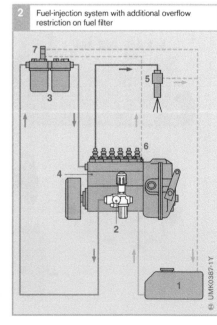

2 Fuel-injection system with additional overflow restriction on fuel filter

required by the fuel-injection pump but also the volume that "bypasses" the fuel-injection pump and returns to the fuel tank.

The following criteria determine the choice of presupply pump:
- The type of fuel-injection pump
- The delivery rate
- The line routing arrangement and
- The available space in the engine compartment

Design and method of operation

A presupply pump draws the fuel from the fuel tank and pumps it under pressure through the fuel filter and into the fuel gallery of the fuel-injection pump (100...350 kPa or 1...3.5 bar). Presupply pumps are generally mechanical plunger pumps that are mounted on the fuel-injection pump (or in rare cases on the engine).

The presupply pump is then driven by an eccentric (Fig. 3, 1) on the fuel-injection pump or engine camshaft (2).

Depending on the fuel delivery rate required, presupply pumps may be single or double-action designs.

Single-action presupply pumps
Single-action presupply pumps (Figs. 3 and 4) are available for fuel-injection pump sizes M, A, MW and P. The drive cam or eccentric (Fig. 3, 1) drives the pump plunger (5) via a push rod (3). The piston is also spring-loaded by a compression spring (7) which effects the return stroke.

The single-action presupply pump operates according to the throughflow principle as follows. The cam pitch on the push rod moves the pump plunger and its integrated suction valve (8) against the force of the compression spring. In the process, the suction valve is opened by the lower pressure created in the fuel gallery (4, Fig. 3a).

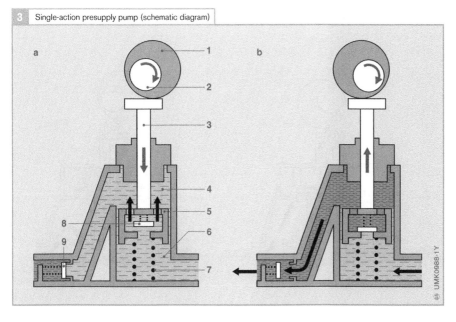

3   Single-action presupply pump (schematic diagram)

a

b

Fig. 3
a   Cam pitch
b   Return stroke

1   Drive eccentric
2   Fuel-injection pump
    camshaft
3   Push rod
4   Pressure chamber
5   Pump plunger
6   Fuel gallery
7   Compression spring
8   Suction valve
9   Delivery valve

UMK0988-1Y

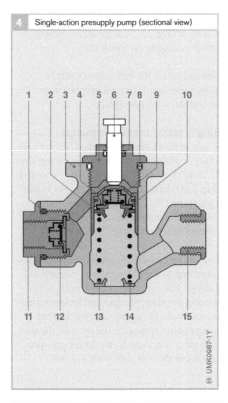

Fig. 4
1 Sealing ring
2 Spring seat
3 Pump housing
  (aluminum)
4 Suction valve
5 Roller-tappet shell
6 Push rod
7 Sealing ring
8 Sealing ring
9 Pump plunger
10 Spacer ring
11 Pressure port
12 Delivery valve
13 Compression spring
14 Spring seat
15 Suction port

As a result, the fuel passes into the chamber between the suction valve and the delivery valve (9). When the pump performs its return stroke under the action of the compression spring, the suction valve closes and the delivery valve opens (Fig. 3b). The fuel then passes under pressure along the high-pressure line to the fuel-injection pump.

Double-action presupply pumps

Double-action presupply pumps (Fig. 5) offer a higher delivery rate and are used for fuel-injection pumps that serve larger numbers of engine cylinders and which consequently must themselves provide greater delivery quantities. This type of presupply pump is suitable for Type P and ZW fuel-injection pumps. As with the single-action version, the double-action presupply pump is driven by a cam or eccentric.

In the double-action plunger pump, fuel is delivered to the fuel-injection pump on both the cam-initiated stroke and the return stroke, in other words there are two delivery strokes for every revolution of the camshaft.

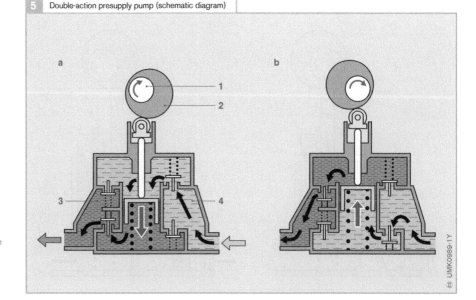

Fig. 5
a Cam pitch
b Return stroke

1 Fuel-injection pump
  camshaft
2 Drive eccentric
3 Pressure chamber
4 Fuel gallery

Manual priming pumps

The priming pump is usually integrated in the presupply pump (Fig. 6, 1). However, it can also be fitted in the fuel line between the fuel tank and the presupply pump. It performs the following functions:

- Priming the suction side of the fuel-injection installation prior to initial operation
- Priming and venting the system after repairs or servicing and
- Priming and venting the system after the fuel tank has been run dry

The latest version of the Bosch priming pump replaces virtually all previous designs. It is backwardly compatible and can therefore be used to replace pumps of older designs. It no longer has to be released or locked in its end position. Consequently, it is easy to operate even in awkward positions.

The priming pump also contains a non-return valve which prevents the fuel flowing back in the wrong direction.

For applications in which the pump has to be fireproof, there is a special version with a steel body.

Preliminary filter

The preliminary filter protects the presupply pump against contamination from coarse particles. In difficult operating conditions, such as where engines are refueled from barrels, it is advisable to fit an additional strainer-type filter inside the fuel tank or in the fuel line to the presupply pump.

The preliminary filter may be integrated in the presupply pump (Fig. 6, 2), mounted on the presupply pump intake or connected to the intake passage between the fuel tank and the presupply pump.

Gravity-feed fuel-tank system

Gravity-feed fuel-tank systems (which operate without a presupply pump) are generally used on tractors and very small diesel engines. The arrangement of the tank and the fuel lines is such that the fuel flows through the fuel filter to the fuel-injection pump under the force of gravity.

With smaller height differences between the fuel tank and the fuel filter or fuel-injection pump, larger-bore lines are better suited to providing an adequate flow of fuel to the fuel-injection pump. In such systems, it is useful to fit a stopcock between the fuel tank and the fuel filter. This allows the fuel inlet to be shut off when carrying out repairs or maintenance so that the fuel tank does not have to be drained.

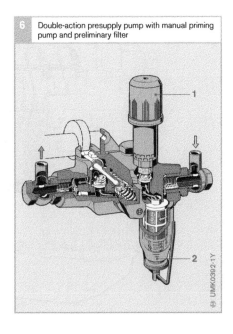

6 Double-action presupply pump with manual priming pump and preliminary filter

Fig. 6
1   Manual priming pump
2   Preliminary filter

Type PE standard in-line fuel-injection pumps

In-line fuel-injection pumps are among the classics of diesel fuel-injection technology. This dependable design has been used on diesel engines since 1927. Over the years they have been continuously refined and adapted to suit their many areas of application. In-line fuel-injection pumps are designed for use on fixed-installation engines, commercial vehicles, and construction and agricultural machinery. They enable high power outputs per cylinder on diesel engines with between 2 and 12 cylinders. When used in conjunction with a governor, a timing device and various auxiliary components, the in-line fuel-injection pump offers considerable versatility. Today in-line fuel-injection pumps are no longer produced for cars.

The power output of a diesel engine is determined essentially by the amount of fuel injected into the cylinder. The in-line fuel-injection pump must precisely meter the amount of fuel delivered to suit every possible engine operating mode.

In order to facilitate effective mixture preparation, a fuel-injection pump must deliver the fuel at the pressure required by the combustion system employed and in precisely the right quantities. In order to achieve the optimum balance between pollutant emission levels, fuel consumption and combustion noise on the part of the diesel engine, the start of delivery must be accurate to within 1 degree of crankshaft rotation.

In order to control start of delivery and compensate for the time taken by the pressure wave to travel along the high-pressure delivery line, standard in-line fuel-injection pumps use a timing device (Fig. 1, 3) which "advances" the start of delivery of the fuel-injection pump as the engine speed increases (see chapter "Governors and control systems for in-line fuel-injection pumps"). In special cases, a load-dependent control system is employed. Diesel-engine load and speed are controlled by varying the injected fuel quantity.

A distinction is made between standard in-line fuel-injection pumps and control-sleeve in-line fuel-injection pumps.

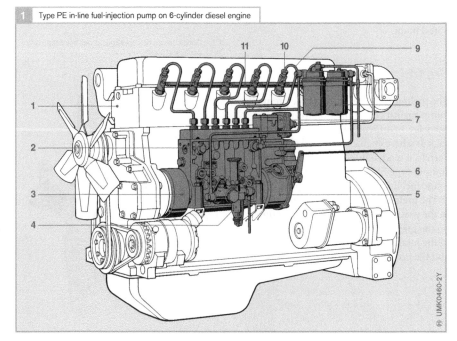

1 Type PE in-line fuel-injection pump on 6-cylinder diesel engine

Fig. 1
1 Diesel engine
2 Standard in-line fuel-injection pump
3 Timing device
4 Presupply pump
5 Governor
6 Control lever with linkage to accelerator
7 Manifold-pressure compensator
8 Fuel filter
9 High-pressure delivery line
10 Nozzle-and-holder assembly
11 Fuel-return line

Fitting and drive system

In-line fuel-injection pumps are attached directly to the diesel engine (Fig. 1). The engine drives the pump's camshaft. On two-stroke engines, the pump speed is the same as the crankshaft speed. On four-stroke engines, the pump speed is half the speed of the crankshaft – in other words, it is the same as the engine camshaft speed.

In order to produce the high injection pressures required, the drive system between the engine and the fuel-injection pump must be as "rigid" as possible.

There is a certain amount of oil inside the fuel-injection pump in order to lubricate the moving parts (e.g. camshaft, roller tappets, etc.). The fuel-injection pump is connected to the engine lube-oil circuit so that oil circulates when the engine is running.

Design and method of operation

Type PE in-line fuel-injection pumps have an internal camshaft that is integrated in the aluminum pump housing (Fig. 2, 14). It is driven either via a clutch unit or a timing device or directly by the engine. Pumps of this type with an integrated camshaft are referred to by the type designation PE.

Above each cam on the camshaft is a roller tappet (13) and a spring seat (12) for each cylinder of the engine. The spring seat forms the positive link between the roller tappet and the pump plunger (8). The pump barrel (4) forms the guide for the pump plunger. The two components together form the pump-and-barrel assembly.

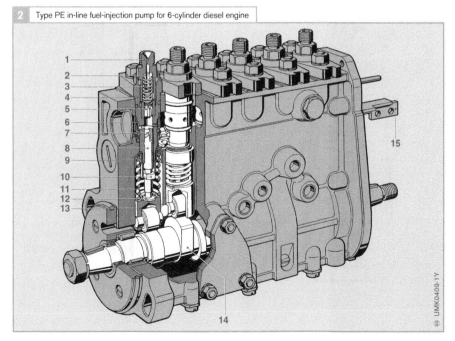

2   Type PE in-line fuel-injection pump for 6-cylinder diesel engine

Fig. 2

1 Pressure-valve holder
2 Filler piece
3 Pressure-valve spring
4 Pump barrel
5 Delivery-valve cone
6 Intake and control port
7 Helix
8 Pump plunger
9 Control sleeve
10 Plunger control arm
11 Plunger spring
12 Spring seat
13 Roller tappet
14 Camshaft
15 Control rack

Design of the pump-and-barrel assembly

In its basic form, a pump-and-barrel assembly consists of a pump plunger (Fig. 3, 9) and a pump barrel (8). The pump barrel has one or two inlet passages that lead from the fuel gallery (1) into the inside of the cylinder. On the top of the pump-and-barrel assembly is the delivery-valve holder (5) with the delivery-valve cone (7). The control sleeve (3) forms the connection between the pump plunger and the control rack (10). The control rack moves inside the pump housing – under the control of the governor as described in the chapter "Governors for in-line fuel-injection pumps" – so as to rotate the positively interlocking "control-sleeve-and-piston" assembly by means of a ring gear or linkage lever. This enables precise regulation of the pump delivery quantity.

The plunger's total stroke is constant. The effective stroke, on the other hand, and therefore the delivery quantity, can be altered by rotating the pump plunger.

In addition to a vertical groove (Fig. 4, 2), the pump plunger also has a helical channel (7) cut into it. The helical channel is referred to as the helix (6).

For injection pressures up to 600 bar, a single helix is sufficient, whereas higher pressures require the piston to have two helixes on opposite sides. This design feature prevents the units from "seizing" as the piston is no longer

Fig. 3
1 Fuel gallery
2 Control-sleeve gear
3 Control sleeve
4 Cover plate
5 Pressure-valve holder
6 Pressure-valve body
7 Delivery-valve cone
8 Pump barrel
9 Pump plunger
10 Control rack
11 Plunger control arm
12 Plunger spring
13 Spring seat
14 Adjusting screw
15 Roller tappet
16 Camshaft

Fig. 4
a Single-port
   plunger-and-barrel
   assembly
b Two-port
   plunger-and-barrel
   assembly

1 Inlet passage
2 Vertical groove
3 Pump barrel
4 Pump plunger
5 Control port
   (inlet and return lines)
6 Helix
7 Helical channel
8 Ring groove for
   lubrication

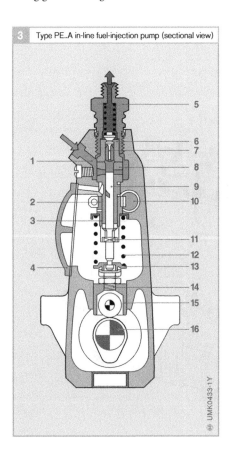

3 Type PE..A in-line fuel-injection pump (sectional view)

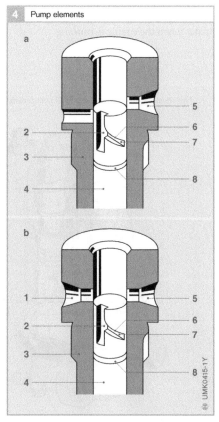

4 Pump elements

forced sideways against the cylinder wall by the injection pressure.

The cylinder then has one or two bores for fuel supply and return (Fig. 4).

The pump plunger is such an exact fit inside the pump barrel that it provides a leakproof seal even at extremely high pressures and at low rotational speeds. Because of this precise fit, pump plungers and barrels can only be replaced as a complete plunger-and-barrel assembly.

The injected fuel quantity possible is dependent on the charge volume of the pump barrel. The maximum injection pressures vary between 400 and 1,350 bar at the nozzle depending on the pump design.

The relative angular positions of the cams on the pump camshaft are such that the injection process is precisely synchronized with the firing sequence of the engine.

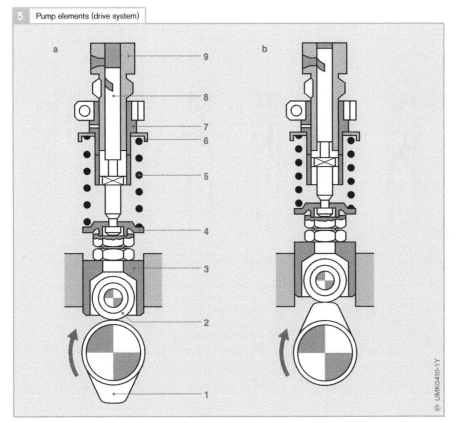

5   Pump elements (drive system)

a

b

Fig. 5
a   BDC position
b   TDC position

1   Cam
2   Tappet roller
3   Roller tappet
4   Lower spring seat
5   Plunger spring
6   Upper spring seat
7   Control sleeve
8   Pump plunger
9   Pump barrel

Method of operation of plunger-and-barrel assembly (stroke phase sequence)

The rotation of the camshaft is converted directly into a reciprocating motion on the part of the roller tappet and consequently into a similar reciprocating action on the part of the pump plunger.

The delivery stroke, whereby the piston moves towards its Top Dead Center (TDC), is assumed by the action of the cam. A compression spring performs the task of returning the plunger to Bottom Dead Center (BDC). It is dimensioned to keep the roller in contact with the cam even at maximum speed, as loss of contact between roller and cam, and the consequent impact of the two surfaces coming back into contact, would

inevitably cause damage to both components in the course of continuous operation.

The plunger-and-barrel assembly operates according to the overflow principle with helix control (Fig. 6). This is the principle adopted on Type PE in-line fuel-injection pumps and Type PF single-plunger fuel-injection pumps.

When the pump plunger is at Bottom Dead Center (BDC) the cylinder inlet passages are open. Under pressure from the presupply pump, fuel is able to flow through those passages from the fuel gallery to the plunger chamber. During the delivery stroke, the pump plunger closes off the inlet passages. This phase of the plunger lift is referred to as

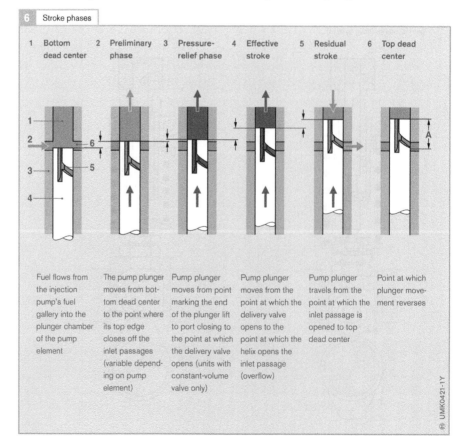

6   Stroke phases

1 Bottom dead center	2 Preliminary phase	3 Pressure-relief phase	4 Effective stroke	5 Residual stroke	6 Top dead center
Fuel flows from the injection pump's fuel gallery into the plunger chamber of the pump element	The pump plunger moves from bottom dead center to the point where its top edge closes off the inlet passages (variable depending on pump element)	Pump plunger moves from point marking the end of the plunger lift to port closing to the point at which the delivery valve opens (units with constant-volume valve only)	Pump plunger moves from the point at which the delivery valve opens to the point at which the helix opens the inlet passage (overflow)	Pump plunger travels from the point at which the inlet passage is opened to top dead center	Point at which plunger movement reverses

Fig. 6
1   Plunger chamber
2   Fuel inlet
3   Pump barrel
4   Pump plunger
5   Helix
6   Fuel return

A   Total stroke

UMK0421-1Y

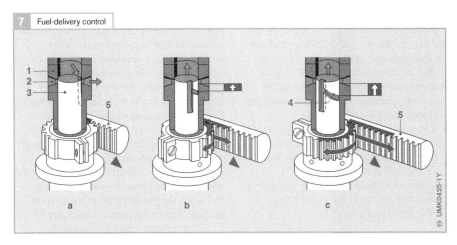

7 Fuel-delivery control

Fig. 7
a   Zero delivery
b   Partial delivery
c   Maximum delivery

1   Pump barrel
2   Inlet passage
3   Pump plunger
4   Helix
5   Geared control rack

the preliminary phase. As the delivery stroke continues, fuel pressure increases and causes the delivery valve at the top of the plunger-and-barrel assembly to open. If a constant-volume valve is used (see section "Delivery valve") the delivery stroke also includes a retraction-lift phase. Once the delivery valve has opened, fuel flows along the high-pressure line to the nozzle for the duration of the effective stroke. Finally, the nozzle injects a precisely metered quantity of fuel into the combustion chamber of the engine.

Once the pump plunger's helix releases the inlet passage again, the effective stroke is complete. From this point on, no more fuel is delivered to the nozzle as, during the residual stroke, the fuel can escape through the vertical groove from the plunger chamber back into the fuel gallery so that pressure in the plunger-and-barrel assembly breaks down.

After the piston reaches Top Dead Center (TDC) and starts to move back in the opposite direction, fuel flows through the vertical groove from the fuel gallery to the plunger chamber until the helix closes off the inlet passage again. As the plunger continues its return stroke, a vacuum is created inside the pump barrel. When the inlet passage is opened again, fuel then immediately flows into the plunger chamber. At this point, the cycle starts again from the beginning.

Fuel-delivery control
Fuel delivery can be controlled by varying the effective stroke (Fig. 7). This is achieved by means of a control rack (5) which twists the pump plunger (3) so that the pump plunger helix (4) alters the point at which the effective delivery stroke ends and therefore the quantity of fuel delivered.

In the final zero-delivery position (a), the vertical groove is directly in line with the inlet passage. With the plunger in this position, the pressure chamber is connected to the fuel gallery through the pump plunger for the entire delivery stroke. Consequently, no fuel is delivered. The pump plungers are placed in this position when the engine is switched off.

For partial delivery (b), fuel delivery is terminated depending on the position of the pump plunger.

For maximum delivery (c), fuel delivery is not terminated until the maximum effective stroke is reached, i.e. when the greatest possible delivery quantity has been reached.

The force transfer between the control rack and the pump plunger, see Figure 7, takes place by means of a geared control rack (PE..A and PF pumps) or via a ball joint with a suspension arm and control sleeve (Type PE..M, MW, P, R, ZW(M) and CW pumps).

Pump unit with leakage return channel

If the fuel-injection pump is connected to the engine lube-oil circuit, leakage fuel can result in thinning of the engine oil under certain circumstances. Assemblies with a leakage return channel to the fuel gallery of the fuel-injection pump largely avoid this problem. There are two designs:

- A ring groove (Fig. 8a, 3) in the plunger collects the leakage fuel and returns it to the fuel gallery via other specially located grooves (2) in the piston.
- Leakage fuel flows back to the fuel gallery via a ring groove in the pump barrel (Fig. 8b, 4) and a hole (1).

Pump plunger design variations

Special requirements such as reducing noise or lowering pollutant emissions in the exhaust gas make it necessary to vary the start of delivery according to engine load. Pump plungers that have an upper helix (Fig. 9, 2) in addition to the lower helix (1) allow load-dependent variation of start of delivery. In order to improve the starting characteristics of some engines, special pump plungers with a starting groove (3) are used. The starting groove – an extra groove cut into the top edge of the plunger – only comes into effect when the plunger is set to the starting position. It retards the start of delivery by 5...10° in terms of crankshaft position.

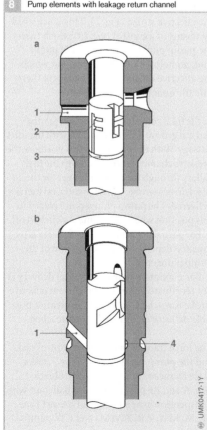

8   Pump elements with leakage return channel

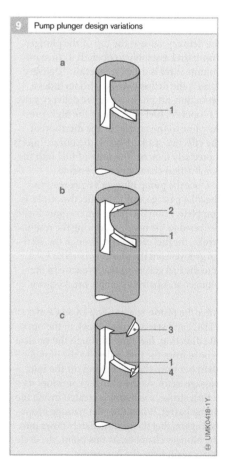

9   Pump plunger design variations

Fig. 8
a   Version with ring groove in plunger
a   Version with ring groove in barrel

1   Leakage return bore
2   Leakage-return slots
3   Ring groove in pump plunger
4   Ring groove in pump barrel

Fig. 9
a   Helix at bottom
b   Helix at top and bottom
c   Helix at bottom and starting groove

1   Bottom helix
2   Top helix
3   Starting groove
4   Start-quantity limitation groove

Cam shapes

Different combustion-chamber geometries and combustion methods demand different fuel-injection parameters. In other words, each individual engine design requires an individually adapted fuel-injection process. The piston speed (and therefore the length of the injection duration) depends on the cam pitch relative to the camshaft angle of rotation. For this reason, there are various different cam shapes according to the specifics of the application. In order to improve injection parameters such as the "rate-of-discharge curve" and "pressure load", special cam shapes can be designed by computer.

The trailing edge of the cam can also be varied (Fig. 10): There are symmetrical cams (a), cams with asymmetric trailing edge (b) and reversal-inhibiting cams (c) which make it more difficult for the engine to start rotating in the wrong direction.

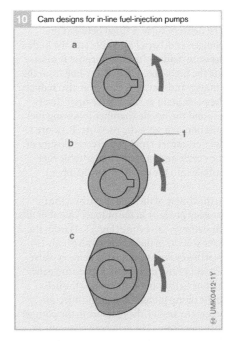

10   Cam designs for in-line fuel-injection pumps

Fig. 10
a   Symmetrical cam
b   Asymmetrical cam
c   Reversal-inhibiting
    cam

1   Trailing edge

History of in-line fuel-injection pumps

No other diesel fuel-injection system can look back on a history as long as the Bosch in-line fuel-injection pump. The very first examples of this famously reliable design came off the production line in Stuttgart as long ago as 1927.

Although the basic method of operation has remained the same, pump and governor design has been continuously adapted and improved to meet new demands. The arrival of electronic diesel control in 1987 and the control-sleeve in-line fuel-injection pump in 1993 opened up new horizons.

Sales figures show that, for a wide range of applications, the in-line fuel-injection pump is far from reaching its "sell-by date" even today. In 2001 roughly 150,000 Type P and Type H pumps left the Bosch factory in Homburg.

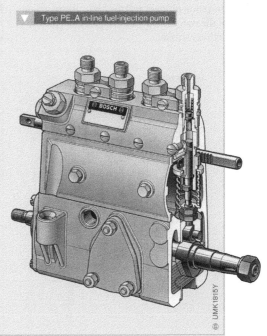

Type PE..A in-line fuel-injection pump

Delivery valve

The delivery valve is fitted between the plunger-and-barrel assembly and the high-pressure delivery line. Its purpose is to isolate the high-pressure delivery line from the plunger-and-barrel assembly. It also reduces the pressure in the high-pressure delivery line and the nozzle chamber following fuel injection to a set static pressure. Pressure reduction causes rapid and precise closure of the nozzle and prevents undesirable fuel dribble into the combustion chamber.

In the course of the delivery stroke, the increasing pressure in the plunger chamber lifts the delivery-valve cone (Fig. 11, 3) from the valve seat (4) in the delivery-valve body (5). Fuel then passes through the delivery-valve holder (1) and into the high-pressure delivery line to the nozzle. As soon as the helix of the pump plunger brings the injection process to an end, the pressure in the plunger chamber drops. The delivery-valve cone is then pressed back against the valve seat by the valve spring (2). This isolates the space above the pump plunger and the high-pressure side of the system from one another until the next delivery stroke.

Constant-volume valve without return-flow restriction

In a constant-volume valve (Bosch designation GRV), part of the valve stem takes the form of a "retraction piston" (Fig. 12, 2). It fits into the valve guide with a minimum degree of play. At the end of fuel delivery, the retraction piston slides into the valve guide and shuts off the plunger chamber from the high-pressure delivery line. This increases the space available to the fuel in the high-pressure delivery line by the charge volume of the retraction piston. The retraction volume is dimensioned precisely to suit the length of the high-pressure delivery line, which means that the latter must not be altered.

In order to achieve the desired fuel-delivery characteristics, torque-control valves are used in some special cases. They have a retraction piston with a specially ground pintle (6) on one side.

Constant-volume valve with return-flow restriction

A return-flow restriction (Bosch designation RDV or RSD) may also be used in addition to the constant-volume valve. Its purpose is to dampen and render harmless returning pressure waves that are produced when the nozzle

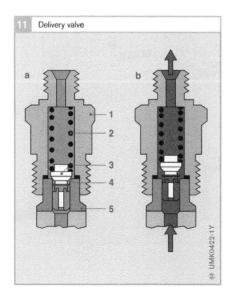

11 Delivery valve

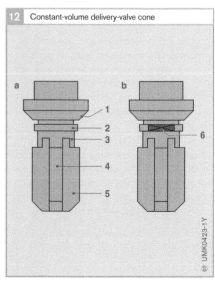

12 Constant-volume delivery-valve cone

closes. This reduces or entirely eliminates wear effects and cavitation in the plunger chamber. It also prevents undesirable secondary injection.

The return-flow restriction is integrated in the upper part of the delivery-valve holder (Fig. 13), in other words between the constant-volume valve and the nozzle. The valve body (4) has a small bore (3) the size of which is dimensioned to suit the application so as to achieve, firstly, the desired flow restriction and, secondly, to prevent reflection of pressure waves as much as possible. The valve opens when fuel is flowing in delivery direction. The delivery flow is therefore not restricted. For pressures up to approx. 800 bar, the valve body shaped like a disk. For higher pressures it is a guided cone.

Pumps with return-flow throttle valves are "open systems", i.e. during the plunger lift to port closing and retraction lift, the static pressure in the high-pressure delivery line is the same as the internal pump pressure. Consequently, this pressure must be at least 3 bar.

Constant-pressure valve
The constant-pressure valve (Bosch designation GDV) is used on fuel-injection pumps

with high injection pressures (Fig. 14). It consists of forward-delivery valve (consisting of delivery valve, 1, 2, 3) and a pressure-holding valve for the return-flow direction (consisting of 2, 5, 6, 7 and 8) which is integrated in the delivery-valve cone (2). The pressure-holding valve maintains a virtually constant static pressure in the high-pressure delivery line between fuel-injection phases under all operating conditions. The advantages of the constant-pressure valve are the prevention of cavitation and improved hydraulic stability which means more precise fuel injection.

During the delivery stroke, the valve acts as a conventional delivery valve. At the end of the delivery stroke, the ball valve (7) is initially open and the valve acts like a valve with a return-flow restriction. Once the closing pressure is reached, the compression spring (5) closes the return-flow valve, thereby maintaining a constant pressure in the fuel line.

However, correct functioning of the constant-pressure valve demands greater accuracy of adjustment and modifications to the governor. It is used for high-pressure fuel-injection pumps (upwards of approx. 800 bar) and for small, fast-revving direct-injection engines.

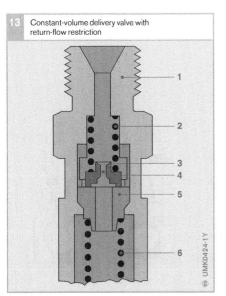

13 Constant-volume delivery valve with return-flow restriction

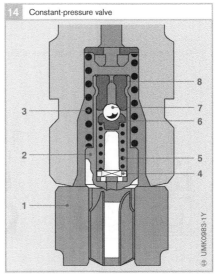

14 Constant-pressure valve

Fig. 13
1 Pressure-valve holder
2 Valve spring
3 Flow throttle
4 Valve body
   (disk in this case)
5 Valve holder
6 Pressure-valve spring

Fig. 14
1 Delivery-valve support
2 Delivery-valve cone
3 Pressure-valve spring
4 Filler piece
5 Compression spring
   (pressure-holding
   valve)
6 Spring seat
7 Ball
8 Flow throttle

Design variations

The range of power outputs for diesel engines with in-line fuel-injection pumps extends from 10 to 200 kW per cylinder. Various pump design variations allow such a wide range of power outputs to be accommodated. The designs are grouped into series whose engine output ranges overlap to some degree. Pump sizes A, M, MW and P are produced in large volumes (Fig. 1).

There are two different designs of the standard in-line fuel-injection pump:
- The open-type design of the Type M and A pumps with a cover plate at the side and
- The closed-type design of the Type MW and P pumps in which the plunger-and-barrel assemblies are inserted from the top

For even higher per-cylinder outputs, there are the pump sizes P10, ZW, P9 and CW.

There are two ways in which the plunger-and-element assemblies can be supplied with fuel (Fig. 2):

With the *longitudinal scavenging* (a), fuel flows from one plunger-and-barrel assembly to the next *in sequence*.

With the *crossflow scavenging* (b), the plunger-and-barrel assemblies are supplied individually from a *common supply channel*. In this way, the fuel-delivery termination pressure does not affect the adjacent cylinder. This achieves tighter quantity tolerances and more precise fuel proportioning.

Fig. 2
a   Longitudinal scavenging
b   Crossflow scavenging (Type P-8000 pump)

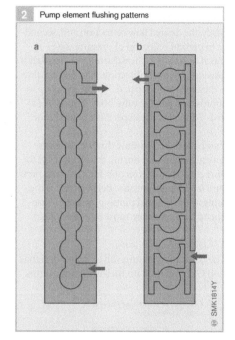

2   Pump element flushing patterns

a   b

SMK1814Y

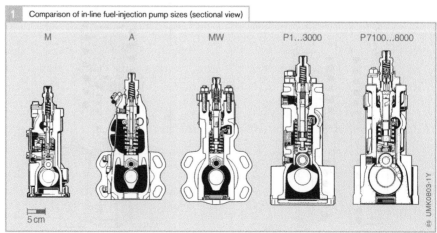

1   Comparison of in-line fuel-injection pump sizes (sectional view)

M   A   MW   P1...3000   P7100...8000

5 cm

UMK0B03-1Y

1978 diesel speed records

In April 1978 the experimental Mercedes-Benz C111-III set nine world speed records, some of which still stand today, and eleven international class records. Some of those records had previously been held by gasoline-engine cars.

The average speed of the record attempts was approximately 325 km/h. The highest speed reached was measured at 338 km/h. The average fuel consumption was only 16 l/100 km.

These considerable achievements were made possible primarily by the highly streamlined plastic body. Its aerodynamic drag coefficient of 0.195 was sensationally low for the time.

The car was powered by a 3-liter, five-cylinder in-line diesel engine with a maximum power output of 170 kW (230 bhp). That meant that it was twice as powerful as its standard production counterpart. The maximum torque of 401 Nm was produced at 3,600 rpm. This performance was made possible by a turbocharger and an intercooler.

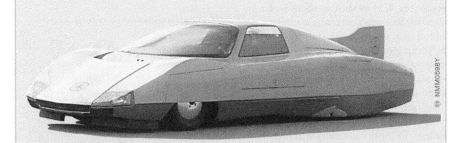

NMM0598Y

Engine compartment of the Mercedes-Benz C111-III

At the engine's nominal speed, the turbocharger was rotating at 150,000 rpm.

Precise fuel delivery and metering was provided by a Bosch Type PE...M in-line fuel-injection pump

NMM0599Y

Size M fuel-injection pumps

The size M in-line fuel-injection pump (Fig. 3 and 4) is the smallest of the Series PE pumps. It has a light-metal (aluminum) body that is attached to the engine by means of a flange.

The size M pump is an open-type in-line fuel-injection pump which has a cover plate on the side and the base. On size M pumps, the peak injection pressure is limited by the pump to 400 bar.

After removal of the side cover plate, the delivery quantities of the plunger-and-barrel assemblies can be adjusted and matched to one another. Individual adjustment is effected by moving the position of the clamp blocks (Fig. 4, 5) on the control rack (4). When the fuel-injection pump is running, the control rack is used to adjust the position of the pump plungers and, as a result, the delivery quantity within design limits. On the size M pump, the control rack consists of a round steel rod that is flatted on one side. Fitted over the control rack are the slotted clamp blocks. Together with its control sleeve, the lever (3), which is rigidly attached to the control sleeve, forms the mechanical link with the corresponding clamp block. This arrangement is referred to as a rod-and-lever control linkage.

The pump plungers sit directly on top of the roller tappets (6). LPC adjustment is achieved by selecting tappet rollers of different diameters.

The size M pump is available in 4, 5 and 6 cylinder versions, and is suitable for use with diesel fuel only.

Fig. 4
1   Delivery valve
2   Pump barrel
3   Control-sleeve
    lever arm
4   Control rack
5   Clamp block
6   Roller tappet
7   Camshaft
8   Cam

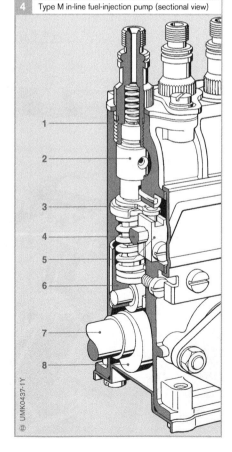

3   Type M in-line fuel-injection pump (external view)

4   Type M in-line fuel-injection pump (sectional view)

Size A fuel-injection pumps

The size A in-line fuel-injection pump (Figs. 5 and 6) is the next size up from the size M pump and offers larger delivery quantities as a result.

It has a light-metal housing and can be either flange-mounted to the engine or attached by means of a cradle mounting.

On the size A fuel-injection pump, which is also an open-type design, the pump barrel (Fig. 6, 2) is inserted directly into the aluminum body from above. It is pressed by the pressure-valve holder against the pump housing via the pressure-valve support. The sealing pressures, which are considerably higher than the hydraulic delivery pressures, must be withstood by the pump housing. For this reason, the peak pressure for a size A pump is internally limited to 600 bar.

In contrast with the size M pump, the size A pump has an adjusting screw (7) for setting the plunger lift to port closing. This simplifies the process of adjusting the basic setting. The adjusting screw is screwed into the roller tappet and fixed by a locking nut.

Another difference with the size M pump is the rack-and-pinion control linkage instead of the rod-and-lever arrangement. This means that the control rack is replaced by a rack (4). Clamped to the control sleeve (5) there is a control-sleeve gear. By loosening the clamp bolt, each control sleeve can be rotated relative to its control-sleeve gear in order to equalize the delivery quantities between individual plunger-and-barrel assemblies.

With this design of pump, all adjustments must be carried out without the pump running and with the housing open. A cover plate is positioned on the side of the pump housing and provides access to the valve-spring chamber.

Size A pumps are available in versions for up to 12 cylinders and, in contrast with the size M models, are suitable for multifuel operation.

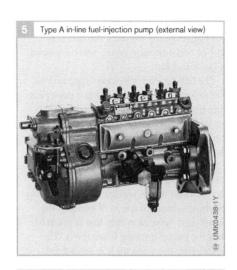

5 Type A in-line fuel-injection pump (external view)

UMK0438-1Y

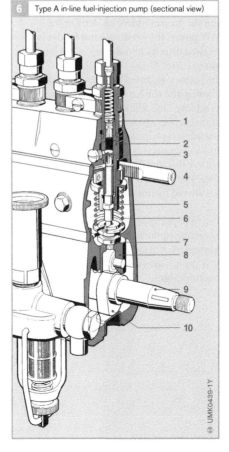

6 Type A in-line fuel-injection pump (sectional view)

UMK0439-1Y

Fig. 6
1 Delivery valve
2 Pump barrel
3 Pump plunger
4 Control rack
5 Control sleeve
6 Plunger spring
7 Adjusting screw
8 Roller tappet
9 Camshaft
10 Cam

Size MW fuel-injection pumps

For higher pump outputs, the size MW in-line fuel-injection pump was developed (Figs. 7 and 8).

The MW pump is a closed-type in-line fuel-injection pump which has a peak pressure limited to 900 bar, it is a lightweight metal design similar to the smaller models, and is attached to the engine by a baseplate, flange or cradle mounting.

Its design differs significantly from that of the Series M and A pumps. The main distinguishing feature of the MW pump is the barrel-and-valve assembly that is inserted into the pump housing from above. The barrel-and-valve assembly is assembled outside the housing and consists of the pump barrel (Fig. 8, 3), the delivery valve (2) and the pressure-valve holder. On the MW pump, the pressure-valve holder is screwed directly into the top of the longer pump barrel. Shims or spacers of varying thicknesses are fitted between the pump housing and the barrel-and-valve assembly to achieve LPC adjustment. The uniformity of fuel delivery between the barrel-and-valve assemblies is adjusted by rotating the barrel-and-valve assembly from the outside. To achieve this, the flange (1) is provided with slots. The position of the pump plunger is not altered by this adjustment.

The MW pump is available with the various mounting options in versions for up to 8 cylinders. It is suitable for diesel fuel only. MW pumps are no longer used for new engine designs.

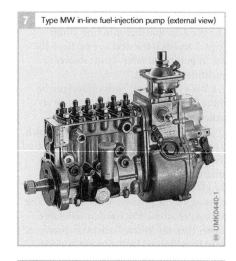

7   Type MW in-line fuel-injection pump (external view)

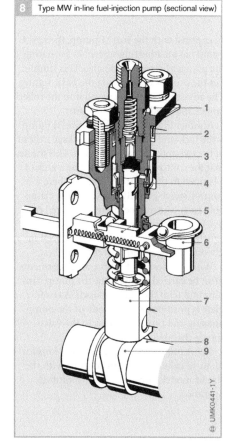

8   Type MW in-line fuel-injection pump (sectional view)

Fig. 8

1   Pump unit mounting flange
2   Delivery valve
3   Pump barrel
4   Pump plunger
5   Control rack
6   Control sleeve
7   Roller tappet
8   Camshaft
9   Cam

Size P fuel-injection pump

The size P in-line fuel-injection pump was similarly developed for higher pump outputs (Figs. 9 and 10). Like the MW pump, it is a closed-type fuel-injection pump and is attached to the engine by its base or by a flange. On size P pumps for peak internal pressures of up to 850 bar, the pump barrel (Fig. 10, 4) is inside an additional flange bushing (3) in which there is an internal thread for the pressure-valve holder. With this design, the sealing forces do not act on the pump housing. LPC adjustment on the P pump takes place in the same way as on the MW pump.

In-line fuel-injection pumps with low injection pressures use conventional fuel gallery flushing whereby the fuel passes through the fuel galleries of the individual barrel-and valve assemblies one after the other from the fuel inlet to the return outlet, traveling along the pump longitudinal axis (longitudinal scavenging). On size P pumps of the type P 8000, which are designed for injection pressures at the pump of 1,150 bar, this flushing method inside the pump would result in a significant temperature difference in fuel temperature (as much as 40°C) between the first and the last cylinder. Consequently, different quantities of energy would be injected into the individual combustion chambers of the engine (the energy density of the fuel decreases with increasing temperature and the associated increase in volume). For this reason, this type of fuel-injection pump has crossflow scavenging (i.e. at right angles to the pump longitudinal axis) whereby the fuel galleries of the individual barrels are isolated from one another by flow throttles and are flushed in parallel with fuel at virtually identical temperatures.

The P-type pump is produced in versions for up to 12 cylinders and is suitable both for diesel-only and for multifuel operation.

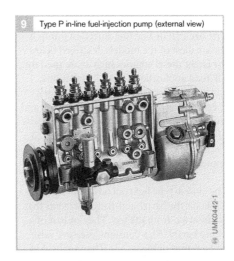

9 Type P in-line fuel-injection pump (external view)

UMK0442-1

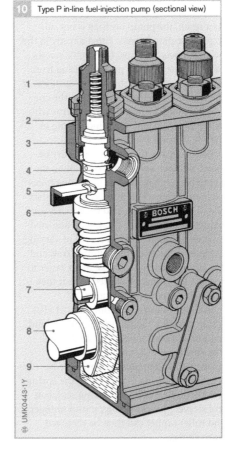

10 Type P in-line fuel-injection pump (sectional view)

UMK0443-1Y

Fig. 10
1 Pressure-valve holder
2 Delivery valve
3 Flange bushing
4 Pump barrel
5 Control rack
6 Control sleeve
7 Roller tappet
8 Camshaft
9 Cam

Size P10 fuel-injection pump

The size P10 in-line fuel-injection pump is the smallest of the models described below for larger diesel engines such as are used for off-road applications, fixed installations, construction and agricultural machinery, specialized vehicles, railway locomotives and ships. It is mounted on the engine by means of a baseplate.

The peak injector pressure is limited to approx. 1,200 bar.

The closed-type light-metal body (Fig. 12, 13) holds the barrel-and-flange elements that are inserted from the top. They consist of a pump barrel (5), a constant-pressure valve and a pump plunger (12). They are held in position by stud bolts (3). A pressure-valve holder (1) seals the constant-pressure valve. As a result, the pump housing is not subjected to sealing stresses. Fitted directly in the pump barrels are impact-deflector screws (4) which protect the pump housing from damage caused by high-energy cutoff jets at the end of the delivery stroke. On the control sleeve (8) there are two link arms with thin cylindrical end lugs which locate in mating slots on the control rack (6).

For balancing the delivery quantity between plunger-and-barrel assemblies, the pump barrels have slotted mounting holes on their flanges. This allows the pump barrels to be suitably adjusted before they are tightened in position. The LPC is adjusted by inserting shims or spacers (2) of varying thicknesses between the pump barrels and the pump housing. To make them easier to replace, the shims are slotted so that they can be inserted from the side.

In order to remove a roller tappet (10) when servicing the pump, the corresponding pump barrel must first be removed. The spring seat (7) above the plunger spring (9) is then pressed downwards. A retaining spring (11) holding the spring seat then releases it. The spring seat, control sleeve, plunger spring, pump plunger and roller tappet can then be removed from above.

Fig. 12

1   Constant-pressure valve socket
2   Shims
3   Stud bolts
4   Impact-deflector screw
5   Pump barrel with mounting flange
6   Control rack
7   Spring seat
8   Control sleeve
9   Plunger spring
10 Roller tappet
11 Spring ring
12 Pump plunger
13 Housing
14 Camshaft

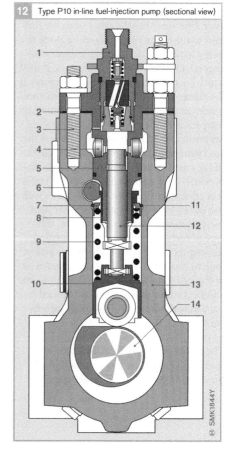

11   Type P10 in-line fuel-injection pump (external view)

SMK1843Y

12   Type P10 in-line fuel-injection pump (sectional view)

SMK1844Y

To refit these components, the plunger spring is compressed using the spring seat and the retaining spring which is snapped into position in the pump housing using a special device.

The camshaft runs on roller elements in the pump housing at each end. In order to obtain a high degree of rigidity, it is also supported by one or two half-shell plain bearings.

The size P10 fuel- injection pump is connected to the engine lube-oil circuit. A throttle bore determines the rate of oil flow. The fuel galleries of the individual plunger-and-barrel assemblies are interconnected and fuel circulates through the pump in a longitudinal direction (longitudinal scavenging). The presupply pump is usually either a gear pump driven by the engine or an electric fuel pump. For effective supply of the fuel-injection pump (and therefore efficient pump cooling), its delivery rate is several times the required fuel quantity.

Size P10 fuel-injection pumps are produced in versions for 6, 8 and 12 cylinders. The standard design is for diesel fuel only, with a special version available for multifuel operation.

Size P9 fuel-injection pump
The size P9 in-line fuel-injection pump is more or less identical in design to the P10 pump. However, it is somewhat larger and therefore positioned between the ZW and CW models.

The P9 fuel-injection pump has a closed-type light-metal housing. As with the P10, the peak nozzle pressure is limited to approx. 1,200 bar. It is attached to the engine by means of a cradle mounting. It is produced in versions for 6, 8 and 12 cylinders. The pump delivery quantity is controlled by a hydraulic or electromechanical governor provided by the engine manufacturer.

Size ZW fuel-injection pump
The size ZW in-line fuel-injection pump (Fig. 13) has an open-style light-metal housing. The pump is attached to the engine by means of a cradle mounting. The peak nozzle pressure is limited to 950 bar.

The pressure-valve holder (Fig. 14 overleaf, 1) screwed into the pump housing (18) provides the seal between the delivery valve and the pump barrel (2) as well as transmitting the hydraulic forces from the plunger. A fixing bolt (14) holds the pump barrel in position.

Two hardened impact-deflector screws (3) fitted in the pump housing opposite the control ports for each cylinder protect the pump housing from damage caused by the high-energy cutoff jet at the end of the delivery stroke.

The delivery quantity is controlled by means of a control rack in the form of a rack (4). This meshes with the control-sleeve gear that is clamped to the control sleeves (6).

For balancing the delivery quantities of the individual plunger-and-barrel assemblies, the clamp bolts (15) are loosened. Each control-sleeve gear can then be rotated relative to its control sleeve. The clamp bolts are then retightened.

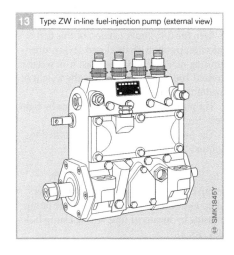

13   Type ZW in-line fuel-injection pump (external view)

SMK1845Y

LPC adjustment takes place by fitting or replacing the LPC disk (9) or a screw in the roller tappet (10).

For the purposes of removing the camshaft (11), the roller tappets can be held at their upper limit of travel by a retaining screw (17) fitted in the side of the pump housing. The camshaft runs on roller elements. For larger numbers of cylinders, there may also be one or two half-shell plain bearings in addition.

The presupply pump used may be a reciprocating piston pump which is flange-mounted on the side of the fuel-injection pump or a separate ring-gear pump or electric fuel pump. The fuel-injection pump is lubricated by the engine lube-oil circuit.

Size ZW fuel-injection pumps are available for engines with 4...12 cylinders. They are suitable for operation with diesel fuel. Fuel-injection pumps with the designation ZW(M) are designed for multifuel operation.

Size CW fuel-injection pump

The size CW in-line fuel-injection pump completes the top end of the Bosch in-line fuel-injection pumps range. The typical area of application for this model is on heavy-duty and relatively slow-revving marine engines and off-highway power units with nominal speeds of up to 1,800 rpm and power outputs of up to 200 kW per cylinder.

Even the 6-cylinder version of this fuel-injection pump with its closed-style pump housing made of nodulized cast iron weighs around 100 kg – this is roughly the weight of medium-sized car engine.

The pump is attached to the engine by eight bolts through its base.

The peak injection pressure is limited to approx. 1,000 bar.

The sealing and retention forces of the pump barrels with their plunger diameters of up to 20 mm are transferred to the pump housing by means of four strong clamp bolts (Fig. 15, 1).

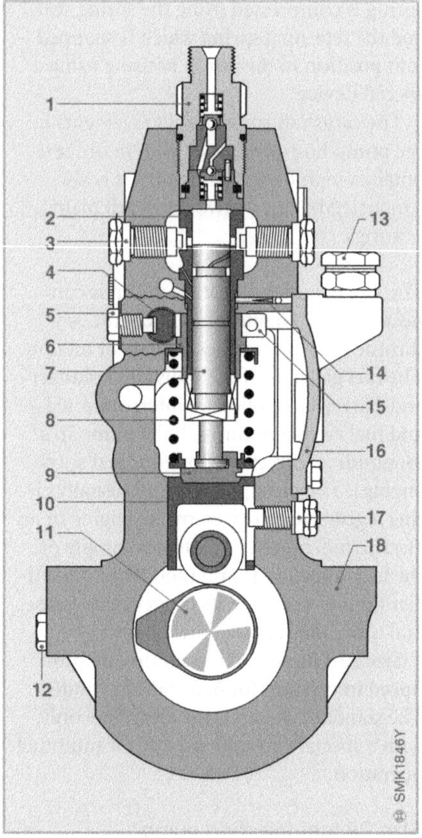

Fig. 14
1 Constant-pressure valve socket
2 Pump barrel
3 Impact-deflector screw
4 Control rack
5 Control rack guide screw
6 Control sleeve
7 Pump plunger
8 Plunger spring
9 LPC disk
10 Roller tappet
11 Camshaft
12 Oil-level checking plug
13 Oil filler plug
14 Pump-unit fixing screw
15 Clamp bolt
16 Cover plate
17 Retaining screw
18 Pump housing

Fig. 15
1 Clamp bolt
2 Impact-deflector screw
2 Screw cap

14 | Type ZW in-line fuel-injection pump (sectional view)

SMK1846Y

15 | Type CW in-line fuel-injection pump (external view)

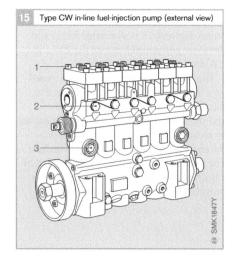

SMK1847Y

The control rack is in the form of a rack. Balancing of the delivery quantity between plunger-and-barrel assemblies is achieved with the aid of small orifices in the side of the pump housing. They are sealed by screw caps (3). LPC adjustment is by inserting shims of varying thicknesses between the roller tappets and the pump plungers.

Fuel supply to the fuel-injection pump is provided by a gear pump driven by the engine or an electric fuel pump.

The fuel-injection pump is controlled by a hydraulic or electromechanical governor provided by the engine manufacturer.

The pump is produced in 6, 8 and 10-cylinder versions and is suitable for use with diesel fuel.

In-line fuel-injection pumps
for special applications

In addition to their use with internal combustion engines, there are a number of specialized applications in which in-line fuel-injection pumps (e.g. driven by an electric motor) are employed. Those include applications in the

● Chemical industry
● Textiles industry
● Machine-tool industry and
● Plant engineering industry

Fuel-injection pumps used in these areas are referred to as *press pumps*. They are mainly Type P and Type ZW(M) designs. Type PE single-plunger fuel-injection pumps without their own camshaft may also be used.

The applications listed above require the delivery or finely and evenly atomized injection of fluids in very small but precisely metered quantities at high pressures. They frequently also demand the ability to vary the delivery quantity quickly, smoothly and as easily as possible.

The fluids pumped must not chemically attack the pump materials (aluminum, copper, steel, perbunane, nylon) to any discernible degree nor contain any solid, i.e. abrasive, components as this is the only way in which premature wear of the pump elements can be prevented. Where necessary, the fluids must be thoroughly filtered before they enter the press pumps. Depending on the fluids involved, special components (e.g. non-corroding compression springs, treated-surface fuel galleries, special seals) may need to be fitted to the press pumps.

High-viscosity fluids must be delivered to the press pump under sufficiently high pressure or made less viscous before passing through the filter by being heated (to max. 80°C).

The viscosity limits for pumped fluids are $v = 7.5 \cdot 10^{-5}\,m^2/s$; or with a higher fuel-gallery pressure of up to 2 bar $v = 38 \cdot 10^{-5}\,m^2/s$.

The fluid pumped should enter the fuel gallery at a pressure of up to 2 bar – depending on viscosity. This can be achieved by a presupply pump mounted on the press pump, a sufficient static head of pressure or a pressurized fluid reservoir.

Delivery capacities are measured using standard commercially available diesel fuels. If fluids of differing viscosities are used, delivery capacities may vary. Precise determination of the maximum delivery quantity is only possible using the actual fluid pumped and in situ in the actual installation.

The permissible *delivery pressure* also depends on whether the pump is operated intermittently or continuously. For Type ZW(M) press pumps, the maximum permissible pressure may be as much as 1,000 bar under certain circumstances (consultation required). If there is a possibility that a peak pressure above the maximum permissible limit may occur during operation, then a safety valve must be fitted in the high-pressure line.

Type PE in-line fuel-injection pumps for alternative fuels

Some specially designed diesel engines can also be run on "alternative" fuels. For such applications, modified versions of the MW and P-type pumps are used.

Multifuel operation

Multifuel engines can be run not only on diesel fuel but also on petrol, paraffin or kerosene. The changeover from one type of fuel to another requires adjustments to the fuel metering system in order to prevent large differences in power output. The most important fuel properties are boiling point, density and viscosity. In order that those properties can be balanced against one another to optimum effect, design modifications to the fuel-injection equipment and the engine are necessary.

Because of the low boiling points of alternative fuels, the fuel has to circulate more rapidly and under greater pressure through the fuel gallery of the fuel-injection pump. There is a special presupply pump available for this purpose.

With low-density fuels (e.g. petrol), the full-load delivery quantity is increased with the aid of a reversible control-rod stop.

In order to prevent leakage losses with low-viscosity fuels, the pump elements have a leakage trap that takes the form of two ring grooves in the pump barrel (see section "Pump unit with leakage return channel"). The upper groove is connected to the fuel gallery by a bore. The fuel that leaks past the plunger during the delivery stroke expands into this groove and flows through the bore back into the fuel gallery.

The lower groove has an inlet passage for the sealing oil. Oil from the engine lube-oil circuit is forced under pressure into this groove via a fine filter. At normal operating speeds, this pressure is greater than the fuel pressure in the fuel gallery, thereby reliably sealing the pump element. A non-return valve prevents crossover of fuel into the lubrication system if the oil pressure drops below a certain level at idle speeds.

Running on alcohol fuels

Suitably modified and equipped in-line fuel-injection pumps can also be used on engines that run on the alcohol fuels methanol or ethanol. The necessary modifications include:
- Fitting special seals
- Special protection for the surfaces in contact with the alcohol fuel
- Fitting non-corroding steel springs and
- Using special lubricants

In order to supply an equivalent quantity of energy, the delivery quantity has to be 2.3 times higher than for diesel fuel in the case of methanol and 1.7 times greater with ethanol. In addition, greater rates of wear must be expected on the delivery-valve and nozzle-needle seats than with diesel fuel.

Running on organic fuels (FAME[1])

For use with FAME, the fuel-injection pump has to be modified in a similar manner to the changes required for alcohol fuels.

RME[2] is one of the varieties of FAME frequently used. With *unmodified fuel-injection pumps*, the present maximum allowable proportion of RME that may be added to the diesel fuel is 5% based on the draft European standard of 2000. If higher proportions or poorer fuel qualities are used, the fuel-injection system may become clogged or damaged. In future there may be other types of FAME that are used either in pure form or as an additive to diesel fuel ($\leq 5\%$).

A definitive standard for FAME is currently in preparation. It will be required to precisely define fuel properties, stability and maximum permissible levels of contamination. Only by such means can trouble-free operation of the fuel-injection system and the engine be ensured.

[1] FAME: Fatty Acid Methyl Ester, i.e. animal or vegetable oil
[2] RME: Rape-Oil Methyl Ester

Operating in-line fuel-injection pumps

In order to operate correctly, a fuel-injection pump must be correctly adjusted, vented to remove all air, connected to the engine lube-oil circuit and its start of delivery must be synchronized with the engine. Only in this way is it possible to obtain the optimum balance between engine fuel consumption and performance and the ever stricter statutory regulations for exhaust-gas emission levels. Consequently a fuel-injection pump test bench is indispensable (see chapter "Service technology").

Venting

Air bubbles in the fuel impair the proper operation of the fuel-injection pump or disable it entirely. The system should therefore always be vented after replacing the filter or any other repair or maintenance work on the fuel-injection pump. While the system is in operation, air is reliably expelled via the overflow valve on the fuel filter (continuous venting). On fuel-injection pumps without an overflow valve, a flow throttle is used.

Lubrication

Fuel-injection pumps and governors are connected to the engine lube-oil circuit. Then the fuel-injection pump is maintenance-free.

On pumps that are attached to the engine through the base or by a cradle mounting, the oil returns to the engine through a lube-oil return (Fig. 1). If the fuel-injection pump is flange-mounted to the engine at its end face, the oil can return directly through the camshaft bearing or special oil bores.

The oil level check takes place at the same time as the regular engine oil changes specified by the engine manufacturer and is performed by removing the oil check plug on the governor. Fuel-injection pumps and governors with separate oil systems have their own dipsticks for checking the oil level.

Shutting down

If the engine, and therefore the fuel-injection pump, is taken out of service for a long period, no diesel fuel may remain inside the fuel-injection pump. Resinification of the diesel fuel would occur, causing the pump plungers and delivery valves to stick and possibly corrode. For this reason, a proportion of up to 10% of a reliable rust-inhibiting oil is added to the diesel fuel in the fuel tank and the fuel is then circulated through the fuel-injection pump for 15 minutes. The same proportion of rust-inhibiting oil is added to the lubricant in the fuel-injection pump's camshaft housing.

New fuel-injection pumps with a "p" in their identification code have been factory-treated with an effective anticorrosive.

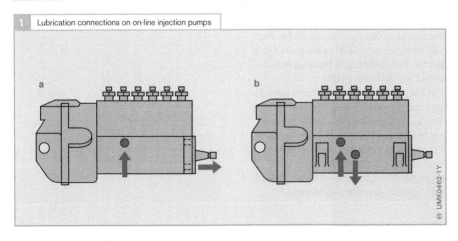

1   Lubrication connections on on-line injection pumps

a

b

Fig. 1
a   Return line via bearing at driven end
b   Return via return line

Governors and control systems for in-line fuel-injection pumps

A diesel fuel-injection pump must reliably supply the engine with precisely the right amount of fuel at exactly the right time under all operating conditions, in all operating statuses and at all engine loads. Even with the control rack in a fixed position, the engine would not maintain an absolutely constant speed. Effective operation of the fuel-injection pump therefore requires a mechanical centrifugal governor or an electronic control system.

The fuel-injection pump delivers precisely metered amounts of fuel at high pressure to the nozzles so that it is injected into the engine's combustion chamber. The fuel-injection system has to ensure that fuel is injected
- In precisely metered quantities according to engine load
- At precisely the right moment
- For a precisely defined length of time and
- In a manner compatible with the combustion method used

It is the job of the fuel-injection pump and governor to ensure that these requirements are met.

The characteristic features of *mechanical governors* are their durability and ease of maintenance. The main topic of this chapter is an examination of the various types of governor and adjustment mechanisms.

    An *Electronic Diesel Control* (EDC) performs a substantially more comprehensive range of tasks than a mechanical governor. The system of electrical actuators for the EDC system is described at the end of this chapter. The structure of the system is described in a separate chapter.

    In the past *pneumatic governors* were also used for smaller fuel-injection pumps. They utilize the intake-manifold pressure (see next page). Because of today's greater demands with regard to control quality, however, the pneumatic governor is no longer produced and therefore not described in any greater detail in this manual.

Open and closed-loop control

Control systems are systems in which one or more input variables (reference variables and disturbance values) govern one or more output variables (Fig. 1).

Open-loop control
In an open-loop control system (Fig. 1a), the effects of control commands are not monitored (open-control loop). This method is used for proportioning the start quantity, for example.

Closed-loop control
The distinguishing feature of a closed-loop control system (Fig. 1b) is the circular nature of the control sequence. The actual value of the controlled variable is constantly compared with the setpoint value. As soon as a discrepancy is detected, an adjustment is made to the settings of the actuators. The advantage of closed-loop control is that external disturbance values on the control process can be detected and compensated for (e.g. changes in engine load). Closed-loop control is used for the engine idle speed, for example.

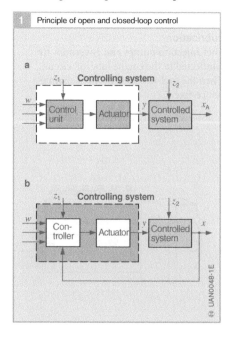

Fig. 1
a   Open-loop control
b   Closed-loop control

w Reference variable(s)
x Controlled variable (closed loop)
x_A Controlled variable (open loop)
y Manipulated variable(s)
z_1, z_2 Disturbance values

1   Principle of open and closed-loop control

a

z_1 Controlling system z_2
w — Control unit → Actuator — y → Controlled system → x_A

b

z_1 Controlling system z_2
w — Con-troller → Actuator — y → Controlled system → x

UAN0048-1E

"Anyone who thinks a diesel engine is a crude machine that will tolerate crude solutions is mistaken!" [1]

A large degree of sensitivity and precision is needed to obtain and maintain the very best performance from a diesel engine.

The specific method by which a diesel engine was governed was originally left to the engine manufacturers themselves. However, in order to do away with the need for a drive system running off the engine, they started to demand fuel-injection pumps with ready-mounted governors.

At the end of the 1920s Bosch took up that new challenge and, as a result of some outstanding engineering work, had a centrifugal idling and maximum-speed governor in volume production by 1931. A variation of that design followed shortly in the guise of a variable-speed governor that was in great demand for tractors and marine engines.

For smaller, faster running diesel engines in motor vehicles, on the other hand, a centrifugal governor did not seem suitable. It wasn't until the pneumatic governor was conceived that new impetus was introduced: "The control rack is attached to a leather diaphragm and the depression in the intake manifold, which is dependent on engine speed, alters the position of the diaphragm and adjusts the delivery quantity according to the position of the control flap" (see illustration). [2]

In the post-war years, an enormous variety of improved designs were used such as floating-pivot governors (1946 to 1948), governors with external springs (1955 onwards) and governors with vibration dampers.

Additional attachments for matching the full-load delivery quantity to the desired engine torque curve were also used as well as devices for automatically adjusting the start quantity.

Today, electronics are as important in diesel fuel injection as in any other branch of technology. An optimized diesel engine controlled by an electronic control system is now virtually taken for granted.

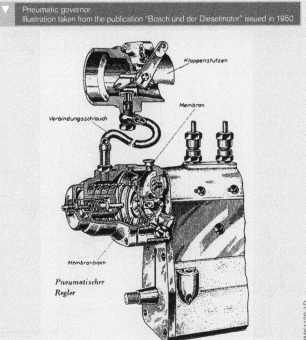

Klappenstutzen

Membran

Verbindungsschlauch

Membranblock

Pneumatischer Regler

UMK1179-1D

[1]   Georg Auer;
"Der Widerspenstigen Zähmung";
Diesel-Report;
Robert Bosch GmbH;
Stuttgart, 1977/78

[2]   Friedrich Schildberger;
Bosch und der Dieselmotor;
Stuttgart, 1950

Action of the governor/control system

All in-line fuel-injection pumps have a pump element consisting of a pump barrel (Fig. 1, 8) and plunger (9) for each engine cylinder. The quantity of fuel injected can be altered by rotating the pump plunger (see chapter "Type PE standard in-line fuel-injection pumps"). The governor/control system adjusts the position of all pump plungers simultaneously by means of the control rack (7) in order to vary the injected fuel quantity between zero and maximum delivery quantity. The control rack travel, s, is proportional to the injected fuel quantity and therefore to the torque produced by the engine.

The helix on the pump plunger can be of various types. Where there is only a bottom helix, fuel delivery always starts at the same point of plunger lift but ends at a variable point dependent on the angle of rotation of the piston. Where there is a top helix, the start of delivery can also be varied. There are also designs which incorporate both a top and bottom helix.

Definitions

No load

No load refers to all engine operating statuses in which the engine is overcoming only its own internal friction. It is not producing any torque output. The accelerator pedal may be in any position. All speed ranges up to and including breakaway speed are possible.

Idle

The engine is said to be idling when it is running at the lowest no-load speed. The accelerator pedal is not depressed. The engine is not generating any output torque. It is overcoming only the internal friction.
Some sources refer to the entire no-load range as idling. The upper no-load speed (breakaway speed) is then called the upper idle speed.

Full load

At full load (Wide-Open Throttle, WOT), the accelerator pedal is fully depressed. Under steady-state conditions, the engine is generating its maximum possible torque. Under non steady-state conditions (limited by turbocharger/supercharger pressure) the

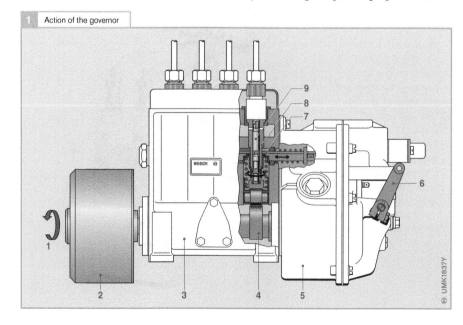

1   Action of the governor

Fig. 1
1 Pump drive
2 Timing device
3 Pump housing
4 Camshaft
5 Governor housing
6 Control lever
7 Control rack
8 Pump barrel
9 Pump plunger

engine develops the maximum possible (lower) full-load torque with the quantity of air available. All engine speeds from idle speed to nominal speed are possible. At breakaway speed the governor/control system automatically reduces the injected fuel quantity and therefore engine torque.

Part load
Part load covers the range between no load and full load. The engine is generating an output between zero and the maximum possible torque.

Part load at idle speed
In this particular case, the governor holds the engine at idle speed. The engine is generating torque output. This may extend to full load.

Overrunning
The engine is said to be overrunning when it is driven by an external force acting through the drivetrain (e.g. when descending an incline).

Steady-state operation
The engine's torque output is equal to the required torque. The engine speed is constant.

Non steady-state operation
The engine's torque output is not equal to the required torque. The engine speed is not constant.

Indices
The indices used in the diagrams and equations in the rest of this chapter have the following meanings:

ı	Idle
n	No load
v	Full load
u	Minimum figure
o	Maximum figure

Some examples:

n_{nu}	Minimum no-load speed ($=$ idle speed n_l)
n_n	Any no-load speed
n_{no}	Maximum no-load speed
n_v	Any full-load speed
n_{vo}	Maximum full-load speed (nominal speed)

Proportional response of the governor

Every engine has a torque curve that corresponds to its maximum load capacity. For every engine speed there is a corresponding maximum torque.

If the load is removed from the engine without the position of the control lever being altered, the engine speed must not be allowed to increase by more than a permissible degree specified by the engine manufacturer (e.g. from full-load speed n_v to no-load speed, n_n, Fig. 2). The increase in engine speed is proportional to the change in engine load, i.e. the greater the amount by which the engine load is reduced, the greater the increase in engine speed. Hence the terms proportional response and proportional characteristics in connection with governors. The proportional response of the governor generally relates to the maximum full-load speed. That is equivalent to the nominal speed.

The proportional response δ is calculated as follows:

$$\delta = \frac{n_{no} - n_{vo}}{n_{vo}}$$

or as a percentage thus:

$$\delta = \frac{n_{no} - n_{vo}}{n_{vo}} \cdot 100\%$$

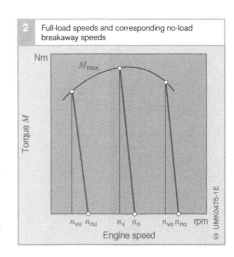

2   Full-load speeds and corresponding no-load breakaway speeds

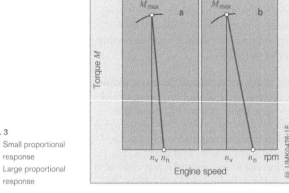

Fig. 3
a Small proportional
   response
b Large proportional
   response

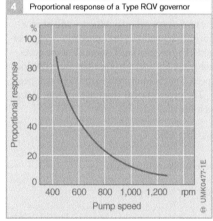

Fig. 4
Curve for varying speeds
set by means of the
control lever

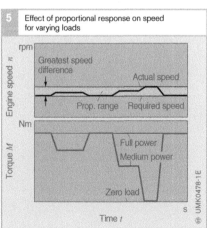

where
n_{no} Upper no-load speed
n_{vo} Upper full-load speed

As the pump speed on four-stroke engines is equivalent to half the engine speed, both the pump speed and the engine speed can be used in the calculation.

Example (pump speeds):
n_{no} = 1,000 rpm, n_{vo} = 920 rpm

$$\delta = \frac{1,000-920}{920} \cdot 100\% = 8.7\%$$

Figure 5 shows the effect of the proportional response based on a practical example: At a constant set speed, the actual speed varies when the engine load alters (e.g. variations in gradient) within the proportional response range.

In general, a greater proportional response allows the achievement of more stable characteristics on the part of the entire control loop consisting of governor, engine and driven machine or vehicle. On the other hand, the proportional response is limited by the operating conditions.

Examples of proportional response:
- Approx. 0...5% for power generators
- Approx. 6...15% for vehicles

Purpose of the governor/control system

The basic task of any governor/control system is to prevent the engine from exceeding the maximum revving speed specified by the engine manufacturer. Since the diesel engine always operates with excess air because the intake flow is not restricted, it would "overrev" if there were no means of limiting its maximum speed. Depending upon the type of governor/control system, its functions may also include holding the engine speed at specific constant levels such as idling or other speeds within a specific band or the entire range between idling and maximum speed. There may be also be other tasks beyond those mentioned, in which case the capabilities of an electronic control system are substantially more extensive than those of a mechanical governor.

The governor/control system is also required to perform control tasks such as

- Automatic enabling or disabling of the greater fuel delivery quantity required for starting (start quantity)
- Variation of the full-load delivery quantity according to engine speed (torque control)
- Variation of the full-load delivery quantity according to turbocharger and atmospheric pressure

Some of those tasks necessitate additional equipment which will be explained in more detail at a later stage.

Maximum speed control function

When the engine is running at maximum full-load speed, n_{vo}, it must not be allowed to exceed the maximum no-load speed, n_{no}, when the load is removed in accordance with the permissible proportional response (Fig. 6). The governor/control system achieves this by moving the control rack back towards the stop setting.

The range $n_{vo} - n_{no}$ is referred to as the maximum speed cutoff range. The greater the proportional response of the governor, the greater the increase in speed from n_{vo} to n_{no}.

Intermediate-speed regulation

If the task so requires (e.g. on vehicles with PTO drives), the governor/control system can also hold the engine at specific constant speeds between idle speed and maximum speed according to the proportional response (Fig. 7). When the intermediate speed regulation function is active, the engine speed, n, thus only varies according to load and within the engine's power band between n_v (at full load) and n_n (at no load).

Idle-speed regulation

The diesel engine's speed can also be controlled at the lower end of the speed range (Fig. 8) – the idle speed range. Without a governor or control system, the engine would either slow down to a standstill or overrev uncontrollably when not under load.

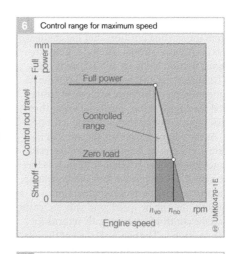

6 Control range for maximum speed

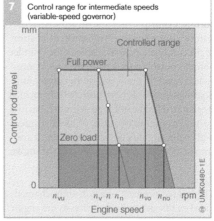

7 Control range for intermediate speeds (variable-speed governor)

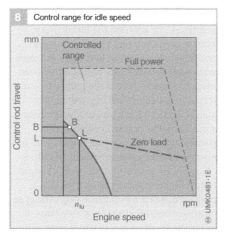

8 Control range for idle speed

When the control rack returns to position B from the starting position after the engine has been started from cold, the engine's internal friction levels are still relatively high. The quantity of fuel required to keep the engine running is therefore somewhat greater and the engine speed somewhat lower than that represented by the idle speed setting L.

As the engine warms up, the internal friction of the engine decreases as do the resistance levels of the external units such as the alternator, air compressor, fuel-injection pump, etc. that are driven by the engine. Consequently, the engine speed gradually increases and the control rack eventually returns to the position L. The engine is then at the idle speed for normal operating temperature.

Torque control

Torque control enables optimum utilization of the combustion air available in the cylinder. Torque control is not a true governor/control system function but is one of the control functions allocated to the governor/control system. It is calibrated for the full-load delivery capacities, i.e. the maximum quantity of fuel delivered within the engine's power output band and combustible without the production of smoke.

Conventionally aspirated engines

The fuel requirement of a non-turbocharged diesel engine generally decreases as the engine speed increases (lower relative air throughput, thermal limits, changes in mixture formation parameters). By contrast, the fuel delivery quantity of a Bosch fuel-injection pump increases with engine speed over a specific range assuming the control rack setting remains unchanged (throttle effect of the pump unit control port). Too much fuel injected into the cylinder, on the other hand, produces smoke or may cause the engine to overheat. The quantity of fuel injected must therefore be adjusted to suit the fuel requirement (Fig. 9).

Governors/control systems with a torque control function move the control rack a specified distance towards the stop setting in the torque control range (Fig. 10). Thus, as the engine speed increases (from n_1 to n_2),

Fig. 9
a Engine fuel
   requirement
b Full-load delivery
   quantity without
   torque control
c Torque-matched
   full-load delivery
   quantity

Fig. 11
a With torque control
b Without torque
   control

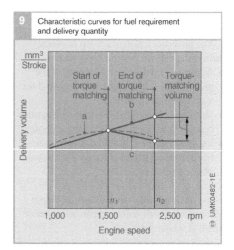

9 Characteristic curves for fuel requirement and delivery quantity

10 Control rack travel curve with positive torque control

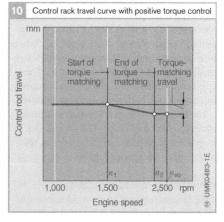

11 Torque curve of a diesel engine

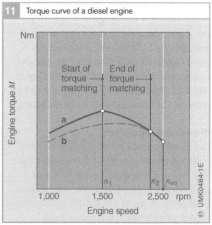

the fuel delivery quantity decreases (positive torque control), and when the engine speed decreases (from n_2 to n_1) the fuel delivery quantity increases.

Torque control mechanisms vary in design and arrangement from one governor/control system to another. Details are provided in the descriptions of the individual governors/control systems.

Figure 11 shows the torque curves of diesel engines with and without torque control. The maximum torque is obtained without exceeding the smoke limit across the entire engine speed range.

Turbocharged engines

In engines with high-compression turbochargers, the full-load fuel requirement at lower engine speeds increases so much that the inherent increase in delivery quantity of the fuel-injection pump is insufficient. In such cases, torque control must be based on engine speed or turbocharger pressure and effected by means of the governor/control system or the manifold pressure compensator alone or by the two in conjunction depending on the circumstances.

This type of torque control is referred to as negative. That means that the delivery quantity increases more rapidly as engine speed rises (Fig. 12). This is in contrast with the usual positive torque control whereby the injection quantity is reduced as engine speed increases.

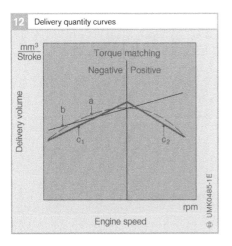

12 Delivery quantity curves

mm³/Stroke

Torque matching

Negative | Positive

Delivery volume

a

b

c_1 c_2

rpm

Engine speed

UMK0485-1E

Types of governor/control system

Continually increasing demands with regard to exhaust-gas emissions, fuel economy and engine smoothness and performance are the defining characteristics of diesel engine development. Those demands are reflected in the requirements placed on the fuel-injection system and in particular the governor or control system.

The various different control tasks required result in the following types of governor:

- *Maximum-speed governors*
  only limit the engine's maximum speed.
- *Minimum/maximum-speed governors*
  also control the idle speed in addition to limiting the maximum speed. They do not control the intermediate range. The injected fuel quantity in that range is controlled by means of the accelerator pedal. This type of governor is used primarily on motor vehicles.
- *Variable-speed governors*
  limit not only the minimum and maximum speeds but also control the intermediate speed range.
- *Combination governors*
  are, as their name suggests, a combination of minimum/maximum-speed governors and variable-speed governors.
- *Generator-engine governors*
  are for use on engines that drive power generators designed to comply with DIN 6280 or ISO 8528.

Mechanical governors

The mechanical governors used with in-line fuel-injection pumps are also referred to as centrifugal governors because of the flyweights they employ. This type of governor is linked to the accelerator pedal by means of a rod linkage and a control lever (Fig. 1 overleaf).

Timing device

In order to control the start of injection and compensate for the time taken by the pressure wave to travel along the high-pressure delivery line, a timing device is used to "advance" the start of delivery of the fuel-injection pump as the engine speed increases.

Fig. 12
a   Engine fuel requirement
b   Full-load delivery quantity without torque control
c   Torque-matched full-load delivery quantity
c_1 Negative torque control
c_2 Positive torque control

Electronic control systems

The Electronic Diesel Control (EDC) satisfies the greater demands placed on modern control systems. It enables electronic sensing of engine parameters and flexible electronic data processing. Closed control loops with electrical actuators offer not only more effective control functions in comparison with mechanical governors but also have additional capabilities such as smooth-running control. In addition, Electronic Diesel Control provide the facility for data exchange with other electronic systems such as the transmission control system and permit comprehensive electronic fault diagnosis. The subsystems and components of the EDC system for in-line fuel-injection pumps are described in the chapter "Electronic Diesel Control (EDC)".

Figures 1 and 2 show schematic diagrams of the control loops for mechanical governors and electronic control systems. Detailed illustrations of the control loops for standard in-line fuel-injection pumps and control-sleeve in-line fuel-injection pumps are displayed on the next double page.

Advantages of electronic control systems

The use of an electronically controlled fuel-injection system offers the following advantages:
- The extensive range of functions and available data enables the achievement of optimum engine response across the entire operating range.
- Clear separation of individual functions: Governor characteristics and fuel rate-of-discharge curves are no longer interdependent; consequently there is wider scope for adaptation to individual applications.
- Greater capacity for manipulating variables that previously could not be included in the equation with mechanical systems (e.g. compensation for fuel temperature, controlling idle speed independently of engine load).
- Higher levels of control precision and consistency throughout engine life by diminishing tolerance effects.
- Improved engine response characteristics: The large volume of stored data (e.g. engine data maps) and parameters allows optimization of the engine-and-vehicle combination.

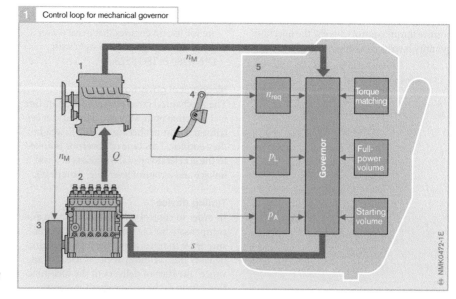

1　Control loop for mechanical governor

Fig. 1
1　Diesel engine
2　In-line fuel-injection pump
3　Timing device
4　Accelerator pedal
5　Governor

n_{req}　Required engine speed
n_M　Engine speed
p_A　Atmospheric pressure
p_L　Turbocharger pressure
Q　Injected fuel quantity
s　Control rack travel

- More extensive range of functions: Additional functions such as cruise control and intermediate-speed regulation can be implemented without major complications.
- Interaction with other electronic systems on the vehicle provides the potential for making future vehicles generally easier to use, more economical, more environmentally friendly and safer (e.g. electronic transmission control EGS, Traction Control System TCS).
- Substantial reduction of space requirements because mechanical attachments to the fuel-injection pump are no longer required.
- Versatility and adaptability: Data maps and stored parameters are programmed individually when the control unit reaches the end of the production line at Bosch or the engine/vehicle manufacturer. This means that a single control unit design can be used for several different engine or vehicle models.

Safety concept

For safety reasons, a compression spring moves the control rack back to the "zero delivery" position whenever the electrical actuators are disconnected from the power supply.

Self-monitoring: The Electronic Diesel Control (EDC) incorporates functions for monitoring the sensors, actuators and the microcontroller in the control unit. Additional safety is provided by extensive use of redundant backup. The diagnostic system provides the facility for obtaining a read-out of recorded faults on a compatible tester or, on older systems, using a diagnostic lamp.

Substitute functions: The system incorporates an extensive array of substitute functions. If, for example, the engine speed sensor fails, the signal from terminal W on the alternator can be used as a substitute for the speed sensor signal. If important sensors fail, a warning lamp lights up.

Fuel shutoff function: In addition to the fuel shutoff function of the control rack when in stop setting, a solenoid valve in the fuel supply line shuts off the fuel supply when disconnected from the power supply. This separate electric or electrohydraulic shutoff valve also shuts off the fuel supply if, for example, the fuel quantity control mechanism fails, thus stopping the engine.

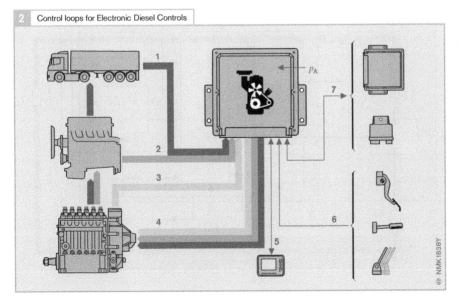

2   Control loops for Electronic Diesel Controls

Fig. 2
1 Vehicle sensors (e.g. for road speed)
2 Engine sensors (e.g. for engine temperature)
3 Injection system sensors (e.g. for start of delivery)
4 Control signals
5 Diagnosis interface
6 Accelerator pedal and desired-value generators (switches)
7 Data communication (e.g. for glow-plug control)

P_A Atmospheric pressure

Control loop configurations

Engine starting, idling, performance, soot emissions and response characteristics are decisively affected by the injected fuel quantity. Accordingly, there are data maps for starting, idling, full load, accelerator characteristics, smoke emission limitation and pump characteristics stored on the control unit.

The control rack travel is used as a substitute variable for injected fuel quantity. For engine response characteristics, an RQ or RQV control characteristic familiar from mechanical governors can be specified.

A pedal-travel sensor detects the driver's torque/engine speed requirements as indicated by the accelerator pedal (Fig. 3). The control unit calculates the required injected fuel quantity (required fuel-injection pump control rack setting), taking account of the stored data maps and the current sensor readings. The required control rack setting is then the reference variable for the control loop. A position control circuit in the control unit detects the actual position of the control rack, and thus the required adjustment, and provides for rapid and precise adjustment of the control rack position.

There are control functions for maintaining various engine speeds: idle speed, a fixed intermediate speed, e.g. for PTO drives, or a set speed for the cruise control function.

Control loop for injected fuel quantity
Based on the calculated required setting, the control unit sends electrical signals to the control rack actuation system for the fuel-injection pump. The required injected fuel quantity specified by the control unit is set using the position control loop: The control unit specifies a required control-rack travel and receives a feedback signal indicating the

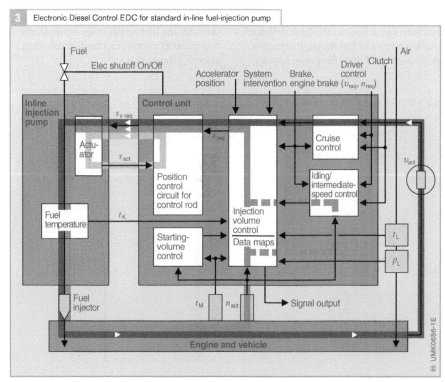

3   Electronic Diesel Control EDC for standard in-line fuel-injection pump

Fig. 3

n_{act} Actual engine speed
n_{req} Required engine speed
p_L Turbocharger pressure
s_{act} Actual control rack travel
s_{req} Required control rack travel
$s_{v\,req}$ Control rack positioning signal
t_K Fuel temperature
t_L Air temperature
t_M Engine temperature
v_{act} Actual road speed
v_{req} Required road speed

actual control rack position from the control rack position sensor. To complete the control loop, the control unit repeatedly recalculates the adjustment required to achieve the required fuel-injection pump setting, thereby continually correcting the actual setting to match the required setting.

Start of delivery control loop
Control-sleeve in-line fuel-injection pumps have a means of adjusting start of delivery as well as the mechanism for adjusting the injected fuel quantity (Fig. 4).

The start of delivery is also adjusted by means of a closed control loop. A needle-motion sensor in one of the nozzle holders signals to the control unit the actual point in time at which injection takes place. Using this information in conjunction with stored data, the control unit then calculates the actual crankshaft position at which injection

starts. Next, it compares the actual start of delivery with the calculated required start of delivery. A signal control circuit in the control unit then operates the timing device on the fuel-injection pump, thus bringing the actual start of delivery into line with the required setting.

Because the timing device actuation mechanism is "structurally rigid", there is no need for an adjustment travel feedback sensor. Structurally rigid means that the lines of action of solenoid and spring always have a precise intersection point so that the travel of the solenoid is proportional to the signal current. That is equivalent to feedback within a closed control loop.

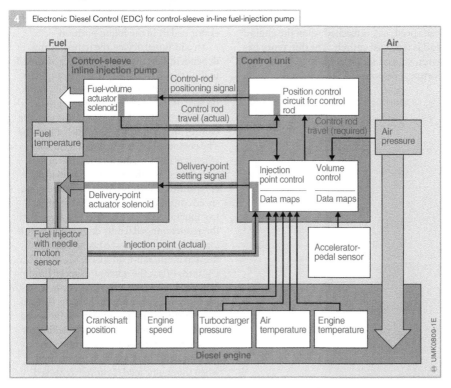

4   Electronic Diesel Control (EDC) for control-sleeve in-line fuel-injection pump

Overview of governor types

Governor type designations
The governor type designation is shown on the identification plate. It indicates the essential features of the governor (e.g. design type, governed speed range, etc.). Figure 3 details the individual components of the governor type designation.

Maximum-speed governors
Maximum-speed governors are intended for diesel engines that drive machinery at their nominal speed. For such applications, the governor's job is merely to hold the engine at its maximum speed; control of idle speed and start quantity are not required. If the engine speed rises above the nominal speed, n_{vo}, because the load decreases, the governor shifts the control rack towards the stop setting, i.e. the control rack travel is shortened and the delivery quantity reduced (Fig. 1). Engine speed increase and control rack travel decrease follow the gradient A – B. The maximum no-load speed, n_{no}, is reached when the engine load is removed entirely. The difference between n_{no} and n_{vo} is determined by the proportional response of the governor.

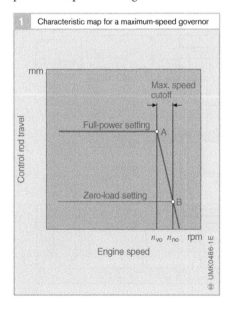

1   Characteristic map for a maximum-speed governor

mm

Control rod travel

Max. speed cutoff

Full-power setting   A

Zero-load setting   B

n_{vo} n_{no} rpm

Engine speed

UMK0486-1E

Minimum/maximum-speed governors
Diesel engines for motor vehicles frequently do not require engine speeds between idling and maximum speed to be governed. Within this range, the fuel-injection pump's control rack is directly operated by the accelerator pedal under the control of the driver so as to obtain the required engine torque. At idle speed, the governor ensures that the engine does not cut out; it also limits the engine's maximum speed. The governor's characteristic map (Fig. 2) shows the following: When the engine is cold, it is started using the start quantity (A). At this point, the driver has fully depressed the accelerator pedal.

If the driver releases the accelerator, the control rack returns to the idle speed setting (B). While the engine is warming up, the idle speed fluctuates along the idle speed curve and finally comes to rest at the point L. Once the engine has warmed up, the maximum start quantity is not generally required when the engine is restarted. Some engines can even start with the control rack actuating lever (accelerator pedal) in the idling position.

An additional device, the temperature-dependent start quantity limiter, can be used to limit the start quantity when the engine is warm even if the accelerator pedal is fully depressed. If the driver fully depresses the accelerator pedal when the engine is running, the control rack is moved to the full-load setting. The engine speed increases as a result and when it reaches n_1, the torque control function comes into effect, i.e. the full-load delivery quantity is slightly reduced. If the engine speed continues to increase, the torque control function ceases to be effective at n_2. With the accelerator pedal fully depressed, the full-load volume continues to be injected until the maximum full-load speed, n_{vo}, is reached. Upwards of n_{vo}, the maximum speed limiting function comes into effect in accordance with the proportional response characteristics so that a further small increase in engine speed results in the control rack travel backing off so as to reduce the delivery quantity. The maximum no-load speed, n_{no}, is reached when the engine load is entirely removed.

2   Characteristic map for a minimum/maximum-speed governor with torque control

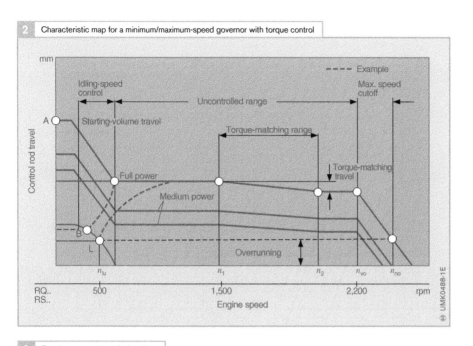

3   Bosch governor type designations

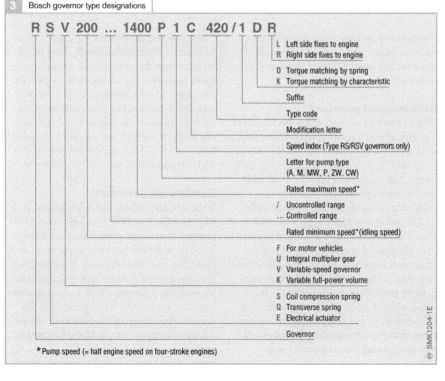

R S V 200 ... 1400 P 1 C 420 / 1 D R

L Left side fixes to engine
R Right side fixes to engine

D Torque matching by spring
K Torque matching by characteristic

Suffix

Type code

Modification letter

Speed index (Type RS/RSV governors only)

Letter for pump type
(A, M, MW, P, ZW. CW)

Rated maximum speed*

/   Uncontrolled range
... Controlled range

Rated minimum speed*(idling speed)

F For motor vehicles
U Integral multiplier gear
V Variable-speed governor
K Variable full-power volume

S Coil compression spring
Q Transverse spring
E Electrical actuator

Governor

*Pump speed (= half engine speed on four-stroke engines)

Fig. 3
With combination
governors, multiple
speeds are specified
(e.g. 300/900...1,200).

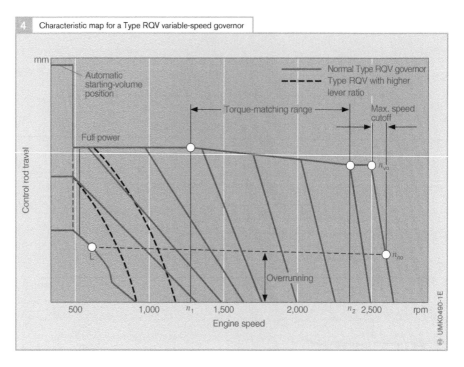

4   Characteristic map for a Type RQV variable-speed governor

When the engine is overrunning (e.g. coasting on a descent) the engine speed may increase further but the control rack travel is backed off to zero.

Variable-speed governors

Vehicles that have to maintain a specific speed (e.g. agricultural tractors, roadsweepers, ships) or have a PTO drive that requires the engine speed to be kept at a constant level (e.g. tank pumps, fire-engine ladders) are fitted with variable-speed governors.

This type of governor controls not only the idling and maximum speeds but also intermediate speeds independently of engine load. The desired speed is set by means of the control lever. The governor characteristic map (Fig. 4) shows the following: the start quantity setting for cold starting, the full-load control characteristic with torque control between n_1 and n_2 up to the maximum speed cutoff band from the maximum full-load speed along the gradient n_{vo}, n_{no}.

The remaining curves show the cutoff characteristics for intermediate speeds. They reveal an increase in the proportional response as speed decreases. The curves shown as broken lines apply to vehicles whose PTO drives operate within the lower engine speed range. As the load increases, the engine speed does not dip as sharply as with normal governors (shallower curves). This is achieved by the use of a higher transmission ratio for the control lever.

Combination governors

If the normal proportional response of a Type RQV or RQUV variable-speed governor at the upper or lower end of the adjustment range is too great for the intended application, and control of intermediate speeds is not required, then the governor movement is designed as a combination governor. With such an arrangement, torque control is not possible in the uncontrolled range of the maximum-speed governor component. On this characteristic map (Fig. 5), the uncontrolled stage is in the lower speed band, while the controlled stage

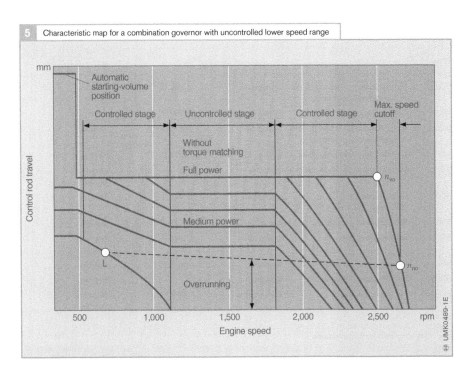

5   Characteristic map for a combination governor with uncontrolled lower speed range

is in the upper speed band. A different type of governor operates in the lower speed band as a variable-speed governor (downward-gradient curves), after which follows an uncontrolled band (horizontal sections of the curves) extending to the maximum speed cutoff band. In both cases, the horizontal sections of the curves represent the control rack travel for varying part-load control lever settings. The lines descending from the full-load curve represent the speed-regulation breakaway characteristics for varying set intermediate speeds. The combination governor differs in design from a variable-speed governor simply by virtue of the use of different governor springs.

Generator governor

For engines driving power generators, German regulations require that governors conform to DIN 6280 (see tables overleaf). Bosch centrifugal governors can be used for design classes 1, 2 and 3. The conditions for design class 4, to which units with a 0% proportional response belong, usually require the use of an

electronic control system. A characteristic map for a generator governor is shown in Figure 6. If parallel operation is not required, the speed setting can be fixed, i.e. a straightforward maximum-speed governor can be used.

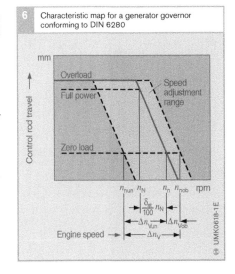

6   Characteristic map for a generator governor conforming to DIN 6280

1   Governor types

Type	Function	Actuator mechanism	Pump size	Torque control
RQ	Minimum/maximum-speed governor or maximum-speed governor	flyweights	A, MW, P	Positive
RQ	Generator governor	flyweights	A, MW, P	None
RQU	Minimum/maximum-speed governor or maximum-speed governor	flyweights[1]	ZW, P9, P10	Positive
RS	Minimum/maximum-speed governor	flyweights	A, MW, P	Positive
RSF	Minimum/maximum-speed governor	flyweights	M	Negative/ positive
RQV	Variable-speed or combination governor	flyweights	A, MW, P	Positive
RQUV	Variable-speed governor	flyweights[1]	ZW, P9, P10	Positive
RQV..K	Variable-speed governor	flyweights	A, MW, P	Negative/ positive
RSV	Variable-speed governor	flyweights	A, M, MW, P	Positive
RSUV	Variable-speed governor	flyweights[1]	P	Positive
RE	Any characteristic	Electromagnet	A, MW, P	Negative/ positive

Table 1
[1] With transmission ratio for slow-running engines

2   Operating limits for design classes

No.	Description	Symbol	Unit	Design class			
				1	2	3	4
4.2.4	Static speed difference or proportional response	δ_{st}	%	8	5	3	STA
4.2.5	Speed fluctuation range	ν_n	%	–	1.5	0.5	STA
4.2.1	Lower speed setting range	$\delta \cdot n_{Vun}$	%	$-(2.5 + \delta_{st})$	$-(2.5 + \delta_{st})$	$-(2.5 + \delta_{st})$	STA
4.2.2	Upper speed setting range	$\delta \cdot n_{Vob}$	%	$+2.5$	$+2.5$	$+2.5$	STA
4.1.6	Frequency regulation time STA	t_{tzu}, t_{tab}	s	–	5	3	STA

Table 2
Applies only to power generator applications
Excerpt from DIN 6280, Part 3

STA Subject to agreement

3	Speed definitions		
No.	**Description**	**Symbol**	**Definition**
4.1	Nominal speed	n_N	The engine speed corresponding to the rated frequency of the generator and to which the generator rated output relates.
4.3	Zero-output speed	n_n	Steady-state speed of the engine under no load. Associated figures for rated-output and intermediate output speeds relate to an unchanged speed setting.
4.7	Minimum variable zero-output speed	n_{nun}	Minimum steady-state engine speed under no load that can be set on the speed setting device or governor.
4.8	Maximum variable zero-output speed	n_{nob}	Maximum steady-state engine speed under no load that can be set on the speed setting device or governor.
4.9	Speed setting range	Δn_V	Range between set minimum and maximum zero-output speeds; the figure for the speed range is obtained by adding the figures for the upper and lower speed setting ranges as per sections 4.9.1 and 4.9.2.
4.9.1	Lower speed setting range	Δn_{Vun}	Range between set minimum zero-output speed and the zero-output speed that results from removal of the engine load at the rated output speed without alteration of the speed setting.
		δn_{Vun}	$\Delta n_{Vun} = n_n - n_{nun}$ The difference between the two speeds is expressed as a percentage of the nominal speed $\delta n_{Vun} = \dfrac{(n_n - n_{nun})}{n_N} \cdot 100$
4.9.2	Upper speed setting range	Δn_{Vob}	Range between set maximum zero-output speed and the zero-output speed that results from removal of the engine load at the rated output speed without alteration of the speed setting. $\Delta n_{Vob} = n_{nob} - n_n$
		δn_{Vob}	The difference between the two speeds is expressed as a percentage of the nominal speed. $\delta n_{Vob} = \dfrac{(n_{nob} - n_n)}{n_N} \cdot 100$
5.1	Static speed difference or proportional response	δ_{St}	Ratio of speed difference between zero-output speed, n_n, and nominal speed, n_N, expressed as a percentage of the nominal speed. $\delta_{St} = \dfrac{(n_n - n_N)}{n_N} \cdot 100$

Table 3
Applies only to power generator applications
Excerpt from DIN 6280, Part 4

Timing devices

The start of delivery (Fig. 1, SD) represents the point at which fuel delivery by the fuel-injection pump commences. The timing of the start of delivery depends on the variables "injection lag" (IL) and "ignition lag" (IGL) which are dependent on the operating status of the engine. The *injection lag* refers to the time delay between the start of delivery (SD) and the start of injection (SI), i.e. the time at which the nozzle opens and starts injecting fuel into the combustion chamber. The *ignition lag* is the time that elapses between the start of delivery and the start of combustion (SC). The combustion start defines the point when air- and-fuel mixture ignites. It can be varied by altering the start of delivery.

Start of delivery, start of injection and combustion start are specified in degrees of crankshaft rotation relative to crankshaft Top Dead Center (TDC).

Engine-speed related adjustment of start of delivery on an in-line fuel-injection pump is best achieved by means of a timing device.

Functions

Strictly speaking, based on its function, the timing device should really be called a deliv-ery start adjuster, as it actually varies the start of delivery directly. It transmits the drive torque for the fuel-injection pump and simul-taneously performs its adjustment function. The torque required to drive the fuel-injec-tion pump depends on the pump size, the number of cylinders, the injected fuel quan-tity, the injection pressure, the plunger di-ameter, and the cam shape used. The drive torque has a retroactive effect on the timing characteristics which must be taken into account in the design as well as the work capacity.

Design

On in-line fuel-injection pumps, the timing device is mounted directly on the injection-pump camshaft. There are basically two types of design – open and closed.

A *closed-type timing device* has its own oil supply outside the housing which is inde-pendent of the engine lube-oil circuit.

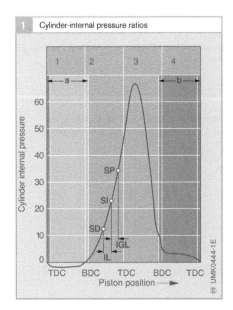

1   Cylinder-internal pressure ratios

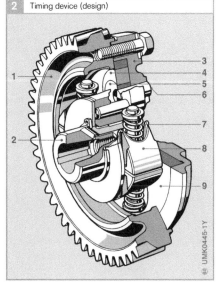

2   Timing device (design)

In the case of the *open-type design*, the timing device is connected directly to the engine lube-oil circuit. Its housing is bolted to a gear wheel. Inside the housing, the adjusting and balancing eccentrics are able to rotate around their respective bearings. They are held by a pin that is rigidly attached to the timing adjuster housing. The advantages of the open-type design are the smaller space requirements, more effective lube-oil supply and lower cost.

Method of operation

The link between the input and output sides of the of the timing device is formed by the nested pairs of eccentrics (Figs. 2 and 3).

The larger eccentrics – the adjusting eccentrics (4) – fit inside the bearing plate (9) that is bolted to the gear wheel that forms the drive input side (1). Fitted inside the adjusting eccentrics are the balancing eccentrics (5). The latter are held by the adjusting eccentrics and the hub pins (6).

The hub pins are attached directly to the hub that forms the drive output side (2). The flyweights (8) locate in the adjusting eccentrics by means of flyweight bolts and are held in their resting position (Fig. 3a) by compression springs (7).

The higher the engine speed – and therefore the speed of the timing device – the further outwards the flyweights move against the action of the compression springs. As a result, the relative position of the input and output sides of the timing adjuster alters by the angle α. Consequently, the engine and pump camshafts are offset by that angle relative to one another and the start of delivery is thus "advanced".

Sizes

By their external diameter and width, the size of the timing device determines the possible mass of the flyweights, the center of gravity separation and the available centrifugal-weight travel. Those three criteria are also the major factors in determining the working capacity and type of application of the timing device.

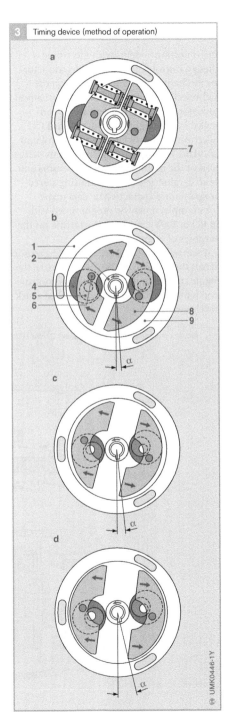

3   Timing device (method of operation)

Fig. 3
a   Resting position
b   Position at low speed
c   Position at medium speed
d   Position at high speed

1   Drive input side
2   Drive output side (hub)
4   Adjusting eccentric
5   Balancing eccentric
6   Hub pin
7   Compression spring
8   Flyweight
9   Bearing plate

α Advance angle

UMK0446-1Y

Electric actuator mechanism

Fuel-injection systems with Electronic Diesel Control (EDC) use an electric actuator mechanism mounted directly on the fuel-injection pump instead of a mechanical governor. The electrical actuator is controlled by the engine control unit or ECU (Electronic Control Unit). The control unit calculates the required control signals on the basis of the input data from the sensors and desired-value generators and using stored programs and characteristic data maps. For example, it may be programmed with an RQ or RQV control characteristic for the purposes of engine response.

A semi-differential short-circuit-ring sensor signals the position of the control rack to the engine control unit so that a closed control loop is formed. The sensor is also called a rack travel sensor.

Design and method of operation

The injected fuel quantity is determined – as with in-line fuel-injection pumps with mechanical governors – by the control rack position and the pump speed.

The linear magnet of the actuator mechanism (Fig. 1, 4) moves the fuel-injection pump's control rack (1) against the action of the compression spring (2). When the magnet is de-energized, the spring pushes the control rack back to the stop setting, thereby cutting off the fuel supply to the engine. As the effective control current increases, the magnet draws the solenoid armature (5) to a higher injected fuel quantity setting. Thus, varying the effective signal current provides a means of infinitely varying the control-rack travel between zero and maximum injected fuel quantity.

The control signal is not a direct-current signal but a Pulse-Width Modulation signal (PWM signal, Fig. 2). This is a square-wave signal with a constant frequency and a vari-

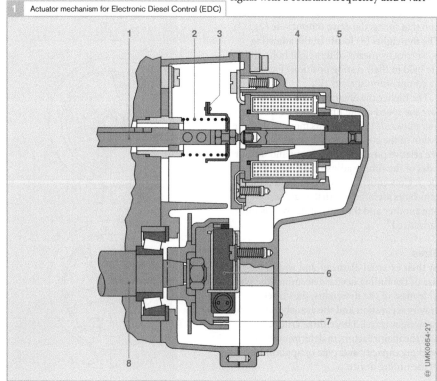

1   Actuator mechanism for Electronic Diesel Control (EDC)

Fig. 1
1   Control rack
2   Compression spring
3   Short-circuiting ring
    for rack-travel sensor
4   Linear magnet
5   Solenoid armature
6   Speed sensor
7   Sensor ring for
    speed sensor or
    marker for start of
    delivery
8   Fuel-injection pump
    camshaft

able pulse duration. The size of the cutin current is always the same. The effective current, which determines the excursion of the armature in the actuator mechanism, depends on the ratio of the pulse duration to the pulse interval. A short pulse duration produces a low effective current, and a long pulse duration a high effective current. The frequency of the signal is chosen to suit the actuator mechanism. This method of control avoids interference problems which low currents might otherwise be susceptible to.

Control-sleeve actuator mechanism

Control-sleeve in-line fuel-injection pumps also have a setting shaft (Fig. 3, 3) for the start of delivery as well as the control rack for the injected fuel quantity (5) (see also the chapter "Control-sleeve in-line fuel-injection pumps"). This shaft is rotated by an additional actuator mechanism (1) by way of

a ball joint (2). A low effective signal current produces a small amount of shaft travel and therefore a retarded start of delivery. As the signal current increases, the start of delivery is shifted towards an "advanced" setting.

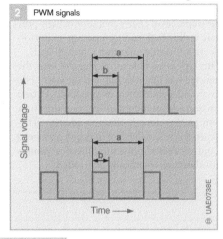

2   PWM signals

Fig. 2
a   Fixed frequency
b   Variable pulse
     duration

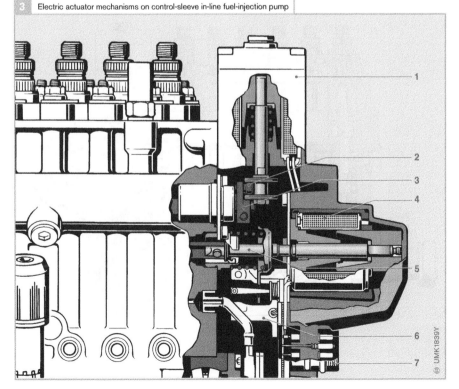

3   Electric actuator mechanisms on control-sleeve in-line fuel-injection pump

Fig. 3
1   Control-sleeve
     actuator mechanism
     (start of delivery
     actuator mechanism)
2   Ball joint
3   Control-collar shaft
4   Linear magnet of
     control-rack travel
     actuator mechanism
5   Control rack
6   Control-track travel
     sensor
7   Connector

Control-sleeve in-line fuel-injection pumps

The reduction of harmful exhaust-gas emissions is a subject to which commercial-vehicle producers are paying increasing attention. On commercial diesel engines, high fuel-injection pressures and optimized start of delivery make a major contribution here. This has led to the development of a new generation of high-pressure in-line fuel-injection pumps – control-sleeve in-line fuel-injection pumps (Fig. 1). This type is capable of varying not only the injected fuel quantity but also the start of delivery independently of engine speed. In comparison with standard in-line fuel-injection pumps, therefore, it offers an additional independently variable fuel-injection parameter. Control-sleeve in-line fuel-injection pumps are always electronically controlled.

The control-sleeve in-line fuel-injection pump is a component of the electric actuation system with which the start of delivery and the injected fuel quantity can be independently varied in response to a variety of determining factors (see chapter "Electronic Diesel Control (EDC)"). This method of control makes it possible to

- Minimize harmful exhaust-gas emissions
- Optimize fuel consumption under all operating conditions
- Precise fuel metering and
- Effective improvement of the starting and warm-up phases

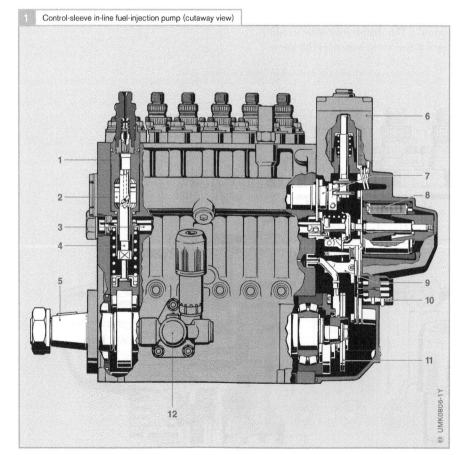

1   Control-sleeve in-line fuel-injection pump (cutaway view)

Fig. 1

1 Pump barrel
2 Control sleeve
3 Control rack
4 Pump plunger
5 Camshaft
   (connection to
   engine)
6 Start of delivery
   actuator mechanism
7 Control-sleeve shaft
8 Actuator solenoid
   for control-rack
   travel
9 Control-rack travel
   sensor
10 Connector
11 Disc for preventing
   fuel delivery which
   is also part of the
   oil-return pump
12 Presupply pump

UMK0806-1Y

A "rigid" pump-mounted timing device designed to cope with high torques is no longer
required.

There are two designs of control-sleeve in-line
fuel-injection pump:
- The Type H1 for 6...8 cylinders and up to
  1,300 bar at the nozzle and
- The Type H1000 which offers a higher
  delivery rate for 5...8 cylinders and up to
  1,350 bar at the nozzle for engines with
  greater fuel-quantity requirements

Design and method
of operation

The control-sleeve in-line fuel-injection pump
differs in design from a standard type by
virtue of a control sleeve (Fig. 3, 4) which
slides over the pump plunger. In all other
aspects, it is the same.

The control sleeve, which slides over the
pump plunger (1) inside a recess (2) in the
pump barrel, provides the facility for varying
the preliminary phase of the delivery stroke
in order to alter the start of delivery and consequently the start of injection. In comparison
with a standard in-line fuel-injection pump,
this provides a second variable fuel-injection
parameter that can be electronically controlled.

A control sleeve in each pump barrel incorporates the conventional spill port (3). A control-
sleeve shaft with control-sleeve levers (6) which
engage in the control sleeves changes the positions of all control sleeves at the same time.
Depending on the position of the control
sleeve (up or down), the start of delivery is
advanced or retarded relative to the position
of the cam. The injected fuel quantity is then
controlled by the helix as on standard in-line
fuel-injection pumps.

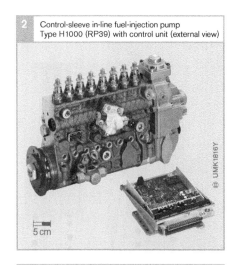

2   Control-sleeve in-line fuel-injection pump
Type H1000 (RP39) with control unit (external view)

5 cm

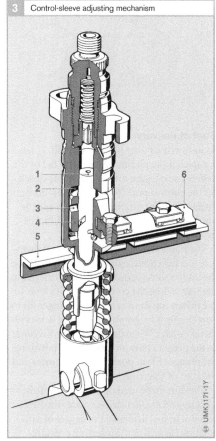

3   Control-sleeve adjusting mechanism

Fig. 3
1 Pump plunger
2 Recess for control
   sleeve
3 Spill port
4 Control sleeve
5 Control rack (injected
   fuel quantity)
6 Control-sleeve shaft

4   Operating cycle of control-sleeve in-line fuel-injection pump

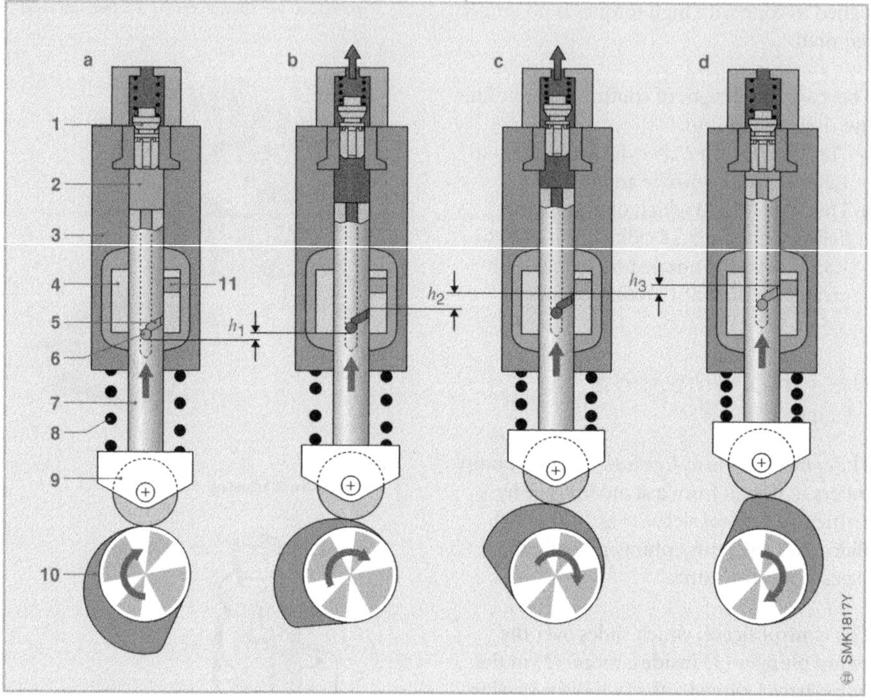

Fig. 4
a Bottom dead center
b Start of delivery
c End of delivery
d Top dead center

1 Delivery valve
2 Plunger chamber
3 Pump barrel
4 Control sleeve
5 Helix
6 Control port
  (start of delivery)
7 Pump plunger
8 Plunger spring
9 Roller tappet
10 Drive cam
11 Spill port

h_1 Plunger lift to port
  closing
h_2 Effective stroke
h_3 Residual travel

Start of delivery

As soon as the pump plunger (Fig. 4b, 7) has completed the preliminary phase (h_1) of the delivery stroke, the control sleeve (4) closes off the control port (6) in the pump plunger. From this point on, the pressure inside the plunger chamber (2) increases and fuel delivery begins.

The point at which fuel delivery, and therefore fuel injection, begins is altered by moving the control sleeve vertically relative to the pump plunger. When the control sleeve is closer to the piston top dead center, the plunger lift to port closing is longer and the start of delivery is therefore later. When the control sleeve is closer to the piston's bottom dead center position, the plunger lift to port closing is shorter and the start of injection is earlier.

The cam shape used determines the delivery velocity and the fuel-delivery rate (theoretical amount of fuel delivered per degree of cam rotation) as well as the injection pressure.

Spill

The piston's effective delivery stroke (h_2) ends when the helix (Fig. 4c, 5) in the pump plunger overlaps the spill port (11) in the control sleeve and allows pressure to escape. Rotating the pump plunger by means of the control rack changes the point at which this occurs and, therefore, the quantity of fuel delivered in the same way as on a standard in-line fuel-injection pump.

Electronic control system

From the input data received from the sensors and desired-value generators described in the chapter "Electronic Diesel Control (EDC)", the control unit (Fig. 5, 5) calculates the required fuel-injection pump settings. It then sends the appropriate electrical signals to the actuator mechanisms for start of delivery (1) and injected fuel quantity (4) on the fuel-injection pump.

Controlling start of delivery

Start of delivery is adjusted by means of a closed control loop. A needle-motion sensor in one of the nozzle holders (generally on no. 1 cylinder) signals to the control unit the actual point at which injection occurs. This information is used to determine the actual start of injection in terms of crankshaft position. This can then be compared with the setpoint value and the appropriate adjustment made by sending a current signal to the electrical start of delivery actuator mechanism.

The start of delivery actuator mechanism is "structurally rigid". For this reason, a separate travel feedback sensor can be dispensed with. Structurally rigid means that the lines of action of solenoid and spring always have a definite point of intersection. This means that the forces are always in equilibrium. Thus, the travel of the linear solenoid is proportional to the signal current. This is equivalent to feedback within a closed control loop.

Controlling injected fuel quantity

The required injected fuel quantity calculated by the microcontroller in the control unit is set using the position control loop: The control unit specifies a required control-rack travel and receives a signal indicating the actual control-rack travel from the control-rack travel sensor (3). The control unit repeatedly recalculates the adjustment needed to achieve the required actuator mechanism setting, thereby continuously correcting the actual setting to match the setpoint setting (closed control loop).

For safety reasons, a compression spring (2) moves the control rack back to the "zero delivery" position whenever the actuator mechanism is de-energized.

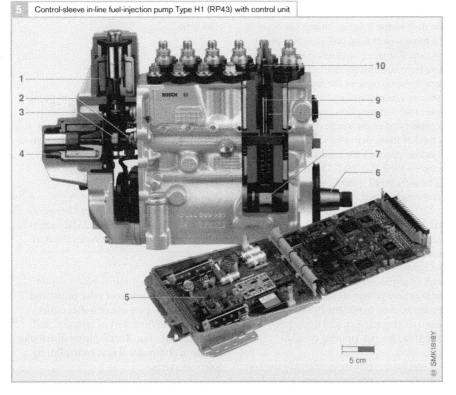

5   Control-sleeve in-line fuel-injection pump Type H1 (RP43) with control unit

5 cm

SMK1818Y

Fig. 5

1 Fuel delivery actuator mechanism
2 Compression spring
3 Control-rack travel sensor
4 Control-rack actuator mechanism (injected fuel quantity)
5 ECU
6 Connection to engine
7 Camshaft
8 Control sleeve
9 Pump plunger
10 Delivery valve

Overview of distributor fuel-injection pump systems

The combustion processes that take place inside a diesel engine are essentially dependent on the way in which the fuel is delivered by the fuel-injection system. The fuel-injection pump plays a decisive role in that connection. It generates the necessary fuel pressure for fuel injection. The fuel is delivered via high-pressure fuel lines to the nozzles, which in turn inject it into the combustion chamber. Small, fast-running diesel engines require a high-performance fuel-injection system capable of rapid injection sequences, that is also light in weight and compact in dimensions. Distributor injection pumps meet those requirements. They consist of a small, compact unit comprising the fuel pump, high-pressure fuel-injection pump and control mechanism.

Areas of application

Since its introduction in 1962, the axial-piston distributor injection pump has become the most widely used fuel-injection pump for cars. The pump and its control system have been continually improved over that period. An increase in the fuel-injection pressure was required in order to achieve lower fuel consumption and exhaust-gas emissions on engines with direct injection. A total of more than 45 million distributor injection pumps were produced by Bosch between 1962 and 2001. The available designs and overall system configurations are accordingly varied.

Axial-piston distributor pumps for engines with indirect injection (IDI) generate pressures of as much as 350 bar (35 MPa) at the nozzle. For direct-injection (DI) engines, both axial-piston and radial-piston distributor injection pumps are used. They produce pressures of up to 900 bar (90 MPa) for slow-running engines, and up to 1,900 bar (190 MPa) for fast-running diesels.

The mechanical governors originally used on distributor injection pumps were succeeded by electronic control systems with electrical actuator mechanisms. Later on, pumps with high-pressure solenoid valves were developed.

Apart from their compact dimensions, the characteristic feature of distributor injection pumps is their versatility of application which allows them to be used on cars, light commercial vehicles, fixed-installation engines, and construction and agricultural machinery (off-road vehicles).

The rated speed, power output and design of the diesel engine determine the type and model of distributor injection pump chosen. They are used on engines with between 3 and 6 cylinders.

Axial-piston distributor pumps are used on engines with power outputs of up to 30 kW per cylinder, while radial-piston types are suitable for outputs of up to 45 kW per cylinder.

Distributor injection pumps are lubricated by the fuel and are therefore maintenance-free.

Designs

Three types of distributor injection pump are distinguished according to the method of fuel-quantity control, type of control system and method of high-pressure generation (Fig. 1).

Method of fuel-quantity control
Port-controlled injection pumps
The injection duration is varied by means of control ports, channels and slide valves. A hydraulic timing device varies the start of injection.

Solenoid-valve-controlled injection pumps
A high-pressure solenoid valve opens and closes the high-pressure chamber outlet, thereby controlling start of injection and injection duration. Radial-piston distributor injection pumps are always controlled by solenoid valves.

Method of high-pressure generation

Type VE axial-piston distributor pumps
These compress the fuel by means of a piston which moves in an axial direction relative to the pump drive shaft.

Type VR radial-piston distributor pumps
These compress the fuel by means of several pistons arranged radially in relation to the pump drive shaft. Radial-piston pumps can produce higher pressures than axial-piston versions.

Type of control system

Mechanical governor
The fuel-injection pump is controlled by a governor linked to levers, springs, vacuum actuators, etc.

Electronic control system
The driver signals the desired torque output/ engine speed by means of the accelerator pedal (sensor). Stored in the control unit are data maps for starting, idling, full load, accelerator characteristics, smoke limits and pump characteristics.

Using that stored information and the actual values from the sensors, specified settings for the fuel-injection pump actuators are calculated. The resulting settings take account of the current engine operating status

and the ambient conditions (e.g. crankshaft position and speed, charge-air pressure, temperature of intake air, engine coolant and fuel, vehicle road speed, etc.). The control unit then operates the actuators or the solenoid valves in the fuel-injection pump according to the required settings.

The EDC (Electronic Diesel Control) system offers many advantages over a mechanical governor:
- Lower fuel consumption, lower emissions, higher power output and torque by virtue of more precise control of fuel quantity and start of injection.
- Lower idling speed and ability to adjust to auxiliary systems (e.g. air conditioning) by virtue of better control of engine speed.
- Greater sophistication (e.g. active surge damping, smooth-running control, cruise control).
- Improved diagnostic functions.
- Additional control functions (e.g. preheating function, exhaust-gas recirculation, charge-air pressure control, electronic engine immobilisation).
- Data exchange with other electronic control systems (e.g. traction control system, electronic transmission control) and therefore integration in the vehicle's overall control network.

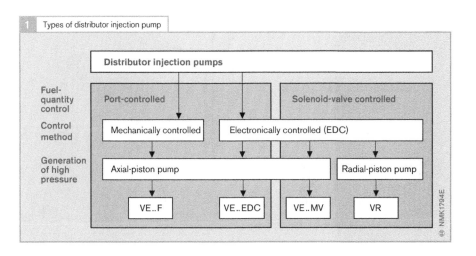

1   Types of distributor injection pump

Helix and port-controlled systems

Mechanically controlled distributor injection pumps

Mechanical control is used only on axial-piston distributor pumps. This arrangement's assets consist of low manufacturing cost and relatively simple maintenance.

Mechanical rotational-speed control monitors the various operating conditions to ensure high quality in mixture formation. Supplementary control modules adapt start of delivery and injected-fuel quantity to various engine operating statuses and load factors:

- Engine speed
- Engine load
- Engine temperature
- Charge-air pressure and
- Barometric pressure

In addition to the fuel-injection pump (Fig. 1, 4), the diesel fuel-injection system includes the fuel tank (11), the fuel filter (10), the pre-supply pump (12), the nozzle-and-holder assembly (8) and the fuel lines (1, 6 and 7). The nozzles and their nozzle-and-holder assemblies are the vital elements of the fuel-injection system. Their design configuration has a major influence on the spray patterns and the rate-of-discharge curves. The solenoid-operated shutoff valve (5) (ELAB) interrupts the flow of fuel to the pump's plunger chamber [1] when the "ignition" is switched off.

A Bowden cable or mechanical linkage (2) relays driver commands recorded by the accelerator pedal (3) to the fuel-injection pump's controller. Specialized control modules are available to regulate idle, intermediate and high-idle speed along. The VE..F series designation stands for "Verteilereinspritzpumpe, fliehkraftgeregelt", which translates as flyweight-controlled distributor injection pump.

[1] The operating concept is reversed on marine engines. On these powerplants, the ELAB shutoff closes under current.

Fig. 1

1 Fuel supply line
2 Linkage
3 Accelerator pedal
4 Distributor injection pump
5 Solenoid-operated shutoff valve (ELAB)
6 High-pressure fuel line
7 Fuel-return line
8 Nozzle-and-holder assembly
9 Sheathed-element glow plug, Type GSK
10 Fuel filter
11 Fuel tank
12 Fuel presupply pump (installed only with extremely long supply lines or substantial differences in relative elevations of fuel tank and fuel-injection pump)
13 Battery
14 Glow-plug and starter switch ("ignition switch")
15 Glow control unit, Type GZS
16 Diesel engine (IDI)

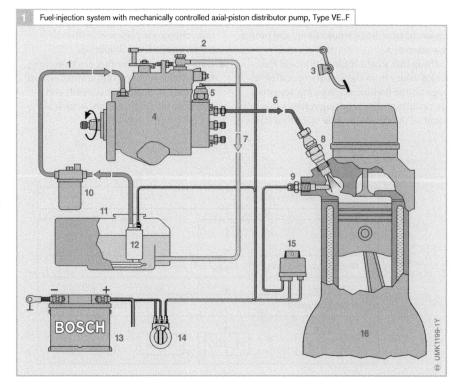

1   Fuel-injection system with mechanically controlled axial-piston distributor pump, Type VE..F

Electronically controlled distributor injection pumps

Electronic Diesel Control (EDC) supports a higher level of functionality than that provided by mechanical control systems. Electric measuring combines with the flexibility contributed by electronic data processing and closed-loop control featuring electric actuators to embed additional operational parameters in the control process.

Figure 2 illustrates the components in a fully-equipped fuel-injection system featuring an electronically-controlled axial-piston distributor pump. Some individual components may not be present in certain applications or vehicle types. The system consists of four sectors:

- Fuel supply (low-pressure circuit)
- Fuel-injection pump
- Electronic Diesel Control (EDC) with system modules for sensors, control unit and final controlling elements (actuators), and
- Peripherals (e.g. turbocharger, exhaust-gas recirculation, glow-plug control, etc.)

The solenoid-controlled actuator mechanism in the distributor injection pump (rotary actuator) replaces the mechanical controller and its auxiliary modules. It employs a shaft to shift the control collar's position and regulate injected fuel quantity. As in the mechanical pump, control collar travel is employed to vary the points at which the port is opened and closed. The ECU uses the stored program map and instantaneous data from the sensors to define the default value for the solenoid actuator position in the fuel-injection pump.

An angle sensor (such as a semidifferential short-circuiting ring sensor) registers the actuator mechanism's angle. This serves as an indicator of control collar travel and this information is fed back to the ECU.

A pulse-controlled solenoid valve compensates for fluctuations in the pump's internal pressure arising from variations in engine speed by shifting the timing device to modify start of delivery.

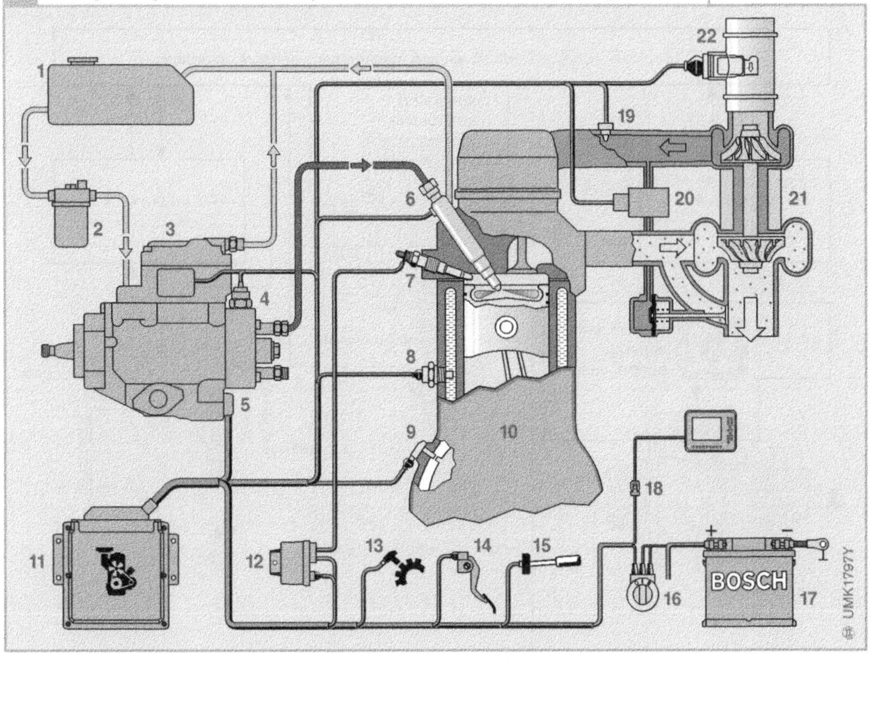

Fig. 2
1 Fuel tank
2 Fuel filter
3 Distributor injection pump with solenoid actuator, rack-travel sensor and fuel-temperature sensor
4 Solenoid-operated shutoff valve, Type ELAB
5 Timing-device solenoid valve
6 Nozzle-and-holder assembly with needle-motion sensor (usually on cylinder no. 1)
7 Glow plug, Type GSK
8 Engine-temperature sensor (in coolant system)
9 Crankshaft-speed sensor
10 Diesel engine (DI)
11 Electronic control unit (MSG
12 Glow control unit, Type GZS
13 Vehicle-speed sensor
14 Accelerator-pedal travel sensor
15 Operator level for cruise control
16 Glow-plug and starter switch ("ignition switch")
17 Battery
18 Diagnostic interface socket
19 Air-temperature sensor
20 Boost-pressure sensor
21 Turbocharger
22 Air-mass meter

2   Fuel-injection system with electronically controlled axial-piston distributor pump, Type VE..EDC

Solenoid-valve-controlled systems

Solenoid-valve controlled fuel-injection systems allow a greater degree of flexibility with regard to fuel metering and variation of injection start than port-controlled systems. They also enable pre-injection, which helps to reduce engine noise, and individual adjustment of injection quantity for each cylinder.

Engine management systems that use solenoid-valve controlled distributor injection pumps consist of four stages (Fig. 1):
- The fuel supply system (low-pressure stage)
- The high-pressure stage including all the fuel-injection components
- The Electronic Diesel Control (EDC) system made up of sensors, electronic control unit(s) and actuators and
- The air-intake and exhaust-gas systems (air supply, exhaust-gas treatment and exhaust-gas recirculation)

Control-unit configuration

Separate control units

First-generation diesel fuel-injection systems with solenoid-valve controlled distributor injection pumps (Type VE..MV [VP30], VR [VP44] for DI engines and VE..MV [VP29] for IDI engines) require two electronic control units (ECUs) – an engine ECU (Type MSG) and a pump ECU (Type PSG). There were two reasons for this separation of functions: Firstly, it was designed to prevent the overheating of certain electronic components by removing them from the immediate vicinity of pump and engine. Secondly, it allowed the use of short control leads for the solenoid valve. This eliminates interference signals that may occur as a result of very high currents (up to 20 A).

While the pump ECU detects and analyzes the pump's internal sensor signals for angle of rotation and fuel temperature in order to adjust start of injection, the engine ECU

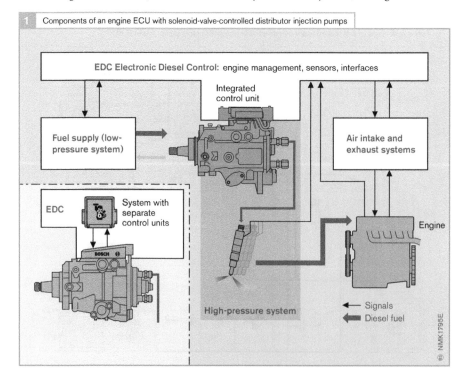

1   Components of an engine ECU with solenoid-valve-controlled distributor injection pumps

EDC Electronic Diesel Control: engine management, sensors, interfaces

Integrated control unit

Fuel supply (low-pressure system)

Air intake and exhaust systems

EDC   System with separate control units

Engine

High-pressure system

← Signals
← Diesel fuel

NMK1795E

processes all engine and ambient data signals from external sensors and uses them to calculate the actuator adjustments on the fuel-injection pump.

The two ECUs communicate over a CAN interface.

Integrated ECU

Heat-resistant printed-circuit boards designed using hybrid technology allow the integration of the engine ECU in the pump ECU on second-generation solenoid-valve-controlled distributor injection pumps. The use of integrated ECUs permits a space-saving system configuration.

Exhaust-gas treatment

There are various means employed for improving emissions and user-friendliness. They include such things as exhaust-gas recirculation, control of injection pattern (e.g. the use of pre-injection) and the use of higher injection pressures. However, in order to meet the increasingly stringent exhaust-gas regulations, some vehicles will require additional exhaut-gas treatment systems.

A number of exhaut-gas treatment systems are currently under development. It is not yet clear which of them will eventually become established. The most important are dealt with in a separate chapter.

---

2 Example of a diesel fuel-injection system with solenoid-valve-controlled radial-piston distributor injection pump and separate control units for engine and pump ECUs

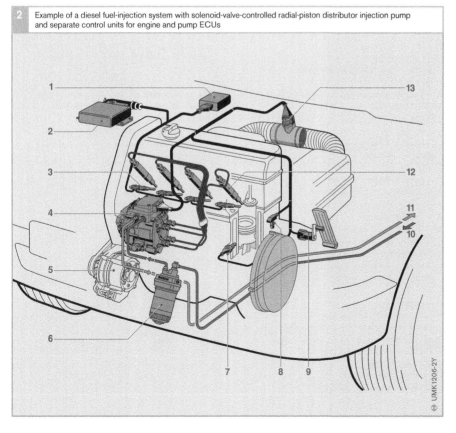

Fig. 2
1 Type GZS glow control unit
2 Type MSG engine ECU
3 Type GSK sheathed-element glow plug
4 Type VP44 radial-piston distributor injection pump with Type PSG5 pump ECU
5 Alternator
6 Fuel filter
7 Engine-temperature sensor (in cooling system)
8 Crankshaft speed sensor
9 Pedal travel sensor
10 Fuel inlet
11 Fuel return
12 Nozzle-and-holder assembly
13 Air-mass meter

System diagram

Figure 3 shows an example of a diesel fuel-injection system using a Type VR radial-piston distributor injection pump on a four-cylinder diesel engine (DI). That pump is fitted with an integrated engine and pump ECU. The diagram shows the full-configuration system. Depending on the nature of the application and the type of vehicle, certain components may not be used.

For the sake of clarity, the sensors and desired-value generators (A) are not shown in their fitted locations. One exception to this is the needle-motion sensor (21).

The CAN bus in the "Interfaces" section (B) provides the means for data exchange with a wide variety of systems and components such as

- The starter motor
- The alternator
- The electronic immobilizer
- The transmission-shift control system
- The Traction Control System (TCS) and
- The Electronic Stability Program (ESP)

The instrument cluster (12) and the air conditioner (13) can also be connected to the CAN bus.

Fig. 3

Engine, engine ECU and high-pressure
fuel-injection components
16 Fuel-injection pump drive
17 Type PSG16 integrated engine/pump ECU
18 Radial-piston distributor injection pump (VP44)
21 Nozzle-and-holder assembly with needle-motion sensor
    (cylinder no. 1)
22 Sheathed-element glow plug
23 Diesel engine (DI)
M Torque

A Sensors and desired-value generators
1 Pedal-travel sensor
2 Clutch switch
3 Brake contacts (2)
4 Vehicle-speed control operator unit
5 Glow-plug and starter switch ("ignition switch")
6 Vehicle-speed sensor
7 Crankshaft-speed sensor (inductive)
8 Engine-temperature sensor (in coolant system)
9 Intake-air temperature sensor
10 Boost-pressure sensor
11 Hot-film air mass-flow sensor (intake air)

B Interfaces
12 Instrument cluster with signal output
    for fuel consumption, rotational speed, etc.
13 Air-conditioner compressor and operator unit
14 Diagnosis interface
15 Glow control unit
CAN Controller Area Network
    (onboard serial data bus)

C Fuel supply system (low-pressure stage)
19 Fuel filter with overflow valve
20 Fuel tank with preliminary filter and presupply pump
    (preliminary pump is only required with long fuel pipes
    or large height difference between fuel tank and fuel-
    injection pump)

D Air supply system
24 Exhaust-gas recirculation positioner and valve
25 Vacuum pump
26 Control valve
27 Exhaust-gas turbocharger with VTG
    (Variable Turbine Geometry)
28 Charge-pressure actuator

E Exhaust-gas treatment
29 Diesel-oxidation catalytic converter DOC
    (Diesel Oxygen Catalyst)

3 Diesel fuel-injection system with Type VP44 solenoid-valve-controlled radial-piston distributor injection pump and Type PSG16 integrated engine and pump ECU

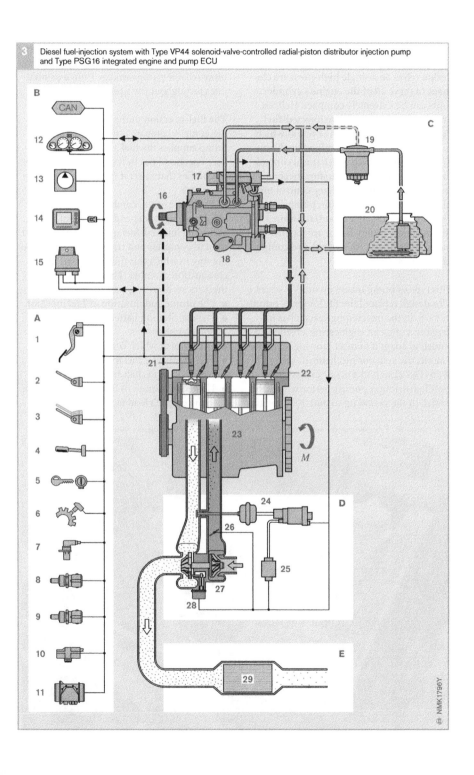

Helix and port-controlled distributor injection pumps

The helix-and-port distributor injection pump is always an axial-piston unit. As the design relies on a single high-pressure element to serve all of the engine's cylinders, units can be extremely compact. Helices, ports and collars modulate injected-fuel quantities. The point in the cycle at which fuel is discharged is determined by the hydraulic timing device. Mechanical control modules or an electric actuator mechanism (refer to section on "Auxiliary control modules for distributor injection pumps") provide flow control. The essential features of this injection-pump design are its maintenance friendliness, low weight and compact dimensions.

This type of pump makes up our VE series. This design replaced the EP/VA series pumps in 1975. In the intervening years it has undergone a range of engineering advances intended to adapt it to meet growing demands. The electric actuator mechanism's advent in 1986 (Fig. 2) started a major expansion in the VE distributor pump's performance potential. In the period up to mid-2002 roughly 42 million VE pumps were manufactured at Bosch. Every year well over a million of these ultra-reliable pumps emerge from assembly lines throughout the world.

The fuel-injection pump pressurizes the diesel fuel to prepare it for injection. The pump supplies the fuel along the high-pressure injection lines to the nozzle-and-holder assemblies that inject it into the combustion chambers.

The shape of the combustion process in the diesel engine depends on several factors, including injected-fuel quantity, the method used to compress and transport the fuel, and the way in which this fuel is injected in the combustion chamber. The critical criteria in this process are:

- The timing and duration of fuel injection
- The distribution pattern in the combustion chamber
- The point at which combustion starts
- The quantity of fuel injected for each degree of crankshaft travel and
- The total quantity of fuel supplied relative to the engine's load factor

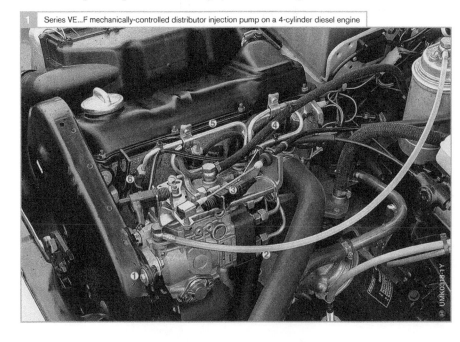

1   Series VE...F mechanically-controlled distributor injection pump on a 4-cylinder diesel engine

Fig. 1
1   Pump drive
2   Fuel inlet
3   Accelerator pedal linkage
4   Fuel return
5   High-pressure fuel line
6   Nozzle-and-holder assembly

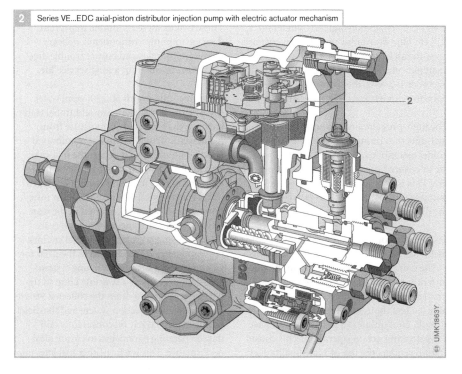

2   Series VE...EDC axial-piston distributor injection pump with electric actuator mechanism

Fig. 2
1 Axial-piston
   distributor
   injection pump
2 Electric actuator
   mechanism

Applications and installation

Fast-turning diesels with limited displacement are one of the applications for helix and port-controlled distributor injection pumps. The pumps furnish fuel in both direct-injection (DI) and prechamber (IDI) powerplants.

The application and the fuel-injection pump's configuration are defined by such factors as nominal speed, power output and design of the individual diesel engine. Distributor injection pumps are fitted in passenger cars, commercial vehicles, construction and agricultural machinery, ships and stationary powerplants to produce power of up to 30 kW per cylinder.

These distributor injection pumps are available with high-pressure spill ports for engines of 3...6 cylinders. The maximum injected-fuel quantity is 125 mm³ per stroke. Requirements for injection pressure vary according to the specific engine's individual demand (DI or IDI). These pressures reach levels of 350...1,250 bar.

The distributor injection pump is flange-mounted directly on the diesel engine (Fig. 1). Motive force from the crankshaft is transferred to the pump by toothed belt, pinion, ring gear, or a chain and sprocket. Regardless of the arrangement selected, it ensures that the pump remains synchronized with the movement of the pistons in the engine (positive coupling).

On the 4-stroke diesel engine, the rotational speed of the pump is half that of the crankshaft. Expressed another way: the pump's rotational speed is the same as that of the camshaft.

Distributor injection pumps are available for both clockwise and counterclockwise rotation[1]. While the injection sequences varies according to rotational direction, the injection sequence always matches the geometrical progression of the delivery ports.

[1] Rotational direction as viewed from the pump drive side

In order to avoid confusion with the designations of the engine's cylinders (cylinder no. 1, 2, 3, etc.) the distributor pump's delivery ports carry the alphabetic designations A, B, C, etc. Example: on a four-stroke engine with the firing order 1–3–4–2, the correlation of delivery ports to cylinders is A-1, B-3, C-4 and D-2.

The high-pressure lines running from the fuel-injection pump to the nozzle-and-holder assemblies are kept as short as possible to ensure optimized hydraulic properties. This is why the distributor injection pump is mounted as close as possible to the diesel engine's cylinder head.

The distributor injection pump's lubricant is fuel. This makes the units maintenance-free.

The components and surfaces in the fuel-injection pump's high-pressure stage and the nozzles are both manufactured to tolerances of just a few thousandths of a millimeter. As a result, contamination in the fuel can have a negative impact on operation. This consideration renders the use of high-quality fuel essential, while a special fuel filter, custom-designed to meet the individual fuel-injec-

tion system's requirements, is another factor. These two elements combine to prevent damage to pump components, delivery valves and nozzles and ensure trouble-free operation throughout a long service life.

Diesel fuel can absorb 50...200 ppm water (by weight) in solution. Any additional water entering the fuel (such as moisture from condensation) will be present in unbound form. Should this water enter the fuel-injection pump, corrosion damage will be the result. This is why fuel filters equipped with a water trap are vital for the distributor injection pump. The water collected in the trap must be drained at the required intervals. The increasing popularity of diesel engines in passenger cars has resulted in the need for an automatic water level warning system. This system employs a warning lamp to signal that it is time to drain the collected water.

Both the fuel-injection system and the diesel engine in general rely on consistently optimized operating parameters to ensure ideal performance. This is why neither fuel lines nor nozzle-and-holder assemblies should be modified during service work on the vehicle.

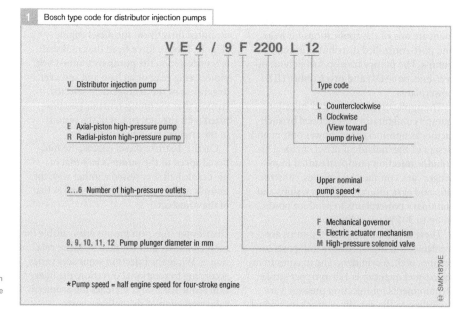

1   Bosch type code for distributor injection pumps

V E 4 / 9 F 2200 L 12

V   Distributor injection pump

E   Axial-piston high-pressure pump
R   Radial-piston high-pressure pump

2...6   Number of high-pressure outlets

8, 9, 10, 11, 12   Pump plunger diameter in mm

Type code

L   Counterclockwise
R   Clockwise
    (View toward
    pump drive)

Upper nominal
pump speed ★

F   Mechanical governor
E   Electric actuator mechanism
M   High-pressure solenoid valve

★Pump speed = half engine speed for four-stroke engine

SMK1879E

Fig. 1
This data code is affixed to the distributor injection pump housing for precise identification

Design

The distributor injection pump consists of the following main assembly groups (Fig. 2):

Low-pressure stage (7)
The vane-type supply pump takes in the diesel fuel and pressurizes the inner chamber of the pump. The pressure-control valve controls this internal pressure (3...4 bar at idle, 10...12 bar at maximum rpm). Air is discharged through the overflow valve. It also returns fuel in order to cool the pump.

High-pressure pump with distributor (8)
High pressure in the helix and port-controlled distributor injection pump is generated by an axial piston. A distributor slot in the pump's rotating plunger distributes the pressurized fuel to the delivery valves (9). The number of these valves is the same as the number of cylinders in the engine.

Control mechanism (2)
The control mechanism regulates the injection process. The configuration of this mechanism is the most distinctive feature of the helix-and-port distributor pump. Here the operative distinction is between:
- The mechanical governor assembly, with supplementary control modules and switches as needed and
- The electric actuator mechanism (VE...EDC), which is controlled by the engine ECU

Both can be equipped with a solenoid-operated shutoff valve (ELAB) (4). This solenoid device shuts off the fuel-injection system by isolating the high-pressure from the low-pressure side of the pump.

Pump versions equipped with a mechanical governor also include a mechanical shutoff device which is integrated in the governor cover.

Hydraulic timing device (10)
The timing device varies the point at which the pump starts to deliver fuel.

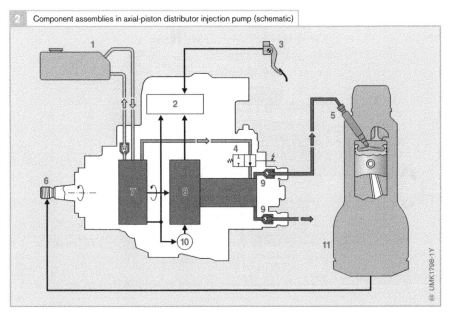

2 Component assemblies in axial-piston distributor injection pump (schematic)

Fig. 2
1 Fuel supply (low-pressure)
2 Control mechanism
3 Accelerator pedal
4 Solenoid-operated shutoff valve (ELAB)
5 Nozzle-and-holder assembly
6 Pump drive
7 Low-pressure stage (vane-type supply pump with pressure-control valve and overflow throttle valve)
8 High-pressure pump with fuel rail
9 Delivery valve
10 Hydraulic timing device
11 Diesel engine

UMK1798-1Y

Fuel flow and control lever

The diesel engine powers an input shaft mounted in two plain bearings in the pump's housing (Fig. 4, 2). The input shaft supports the vane-type supply pump (3) and also drives the high-pressure pump's cam plate (6). The cam plate's travel pattern is both rotational and reciprocating; its motion is transferred to the plunger (11).

On injection pumps with mechanical control, the input shaft drives the control assembly (9) via a gear pair (4) with a rubber damper.

At the top of the control assembly is a control-lever shaft connected to the external control lever (1) on the governor cover. This control-lever shaft intervenes in pump operation based on the commands transmitted to it through the linkage leading to the accelerator pedal. The governor cover seals the top of the distributor injection pump.

Fuel supply

The fuel-injection pump depends on a continuous supply of pressurized bubble-free fuel in the high-pressure stage. On passenger cars and light trucks, the difference in the elevations of fuel-injection pump and fuel tank is usually minimal, while the supply lines are short with large diameters. As a result, the suction generated by the distributor injection pump's internal vane-type supply pump is usually sufficient.

Vehicles with substantial elevations differences and/or long fuel lines between fuel tank and fuel-injection pump need a presupply pump to overcome resistance in lines and filters. Gravity-feed fuel-tank operation is found primarily in stationary powerplants.

Fig. 3
1   Vane supply pump
    with pressure-
    control valve:
    Fuel induction and
    generation of inter-
    nal pump pressure
2   High-pressure
    pump with
    distributor head:
    Generation of
    injection pressure,
    fuel delivery and dis-
    tribution to individual
    engine cylinders
3   Mechanical
    governor:
    Controls rotational
    speed; control
    mechanisms vary
    delivery quantity
    in control range
4   Solenoid-operated
    shutoff valve
    (ELAB): Interrupts
    fuel supply to shut
    off engine
5   Timing device:
    Determines start of
    delivery as function
    of engine speed and
    (partially) engine
    load

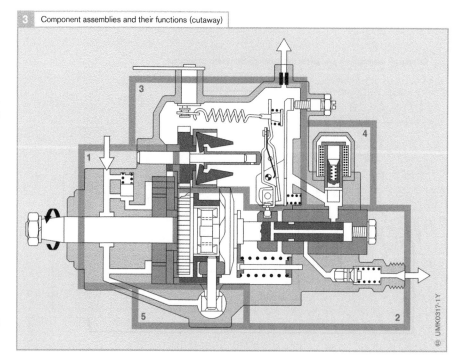

3   Component assemblies and their functions (cutaway)

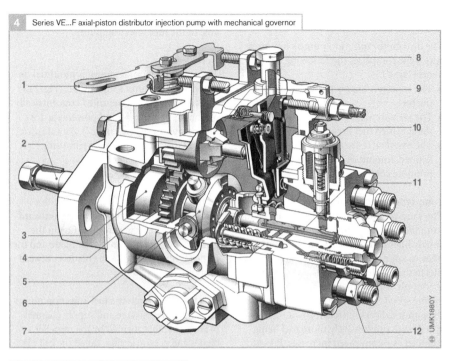

4   Series VE...F axial-piston distributor injection pump with mechanical governor

Fig. 4
1 Flow-control lever
   (linked to accelera-
   tor pedal)
2 Input shaft
3 Vane-type supply
   pump
4 Governor drive gear
5 Roller on roller ring
6 Cam plate
7 Hydraulic timing
   device
8 Overflow restriction
9 Governor assembly
   (mechanical
   governor)
10 Solenoid-operated
   shutoff valve (ELAB)
11 Distributor plunger
12 Delivery valve

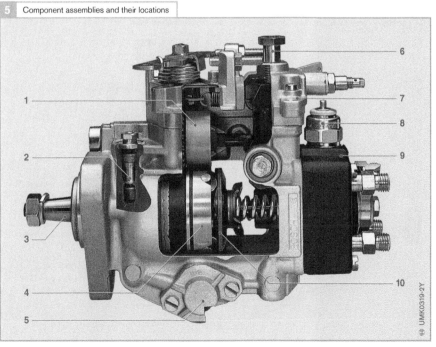

5   Component assemblies and their locations

Fig. 5
1 Governor assembly
2 Pressure-control
   valve
3 Input shaft
4 Roller ring
5 Hydraulic timing
   device
6 Overflow restriction
7 Governor cover
8 Solenoid-operated
   shutoff valve (ELAB)
9 Distributor head with
   high-pressure pump
10 Cam plate

Low-pressure stage

The distributor injection pump's low-pressure stage comprises the following components (Fig. 1):
- The *vane-type supply pump* (4) supplies the fuel
- The *pressure-control valve* (3) maintains the specified fuel pressure in the system
- The *overflow restriction* (9) returns a defined amount of fuel to the pump to promote cooling

Vane-type supply pump

The vane-type supply pump extracts the fuel from the tank and conveys it through the supply lines and filters. As each rotation supplies an approximately constant amount of fuel to the inside of the fuel-injection pump, the supply volume increases as a function of engine speed. Thus the volume of fuel that the pump delivers reflects its own rotational speed, with progressively more fuel being supplied as pump speed increases. Pressur-

ized fuel for the high-pressure side is available in the fuel-injection pump.

Design

The vane-type supply pump is mounted on the main pump unit's input shaft (Fig. 2). The impeller (10) is mounted concentrically on the input shaft (8) and powers the former through a disk spring (7). An eccentric ring (2) installed in the pump housing (5) surrounds the impeller.

Operating concept

As the impeller rotates, centrifugal force presses the four floating blades (9) outward against the eccentric ring. The fuel in the gap between the bottom of the blade and the impeller body supports the blade's outward motion.

Fuel travels through the fuel-injection pump housing's inlet passage and supply channel (4) to a chamber formed by the impeller, blades and eccentric ring, called the cell (3). The rotation presses the fuel from between

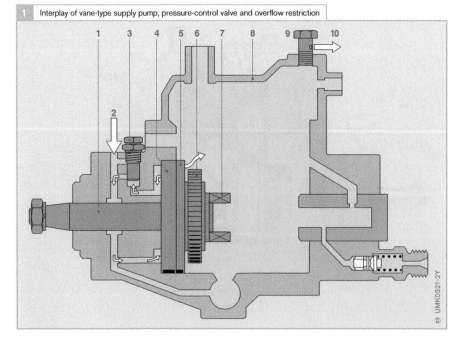

1   Interplay of vane-type supply pump, pressure-control valve and overflow restriction

Fig. 1
1   Pump drive
2   Fuel supply
3   Pressure-control valve
4   Eccentric ring on vane supply pump
5   Support ring
6   Governor drive gear
7   High-pressure pump drive claw
8   Pump housing
9   Overflow restriction
10 Fuel return

UMK0321-2Y

the blades toward the spill port (6), from where it proceeds through a bore to the pump's inner chamber. The eccentric shape of the ring's inner surface decreases the volume of the cell as the vane-type supply pump rotates to compress the fuel. A portion of the fuel proceeds through a second bore to the pressure-control valve (see Fig. 1).

The inlet and discharge sides operate using suction and pressure cells and have the shape of kidneys.

Pressure-control valve

As fuel delivery from the vane-type supply pump increases as a function of pump speed, the pump's internal chamber pressure is proportional to the engine's rotational speed. The hydraulic timing device relies on these higher pressurization levels to operate (see section on "Auxiliary control modules for distributor injection pumps"). The pressure-control valve is needed to govern pressurization and ensure that pressures correspond to the levels required for optimized operation of both the timing device and the

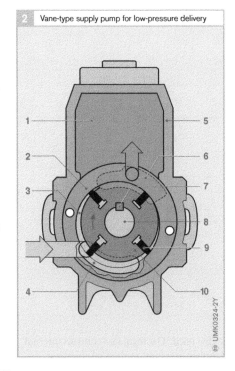

2   Vane-type supply pump for low-pressure delivery

Fig. 2
1   Pump inner chamber
2   Eccentric ring
3   Crescent-shaped cell
4   Fuel inlet (suction cells)
5   Pump housing
6   Fuel discharge (pressure cells)
7   Woodruff key
8   Input shaft
9   Blade
10 Impeller

UMK0324-2Y

3   Components of vane-type supply pump on input shaft

Fig. 3
1   Input shaft
2   Impeller
3   Blade
4   Eccentric ring
5   Support ring
6   Governor drive gear
7   High-pressure pump drive claw

UMK0320-2Y

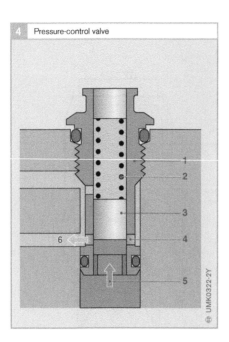

4   Pressure-control valve

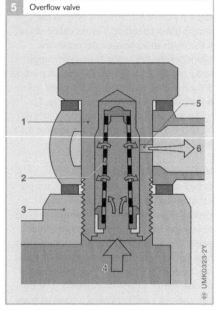

5   Overflow valve

Fig. 4
1   Valve body
2   Compression spring
3   Valve plunger
4   Bore
5   Supply from vane-
    type supply pump
6   Return to vane-type
    supply pump

Fig. 5
1   Housing
2   Filter
3   Governor cover
4   Fuel supply
5   Throttle bore
6   Return line to fuel
    tank

engine itself. The regulator controls internal chamber pressures in the fuel-injection pump according to the quantity of fuel supplied by the vane-type supply pump. At a specific rotational speed, a specific internal pump pressure occurs, which then induces a defined shift in start of delivery.

A passage connects the pressure-control valve to the pressure cell (Fig. 2). It is directly adjacent to the vane-type supply pump.

The pressure-control valve is a spring-loaded slide valve (Fig. 4). When fuel pressure rises beyond a specified level, it pushes back the valve plunger (3), compressing the spring (2) and simultaneously exposing the return passage. The fuel can now proceed through the passage to the vane-type supply pump's suction side (6). When fuel pressure is low, the spring holds the return passage closed. The actual opening pressure is defined by the adjustable tension of the spring.

Overflow restriction

The distributor injection pump is cooled by fuel that flows back to the tank via an overflow restriction screwed to the governor cover (Fig. 5). The overflow restriction is located at the fuel-injection pump's highest point to bleed air automatically. In applications where higher internal pump pressures are required for low-speed operation, an overflow valve can be installed in place of the overflow restriction. This spring-loaded ball valve functions as a pressure-control valve.

The amount of fuel that the overflow restriction's throttle port (5) allows to return to the fuel tank varies as a function of pressure (6). The flow resistance furnished by the port maintains the pump's internal pressure. Because a precisely defined fuel pressure is required for each individual engine speed, the overflow restriction and the pressure-control valve must be matched.

High-pressure pump with fuel distributor

The fuel-injection pump's high-pressure stage pressurizes the fuel to the levels required for injection and then distributes this fuel to the cylinders at the specified delivery quantities. The fuel flows through the delivery valve and the high-pressure line to the nozzle-and-holder assembly, where the nozzle injects it into the engine's combustion chamber.

Distributor plunger drive
A power-transfer assembly transmits the rotational motion of the input shaft (Fig. 1, 1) to the cam plate (6), which is coupled to the distributor plunger (10). In this process, the claws from the input shaft and the cam plate engage in the intermediate yoke (3).

The cam plate transforms the input shaft's rotation into a motion pattern that combines rotation with reciprocation (total stroke of 2.2...3.5 mm, depending on pump version). The shaft's motion is translated by the motion of the cams on the cam plate (4) against the rollers on the roller ring (5). While the latter runs on bearing surfaces in the housing, there is no positive connection joining it to the input shaft. Because the profiles of the cams on the cam plate extend along the plane defined by the input shaft, they are sometimes referred to as "axial cams".

The distributor plunger's base (10) rests in the cam plate, where its position is maintained by a locating stud. Plunger diameters range from 8...12 mm, depending on the desired injected-fuel quantity.

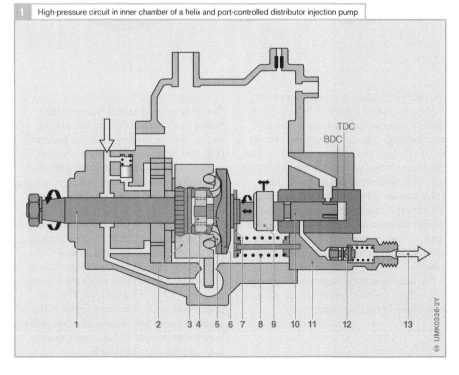

1   High-pressure circuit in inner chamber of a helix and port-controlled distributor injection pump

TDC
BDC

Fig. 1
1 Input shaft
2 Roller ring
3 Yoke
4 Cams
5 Roller
6 Cam plate
7 Spring-loaded
   cross brace
8 Plunger return
   spring
9 Control collar
10 Distributor plunger
11 Distributor head
12 Delivery valve
13 Discharge to high-
    pressure line

TDC **T**op **D**ead **C**enter
    for pump plunger
BDC **B**ottom **D**ead
     Center for pump
     plunger

1   2   3 4   5   6 7   8 9   10 11   12   13

UMK0326-2Y

2   Pump assembly with distributor head

The cam plate's cams drive the plunger toward its **Top Dead Center** (TDC) position. The two symmetrically arranged plunger return springs (Fig. 2, 13) push the plunger back to **Bottom Dead Center** (BDC). At one end, these springs rest against the distributor body (15), while the force from the other end is transferred to the distributor plunger (10) through a spring coupling (11). The plunger return springs also prevent the cam plate (7) from slipping off the roller ring's rollers (5) in response to high rates of acceleration.

The heights of the plunger return springs are precisely matched to prevent the plunger from tilting in the distributor body.

Cam plates and cam profiles
The number of cams and rollers is determined by the number of cylinders in the engine and the required injection pressure (Fig. 3). The cam profile affects injection pressure as well as the maximum potential injection duration. Here the primary criteria are cam pitch and stroke velocity.

The conditions of injection must be matched to the combustion chamber's configuration and the engine's combustion process (DI or IDI). This is reflected in the cam profiles on the face of the cam plate, which are specially calculated for each engine type. The cam plate serves as a custom component in the specified pump type. This is why cam plates in different types of VE injection types are not mutually interchangeable.

Distributor body

The plunger (5) and the plunger barrel (2) are precisely matched (lapped assembly) in the distributor body (Fig. 4, 3) which is screwed to the pump housing. The control collar (1) is also part of a custom-matched assembly with the plunger. This allows the components to provide a reliable seal at extremely high pressures. At the same time, a slight pressure loss is not only unavoidable, it is also desirable as a source of lubrication for the plunger. Precise mutual tolerances in these assemblies mean that the entire distributor group must be replaced as a unit; no attempt should ever be made to replace the plunger, distributor body or control collar as individual components.

Also mounted in the distributor body are the solenoid-operated shutoff valve (ELAB) (not shown in this illustration) used to interrupt the fuel supply along with the screw cap (4) with vent screw (6) and delivery valves (7).

3 Various roller and cam combinations

Roller ring   Cam plate

a

b

c

d

SMK1881E

Fig. 3

a **Three-cylinder engine**
   Six-cylinder version (d) is also available. In this version, every second spill port is rerouted to the inner chamber of the pump.

b **Four-cylinder**
   On two-cylinder engines, every second spill port is rerouted to the inner chamber of the pump.

c **Five-cylinder engine**
   The rollers in blue are not fitted to pumps for IDI engines as these operate at lower fuel-injection pressures and reduced physical loads.

d **Six-cylinder engine**
   Only four rollers used in this application.

4 Distributor head component assembly

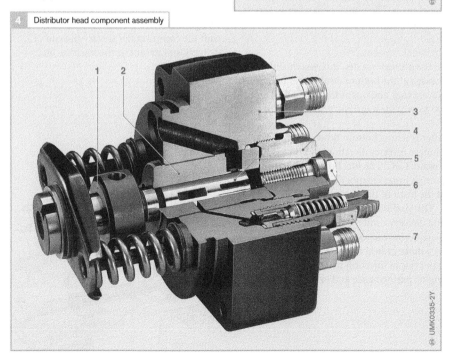

UMK0335-2Y

Fig. 4

1 Control collar
2 Plunger barrel
3 Distributor head
4 Screw cap
5 Distributor plunger
6 Vent screw
7 Delivery valve

Fuel metering

The distributor head assembly generates the pressure required for injection. It also distributes the fuel to the various engine cylinders. This dynamic process correlates with several different phases of the plunger stroke, called delivery phases.

The plunger's stroke phases as illustrated in Figure 6 show the process of metering fuel for a single cylinder. Although the plunger's motion is horizontal (as in the in-line fuel-injection pump), the extremes of its travel are still refereed to as top and bottom dead center (TDC and BDC). On four-cylinder engines the plunger rotates by one fourth of a turn during a delivery phase, while a sixth of a turn is available on six cylinder engines.

Suction (6a)

As the plunger travels from top to bottom dead center, fuel flows from the pump chamber and through the exposed inlet passage (2) into the plunger chamber (6) above the plunger. The plunger chamber is also called the element chamber.

Prestroke (option, 6b)

As the plunger rotates, it closes the inlet passage at the bottom dead center end of its travel range and opens the distributor slot (8) to provide a specific discharge.

After reaching bottom dead center, the plunger reverses direction and travels back toward TDC. The fuel flows back to the pump's inner chamber through a slot (7) at the front of the plunger.

This preliminary stroke delays fuel supply and delivery until the profile on the cam plate's cam reaches a point characterized by a more radical rise. The result is a more rapid rise in injection pressure for improved engine performance and lower emissions.

Fuel delivery (effective stroke, 6c)

The plunger continues moving toward TDC, closing the prestroke passage in the process. The enclosed fuel is now compressed. It travels through the distributor slot (8) to the delivery valve on one of the delivery ports (9). The port opens and the fuel is propelled through the high-pressure line to the nozzle-and-holder assembly.

Residual stroke (6d)

The effective stroke phase terminates when the lateral control passage (11) on the plunger reaches the control collar's helix (10). This allows the fuel to escape into the pump's inside chamber, collapsing the pressure in the element chamber. This terminates the delivery process. No further fuel is delivered to the nozzle (end of delivery). The delivery valve closes off the high-pressure line.

Fuel flows through the connection to the inside of the pump for as long as the plunger continues to travel toward TDC. The inlet passage is opened once again in this phase.

The governor or actuator mechanism can shift the control collar to vary the end of delivery and modulate the injected-fuel quantity.

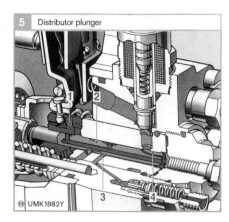

5   Distributor plunger

⊕ UMK1882Y

Fig. 5
1   Distributor plunger
2   Control collar
3   Distributor slot
4   Inlet metering slot
    (a metering slot is
    provided for each of
    engine cylinders)

6 Stroke phases on helix and port-controlled distributor injection pump

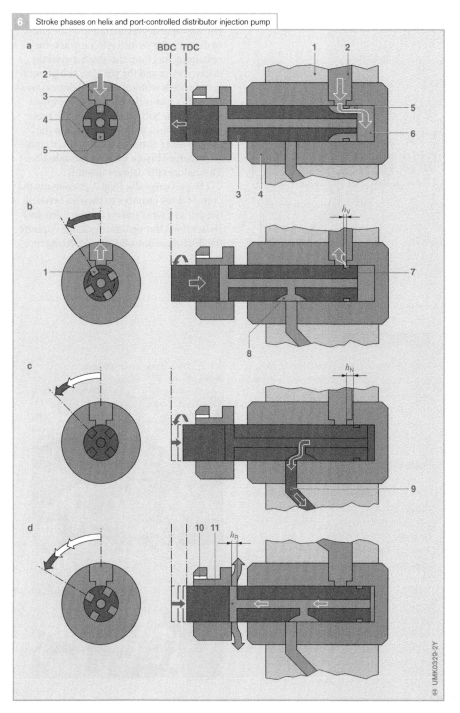

Fig. 6
a Suction
b Prestroke
c Effective stroke
d Residual stroke

1 Distributor head
2 Inlet passage
  (fuel supply)
3 Distributor plunger
4 Plunger barrel
5 Inlet metering slot
6 Plunger chamber
  (element chamber)
7 Prestroke groove
8 Distributor slot
9 Inlet passage to
  delivery valve
10 Control collar
11 Control bore

h_N Effective stroke
h_R Residual stroke
h_V Prestroke

TDC Top Dead Center
    on pump plunger
BDC Bottom Dead
    Center on pump
    plunger

UMK0329-2Y

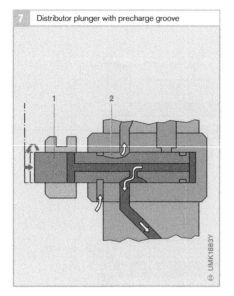

7　Distributor plunger with precharge groove

1
2

UMK1883Y

Fig. 7
1　Distributor plunger
2　Precharge groove

Precharge slot

During rapid depressurization at the end of delivery, the venturi effect extracts the remaining fuel from the area between the delivery valve and the plunger. At high rotational speeds with substantial delivery quantities, this area is closed off before enough fuel can flow back in. As a result, the pressure in this area is lower than that in the pump's inner chamber. The slot must thus be recharged before the next fuel injection. This reduces the delivery quantity.

The precharge slot (Fig. 7, 2) connects the pump's inner chamber to the area between the delivery valve and the plunger. The fuel always flows through the discharge opposite the discharge controlled for delivering fuel.

▷　Off-road applications

In addition to their use in on-road vehicles, helix and port-controlled distributor injection pumps are also used in a multitude of off-road applications. This sector includes stationary powerplants as well as construction and agricultural machines. Here, the primary requirements are robust construction and ease of maintenance.

The conditions that fuel-injection systems meet in off-road use can be exceptionally challenging (for instance, when exposed engines are washed down with high-pressure steam cleaners, poor fuel quality, frequent refuelings from canisters, etc.). Special filtration systems operate in conjunction with water separators to protect fuel-injection pumps against damage from fuel of poor quality).

PTOs of the kind employed to drive pumps and cranes rely on constant engine speeds with minimum fluctuation in response to load shifts (low speed droop). Heavy flyweights in the governor assembly provide this performance.

Excavator with 112 kW (152 bhp)

Combine harvester with 125 kW (170 bhp)

Harvester with 85 kW (116 bhp)
Tractor with 98 kW (133 bhp)

1　Off-road vehicles with VE pump

SAV0058Y

Delivery valve

The delivery valve closes the high-pressure line to the pump in the period between fuel supply phases. It insulates the high-pressure line from the distributor head.

The delivery valve (Fig. 8) is a slide valve. The high pressure generated during delivery lifts the valve plunger (2) from its seat. The vertical grooves (8) terminating in the ring groove (6) carry the fuel through the delivery-valve holder (4), the high-pressure line and the nozzle holder to the nozzle.

At the end of delivery, the pressure in the plunger chamber above the plunger and in the high-pressure lines falls to the level present in the pump's inner chamber. The valve spring (3) and the static pressure in the high-pressure line push the valve plunger back against its seat.

Yet another function of the delivery valve is to relieve injection pressure from the injection line after completion of the delivery phase by increasing a defined volume on the line side. This function is discharged by the retraction piston (7) that closes the valve before the valve plunger (2) reaches its seat. This pressure relief provides precisely calibrated termination of fuel discharge through the nozzle at the end of the injection process. It also stabilizes pressure in the high-pressure lines between injection processes to compensate for shifts in injected-fuel quantities.

Delivery valve with torque control

The dynamic response patterns associated with high-pressure delivery processes in the fuel-injection pump cause flow to increase as a function of rotational speed. However, the engine needs less fuel at high speeds. A positive torque control capable of reducing flow rates as rotational speed increases is thus required in many applications. This function is usually executed by the governor. Another option available on units designed for low injection pressures (IDI powerplants) is to use the pressure-control valve for this function.

Pressure-control valves with torque-control functions feature a torque-control collar (2) adjacent to the retraction piston (Fig. 9, 1) with either one or two polished recesses (3), according to specific requirements. The resulting restricted opening reduces delivery quantity at high rpm.

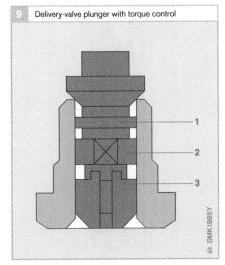

8   Delivery valve

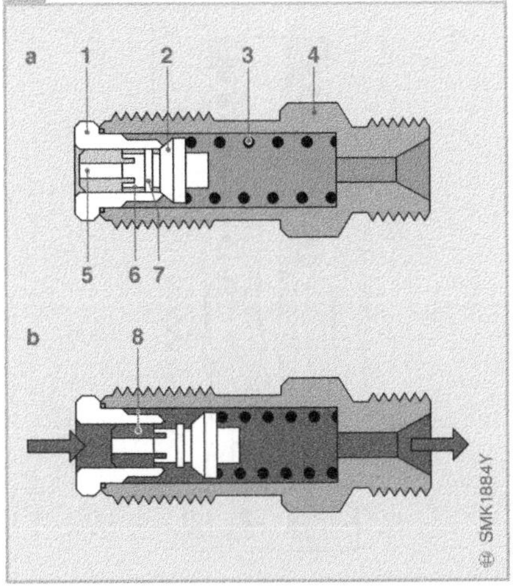

a
1 2 3 4
5 6 7
b
8

SMK1884Y

9   Delivery-valve plunger with torque control

1
2
3

SMK1885Y

Fig. 8
a   Closed
b   Open

1   Valve holder
2   Valve plunger
3   Valve spring
4   Delivery-valve holder
5   Stem
6   Ring groove
7   Retraction piston
8   Vertical groove

Fig. 9
1   Retraction piston
2   Control-torque collar
3   Specially ground
     recess

Cutoff bores with flattened surfaces in the control collar, specially designed for the individual engine application, can also provide a limited degree of torque control.

Delivery valve with return-flow restriction

The precise pressure relief required at the end of the injection event generates pressure waves. These are reflected by the delivery valve. At high injection pressures, these pulses have the potential to reopen the nozzle needle or induce phases of negative pressure in the high-pressure line. These processes cause post-injection dribble, with negative consequences on emission properties and/or cavitation and wear in the high-pressure line or at the nozzle.

Harmful reflections are inhibited by a calibrated restriction mounted upstream of the delivery valve, where it affects return flow only. This calibrated restriction attenuates pressure waves but is still small enough to allow maintenance of static pressure between injection events.

The return-flow restriction consists of a valve plate (Fig. 10, 4) with a throttle bore (3) and a spring (2). The valve plate lifts to prevent the throttle from exercising any effect on delivery quantity. During return flow, the valve plate closes to inhibit pulsation.

Constant-pressure valve

On high-speed diesel engines, the volumetric relief provided by the retraction piston and delivery valve is often not enough to prevent cavitation, dribble and blowback of combustion gases into the nozzle-and-holder assembly under all conditions.

Under these conditions, constant-pressure valves (Fig. 11) are installed. These valves use a single-action non-return valve with adjustable pressure (such as 60 bar) to relieve pressure on the high-pressure system (line and nozzle-and-holder assembly).

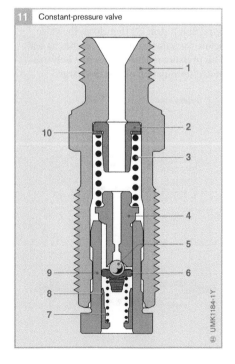

10 Delivery valve with return-flow restriction

11 Constant-pressure valve

Diesel records in 1972

In 1972 a modified Opel GT set a total of 20 international records for diesel-powered vehicles. The original roof structure was replaced to reduce aerodynamic resistance, while the 600-liter fuel tank was installed in place of the passenger seat. This vehicle was propelled by a 2.1-liter 4-cylinder diesel engine fitted with combustion swirl chambers. A distinguishing factor relative to the standard factory diesel was the exhaust-gas turbocharger. This engine produced 95 DIN bhp (approx. 70 kW) at 4,400 rpm while consuming fuel at the rate of 13 liters per hundred kilometers. Optimized fuel injection was provided by an EP/VA CL 163 rotary axial-piston pump with mechanical injection from Bosch.

Unlike today's diesel engines, this unit was designed around an existing gasoline engine.

As a result, the distributor injection pump had to be installed in the location originally intended for the ignition distributor. This led to a virtually vertical installation in the engine compartment. Numerous issues awaited resolution: control variables based on horizontal installation had to be recalculated for vertical operation, problems with air in the fuel led to difficult starting, the pump tended to cavitate, etc.

Here is a sampling of the records posted for this vehicle:

- It covered 10,000 kilometers in 52 hours and 23 minutes. This translates into an average speed of 190.9 km/h. The car was driven by six drivers relieving each other every three to four hours.
- Absolute world records for all vehicle classes were posted for 10 km (177.4 km/h) and 10 miles (184.5 km/h) from standing start.

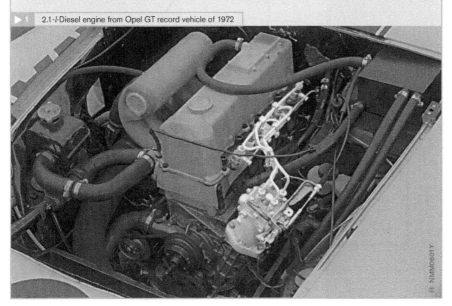

▶1 2.1-*l*-Diesel engine from Opel GT record vehicle of 1972

Auxiliary control modules for distributor injection pumps

The auxiliary control modules installed on distributor injection pumps with axial pistons govern start of delivery and regulate the volume of fuel discharged into the combustion chamber during injection. These control modules are composed of mechanical control elements, or actuators. They respond to variations in operating conditions (load factor, rotational speed, charge-air pressure, etc.) with precise adjustments for ideal performance. On distributor injection pumps with Electronic Diesel Control (EDC), an electric actuator mechanism replaces the mechanical actuators.

Overview

Auxiliary control modules adapt the start of delivery and delivery period to reflect both driver demand and the engine's instantaneous operating conditions (Fig. 1).

In the period since the distributor injection pump's introduction in 1962, an extensive range of controllers has evolved to suit a wide range of application environments. Numerous component configurations are also available to provide ideal designs for a variety of engine versions. Any attempt to list all possible versions of auxiliary control modules would break the bounds of this section. Instead, we will concentrate on the most important control modules. The individual units are the:

- Speed governor
- Timing device
- Adjustment and torque control devices
- Switches and sensors
- Shutoff devices
- Electric actuator mechanisms and
- Diesel immobilizers (component in the electronic vehicle immobilizer)

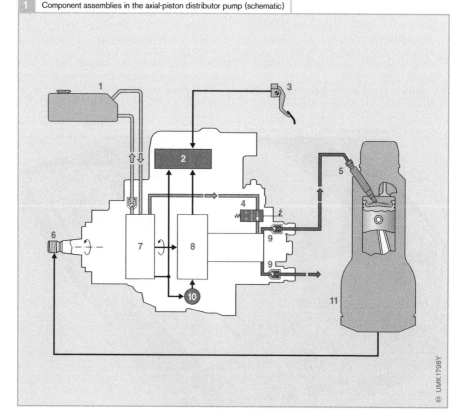

1 Component assemblies in the axial-piston distributor pump (schematic)

Fig. 1

1 Fuel supply
   (low-pressure)
2 Controlling system
3 Accelerator pedal
4 ELAB Electric
   shutoff device
5 Nozzle-and-holder
   assembly
6 Pump drive
   assembly
7 Low-pressure stage
   (vane-type supply
   pump with pressure-
   control valve and
   overflow throttle
   valve)
8 High-pressure pump
   with fuel rail
9 Delivery valve
10 Hydraulic timing
   device
11 Diesel engine

UMK1798Y

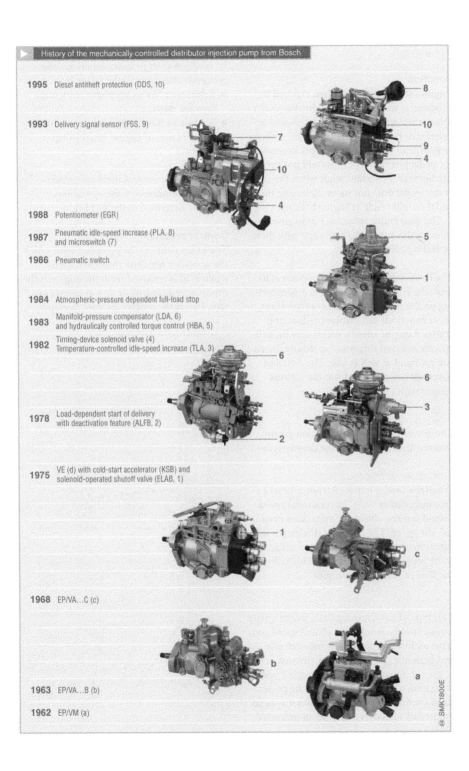

1995 Diesel antitheft protection (DDS, 10)

1993 Delivery signal sensor (FSS, 9)

1988 Potentiometer (EGR)

1987 Pneumatic idle-speed increase (PLA, 8) and microswitch (7)

1986 Pneumatic switch

1984 Atmospheric-pressure dependent full-load stop

1983 Manifold-pressure compensator (LDA, 6) and hydraulically controlled torque control (HBA, 5)

1982 Timing-device solenoid valve (4) Temperature-controlled idle-speed increase (TLA, 3)

1978 Load-dependent start of delivery with deactivation feature (ALFB, 2)

1975 VE (d) with cold-start accelerator (KSB) and solenoid-operated shutoff valve (ELAB, 1)

1968 EP/VA...C (c)

1963 EP/VA...B (b)

1962 EP/VM (a)

SMK1800E

Governors

Function

Engines should not stall when exposed to progressive increases in load during acceleration when starting off. Vehicles should accelerate or decelerate smoothly and without any surge in response to changes in accelerator pedal position. Speed should not vary on gradients of a constant angle when the accelerator pedal does not move. Releasing the pedal should result in engine braking.

The distributor injection pump governor offers active control to help cope with these operating conditions.

The basic function of every governor is to limit the engine's high idle speed. Depending on the individual unit's configuration, other functions can involve maintaining constant engine speeds, which may include specific selected rotational speeds or the entire rev band as well as idle. The different governor designs arise from the various assignments (Fig. 1):

Idle controller

The governor in the fuel-injection pump controls the idle speed of the diesel engine.

Maximum-speed governor

When the load is removed from a diesel engine operated at WOT, its rotational speed should not rise above the maximum permitted idle speed. The governor discharges this function by retracting the control collar toward "stop" at a certain speed. This reduces the flow of fuel into the engine.

Variable-speed governor

This type of governor regulates intermediate engine speeds, holding rotational speed constant within specified limits between idle and high idle. With this setup, fluctuations in the rotational speed n of an engine operating under load at any point in the power band are limited to a range between a speed on the full-load curve n_{VT} and an unloaded engine speed n_{LT}.

Other requirements

In addition to its primary functions, the governor also controls the following:

- Releases or blocks supplementary fuel flow for starting
- Modulates full-throttle delivery quantity as a function of engine speed (torque control)

Special torque-control mechanisms are required for some of these operations. They are described in the text below.

Control precision

The index defining the precision with which the governor regulates rotational speed when load is removed from the engine is the droop-speed control, or droop-speed control. This is the relative increase in engine

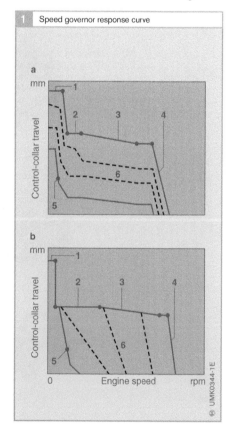

1   Speed governor response curve

a
mm
Control-collar travel

b
mm
Control-collar travel

0   Engine speed   rpm

UMK0344-1E

Fig. 1

a   Idle/maximum-speed governor
b   Variable-speed governor

1   Start quantity
2   Full-load delivery quantity
3   Torque control (positive)
4   Full-load speed regulation
5   Idle
6   Intermediate speed

speed that occurs when load is removed from the diesel engine while control lever position (accelerator-pedal travel) remains constant. The resulting rise in engine speed should not exceed a certain level in the controlled range. The maximum rpm specified for an unloaded engine represents the high-idle speed. This figure is encountered when the load factor of a diesel engine operating at Wide-Open Throttle (WOT) decreases from 100% to 0%. The rise in rotational speed is proportional to the variation in load factor. Larger load shifts produce progressively larger increases in rotational speed.

$$\delta = \frac{n_\mathrm{no} - n_\mathrm{vo}}{n_\mathrm{vo}}$$

or in %:

$$\delta = \frac{n_\mathrm{no} - n_\mathrm{vo}}{n_\mathrm{vo}} \cdot 100\%$$

where: δ droop-speed control, n_no upper no-load speed, n_vo upper full-load speed (some sources refer to the upper no-load speed as the high-idle speed).

The desired droop-speed control is defined by the diesel engine's intended application environment. Thus low degrees of proportionality (on the order of 4%) are preferred for electric power generators, as they respond to fluctuations in load factor by holding changes and engine speed and the resultant frequency shifts to minimal levels. Larger degrees of proportionality are better in motor vehicles because they furnish more consistent control for improved driveability under exposure to minor variations in load (vehicle acceleration and deceleration). In vehicular applications, limited droop-speed control would lead to excessively abrupt response when load factors change.

Design structure

The governor assembly (Fig. 2) consists of the mechanical governor (2) and the lever assembly (3). Operating with extreme precision, it controls the position of the control collar (4) to define the effective stroke and with it the injected-fuel quantity. Different versions of the control lever assembly can be employed to vary response patterns for various applications.

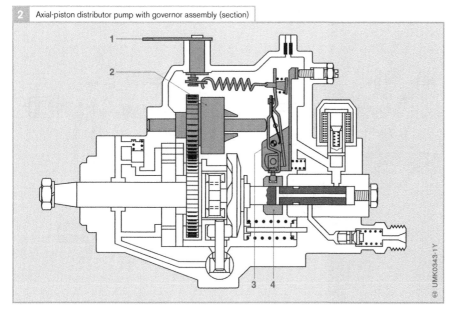

2 Axial-piston distributor pump with governor assembly (section)

UMK0343-1Y

Fig. 2

1 Rotational-speed control lever (accelerator pedal)
2 Mechanical governor
3 Lever assembly
4 Control collar

Variable-speed governor

The variable-speed governor regulates all engine speeds from start or high idle. In addition to these two extremes, it also controls operation in the intermediate range. The speed selected at the rotational-speed control lever (accelerator pedal or supplementary lever) is held relatively constant, with actual consistency varying according to the droop-speed control (Fig. 4).

This sort of adjustment is required when ancillary equipment (winch, extinguisher water pump, crane, etc.) is used on commercial vehicles or in stand-alone operation. This function is also frequently used in commercial vehicles and with agricultural machinery (tractors, combine harvesters).

Design

The governor assembly, consisting of flyweight housing and flyweights, is powered by the input shaft (Fig. 2). A governor shaft mounted to allow rotation in the pump housing supports the assembly (Fig. 3, 12). The radial travel of the flyweights (11) is translated into axial movement at the sliding sleeve (10). The force and travel of the sliding sleeve modify the position of the governor mechanism. This mechanism consists of the control lever (3), tensioning lever (2) and starting lever (4).

The control lever's mount allows it to rotate in the pump housing, where its position can be adjusted using the WOT adjusting screw (1). The starting and tensioning levers are also mounted in bearings allowing them to rotate in the control lever. At the lower end of the starting lever is a ball pin that engages with the control collar (7), while a starting spring (6) is attached to its top. Attached to a retaining stud (18) on the upper side of the tensioning lever is the idle-speed spring (19). The governor spring (17) is also mounted in the retaining stud. A lever (13) combines with the control-lever shaft (16) to form the link with the rotational-speed control lever (14).

The spring tension operates together with the sliding sleeve's force to define the position of the governor mechanism. The adjustment travel is transferred to the control collar to determine the delivery quantity h_1 and h_2, etc.).

Fig. 3
a   Start position (rotational-speed control lever can be at WOT or idle position for starting)
b   Idle position

1   Full-load screw
2   Tensioning lever
3   Control lever
4   Starting lever
5   Stop pin in housing
6   Starting spring
7   Sliding sleeve
8   Distributor plunger spill port
9   Distributor plunger
10 Sliding sleeve
11 Flyweight
12 Controller base
13 Lever
14 Rotational-speed control lever
15 Idle-speed adjusting screw
16 Control-lever shaft
17 Governor spring
18 Retaining pin
19 Idle-speed spring

a Starting-spring travel
c Idle-speed spring travel
h_1 Max. effective stroke (start)
h_2 Min. effective stroke (idle)
M_2 Pivot point for 4 and 5

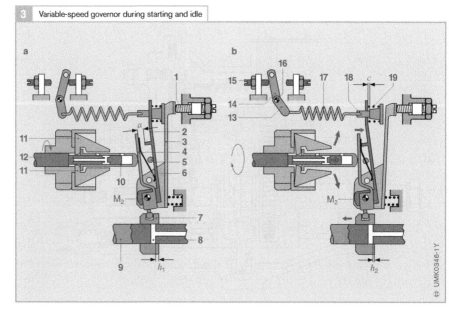

3   Variable-speed governor during starting and idle

Starting response

When the distributor injection pump is stationary, the flyweights and the sliding sleeve are at their base positions (Fig. 3a). The starting spring pushes the starting lever into the position for starting by rotating it about its axis M_2. The starting lever simultaneously transfers force through the ball pin onto the distributor plunger to shift the control collar to its starting position. This causes the distributor plunger (9) to execute substantial effective stroke. This provides maximum fuel start quantity.

Low rotational speeds are enough to overcome the compliant starting spring's tension and shift the sliding sleeve by the distance a. The starting lever again rotates about its pivot axis M_2, and the initial enhanced start quantity is automatically reduced to idle delivery quantity.

Idle-speed control

After the diesel engine starts and the accelerator pedal is released, the control lever returns to its idle position (Fig. 3b) as defined by the full-load stop on the idle adjusting screw (15). The selected idle speed ensures that the engine consistently continues to turn over smoothly when unloaded or under minimal loads. The control mechanism regulates the idle-speed spring mounted on the retaining stud. It maintains a state of equilibrium with the force generated by the flyweights.

By defining the position of the control collar relative to the control passage in the plunger, this equilibrium of forces determines the effective stroke. At rotational speeds above idle, it traverses the spring travel c to compress the idle-speed spring. At this point, the effective spring travel c is zero.

The housing-mounted idle-speed spring allows idle adjustments independent of accelerator pedal position while accommodating increases to compensate for temperature and load factor shifts (Fig. 5).

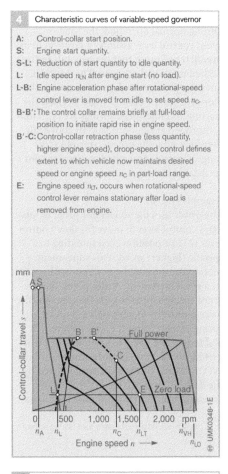

4   Characteristic curves of variable-speed governor

A: Control-collar start position.
S: Engine start quantity.
S-L: Reduction of start quantity to idle quantity.
L: Idle speed n_{LN} after engine start (no load).
L-B: Engine acceleration phase after rotational-speed control lever is moved from idle to set speed n_C.
B-B':The control collar remains briefly at full-load position to initiate rapid rise in engine speed.
B'-C:Control-collar retraction phase (less quantity, higher engine speed), droop-speed control defines extent to which vehicle now maintains desired speed or engine speed n_C in part-load range.
E: Engine speed n_{LT}, occurs when rotational-speed control lever remains stationary after load is removed from engine.

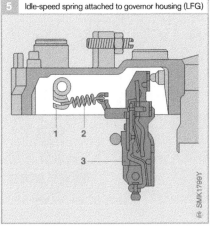

5   Idle-speed spring attached to governor housing (LFG)

Fig. 5
1   Rocker (fixed to pump housing)
2   Idle-speed spring
3   Lever assembly

Operation under load

During normal operation, the rotational-speed control lever assumes a position in its overall travel range that corresponds to the desired rotational or speed. The driver dictates this position by depressing the accelerator pedal to the desired angle. Because the starting and idle-speed springs are both fully compressed at rotational speeds above idle, they have no influence on control in this range. Control is exercised by the governor spring.

Example (Fig. 6):
The driver uses the accelerator pedal or auxiliary control lever to move the flow-control lever (2) to a position corresponding to a specific (higher) speed. This adjustment motion compresses the governor spring (4) by a given increment. At this point, the governor spring's force exceeds the centrifugal force exerted by the flyweights (1). The spring's force rotates the starting lever (7) and tensioning lever (6) around their pivot axis M_2 to shift the control collar toward its WOT position by the increment defined by the design's ratio of conversion. This increases delivery quantity to raise the engine

speed. The flyweights generate additional force, which the sliding sleeve (11) then transfers to oppose the spring force.

The control collar remains in its WOT position until the opposed forces achieve equilibrium. From this point onward, any additional increase in engine speed will propel the flyweights further outwards and the force exerted by the sliding sleeve will predominate. The starting and tensioning levers turn about their shared pivot axis (M_2) to slide the control collar toward "stop" to expose the cutoff bore earlier. The mechanism ensures effective limitation of rotational speeds with its ability to reduce delivery capacity all the way to zero. Provided that the engine is not overloaded, each position of the flow-control lever corresponds to a specific rotational range between WOT and no-load. As a result, the governor maintains the selected engine speed with the degree of intervention defined by the system's droop-speed control (Fig. 4).

Once the imposed loads have reached such a high order of magnitude as to propel the control collar all the way to its WOT position (hill gradients, etc.), no additional increases in fuel quantity will be available,

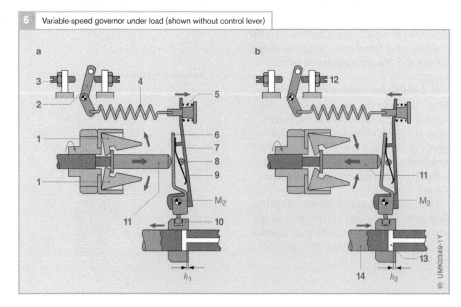

6   Variable-speed governor under load (shown without control lever)

even in response to lower rotational speed. The engine is overloaded, and the driver should respond by downshifting to a lower gear.

Overrun

When the vehicle descends a steep gradient, or when the accelerator pedal is released at high speeds (overrun), the engine is driven by the vehicle's inertia. The sliding sleeve responds by pressing against the starting and tensioning levers. Both levers move to shift the control collar to decrease delivery quantity; this process continues until the fuel-delivery quantity reflects the requirements of the "new" load factor, or zero in extreme cases. The response pattern of the variable-speed governor described here is valid at all flow-control lever positions, and occurs whenever any factor causes load or rpm to vary so substantially as to shift the control collar all the way to its WOT or "stop" end position.

Minimum/maximum speed governors

This governor regulates the idle and high-idle engine speeds only. Response in the intermediate range is regulated exclusively by

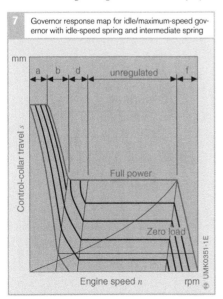

7 Governor response map for idle/maximum-speed governor with idle-speed spring and intermediate spring

mm

a b d   unregulated   f

Control-collar travel s

Full power

Zero load

Engine speed n rpm

© UMK0351-1E

the accelerator pedal (Fig. 6). This reduces surge, but is not suitable for use in light commercial vehicles equipped with ancillaries.

Design

The governor assembly with its flyweights and assembly of control levers is essentially comparable to the variable-speed governor described above. One difference in the idle and maximum-speed governor's design is the governor spring (Fig. 7 on next page, 4) and its installation. The compression spring applies pressure from its position in a guide sleeve (5). A retaining stud (7) provides the link between the tensioning lever and the governor spring.

Starting response

Because the flyweights (1) are at rest, the sliding sleeve (15) is in its base position. This allows the starting spring (12) to transfer force through the starting lever (9) and the sliding sleeve to push the flyweights to the inside. The control collar (13) on the distributor plunger is positioned to provide the delivery quantity prescribed for starting.

Idle-speed control

Once the engine starts and the accelerator pedal is released, the rotational-speed control lever (2) responds to the force exerted by the return spring against the pump housing by returning to the idle position. As engine speed climbs, the centrifugal force exerted by the flyweights (Fig. 8a, next page) rises, and their slots press the sliding sleeve back against the starting lever. The control process relies on the idle-speed spring (8) mounted on the tensioning lever (10). The starting lever's rotation moves the control collar to reduce the delivery quantity. The control collar's position is regulated by the combined effects of centrifugal and spring force.

Operation under load

When the driver varies the position of the accelerator pedal the rotational-speed control lever rotates by a given angle. This action negates the effect of the starting and idle-

Fig. 7
a Starting spring range
b Starting and idle-speed spring range
d Intermediate spring range
f Governor spring range

speed spring and engages the intermediate spring (6), which smoothes the transition to the unregulated range on engines with idle/maximum-speed governors. Rotating the rotational-speed control lever further toward WOT allows the intermediate spring to expand until the retainer's shoulder is against the tensioning lever (Fig. 8b). The assembly moves outside the intermediate spring's effective control sector to enter the uncontrolled range. The uncontrolled range is defined by the tension on the governor spring, which can be considered as rigid in this speed range. The driver's adjustments to the rotational-speed control lever (accelerator pedal) can now be transferred directly through the control mechanism to the control collar. Delivery quantity now responds directly to movement at the accelerator pedal.

The driver must depress the pedal by an additional increment in order to raise the vehicle speed or ascend a gradient. If the object is to reduce engine output, the pressure on the accelerator pedal is reduced.

If the load on the engine decreases while the control lever position remains unchanged, the delivery quantity will remain constant, and rotational speed will increase.

This leads to greater centrifugal force, and the flyweights push the sliding sleeve against the starting and tensioning levers with increased force. The full-load speed governor assumes active control only once the sleeve's force overcomes the spring's tension.

When the load is completely removed from the engine, it accelerates to its high-idle speed, where the system protects it against further increases in speed.

Part-load governor

Passenger vehicles are usually equipped with a combination of variable-speed and idle/maximum-speed governors. This type of part-load governor features supplementary governor springs allowing it to function as a variable-speed governor at the low end of the rev band while also operating as an idle/maximum-speed governor when engine speed approaches the specified maximum. This arrangement provides a stable rotational speed up to roughly 2,000 rpm, after which no further constraints are imposed on further rises if the engine is not under load.

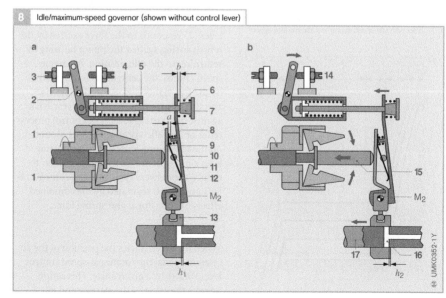

8   Idle/maximum-speed governor (shown without control lever)

UMK0352-1Y

Timing device

Function
The injection event must be initiated at a specific crankshaft angle (piston position) to ensure efficient combustion and optimal power generation. The pump's start of delivery and the resultant start of delivery must thus vary as rotational speed changes in order to compensate for two primary factors:

Injection lag
Fuel delivery to the nozzles starts when the plunger in the distributor head closes the inlet passage (start of delivery). The resulting pressure wave initiates the injection event when it reaches the injector (start of delivery). The pulse propagates through the injection line at the speed of sound. The travel duration remains essentially unaffected by rotational speed. The pressure wave's propagation period is defined by the length of the injection line and the speed of sound, which is approximately 1,500 m/s in diesel fuel. The period that elapses between the start of delivery and the start of injection is the injection lag.

Thus the actual start of injection always occurs with a certain time offset relative to the start of delivery. As a result of this phenomenon, the nozzle opens later (relative to engine piston position) at high than at low rotational speeds. To compensate for this, start of delivery must be advanced, and this forward shift is dependent on pump and engine speed.

Ignition lag
Following injection, the diesel fuel requires a certain period to evaporate into a gaseous state and form a combustible mixture with the air. Engine speed does not affect the time required for mixture formation. The time required between the start of injection and the start of combustion in diesel engines is the ignition lag.

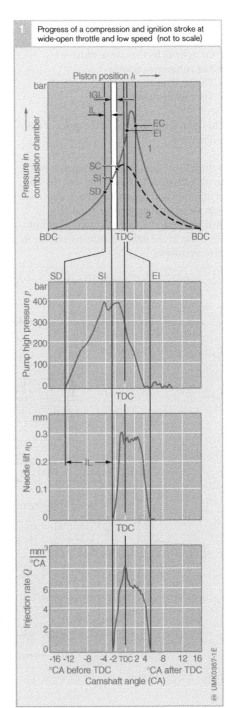

1 Progress of a compression and ignition stroke at wide-open throttle and low speed (not to scale)

Fig. 1
1 Combustion pressure
2 Compression

SD Start of Delivery
SI   Start of Injection
IL   Injection Lag
SC Start of Combustion
IGL Ignition Lag
EI   End of Injection
EC End of Combustion
BDC Engine piston at Bottom Dead Center
TDC Engine piston at Top Dead Center

Factors influencing ignition lag include
- The ignitability of the diesel fuel
  (as defined by the cetane number)
- The compression ratio
- The air temperature and
- The fuel discharge process

The usual ignition lag is in the order of one millisecond.

If the start of injection remains constant, the crankshaft angle traversed between the start of injection and the start of combustion will increase as rotational speed rises. This, in turn, prevents the start of combustion from occurring at the optimal moment relative to piston position.

The hydraulic timing device compensates for lag in the start of injection and ignition by advancing the distributor injection pump's start of delivery by varying numbers of degrees on the diesel engine's crankshaft. This promotes the best-possible combustion and power generation from the diesel engine at all rotational speeds.

Spill and end of combustion
Opening the cutoff bore initiates a pressure drop in the pump's high-pressure system (spill), causing the nozzle to close (end of injection). This is followed by the end of combustion. Because spill depends on the start of delivery and the position of the control collar, it is indirectly adjusted by the timing device.

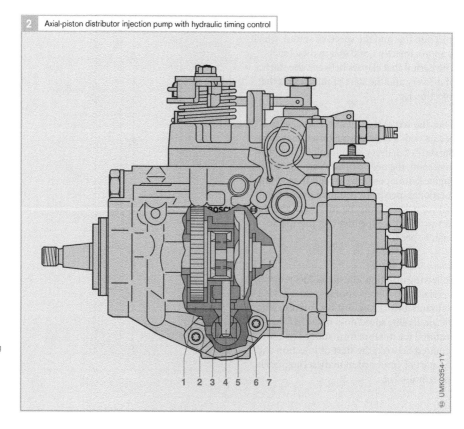

2 Axial-piston distributor injection pump with hydraulic timing control

1 2 3 4 5 6 7

Fig. 2
1 Roller ring
2 Rollers on roller ring
3 Sliding block
4 Pin
5 Timing plunger
6 Cam plate
7 Distributor plunger

UMK0354-1Y

Design

The hydraulic timing device is located in the distributor injection pump's underside at a right angle to the pump's longitudinal axis (Figs. 2 and 3).

The timing plunger (Fig. 3, 7) is guided in the pump housing (1). Both ends of this housing are sealed by covers (6). A bore (5) in the timing plunger allows fuel inlet, and a spring (9) is installed on the opposite side.

A sliding block (8) and pin (4) connect the timing plunger to the roller ring (2).

Operating concept

The timing plunger in the distributor injection pump is held in its base position by the pre-loaded tension of the compressed timing spring (Fig. 3a). As the pump operates, the pressure-control valve regulates the pressure of the fuel in its inner chamber in proportion to rotational speed. The fuel thus acts on the side of the plunger opposite the timing spring at a pressure proportional to the pump's rotational speed.

The pump must reach a specific rotational speed, such as 300 rpm, before the fuel pressure (pressure in inner chamber) overcomes the spring's tension to compress it and shift the timing plunger (to the left in Fig. 3b). The sliding block and pin transfer the plunger's axial motion to the rotating roller ring. This modifies the relative orientation of the cam disk and roller ring, and the rollers in the roller ring lift the turning cam disk earlier. Thus the rollers and roller ring rotate relative to the cam disk and plunger by a specific angular increment (a) determined by rotational speed. (α). The maximum potential angle is usually twelve camshaft degrees (24 degrees crankshaft).

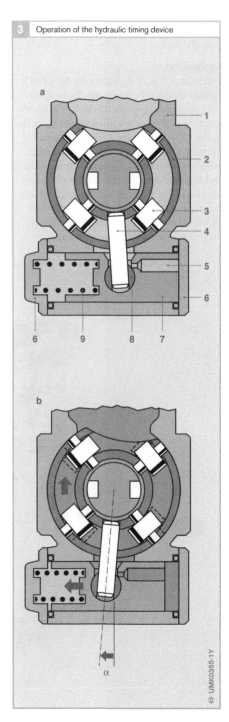

3 Operation of the hydraulic timing device

a

b

α

UMK0355-1Y

Fig. 3
a Passive state
b In operation

1 Pump housing
2 Roller ring
3 Rollers on roller ring
4 Pin
5 Bore in timing plunger
6 Cover plate
7 Timing plunger
8 Sliding block
9 Timing spring

α Roller ring pivot angle

Mechanical torque-control modules

Application

Distributor injection pumps are designed around modular building blocks, with various supplementary control devices available to satisfy individual engine requirements (as in Fig. 1). This concept affords an extended range of torque-control options allowing pumps to provide maximum torque, power and fuel economy along with low emissions. This overview serves as a compilation of the various torque-control modules and their effects on the diesel engine. The schematic diagram illustrates the interrelationships between the basic distributor injection pump and the different torque-control modules (Fig. 2).

Torque control

The torque-control process is the operation in which delivery quantity is adjusted to the engine's full-load requirement curve in response to changes in rotational speed.

This adaptive strategy may be indicated when special demands on full-load characteristics (improved exhaust-gas composition, torque generation and fuel economy) are encountered.

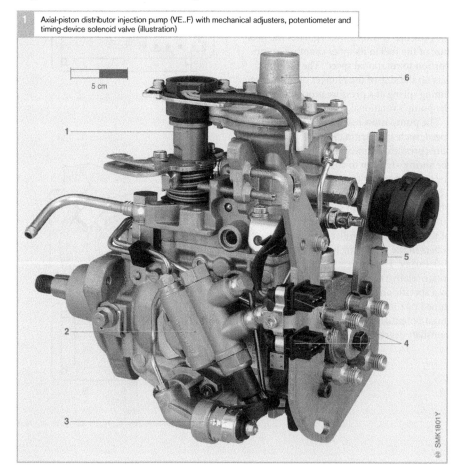

1   Axial-piston distributor injection pump (VE..F) with mechanical adjusters, potentiometer and timing-device solenoid valve (illustration)

5 cm

Fig. 1
1   Potentiometer
2   Hydraulic cold-start accelerator KSB
3   Load-dependent start of delivery with deactivation feature-ALFB
4   Connector
5   Pneumatic idle-speed increase PLA
6   Hydraulically controlled torque-control HBA

SMK1801Y

2   Block diagram of VE distributor injection pump with mechanical/hydraulic full-load torque control

Charge-air pressure-dependent manifold-pressure compensator LDA
Delivery-quantity control relative to charge-air pressure (engine with turbocharger).

Hydraulically controlled torque control device HBA
Control of delivery quantity based on engine speed (not on turbocharged engines with LDA).

Load-dependent start of delivery LFB
Adjust start of delivery to reflect load to reduce noise emissions and, as primary aim, to reduce exhaust-gas emissions.

Atmospheric-pressure dependent full-load stop ADA
Control of delivery quantity based on atmospheric pressure.

Cold-start accelerator KSB
Adjusts start of delivery for improved performance during cold starts.

Graduated (or adjustable) start quantity GST
Avoid start enrichment during hot starts.

Temperature-controlled idle-speed increase TLA
Improved post-start warm-up and smoother operation with idle-speed increase on cold engine.

Solenoid-operated shutoff valve ELAB
Makes it possible to shut down engine with "ignition key".

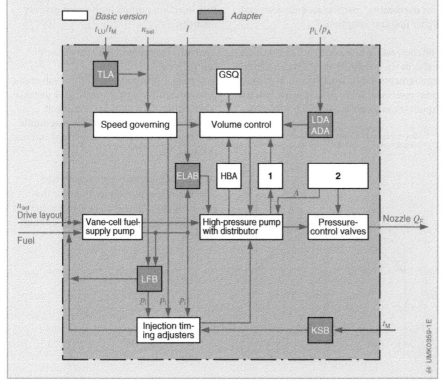

Fig. 2
1   Full-load torque
    control with control-
    lever assembly
2   Hydraulic full-load
    torque control

A Cutoff bore
n_{act} Actual speed
    (control parameter)
n_{set} Set speed (refer-
    ence parameter)
Q_F Delivery quantity
t_M Engine temperature
t_{LU} Ambient air
    temperature
p_L Charge-air pressure
p_A Atmospheric
    pressure
p_i Pump inner chamber
    pressure
I Current at ELAB
    (PWM signal)

The object is to inject precisely the amount of fuel required by the engine. Following an initial rise, the engine's fuel requirement falls slightly as its speed increases. Fig. 3 shows the curve for fuel delivery on an fuel-injection pump without torque control (1).

As the curve indicates, the distributor injection pump delivers somewhat more fuel at high rotational speeds, while the control collar remains stationary relative to the plunger. This increase in the pump's delivery quantity is traceable to the venturi effect at the cutoff bore on the plunger.

Permanently defining the fuel-injection pump's delivery quantity for maximum torque generation at low rotational speeds results in an excessive rate of high-speed fuel injection, and the engine is unable to burn the fuel without producing smoke. The results of excessive fuel injection include engine overheating, particulate emissions and higher fuel consumption.

The contrasting case occurs when maximum fuel delivery is defined to reflect the engine's requirements in full-load operation at maximum rotational speed; this strategy fails to exploit the engine's low-speed power-pro-

duction potential. Again, delivery quantity decreases as engine speed increases. In this case, power generation is the factor that is less than optimal. Conclusion: A means is required to adjust injected fuel quantities to reflect the engine's actual instantaneous requirements.

On distributor injection pumps, this torque-control can be executed by the delivery valve, the cutoff bore, an extended control lever assembly or hydraulically controlled torque control (HBA). The control lever assembly is employed when negative full-load torque control is required.

Positive torque control

Positive full-load torque control is required on fuel-injection pumps that would otherwise deliver too much fuel at the top of the speed range. This type of system is employed to avoid this issue by reducing the fuel-injection pump's high-speed delivery quantity.

Positive torque control with the pressure-control valve
in certain limits, pressure-control valves can provide positive torque control, for instance, when equipped with more compliant springs. This strategy limits the magnitude of the high-speed rise in the pump's internal pressure.

Positive torque control using cutoff bore
Selected shapes and dimensions for the plunger's cutoff bore can be selected to reduce delivery quantities delivered at high speed.

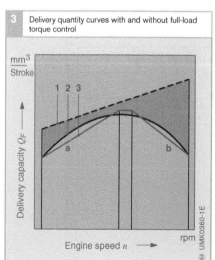

3   Delivery quantity curves with and without full-load torque control

Fig. 3
a   Negative torque control
b   Positive torque control

1   Full-load delivery quantity with no torque control
2   Engine fuel requirement
3   Full-load delivery quantity with torque control

Shaded area:
Excessive fuel injected

Negative torque control

Negative full-load torque control may be
indicated with engines that suffer from a
low-speed tendency to generate black smoke
or need special torque-rise characteristics.
Turbocharged engines often require negative
torque control to take over from the manifold-pressure compensator (LDA). The response to these scenarios is to increase delivery quantity as engine speed rises (Fig. 3,
Sector a).

*Negative torque control with control lever
assembly (Fig. 4)*
The starting spring (4) compresses and the
torque-control lever (9) presses against the
tensioning lever (2) via the stop pin (3). The
torque-control shaft (11) is also pressed
against the tensioning lever. When speed
rises, increasing the sleeve force F_M, the
torque-control lever presses against the
torque-control spring. If the sleeve force exceeds the torque-control spring's force, the
torque-control lever (9) is pressed toward
the pin shoulder (5). This shifts the pivot
axis M_4 shared by the starting lever and the
torque-control lever. The starting lever simultaneously rotates about M_2 to slide the
control collar (8) for increased delivery
quantity. The torque-control process terminates with the torque-control lever coming
to rest against the pin shoulder.

*Negative torque control with hydraulically
controlled torque control (HBA)*
A torque control mechanism similar to the
manifold-pressure compensator (LDA) can
be used to define the full-load delivery characteristics as a function of engine speed
(Fig. 5). As engine speed increases, rising
pressure in the pump's internal chamber p_i
is transferred to the control plunger (6).
This system differs from spring-based adjustment by allowing use of a cam profile on
the control pin to define full-throttle curves
(to a limited degree).

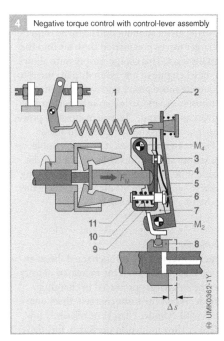

4   Negative torque control with control-lever assembly

Fig. 4
1 Governor spring
2 Tensioning lever
3 Full-load stop pin
4 Starting spring
5 Pin shoulder
6 Impact point
7 Starting lever
8 Control collar
9 Torque-control lever
10 Torque-control
    spring
11 Torque-control shaft

M_2 Pivot point for
    2 and 7
M_4 Pivot axis for
    7 and 9

F_M Sleeve force
Δs Control-collar travel

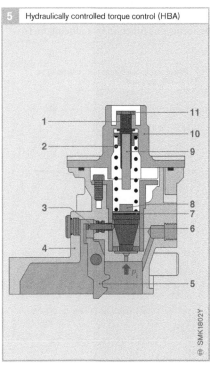

5   Hydraulically controlled torque control (HBA)

Fig. 5
1 Adjusting screw
2 Spring
3 Pin
4 Pump cover
5 Reverse-transfer
   lever with full-load
   stop
6 Adjustment piston
7 Shim
8 Base disk
9 Stop pin
10 Cover plate
11 Locknut

p_i Pump inner chamber
   pressure

Charge-air pressure torque control

During artificial induction a (turbo)super-charger forces pressurized fresh air into the intake tract. This charge-air pressure allows a diesel engine of any given displacement to generate more power and torque than its atmospheric-induction counterpart in any given speed band. The rise in effective power corresponds to the increase in air mass (Fig. 6). In many cases, it proves possible to reduce specific fuel consumption at the same time. A standard means of generating charge-air pressure for diesel engines is the exhaust-gas turbocharger.

Manifold-pressure compensator (LDA)
Function

Fuel delivery in the turbocharged diesel engine is adapted to suit the increased density of the air charges produced in charging mode. When the turbocharged diesel operates with cylinder charges of relatively low density (low induction pressure), fuel delivery must be adjusted to reflect the lower air mass. The manifold-pressure compensator performs this function by reducing fuel-throttle fuel flow below a specific (selected) charge-air pressure (Fig. 7).

Design

The manifold-pressure compensator is mounted on top of the distributor injection pump (Figs. 8 and 9). On top of this mechanism are the charge-air-pressure connection (7) and the vent port (10). A diaphragm (8) separates the inside chamber into two air-tight and mutually isolated sections. A compression spring (9) acts against the diaphragm, while the adjusting screw (5) holds the other side. The adjusting screw is used to adjust the spring's tension. This process adapts the manifold-pressure compensator to the charge-air pressure generated by the turbocharger. The diaphragm is connected to the sliding bolt (11). The sliding bolt features a control cone (12) whose position is monitored by a probe (4). The probe transfers the sliding bolt's motion through the reverse-transfer lever (3) to vary the full-load stop. The adjusting screw (6) on top of the LDA defines the initial positions of the diaphragm and sliding bolt.

Fig. 6
— Naturally aspirated
    engine
— Turbocharged
    engine

Fig. 7
a   with turbocharger
b   naturally aspirated

p_1 Lower charge-air
    pressure
p_2 Upper charge-air
    pressure

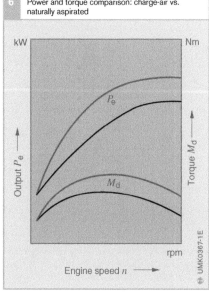

6   Power and torque comparison: charge-air vs. naturally aspirated

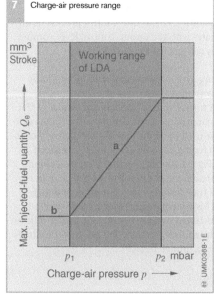

7   Charge-air pressure range

Operating concept
At the lower end of the speed range, the charge-air pressure generated by the turbocharger is not powerful enough to compress the spring. The diaphragm remains in its initial position. Once the rising charge-air pressure p_L starts to deflect it, the diaphragm pushes the sliding bolt and control cone down against the pressure of the spring.

This vertical movement in the sliding bolt shifts the position of the probe, causing the reverse-transfer lever to rotate about its pivotal point M_1. The tensile force exerted by the governor spring creates a positive connection between tensioning lever, reverse-transfer lever, probe and control cone.

The tensioning lever thus mimics the reverse-transfer lever's rotation, and the starting and tensioning levers execute a turning motion around their shared pivot axis to shift the control lever and raise delivery quantity. This process adapts fuel flow to meet the demands of the greater air mass with the engine's combustion chambers.

8   Axial-piston distributor injection pump with manifold-pressure compensator

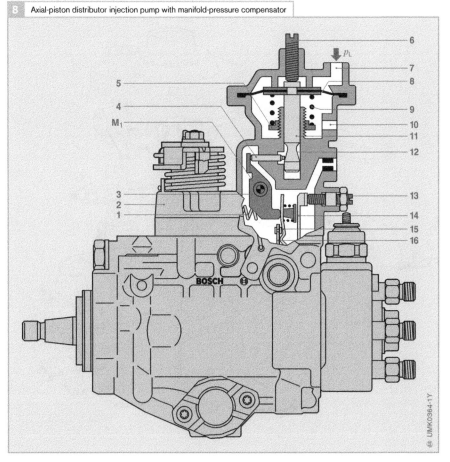

Fig. 8
1 Governor spring
2 Governor cover
3 Reverse-transfer lever
4 Sensor pin
5 Adjustment nut
6 Adjustment pin
7 Charge-air pressure connection
8 Diaphragm
9 Spring
10 Vent
11 Sliding bolt
12 Control cone
13 Adjusting screw for full-load delivery
14 Control lever
15 Tensioning lever
16 Starting lever

p_L Charge-air pressure

M_1 Pivot axis for 3

As charge-air pressure falls, the spring beneath the diaphragm presses the sliding bolt back up. The control mechanism now acts in the opposite direction, and fuel quantity is reduced to reflect the needs of the lower charge-air pressure.

The manifold-pressure compensator responds to turbocharger failure by reverting to its initial position, and limits full-throttle fuel delivery to ensure smoke-free combustion. Full-load fuel quantity with charge-air pressure is adjusted by the full-throttle stop screw in the governor cover.

Atmospheric pressure-sensitive torque control

Owing to the lower air density, the mass of inducted air decreases at high altitudes. If the standard fuel quantity prescribed for full-load operation is injected, there will not be enough air to support full combustion. The immediate results are smoke generation and rising engine temperatures. The atmospheric pressure-sensitive full-load stop can help prevent this condition. It varies full-load fuel delivery in response to changes in barometric pressure.

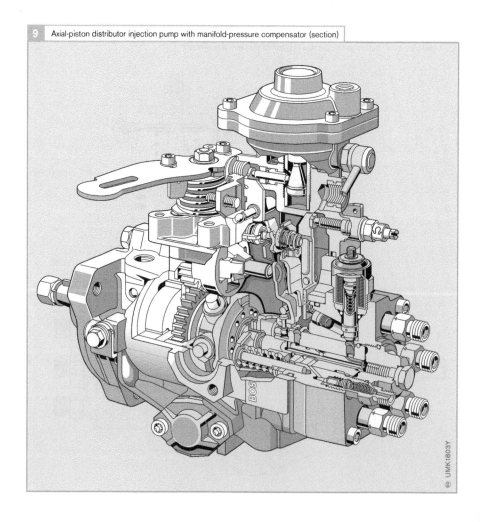

9   Axial-piston distributor injection pump with manifold-pressure compensator (section)

UMK1803Y

Atmospheric-pressure sensitive full-load stop (ADA)

Design

The basic structure of the atmospheric pressure-sensitive full-load stop is identical to that of the charge-air pressure-sensitive full-load stop. In this application, it is supplemented by a vacuum unit connected to a vacuum-operated pneumatic device (such as the power brake system). The vacuum unit provides a constant reference pressure of 700 mbar (absolute).

Operating concept

Atmospheric pressure acts on the upper side of the ADA's internal diaphragm. On the other side is the constant reference pressure supplied by the vacuum unit.

Reductions in atmospheric pressure (as encountered in high-altitude operation) cause the adjustment piston to rise away from the lower full-load stop. As with the LDA, a reverse-transfer lever then reduces the injected fuel quantity.

Load-sensitive torque control

Load-sensitive start of delivery (LFB)

Function

As the diesel engine's load factor changes, the start of injection – and thus the start of delivery – must be advanced or retarded accordingly.

The load-sensitive start of delivery is designed to react to declining loads (from full-load to part throttle, etc.) at constant control lever positions by retarding start of delivery. It responds to rising load factors by shifting the start of delivery forward. This adaptive process provides smoother engine operation along with cleaner emissions at part throttle and idle.

Fuel-injection pumps with load-sensitive start of delivery can be recognized by the press-fit ball plug inserted during manufacture (Fig. 10, 10).

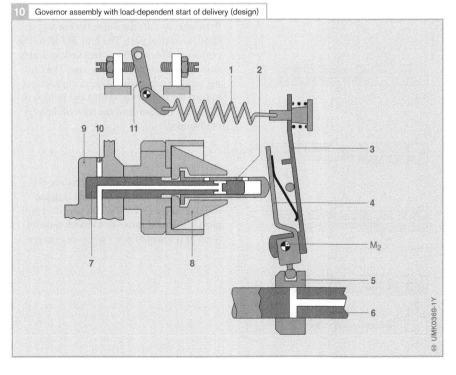

10 Governor assembly with load-dependent start of delivery (design)

Fig. 10
1 Governor spring
2 Sliding sleeve
3 Tensioning lever
4 Starting lever
5 Control collar
6 Distributor plunger
7 Controller base
8 Flyweight
9 Pump housing
10 Cone
11 Rotational-speed
   control lever

M₂ Pivot for 3 and 4

Design

Load-sensitive start of delivery torque control (Fig. 10) relies on modifications to the sliding sleeve (2), governor base (7) and pump housing (9). In this configuration, the sliding sleeve is equipped with a supplementary control port while the base includes a ring groove, one longitudinal and two transverse passages. The pump housing contains an additional bore allowing this assembly to link the pump's inner chamber with the suction-side of the vane-type supply pump.

Operating concept

Rises in engine speed and the accompanying increases in the pump's internal pressure advance the start of delivery. The reduction in the pump's inner chamber supplied by the LFB provide a (relative) retardation of the timing. The control function is regulated by the ring groove on the base and the control port on the sliding sleeve. Individual full-load speeds can be specified by the rotational-speed control lever (11).

If the engine attains this speed without reaching wide-open throttle, the rotational speed can continue to rise. The flyweights (8) respond by spinning outward to shift the position of the sliding sleeve. Under normal control conditions, this reduces delivery quantity, while the sliding sleeve's control port is exposed by the ring groove in the base (open, Fig. 11). At this point, a portion of the fuel flows through the base's longitudinal and transverse passages to the suction side, reducing the pressure in the pump's inner chamber.

This pressure reduction moves the timing plunger to a new position. This forces the roller ring to rotate in the pump's rotational direction, retarding the start of delivery.

The engine speed will drop if load factor continues to rise while the control lever's position remains constant. This retracts the flyweights to shift the sliding sleeve and re-close its control port. The flow of fuel to the pump's inside chamber is blocked, and pressure in the chamber starts to rise. The timing plunger acts against the spring pressure to turn the roller ring against the pump's direction of rotation and the start of delivery is advanced.

Load-dependent start of delivery with deactivation feature (LFB)

The LFB can be deactivated to reduce HC emissions generated by the diesel engine when it is cold (< 60°C). This process employs a solenoid valve (8) to block the fuel flow. This solenoid open when de-energized.

Fig. 11

a Start (base) position

b Full-load position just before activation

c Activation (pressure reduction in inner chamber)

d Load-dependent start of delivery with deactivation feature ALFB

1 Longitudinal passage in plunger

2 Controller axis

3 Sliding sleeve control port

4 Sliding sleeve

5 Transverse passage in controller axis

6 Control helix on ring groove on controller axis

7 Transverse passage in controller axis

8 Solenoid valve

9 Vane-type supply pump

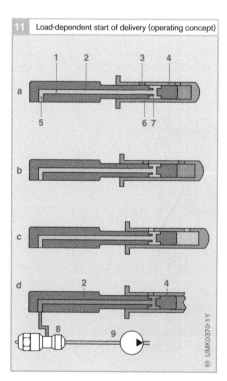

11 Load-dependent start of delivery (operating concept)

Cold-start compensation

The cold-start compensation device improves the diesel engine's cold-start response by advancing the start of delivery. This feature is controlled by a driver-operated cable or by an automatic temperature-sensitive control device (Fig. 13).

Mechanical cold-start accelerator (KSB) on roller ring

Design

The KSB is mounted on the pump's housing. In this assembly, a shaft (Fig. 12, 12) connects the stop lever (Fig. 13, 3) with an inner lever featuring an eccentrically mounted ball head (3). This lever engages with the roller ring. The stop lever's initial position is defined by the full-load stop and the leg spring (13). The control cable attached to the upper end of the stop lever serves as the link to the adjustment mechanism which is either manual or automatic. The automatic adjuster is installed in a bracket on the distributor injection pump (Fig. 13), while the manual adjustment cable terminates in the passenger compartment.

There also exists a version in which the adjuster intervenes through the timing plunger.

Operating concept

The only difference between the manual and automatic versions of the cold-start accelerator is the external control mechanisms. The key process is always the same. When the control cable is not tensioned, the leg spring presses the stop lever against its full-load stop. Both ball head and roller ring (6) remain in their initial positions. Tension on the control cable causes the stop lever, shaft and inside lever to turn with the ball head.

This rotation changes the position of the roller ring to advance the start of delivery. The ball head engages a vertical groove in the roller ring. This prevents the timing plunger from advancing the start of delivery further before a specified engine speed is reached.

When the driver activates the cold-start acceleration device (KSB timing device), a position shift of approximately 2.5° camshaft (b) remains, regardless of the adjustment called for by the timing device (Fig. 14a).

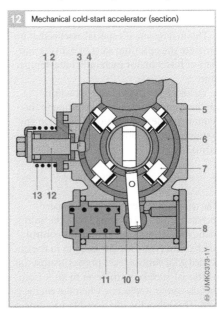

12 Mechanical cold-start accelerator (section)

Fig. 12
1 Lever
2 Adjustment window
3 Ball head
4 Vertical groove
5 Pump housing
6 Roller ring
7 Rollers in roller ring
8 Timing plunger
9 Pin
10 Sliding block
11 Timing spring
12 Shaft
13 Leg spring

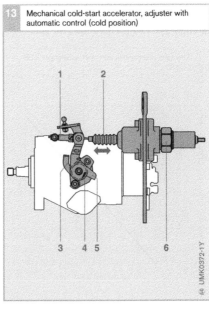

13 Mechanical cold-start accelerator, adjuster with automatic control (cold position)

Fig. 13
1 Retainer
2 Bowden cable
3 Stop lever
4 Leg spring
5 Control lever KSB
6 Adjuster dependent on coolant and ambient temperatures

In cold-start systems with automatic control, the actual increment depends on engine temperature and/or ambient temperature.

Automatic adjustment relies on a control mechanism in which a temperature-sensitive expansion element translates variations in engine temperature into linear motion.

This arrangement's special asset is that it provides the optimum start of delivery and start of injection for each individual temperature.

Various lever layouts and actuation mechanisms are available according to the mounting side and rotational direction encountered in individual installation environments.

Temperature-controlled idle-speed increase (TLA)

The temperature-controlled idle-speed increase, which is combined with the automatic KSB, is also operated by the control mechanism (Fig. 15). The ball pin in the extended KSB control lever presses against the rotational-speed control lever to lift it from the idle-speed stop screw when the engine is cold. This raises the idle speed to promote smoother engine operation. The KSB control lever rests against its full-load stop when the engine is warm. This allows the rota-

tional-speed control lever to return to its own full-load stop, at which point the temperature-controlled idle-speed increase system is no longer active.

Hydraulic cold-start accelerator

There are inherent limits on the use of strategies that shift the timing plunger to advance start of injection. Hydraulic start of injection advance applies speed-controlled pressure in the pump's inside chamber to the timing plunger. The system employs a bypass valve in the pressure-control valve to modify the inner chamber's automatic pressure control, automatically increasing internal pressure to obtain additional advance extending beyond the standard advance curve.

Design

The hydraulic cold-start accelerator comprises a modified pressure-control valve (Fig. 17, 1), a KSB ball valve (7), an electrically heated expansion element (6) and a KSB control valve (9).

Fig. 14

a   Injection adjusted by timing device

b   Minimum adjustment (approximately 2.5° camshaft)

Fig. 15

1   Rotational-speed control lever

2   Ball pin

3   Control lever KSB

4   Full-load stop

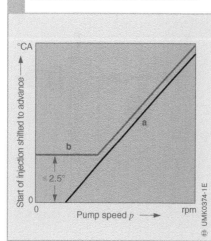
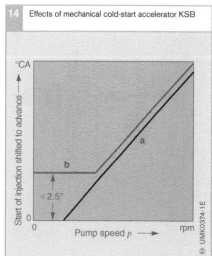

14   Effects of mechanical cold-start accelerator KSB

Start of injection shifted to advance (°CA) vs. Pump speed p (rpm)

a

b

≤ 2.5°

UMK0374-1E

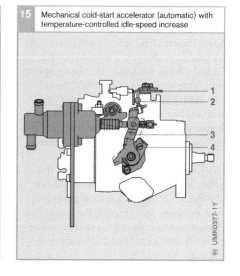

15   Mechanical cold-start accelerator (automatic) with temperature-controlled idle-speed increase

1
2
3
4

UMK0377-1Y

Operating concept

The fuel supplied by the supply pump (5) flows through the distributor injection pump's inner chamber and to the end of the timing plunger (11), which compresses the return spring (12) and shifts position in response to the inner chamber pressure to vary the start of injection. The pressure-control valve controls the pressure in the inner chamber, raising pressure for greater delivery quantity as engine speed increases (Fig. 16).

The throttle port in the pressure-control valve's plunger (Fig. 17, 3) supplies the added pressure that allows the KSB to provide the forward offset in start of injection advance (Fig. 16, blue curve). This conveys an equal pressure to the spring side of the pressure-control valve. The KSB ball valve, with its higher pressure setting, regulates activation and deactivation (with the thermal element) while also serving as a safety release. An adjusting screw on the integrated KSB control valve is available for adjusting KSB operation to a specific engine speed. Pressure from the supply pump presses the KSB control-valve plunger (10) against a spring. A damping throttle inhibits pulsation against the control plunger. The KSB

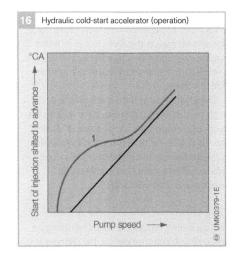

16   Hydraulic cold-start accelerator (operation)

Fig. 16
1   Start of delivery advanced

pressure curve is controlled by the timing edge on the control plunger and the opening on the valve holder. The spring rate on the control valve and the control port configuration can be modified to match KSB functionality to the individual application. The ambient temperature will act on the expansion element to open the cold-start accelerator ball valve before starting the hot engine.

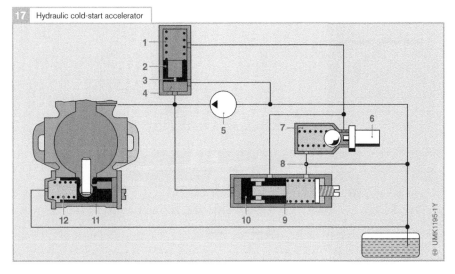

17   Hydraulic cold-start accelerator

Fig. 17
1   Pressure-control valve
2   Valve plunger
3   Throttled bore
4   Inner chamber pressure
5   Vane-type supply pump
6   Electrically heated expansion element
7   Ball valve KSB
8   Fuel drains without pressure
9   Adjustable KSB control valve
10   Control plunger
11   Timing device
12   Return spring

Soft-running device

To achieve the desired emission properties, the system injects the fuel charge into the engine's combustion chamber in the briefest possible time span; it operates at high fuel-delivery rates.

Depending on system configuration, these high fuel-delivery rates can have major consequences, especially in the form of diesel knock at idle. Remedial action is available through the use of extended injection periods to achieve smoother combustion processes at and near idle.

Design and operating concept

In distributor injection pumps featuring an integral soft-running device, the plunger is equipped with two longitudinal passages (Fig. 19, 3 and 5) connected by a ring groove (6). The longitudinal passage 3 is connected to a cutoff bore (7) with a restrictor in the area adjacent to the control collar (1).

The plunger travels through the stroke h_1 on its path toward TDC. The cutoff bore (7) connected to passage 3 emerges from the control collar earlier than the cutoff bore (2) on passage 5.

Because the ring groove (6) links passages 3 and 5, this causes a portion of the fuel to seep from the plunger chamber back to the pump's inner chamber. This reduces the fuel-delivery rate (less fuel discharged for

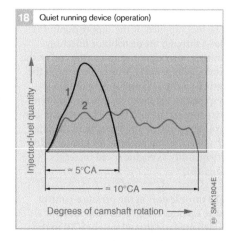

SMK1804E

18   Quiet running device (operation)

Degrees of camshaft rotation ⟶

Injected-fuel quantity

≈ 5°CA

≈ 10°CA

each degree of camshaft travel). The travel that the camshaft executes for any given quantity of fuel injected is roughly doubled (Fig. 18).

Under extreme loads, the control collar is closer to the distributor head. This makes the distance h_2 smaller than h_1. When the plunger now moves toward TDC, the ring groove (6) is covered before the cutoff bore (7) emerges from the control collar. This cancels the link joining passages 3 and 5, deactivating the soft-running mechanism in the high-load range.

Fig. 18
1   without quiet
    running
2   with quiet running

Fig. 19
1   Control collar
2   Cutoff bore
3   Port 3
4   1-way non-return
    valve for nozzle
5   Port 5
6   Ring groove
7   Cutoff bore

h_1 Lift 1
h_2 Lift 2

TDC **T**op **D**ead **C**enter
    on distributor
    plunger

19   Quiet running device (section)

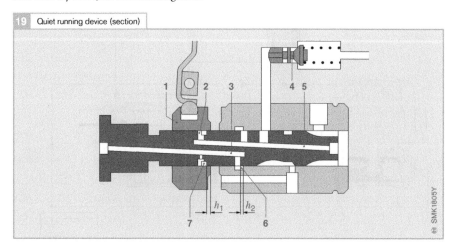

SMK1805Y

Load switch

The load switches are installed at the distributor injection pump's main rotational-speed control lever. These switches control operation of assemblies outside of the fuel-injection pump. They rely on microswitches for electronic control, or valves for pneumatic switching of these external units. Their primary application is in triggering the exhaust-gas recirculation valve. In this application, the switch opens and closes the valve in response to variations in the main control lever's position.

Two different elements are available to control microswitches and pneumatic valves:
● Angular stop plate with control recess and
● Cast aluminum control cams

The control plate or cam attached to the main control lever initiates activation or deactivation based on the lever's travel. The switching point corresponds to a specific coordinate on the pump's response curve (rotational speed vs. delivery quantity).
    Either two microswitches or one pneumatic valve can be installed on the fuel-injection pump due to differences in the respective layouts.

Microswitch
The electric microswitch is a bipolar make-and-break switch. It consists of leaf spring elements with a rocker arm. A control pin presses on the springs. A supplementary control element (Fig. 1, 6) limits mechanical wear and provides a unified control stroke for all applications on distributor injection pumps.

Pneumatic valve
The pneumatic valve (Fig. 2) blocks air flow in a vacuum line.

Potentiometer

The potentiometer is an option available for cases in which the object is to control several different points on the curve for rotational speed vs. load factor. It is mounted on top of the rotational-speed control lever, to which it is attached by clamps (Fig. 1). When energized, the potentiometer transmits a continuous electrical voltage signal that reflects the control lever's position with a linear response curve. A suitable governor design then provides useful information in the defined range.

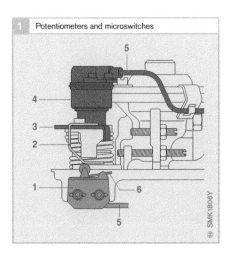

1 Potentiometers and microswitches

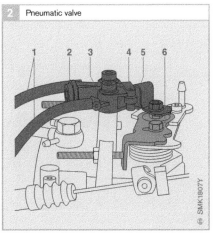

2 Pneumatic valve

Fig. 1
1 Microswitch
2 Angular stop bracket
3 Rotational-speed control lever
4 Potentiometer
5 Electrical connection
6 Auxiliary controller

Fig. 2
1 Pneumatic connections
2 Vacuum adjusting screw
3 Pneumatic valve
4 Control lever
5 Control roller
6 Rotational-speed control lever

Delivery-signal sensor

Application

The FSS delivery-signal sensor is a dynamic pressure sensor installed in the threaded center plug of the diesel fuel-injection pump in place of the vent screw (Fig. 2). It monitors pressure in the element chamber. The sensor signal can be employed to register the start of delivery, the delivery period and the pump speed n.

The pressure-signal measurement range extends from 0...40 MPa, or 0...400 bar, at which it is suitable for use with IDI fuel-injection pumps. This device is a "dynamic sensor". It registers pressure variations instead of monitoring static pressure levels.

Design and operating concept

The delivery-signal sensor relies on piezo-electric principles for operation. The pressure in the element chamber acts on a sensor measurement cell (Fig. 2, 13). This contains the piezoelectric ceramic layer. Pressure variations modify electrical charge states in this layer. These charge shifts generate minute electric voltages, which the integrated circuit (11) in the sensor converts to a square-wave signal (Fig. 1).

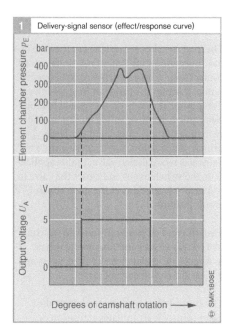

1 Delivery-signal sensor (effect/response curve)

Degrees of camshaft rotation →

Fig. 1
a Pressure curve for element chamber
b Signal from delivery-signal sensor

Fig. 2
1 Disk
2 Spring
3 Seal
4 Contact pin
5 Housing
6 Contact spring
7 Cable housing
8 Power supply connection (yellow/white)
9 Signal connection (gray/white)
10 O-ring
11 Circuit board with integrated circuit
12 Insulator
13 Sensor measurement cell (piezoelectric ceramic material)
14 Fuel-injection pump element chamber

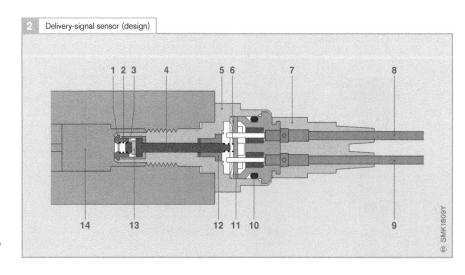

2 Delivery-signal sensor (design)

Shutoff devices

Shutoff

One inherent property of the diesel engine's auto-ignition concept is that interrupting the fuel supply represents the only way to switch off the engine.

The mechanically controlled distributor injection pump is usually deactivated by an electric shutoff valve (ELAB). It is only in rare special application environments that these pumps are equipped with mechanical shutoff devices.

Solenoid-operated shutoff valve (ELAB)

The electric shutoff device is a solenoid valve. Operation is controlled by the "ignition" key to offer drivers a high level of convenience.

The solenoid valve employed to interrupt fuel delivery in the distributor injection pump is mounted on top of the distributor head (Fig. 1). When the diesel engine is running, the solenoid (4) is energized, and the armature retracts the sealing cone (6) to hold the inlet passage to the plunger chamber open. Switching off the engine with the key interrupts the current to the solenoid coil. The magnetic field collapses and the spring (5) presses the armature and sealing cone back against the seat. By blocking fuel flow through the plunger chamber's inlet passage, the seal prevents delivery from the plunger. The solenoid can be designed as either pulling or pushing electromagnet.

The ELABs used on marine engines open when de-energized. This allows the vessel to continue in the event of electrical system failure. Minimizing the number of electrical consumers is a priority on ships, as the electromagnetic fields generated by continuous current flow promote salt-water corrosion.

Mechanical shutoff device

The mechanical shutoff devices installed on distributor injection pumps consist of lever assemblies (Fig. 2). This type of assembly consists of an inside and outside stop lever mounted on the governor cover (1, 5). The driver operates the outer stop lever from the

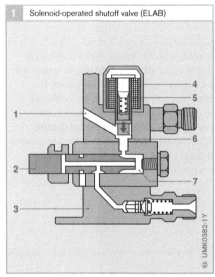

1 Solenoid-operated shutoff valve (ELAB)

Fig. 1
1   Inlet passage
2   Distributor plunger
3   Distributor head
4   Solenoid (pulling electromagnet in this instance)
5   Spring
6   Armature with sealing cone
7   Plunger chamber

2 Mechanical shutoff device

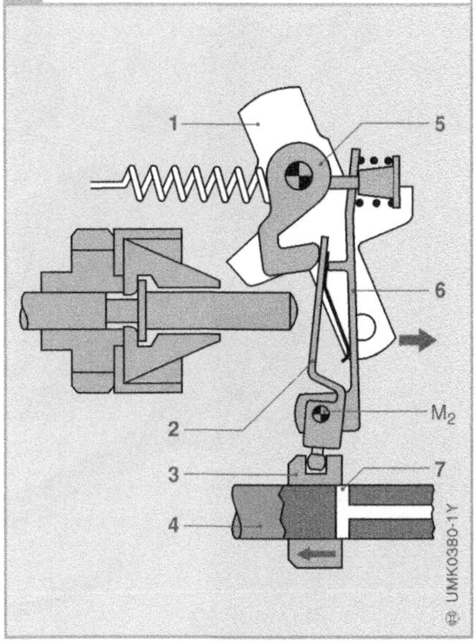

Fig. 2
1   External stop lever
2   Starting lever
3   Control collar
4   Distributor plunger
5   Inner stop lever
6   Tensioning lever
7   Control port

M_2 Pivot axis for 2 and 6

vehicle's interior via Bowden cable, etc. Cable operation causes the two stop levers to rotate about their pivot axis, with the inside stop lever pressing against the control mechanism's (2) starting lever. The starting lever pivots about its axis M_2 to push the control collar (3) into its stop position. This acts on the control passage (7) in the plunger to prevent fuel delivery.

Electric actuator mechanism
Systems with electronic diesel control employ the fuel-delivery actuator to switch off the engine (control parameter from electronic control unit (ECU): injected-fuel quantity = zero; refer to following section). The separate solenoid-operated shutoff valve (ELAB) serves only as a backup system in the event of a defect in the actuator mechanism.

Electronic Diesel Control

Systems with mechanical rotational-speed control respond to variations in operating conditions to reliably guarantee high-quality mixture formation.

EDC (Electronic Diesel Control) simultaneously satisfies an additional range of performance demands. Electric monitoring, flexible electronic data processing and closed-loop control circuits featuring electric actuators extend performance potential to include parameters where mechanical systems have no possibility of supply regulation. At the core of EDC is the electronic control unit that governs distributor injection pump operation.

As Electronic Diesel Control supports data communications with other electronic systems (traction control, electronic transmission-shift control, etc.) it can be integrated in an all-encompassing vehicle system environment.

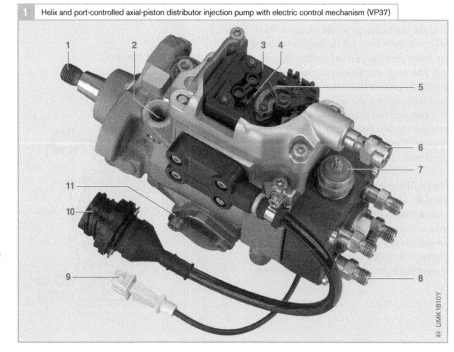

1   Helix and port-controlled axial-piston distributor injection pump with electric control mechanism (VP37)

Fig. 1
1 Pump drive
2 Fuel inlet
3 Fuel-injection pump delivery positioner
4 Fuel temperature sensor
5 Angle sensor
6 Fuel return
7 Type ELAB electric shutoff valve
8 Delivery valve (discharge to nozzle)
9 Connection for start of injection solenoid valve
10 Connection for fuel-injection pump delivery positioner
11 Timing device

UMK1810Y

Helix and port-controlled electronic distributor injection pumps use a fuel-delivery actuator mechanism to regulate injected-fuel quantities and a solenoid valve to control start of injection.

Solenoid actuator for fuel-delivery control

The solenoid actuator (rotary actuator, Fig. 2, 2) acts on the control collar via shaft. On mechanically controlled fuel-injection pumps, the position determines the point at which the cutoff bores are exposed.

Injected fuel quantity can be progressively varied continuously from zero to the system's maximum capacity (for instance, during cold starts). This valve is triggered by a Pulse-Width Modulated (PWM) signal. When current flow is interrupted, the rotary-actuator return springs reduce the fuel-delivery quantity to zero.

A sensor with a semidifferential short-circuit ring (1) relays the actuator-mechanism rotation angle, and thus the control collar position, back to the control unit [1]. This action determines the injected-fuel quantities for the instantaneous engine speed.

Solenoid valve for start of injection timing

As with mechanical timing devices, the timing plunger reacts to pressure in the pump's inner chamber, which rises proportionally to rotational speed. This force against the timing device's pressure-side is controlled by the timing-device solenoid valve (5). This valve is triggered by a PWM signal.

When the solenoid valve remains constantly open, pressure falls to retard start of injection, while closing the valve advances the timing. The electronic control unit can continuously vary the PWM signal's pulse-duty factor (ratio of open to closed periods at the valve) in the intermediate range.

[1] On 1st generation pumps, a potentiometer was employed to register rotation angle.

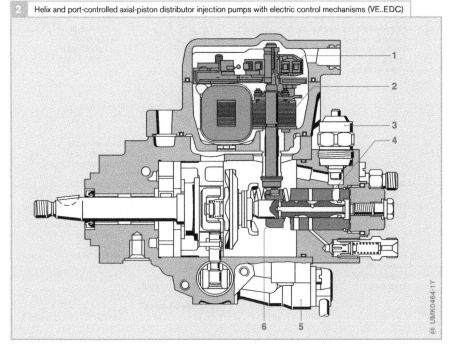

2   Helix and port-controlled axial-piston distributor injection pumps with electric control mechanisms (VE..EDC)

UMK0464-1Y

Fig. 2
1   Semi-differential short-circuiting ring sensor
2   Solenoid control mechanism
3   Electric shutoff valve ELAB
4   Distributor plunger
5   Timing-device solenoid valve
6   Control collar

Continuing efforts to improve the performance of diesel engines while simultaneously reducing harmful exhaust-gas emissions and fuel consumption mean that the number of sensors on and around the engine is constantly growing. The illustration below provides an overview of the parameters and variables that can be measured on the engine while it is running.

Some of this information is only collected and analyzed in the course of engine development or when it is being serviced (*). Of the remaining data, only a certain proportion is recorded when the engine is operational. The specific data that is required depends on the engine design, the fuel-injection system and the equipment fitted on the vehicle.

The parameters and variables are detected by sensors. The degree of accuracy and rate of detection required are determined by the type of application for which the engine is intended.

Measured variables on diesel engines

p **Pressures**
– Intake air,
– Air upstream/downstream of turbocharger,
– Recirculated exhaust upstream/downstream of cooler*,
– Exhaust upstream/downstream of turbocharger*,
– Exhaust upstream of catalytic converter,
– Exhaust downstream of catalytic converter*,
– Combustion chamber*,
– High-pressure fuel pipe*,
– Fuel supply,
– Fuel return*,
– Engine coolant*,
– Engine oil.

s **Travel**
– Needle stroke (for injection point),
– Governor settings,
– Injection timing adjuster setting,
– Valve positions.

t **Times**
– Injection period *,
– Delivery point,
– Delivery period.

U, I **Control signals**
– Injectors,
– Actuators,
– Valves (e.g. exhaust recirculation, wastegate),
– Flaps,
– Auxiliary systems.

Noise emission*

n **Speeds**
– Crankshaft
– Camshaft
– Turbocharger*,
– Auxiliary units*.

M **Torque***

Exhaust constituents
– Carbon dioxide (CO_2)*,
– Carbon monoxide (CO)*,
– Methane (CH_4)*,
– Nitrogen oxides (NO_x)*,
– Oxygen (O_2),
– Aldehydes*,
– Hydrocarbons (HC)*,
– Particulates (smoke index, soot concentration, exhaust opacity)*,
– Sulfur dioxide (SO_2)*.

T **Temperatures**
– Intake air,
– Air upstream/downstream of turbocharger,
– Recirculated exhaust upstream/downstream of cooler*,
– Exhaust upstream/downstream of turbocharger,
– Exhaust upstream/downstream of catalytic converter,
– Fuel supply,
– Fuel return,
– Engine coolant,
– Engine oil.

m **Masses**
– Intake air,
– Fuel,
– Recirculated exhaust*,
– Blow-By ([piston-ring] blow-by)*

a **Acceleration (vibration) of components***

UAE0754-1E

Diesel-engine immobilizers

Diesel immobilizers is a component in the electronic vehicle immobilizer installed on vehicles with a mechanically-controlled distributor injection pump. The diesel immobilizer is mounted above the solenoid-operated shutoff valve (ELAB) of the distributor injection pump. It responds to signals from the immobilizer control unit by switching current flow to the ELAB off and on to allow or block the flow of diesel fuel to the engine. The diesel immobilizer unit and ELAB are mounted behind a metal plate which must be destroyed for removal.

The diesel-engine immobilizer's input wiring usually comprises three terminals, consisting of power supply, data line and ground wire. The single exit wire carries the current to the ELAB.

Control units
The DDS 1.1. and DDS 3.1 control units are available for diesel-engine immobilizers.

DDS1.1
DDS 1.1 is a PCB-equipped ECU featuring an integrated protective housing made of plastic with injected sheet metal. On its own, DDS 1.1 provides only the lowest degree of protection. The security rating can be raised by adding various external protection features to the pump.

DDS3.1
DDS 3.1 is a hybrid control unit with a supplementary protective housing made of manganese steel. DDS 3.1 can provide optimum security without any supplements.

The DDS system must be released on the pump for operation on the test bench.

1 Helix and port-controlled axial-piston distributor injection pump with diesel antitheft protection

5 cm

SMK1B11Y

Fig. 1
1 DDS 1
    (via ELAB)
2 Plug connection

Solenoid-valve controlled distributor injection pumps

Ever stricter emission limits for diesel engines and the demand for further reductions in fuel consumption have resulted in the continuing refinement of electronically controlled distributor injection pump. High-pressure control using a solenoid valve permits greater flexibility in the variation of start and end of delivery and even greater accuracy in the metering of the injected-fuel quantity than with port-controlled fuel-injection pumps. In addition, it permits pre-injection and correction of injected-fuel quantity for each cylinder.

The essential differences from port-controlled distributor injection pumps are the following:
- A control unit mounted on the pump
- Control of fuel injection by a high-pressure solenoid valve and
- Timing of the high-pressure solenoid valve by an angle-of-rotation sensor integrated in the pump

As well as the traditional benefits of the distributor injection pump such as light weight and compact dimensions, these characteristics provide the following additional benefits:
- A high degree of fuel-metering accuracy within the program map
- Start of injection and injection duration can be varied independently of factors such as engine speed or pump delivery quantity
- Injected-fuel quantity can be corrected for each cylinder, even at high engine speeds
- A high dynamic volume capability
- Independence of the injection timing device range from the engine speed and
- The capability of pre-injection

Areas of application

Solenoid-valve controlled distributor injection pumps are used on small and medium-sized diesel engines in cars, commercial vehicles and agricultural tractors. They are fitted both on direct-injection (DI) and indirect-injection (IDI) engines.

The nominal speed, power output and design of the diesel engine determine the type and size of the fuel-injection pump. Distributor injection pumps are used on car, commercial-vehicle, agricultural tractor and fixed-installation engines with power outputs of up to 45 kW per cylinder. Depending on the type of control unit and solenoid valve, they can be run off either a 12-volt or a 24-volt electrical system.

Solenoid-valve controlled distributor injection pumps are available with high-pressure outlets for either four or six cylinders. The maximum injected-fuel capacity per stroke is in the range of $70...175$ mm^3. The required maximum injection pressures depend on the requirements of the engine (DI or IDI). They range from $800...1,950$ bar.

All distributor injection pumps are lubricated by the fuel. Consequently, they are maintenance-free.

Designs

There are two basic designs:
- Axial-piston distributor pumps (Type VE..MV or VP29/VP30) and
- Radial-piston distributor pumps (Type VR or VP44)

There are a number of design variations (e.g. number of outlets, pump drive by ring gear) according to the various types of application and engine.

The hydraulic performance capacity of the axial-piston pump with injection pressures up to 1,400 bar at the nozzle will remain sufficient for many direct-injection and indirect-injection engines in the future.

Where higher injection pressures are required for direct-injection engines, the radial-piston distributor pump introduced in 1996 is the more suitable choice. It has the capability of delivering injection pressures of up to 1,950 bar at the nozzle.

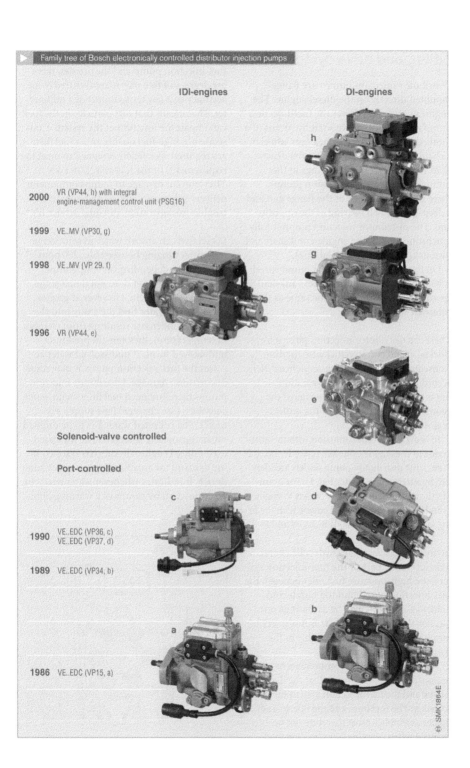

Family tree of Bosch electronically controlled distributor injection pumps

IDI-engines   DI-engines

h

2000   VR (VP44, h) with integral
engine-management control unit (PSG16)

1999   VE..MV (VP30, g)

1998   VE..MV (VP 29, f)

f   g

1996   VR (VP44, e)

e

Solenoid-valve controlled

Port-controlled

c   d

1990   VE..EDC (VP36, c)
VE..EDC (VP37, d)

1989   VE..EDC (VP34, b)

b

a

1986   VE..EDC (VP15, a)

SMK1864E

Fitting and drive system

Distributor injection pumps are flange-mounted directly on the diesel engine. The engine's crankshaft drives the fuel-injection pump's heavy-duty drive shaft by means of a toothed-belt drive, a pinion, a gear wheel or a chain. In axial-piston pumps, the drive shaft runs on two plain bearings in the pump housing. In radial-piston pumps, there is a plain bearing at the flange end and a deep-groove ball bearing at the opposite end. The distributor injection pump is fully synchronous with the engine crankshaft and pistons (positive mechanical link).

On four-stroke engines, the pump speed is half that of the diesel-engine crankshaft speed. In other words, it is the same as the camshaft speed.

There are distributor injection pumps for clockwise and counterclockwise rotation [1]. Consequently, the injection sequence differs according to the direction of rotation – but is always consecutive in terms of the geometrical arrangement of the outlets (e.g. A-B-C or C-B-A).

In order to prevent confusion with the numbering of the engine cylinders (cylinder no. 1, 2, 3, etc.), the distributor-pump outlets are identified by letters (A, B, C, etc., Fig. 1). For example, the assignment of pump outlets to engine cylinders on a four-cylinder engine with the firing sequence 1-3-4-2 is A-1, B-3, C-4 and D-2.

In order to achieve good hydraulic characteristics on the part of the fuel-injection system, the high-pressure fuel lines between the fuel-injection pump and the nozzle-and-holder assemblies must be as short as possible. For this reason, the distributor injection pump is fitted as close as possible to the diesel-engine cylinder head.

The interaction of all components has to be optimized in order to insure that the diesel engine and the fuel-injection system function correctly. The fuel lines and the nozzle-and-holder assemblies must therefore not be interchanged when carrying out servicing work.

In the high-pressure system between the fuel-injection pump and the nozzles, there are degrees of free play manufactured to an accuracy of a few thousandths of a millimeter. That means that dirt particles in the fuel can impair the function of the system. Consequently, high fuel quality and a fuel filter are required specifically designed to meet the requirements of the fuel-injection system. They prevent damage to pump components, delivery valves and nozzles and guarantee trouble-free operation and long service life.

Diesel fuel can absorb water in solution in quantities ranging between 50...200 ppm (by weight), depending on temperature. This dissolved water does not damage the fuel-injection system. However, if greater quantities of water find their way into the fuel (e.g. condensate resulting from changes in temperature), they remain present in undissolved form. If undissolved water accesses the fuel-injection pump, it may result in corrosion damage. Distributor injection pumps therefore need fuel filters with water chambers (see chapter "Fuel supply system"). The collected water has to be drained off at appropriate intervals. The increased use of diesel engines in cars has resulted in the demand for an automatic water warning device. It indicates whenever water needs to be drained off by means of a warning light.

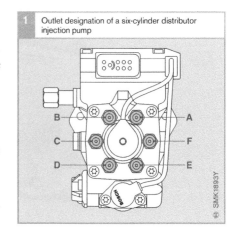

1   Outlet designation of a six-cylinder distributor injection pump

[1] Direction of rotation as viewed from drive-shaft end of the pump

Fig. 1
Outlets are always numbered counterclockwise. Numbering starts at top right.

The 24-hour race on Germany's famous "Nürburgring" race track is not just about speed but about the durability of automotive technology. On 14th June 1998 the race was won for the first time by a diesel-engined car. The BMW 320 d left its gasoline-engined rivals trailing.

The average speed achieved during practice was approximately 160 km/h. The maximum speed was around 250 km/h. The 1,040-kg vehicle accelerated from 0 to 100 km/h in 4.5 seconds.

The racing car was powered by a four-cylinder direct-injection diesel engine with a capacity of 1,950 cc and a maximum power output of more than 180 kW (245 bhp) at 4,200...4,600 rpm. The engine produced its maximum torque of 430 N·m at 2,500...3,500 rpm.

The car's performance was due in no small part to its high-performance fuel-injection system designed by Bosch. The central component of the system is the Type VP44 radial-piston distributor injection pump that has been in volume production since 1996.

Compared with normal production engines, the racing diesel is an "enhanced-performance" model achieved by larger injected-fuel quantities and even higher injection pressures and engine speeds. For this purpose, a new high-pressure stage with 4 delivery plungers instead of 2 was fitted in combination with new nozzles. The software for the control units was also modified.

These design modifications and the fact that the engines are run continuously at full power obviously reduce the life of the components (designed for 24...48 hours of service), and increase the concentration levels of the noxious constituents and fuel consumption. The car used around 23 l/100 km (a comparable gasoline racing engine uses almost twice as much).

That impressive victory demonstrates once again that the diesel engine is no longer the "lame duck" that it used to be.

▶ BMW 320 d: Winner of the 1998 24-hour race

SMM0607Y

Design and method of operation

Assemblies

Solenoid-valve controlled distributor injection pumps are modular in design. This section explains the interaction between the individual modular components. They are described in more detail later on. The modular components referred to are the following (Figs. 1 and 2):

- *The low-pressure stage* (7), consisting of the vane-type supply pump, pressure-control valve and overflow throttle valve
- *The high-pressure stage* (8)
- *The delivery valves* (11)
- *The high-pressure solenoid valve* (10)
- *The timing device* (9) with timing-device solenoid valve and angle-of-rotation sensor and
- *The pump ECU* (4)

The combination of these modular components in a compact unit allows the interaction of individual functional units to be very precisely coordinated. This allows compli-

ance with strict tolerances and fully meets demands regarding performance characteristics.

The high pressures and the associated physical stresses require a wide variety of fine-tuning adjustments in the design of the components. Lubrication is carried out by the fuel itself. In some cases, lubricant-film thickness is less than 0.1 µm and is much less than the minimum achievable surface roughness. For this reason, the use of special materials and manufacturing methods is necessary in addition to special design features.

Fuel supply and delivery

Low-pressure stage

The vane-type supply pump in the low-pressure stage (7) pumps fuel from the fuel tank and produces a pressure of 8...22 bar inside the fuel-injection pump depending on pump type and speed.

The vane-type supply pump delivers more fuel than is required for fuel injection. The excess fuel flows back to the fuel tank.

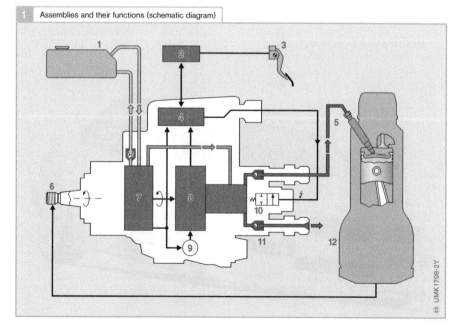

1 Assemblies and their functions (schematic diagram)

High-pressure stage

The high-pressure pump (8) generates the high pressure required for fuel injection and controls the injected-fuel quantity as required. At the same time the fuel rail opens the outlet for the appropriate engine cylinder so that the fuel is delivered via the delivery valve (11) to the nozzle-and-holder assembly (5). The assembly then injects the fuel into the combustion chamber (12) of the engine. It is in the design of the high-pressure pump and fuel distributor that the axial-piston and radial-piston distributor pumps differ the most.

The high-pressure pump is able to deliver fuel as long the high-pressure solenoid valve (10) keeps the pump's plunger chamber sealed. In other words, it is the high-pressure solenoid valve that determines the delivery period and therefore, in conjunction with pump speed, the injected-fuel quantity and injection duration.

The injection pressure increases during the period of fuel injection. The maximum pressure depends on the pump speed and the injection duration.

Timing

As with port-controlled distributor injection pumps, the timing device (9) varies the start of injection. It alters the cam position in the high-pressure pump.

An injection-timing solenoid valve controls the supply-pump pressure acting on the spring-loaded timing-device piston. This controls the start of injection independently of engine speed.

The integrated angle-of-rotation sensor detect and, as necessary, correct the pump speed and, together with the speed sensor on the engine crankshaft, the position of the timing device.

Electronic control unit

The pump ECU (4) calculates the triggering signals for the high-pressure solenoid valve and the timing-device solenoid valve on the basis of a stored map.

2 Assemblies and their functions (schematic section diagram of a radial-piston distributor injection pump)

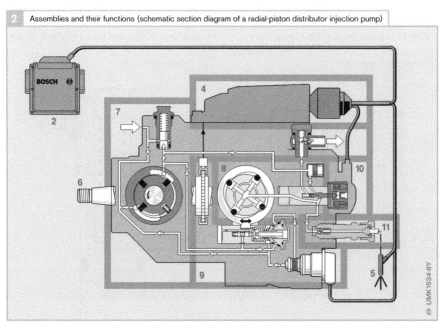

Fig. 2

For the sake of clarity, various components are shown end-on rather than side-on. The index figures are the same as for Fig. 1.

2 Engine ECU
4 Pump ECU
5 Nozzle-holder assembly
6 Pump drive shaft
7 Low-pressure stage (vane-type supply pump with pressure-control valve and overflow throttle valve)
8 High-pressure pump with fuel rail
9 Timing device with timing-device solenoid valve and angle-of-rotation sensor
10 High-pressure solenoid valve
11 Delivery valve

Low-pressure stage

The low-pressure stage delivers sufficient fuel for the high-pressure stage and generates the pressure for the high-pressure pump and the timing device (8...25 bar depending on pump type). Its basic components include the vane-type supply pump, the pressure-control valve and the overflow valve.

Vane-type supply pump

The purpose of the vane-type supply pump (Fig. 1) is to draw in a sufficient quantity of fuel and to generate the required internal pressure.

The vane-type supply pump is positioned around the drive shaft (4) in the distributor injection pump. Between the inner surface of the pump housing and a support ring acting as the end plate is the retaining ring (2) which forms the inner surface of the vane-pump stator. On the inner surface of the pump housing, there are two machined recesses which form the pump inlet (5) and outlet (6). Inside the retaining ring is the impeller (3) which is driven by an interlocking gear (Type VP44) or a Woodruff key (VP29/30) on the drive shaft. Guide slots

in the impeller hold the vanes (8) which are forced outwards against the inside of the retaining ring by centrifugal force. Due to the higher delivery pressures involved in radial-piston distributor injection pumps, the vanes have integrated springs which also help to force the vanes outwards. The compression-chamber "cells" (7) are formed by the following components:

● The inner surface of the pump housing ("base")
● The support ring ("cap")
● The shaped inner surface of the retaining ring
● The outer surface of the impeller and
● Two adjacent vanes

The fuel that enters through the inlet passage in the pump housing and the internal passages in the compression-chamber cell is conveyed by the rotation of the impeller to the compression-chamber outlet. Due to the eccentricity (VP29/30) or profile (VP44) of the inner surface of the retaining ring, the cell volume reduces as the impeller rotates. This reduction in volume causes the fuel pressure to rise sharply until the fuel escapes through the compression-chamber outlet – in other words, the fuel is compressed. From

Fig. 1
1   Pressure-control valve
2   Eccentric retaining ring
3   Impeller
4   Pump drive shaft
5   Fuel inlet
6   Fuel outlet to pump intake chamber
7   Compression-chamber cell
8   Vane

Fig. 2
1   Valve body
2   Compression spring
3   Valve plunger
4   Outlet to pump intake
5   Inlet from pump outlet
6   Bore

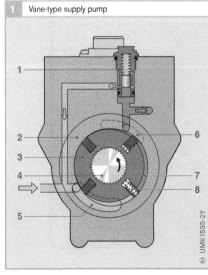

1   Vane-type supply pump

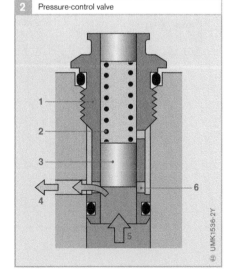

2   Pressure-control valve

the compression-chamber outlet, the various components are supplied with pressurized fuel via internal passages in the pump housing.

The pressure level required in a radial-piston distributor injection pump is relatively high compared with other types of distributor injection pump. Due to high pressure, the vanes have a bore in the center of the end face so that only one of the end-face edges is in contact with the inner surface of the retaining ring at one time. This prevents the entire end face of the vane from being subjected to pressure, which would result in an undesirable radial movement. At the point of changeover from one edge to the other (e.g. when changing over from inlet to outlet), the pressure acting on the end face of the vane can transfer to the other side of the vane through the bore. The opposing pressures balance each other out to a large extent and the vanes are pressed against the inner surface of the retaining ring by centrifugal force and the action of the springs as described above.

Pressure-control valve

The fuel pressure created at the pressure outlet of the vane-type supply pump depends on the pump speed. To prevent the pressure from reaching undesirably high levels at high pump speeds, there is a pressure-control valve in the immediate vicinity of the vane-type supply pump which is connected to the pressure outlet by a bore (Fig. 1, 1). This spring-loaded slide valve varies the delivery pressure of the vane-type supply pump according to the fuel quantity delivered. If the fuel pressure rises beyond a certain level, the valve plunger (Fig. 2, 3) opens radially positioned bores (6) through which the fuel can flow via an outlet (4) back to the intake port of the vane-type supply pump. If the fuel pressure is too low, the pressure-control valve remains closed and the entire fuel quantity is pumped into the distributor-pump intake chamber. The adjustable tension on the compression spring (2) determines the valve opening pressure.

Overflow valve

In order to vent and, in particular, cool the distributor injection pump, excess fuel flows back to the fuel tank through the overflow valve (Fig. 3) screwed to the pump housing.

The overflow valve is connected to the overflow valve (4). Inside the valve body (1), there is a spring-loaded ball valve (3) which allows fuel to escape when the pressure exceeds a preset opening pressure.

In the overflow channel to the ball valve, there is a bore that is connected to the pump overflow via a very small throttle bore (5). Since the overflow valve is mounted on top of the pump housing, the throttle bore facilitates automatic venting of the fuel-injection pump.

The entire low-pressure stage of the fuel-injection pump is precisely coordinated to allow a defined quantity of fuel to escape through the overflow valve and return to the fuel tank.

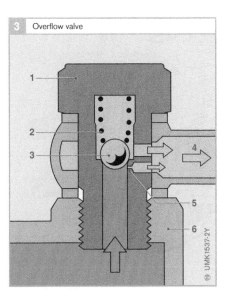

3 Overflow valve

UMK1537-2Y

Fig. 3
1 Valve body
2 Compression spring
3 Valve ball
4 Fuel overflow
5 Throttle bore
6 Pump housing

High-pressure stage of the axial-piston distributor injection pump

Solenoid-valve controlled (Fig. 1) and port-controlled distributor injection pumps have essentially the same dimensions, fitting requirements and drive system including cam drive.

Design and method of operation

A clutch unit transmits the rotation of the drive shaft (Fig. 2 overleaf, 1) to the cam plate (5). The claws on the drive shaft and the cam plate engage in the yoke (3) positioned between them.

The cam plate converts the purely rotational movement of the drive shaft into a combined rotating-reciprocating movement. This is achieved by the fact that the cams on the cam plate rotate over rollers held in the

roller ring (2). The roller ring is mounted inside the pump housing but has no connection to the drive shaft.

The cam plate is rigidly connected to the distributor plunger (8). Consequently, the distributor plunger also performs the rotating-reciprocating movement described by the cam plate. The reciprocating movement of the distributor plunger is aligned axially with the drive shaft (hence the name axial-piston pump).

The movement of the plunger back to the roller ring takes place the symmetrically arranged plunger return springs (7). They are braced against the distributor body (9) and act against the distributor plunger by means of a thrust plate (6). The piston return springs also prevent the cam plate from jumping away from the rollers in the roller ring when subjected to high acceleration forces.

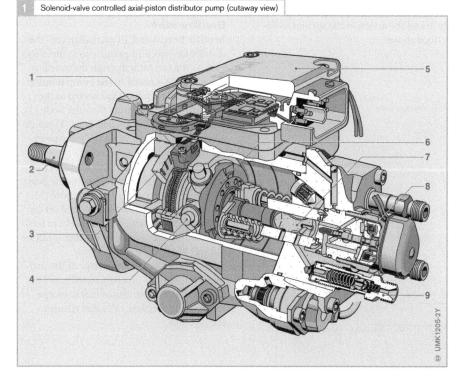

1 Solenoid-valve controlled axial-piston distributor pump (cutaway view)

Fig. 1
1   Angle-of-rotation
    sensor
2   Pump drive shaft
3   Support ring of
    vane-type supply
    pump
4   Roller ring
5   Pump control unit
6   Cam plate
7   Distributor plunger
8   High-pressure
    solenoid valve
9   High-pressure outlet

UMK1205-2Y

The lengths of the plunger return springs are precisely matched to one another in order to prevent lateral forces from acting on the distributor plunger.

Although the distributor plunger moves horizontally, the limits of its travel are still referred to as Top Dead Center (TDC) and Bottom Dead Center (BDC). The length of the plunger stroke between bottom and top dead center is application-specific. It can be up to 3.5 mm.

The number of cams and rollers is determined by the number of cylinders in the engine. The cam shape affects injection pressure (injection pattern and maximum injection pressure) and the maximum possible injection duration. The factors determining this connection are cam lift and the speed of movement.

When designing the fuel-injection pump, the injection parameters must be individually adapted to suit the design of the combustion chamber and the nature of the combustion process (DI or IDI) employed by the engine on which the pump is to be used. For this reason, a specific cam profile is calculated for each type of engine and is then machined on the end face of the cam plate. The cam plate produced in this way is an application-specific component of the distributor injection pump. Cam plates are not interchangeable between different types of pump.

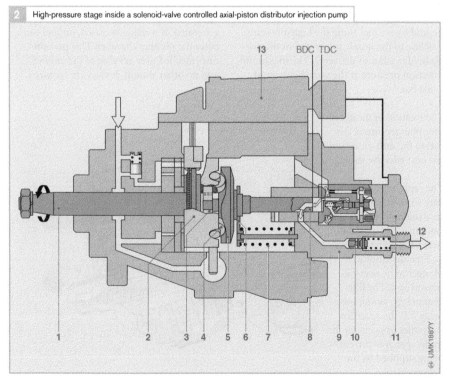

2   High-pressure stage inside a solenoid-valve controlled axial-piston distributor injection pump

Fig. 2
1 Drive shaft
2 Roller ring
3 Yoke
4 Roller
5 Cam plate
6 Spring plate
7 Piston return spring
   (only one shown)
8 Distributor plunger
9 Distributor body
   (also called
   distributor head
   or distributor-head
   flange)
10 Delivery valve
11 High-pressure
   solenoid valve
12 Outlet to high-
   pressure delivery
   line
13 Pump ECU

TDC Distributor-plunger
    Top Dead Center
BDC Distributor-plunger
    Bottom Dead
    Center

Delivery phases

Induction (Fig. 4a)

As the distributor plunger (4) moves towards Bottom Dead Center (BDC), it draws fuel into the plunger chamber (6) through the fuel inlet (1) in the distributor body (3) and the open high-pressure solenoid valve (7).

Effective stroke (Fig. 4b)

At bottom dead center, before the cam lobes are in contact with the rollers, the pump ECU sends a control signal to the high-pressure solenoid valve. The valve needle (9) is pressed against the valve seat (7). The high-pressure solenoid valve is then closed.

When the distributor plunger then starts to move towards Top Dead Center (TDC), the fuel cannot escape. It passes through channels and passages in the distributor plunger to the high-pressure outlet (10) for the appropriate cylinder. The rapidly developed high pressure opens the orifice check valve (DI) or delivery valve (IDI), as the case may be, and forces fuel along the high-pressure fuel line to the nozzle integrated in the nozzle holder (start of delivery). The maximum injection pressure at the nozzle is around 1,400 bar.

The rotation of the distributor plunger directs the fuel to the next outlet on the next effective stroke.

The rapid release of pressure when the high-pressure solenoid valve opens can cause the space between the delivery valve and the distributor plunger to be over-depressurized. The filler channel (5) simultaneously fills the space for the outlet opposite to the outlet which is currently being supplied by the delivery stroke.

Residual stroke (Fig. 4c).

Once the desired injected-fuel quantity has been delivered, the ECU cuts off the power supply to the solenoid coil. The high-pressure solenoid valve opens again and pressure in the high-pressure stage collapses (end of delivery). As the pressure drops, the nozzle and the valve in the high-pressure outlet close again and the injection sequence comes to an end. The point of closure and the open duration of the high-pressure solenoid valve, the cam pitch during the delivery stroke and the pump speed determine the injected-fuel quantity.

The remaining travel of the pump plunger to top dead center forces the fuel out of the plunger chamber and back into the pump intake chamber.

As there are no other inlets, failure of the high-pressure solenoid valve prevents fuel injection altogether. If the valve remains open, the required high pressure cannot be generated. If it remains closed, no fuel can enter the plunger chamber. This prevents uncontrolled over-revving of the engine and no other shutoff devices are required.

Fig. 3

The index figures are the same as for Fig. 4.

2 Filter
4 Distributor plunger
6 Plunger chamber
7 Valve seat
9 Solenoid-valve needle
10 Outlet to high-pressure delivery line

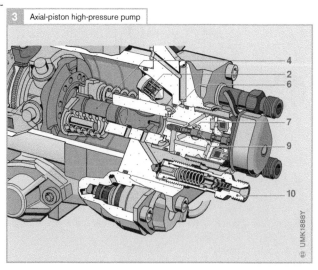

3   Axial-piston high-pressure pump

UMK1888Y

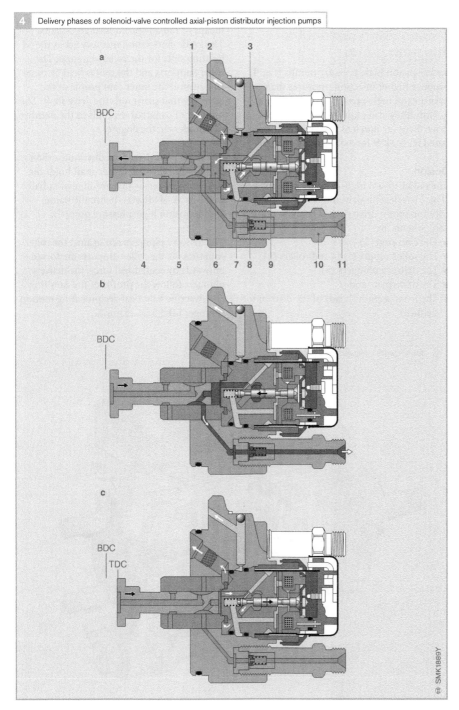

4 Delivery phases of solenoid-valve controlled axial-piston distributor injection pumps

Fig. 4
a Induction
b Effective stroke
c Residual stroke

1 Inlet passage
   (fuel inlet)
2 Filter
3 Distributor body
4 Distributor plunger
5 Filler groove
6 Plunger chamber
7 Valve seat
8 Delivery valve
9 Solenoid-valve
   needle
10 High-pressure outlet
11 Outlet to high-
   pressure delivery
   line

TDC Pump-plunger
    Top **D**ead **C**enter
BDC Pump-plunger
    Bottom **D**ead
    Center

High-pressure stage of the radial-piston distributor injection pump

Radial-piston high-pressure pumps (Fig. 1) produce higher injection pressures than axial-piston high-pressure pumps. Consequently, they also require more power to drive them (as much as 3.5...4.5 kW compared with 3 kW for axial-piston pumps).

Design

The radial-piston high-pressure pump (Fig. 2 overleaf) is driven directly by the distributor-pump drive shaft. The main pump components are
- The cam ring (1)
- The roller supports (4) and rollers (2)
- The delivery plungers (5)
- The drive plate and
- The front section (head) of the distributor shaft (6)

The drive shaft drives the drive plate by means of radially positioned guide slots. The guide slots simultaneously act as the locating slots for the roller supports. The roller supports and the rollers held by them run around the inner cam profile of the cam ring that surrounds the drive shaft. The number of cams corresponds to the number of cylinders in the engine.

The drive plate drives the distributor shaft. The head of the distributor shaft holds the delivery plungers which are aligned radially to the drive-shaft axis (hence the name "radial-piston high-pressure pump").

The delivery plungers rest against the roller supports. As the roller supports are forced outwards by centrifugal force, the delivery plungers follow the profile of the cam ring and describe a cyclical-reciprocating motion (plunger lift 3.5...4.15 mm).

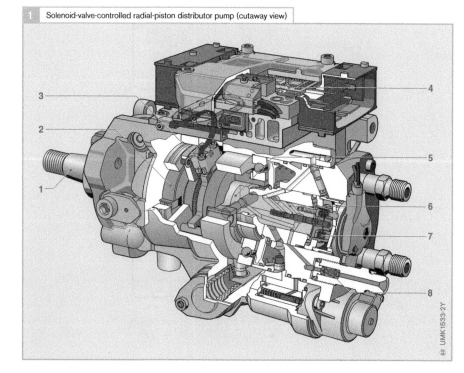

1   Solenoid-valve-controlled radial-piston distributor pump (cutaway view)

Fig. 1
1 Pump drive shaft
2 Vane-type supply pump
3 Angle-of-rotation sensor
4 Pump ECU
5 Radial-piston high-pressure pump
6 Distributor shaft
7 High-pressure solenoid valve
8 Delivery valve

UMK1533-2Y

When the delivery plungers are pushed inwards by the cams, the volume in the central
plunger chamber between the delivery
plungers is reduced. This compresses and
pumps the fuel. Pressures of up to 1,200 bar
are achievable at the pump.

Through passages in the distributor shaft,
the fuel is directed at defined times to the
appropriate outlet delivery valves (Fig. 1, 8
and Fig. 3, 5).

There may be 2, 3 or 4 delivery plungers depending on the number of cylinders in the
engine and the type of application (Fig. 2).
Sharing the delivery work between at least
two plungers reduces the forces acting on
the mechanical components and permits the
use of steep cam profiles with good delivery
rates. As a result, the radial-piston pump
achieves a high level of hydraulic efficiency.

The direct transmission of force within
the cam-ring drive gear minimizes the
amount of "give", which also improves the
hydraulic performance of the pump.

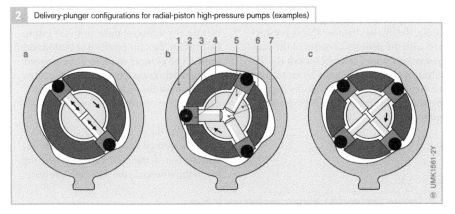

2   Delivery-plunger configurations for radial-piston high-pressure pumps (examples)

a
b
1 2 3 4   5   6 7
c

UMK1561-2Y

Fig. 2
a   For 4- or 6-cylinder
     engines
b   For 6-cylinder
     engines
c   For 4-cylinder
     engines

1   Cam ring
2   Roller
3   Guide slot in drive
     shaft
4   Roller support
5   Delivery plunger
6   Distributor shaft
7   Plunger chamber

3   High-pressure stage within a solenoid-valve controlled radial-piston distributor injection pump

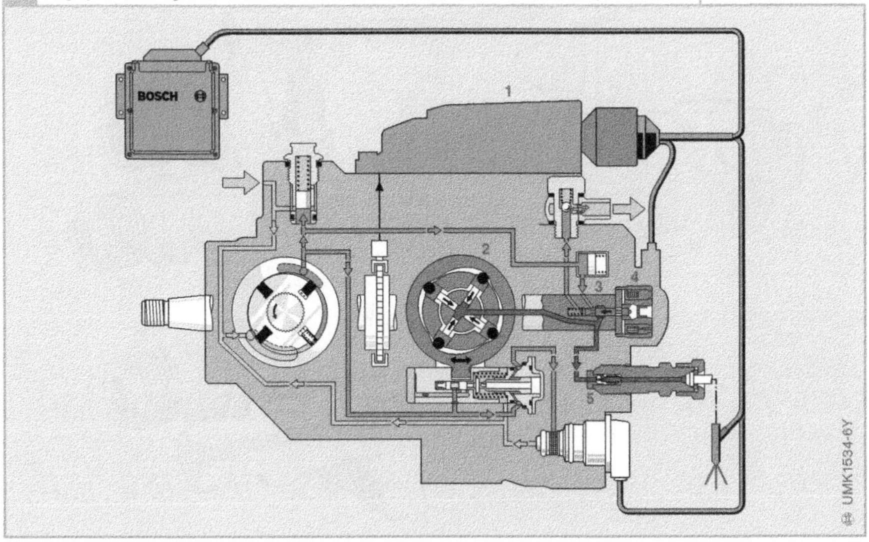

BOSCH

1

2

3

4

5

UMK1534-6Y

Fig. 3
For the sake of clarity,
various components are
shown end-on rather
than side-on.

1   Pump ECU
2   Radial-piston high-
     pressure pump
     (end-on view)
3   Distributor shaft
4   High-pressure
     solenoid valve
5   Delivery valve

Distributor-body assembly

The distributor-body assembly (Fig. 4) consists of the following:

- The distributor body (2)
- The control sleeve (5) which is shrink-fitted in the distributor body
- The rear section of the distributor shaft (4) which runs in the control sleeve
- The valve needle (6) of the high-pressure solenoid valve
- The accumulator diaphragm (1) and
- The delivery valve (7) with orifice check valve

In contrast with the axial-piston distributor pump, the intake chamber that is pressurized with the delivery pressure of the vane-type supply pump consists only of the diaphragm chamber enclosed by an accumulator diaphragm (1). This produces higher pressures for supplying the high-pressure pump.

Delivery phases (method of operation)

Induction

During the induction phase (Fig. 5a), the delivery plungers (1) are forced outwards by the supply-pump pressure and centrifugal force. The high-pressure solenoid valve is open. Fuel flows from the diaphragm chamber (12) past the solenoid-valve needle (4) and through the low-pressure inlet (13) and the annular groove (10) to the plunger chamber (8). Excess fuel escapes via the return passage (5).

Effective stroke

The high-pressure solenoid valve (Fig. 5b, 7) is closed by a control pulse from the pump ECU when the cam profile is at bottom dead center. The plunger chamber is now sealed and fuel delivery starts as soon as the cams start to move the pistons inwards (start of delivery).

Residual stroke

Once the desired injected-fuel quantity has been delivered, the ECU cuts off power supply to the solenoid coil. The high-pressure

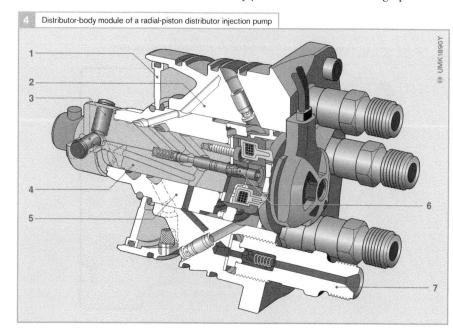

4   Distributor-body module of a radial-piston distributor injection pump

UMK1890Y

1
2
3
4
5
6
7

Fig. 4
1   Accumulator
    diaphragm
2   Distributor body
3   Delivery plunger
4   Distributor shaft
5   Control sleeve
6   Valve needle
7   Delivery valve

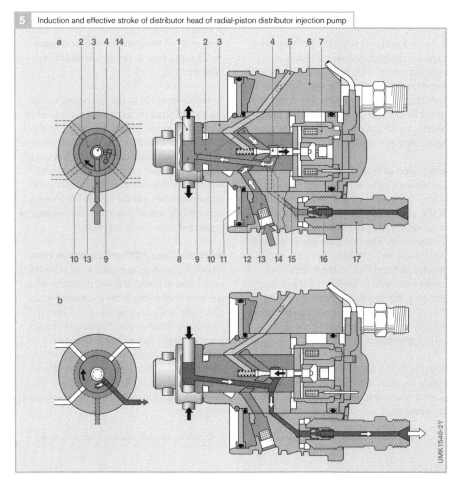

5 Induction and effective stroke of distributor head of radial-piston distributor injection pump

Fig. 5
For the sake of clarity, the pump plungers are shown end-on rather than side-on.

a Induction
b Effective stroke

1 Delivery plunger
2 Distributor shaft
3 Control sleeve
4 Valve needle
5 Fuel return
6 Distributor body
7 Solenoid coil
8 Plunger chamber
9 High-pressure fuel
10 Annular groove
11 Accumulator diaphragm
12 Diaphragm chamber
13 Low-pressure inlet
14 Distributor slot
15 High-pressure outlet
16 Orifice check valve
17 Delivery-valve body

solenoid valve opens again and pressure in the high-pressure stage collapses (end of delivery). The pressure drop closes the nozzle and the delivery valve again and the injection sequence comes to an end.

The excess fuel that is delivered by the pump while the pistons continue to move toward the cam top dead center is diverted back to the diaphragm chamber (12). The high pressure peaks that are thus produced in the low-pressure stage are damped by the accumulator diaphragm (11). In addition, the fuel stored in the diaphragm chamber helps to fill the plunger chamber for the next injection cycle.

Fuel metering takes place between the start of the cam lift and opening of the high-pressure solenoid valve. This phase is referred to as the delivery period. It determines the injected-fuel quantity in conjunction with the pump speed.

The high-pressure solenoid valve can completely shut off high-pressure fuel delivery in order to stop the engine. For this reason, an additional shutoff valve as used with port-controlled distributor injection pumps is not necessary.

Delivery valves

Between injection cycles, the delivery valve shuts off the high-pressure delivery line from the pump. This isolates the high-pressure delivery line from the outlet port in the distributor head. The residual pressure retained in the high-pressure delivery line ensures rapid and precise opening of the nozzle during the next injection cycle.

Integrated orifice check valve

The delivery valve with integral Type RSD orifice check valve (Fig. 1) is a piston valve. At the start of the delivery sequence, the fuel pressure lifts the valve cone (3) away from the valve seat. The fuel then passes through the delivery-valve holder (5) to the high-pressure delivery line to the nozzle-and-holder assembly. At the end of the delivery sequence, the fuel pressure drops abruptly. The valve spring (4) and the pressure in the delivery line forces the valve cone back against the valve seat (1).

With the high pressures used for direct-injection engines, reflected pressure waves can occur at the end of the delivery lines. They can cause the nozzle to reopen, resulting in undesired dribble which adversely affects pollutant levels in the exhaust gas. In addition, areas of low pressure can be created and this can lead to cavitation and component damage.

A throttle bore (2) in the valve cone dampens the reflected pressure waves to a level at which they are no longer harmful. The throttle bore is designed so that the static pressure in the high-pressure delivery line is retained between injection cycles. Since the throttle bore means that the space between the fuel-injection pump and the nozzle is no longer hermetically sealed, this type of arrangement is referred to as an open system.

Separate Type RDV orifice check valve

On axial-piston pumps and some types of radial-piston pump, a delivery valve with a separate orifice check valve is used (Fig. 2). This valve is also referred to as a Type GDV constant-pressure valve. It creates dynamic pressure in the delivery line. During the injection sequence, the valve plate (5) opens so that the throttle bore (6) has no effect. When fuel flows in the opposite direction, the valve plate closes and the throttle comes into action.

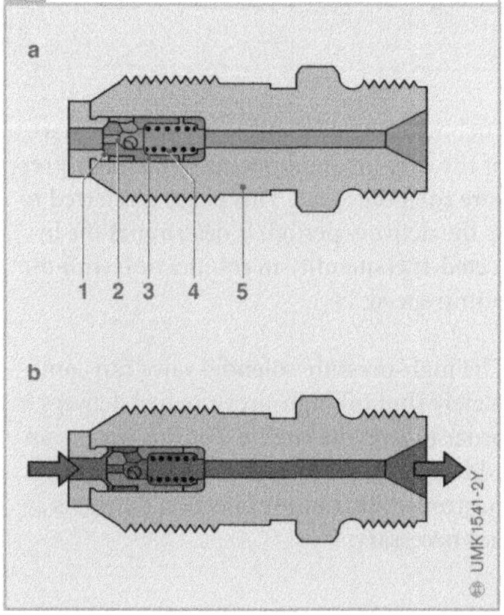

Fig. 1
a Valve closed
b Valve open

1 Valve seat
2 Throttle bore
3 Valve cone
4 Valve spring
5 Delivery-valve holder

Fig. 2
1 Valve holder
2 Pressure-valve stem
3 Retraction piston
4 Valve spring
  (delivery valve)
5 Valve plate
6 Throttle bore
7 Valve spring
  (valve plate)
8 Delivery-valve holder

1 Delivery valve with integrated orifice check valve

a

b

1 2 3 4 5

UMK1541-2Y

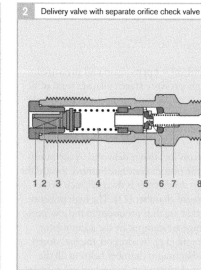

2 Delivery valve with separate orifice check valve

1 2 3 4 5 6 7 8

UMK1183-3Y

Micromechanics

Micromechanics is defined as the application of semiconductor techniques in the production of mechanical components from semiconductor materials (usually silicon). Not only silicon's semiconductor properties are used but also its mechanical characteristics. This enables sensor functions to be implemented in the smallest-possible space. The following techniques are used:

Bulk micromechanics

The silicon wafer material is processed at the required depth using anisotropic (alkaline) etching and, where needed, an electrochemical etching stop. From the rear, the material is removed from inside the silicon layer (Fig. 1, 2) at those points underneath an opening in the mask. Using this method, very small diaphragms can be produced (with typical thicknesses of between 5 and 50 µm, as well as openings (b), beams and webs (c) as are needed for instance for acceleration sensors.

Surface micromechanics

The substrate material here is a silicon wafer on whose surface very small mechanical structures are formed (Fig. 2). First of all, a "sacrificial layer" is applied and structured using semiconductor processes such as etching (a). An approx. 10 µm polysilicon layer is then deposited on top of this and structured vertically using a mask and etching. In the final processing step, the "sacrificial" oxide layer underneath the polysilicon layer is removed by means of gaseous hydrogen fluoride. In this manner, the movable electrodes for acceleration sensors (Fig. 3) are exposed.

Wafer bonding

Anodic bonding and sealglass bonding are used to permanently join together (bonding) two wafers by the application of tension and heat or pressure and heat. This is needed for the hermetic sealing of reference vacuums for instance, and when protective caps must be applied to safeguard sensitive structures.

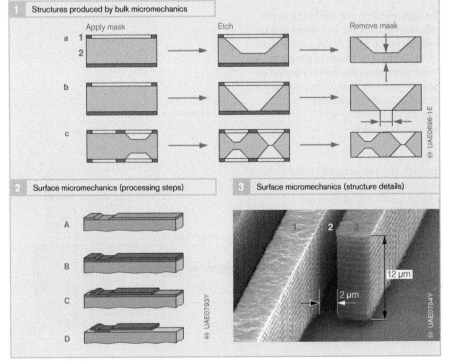

1 Structures produced by bulk micromechanics

Apply mask   Etch   Remove mask

a 1
   2

b

c

UAE0698-1E

2 Surface micromechanics (processing steps)

A

B

C

D

UAE0793Y

3 Surface micromechanics (structure details)

12 µm

2 µm

UAE0794Y

Fig. 1
a Diaphragms
b Openings
c Beams and webs

1 Etching mask
2 Silicon

Fig. 2
A Cutting and structuring the sacrificial layer
B Cutting the polysilicon
C Structuring the polysilicon
D Removing the sacrificial layer

Fig. 3
1 Fixed electrode
2 Gap
3 Spring electrodes

High-pressure solenoid valve

Design and operating concept

The high-pressure solenoid valve (Fig. 1) is a 2/2-way valve, i.e. it has two hydraulic ports and two possible settings. The valve is built into the distributor-body assembly.

The main components are
- The valve body consisting of the housing (4), and attachments
- The valve needle (3) and solenoid armature (6)
- The solenoid plate and
- The electromagnet (7) with its electrical connection to the pump ECU (8)

The valve needle (3) protrudes into the distributor shaft (5) and rotates in synchronization with it. Arranging the solenoid coil (7) concentrically with the valve needle creates a compact combination of high-pressure solenoid valve and distributor body.

The high-pressure solenoid valve opens and closes in response to control signals from the pump ECU (valve-needle stroke 0.3...0.4 mm). The length of time it remains closed determines the delivery period of the high-pressure pump. This means that the fuel quantity can be very precisely metered for each individual cylinder.

Requirements

The high-pressure solenoid valve has to meet the following criteria:
- Large valve cross-section for complete filling the plunger chamber, even at high speeds
- Light weight (small mass movements) in order to minimize component stress
- Rapid switching times for precise fuel metering and
- Generation of high magnetic forces in keeping with the loads encountered at high pressures

The high-pressure solenoid valve is controlled by regulating the current. Steep current-signal edges must be used to achieve a high level of reproducibility on the part of the injected-fuel quantity. In addition, control must be designed in such a way as to minimize power loss in the ECU and the solenoid valve. This is achieved by keeping the control currents as low as possible, for example. Therefore, the solenoid valve reduces the current to the holding current (approx. 10 A) after the pickup current (approx. 18 A) has been applied.

The pump control unit can detect when the solenoid-valve needle meets the valve seat by means of the current pattern (BIP signal; Beginning of Injection Period). This allows the exact point at which fuel delivery starts to be calculated and the start of injection to be controlled very precisely.

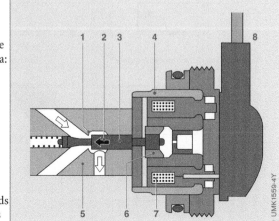

1   High-pressure solenoid valve

Fig. 1

1   Valve seat
2   Closing direction
3   Valve needle
4   Housing
5   Distributor shaft
6   Solenoid armature
7   Solenoid coil (electromagnet)
8   Electrical connection

Solenoid-valve actuation

The high-pressure fuel-quantity solenoid valve is triggered via current control (Fig. 2) which subdivides the triggering process into a pickup-current phase (a) of approx. 18 A, and a holding-current phase (c) of approx. 10 A. At the beginning of the controlled holding-current phase (after 200...250 μs), the BIP evaluation circuit detects the solenoid-valve needle closing against the valve seat (BIP = Beginning of the Injection Period).

The latest generation uses the PSG16 pump ECU, and is also provided with a BIP current (b) between the pickup and holding-current phases which is at the optimum level for the BIP detection function.

In order that the pump's injection characteristics are always reproducible irrespective of operating conditions, the triggering circuitry as a whole, and the current control, must be extremely accurate. Furthermore, they must keep the power loss in ECU and solenoid valve down to a minimum.

Defined, rapid opening of the solenoid valve is required at the end of the injection process. To this end, high-speed quenching (d) using a high quenching voltage (1) is applied at the valve to dissipate the energy stored in its solenoid.

The solenoid valve can also be used to control Pilot Injection (PI) for reduction of combustion noise. Here, the solenoid valve is operated ballistically between the PI point and the MI (Main Injection) point. In other words it is only partially opened, which means that it can be closed again very quickly. The resulting injection spacing is very short so that even at high rotational speeds, adequate cam pitch remains for the main injection process.

The subdivision into the individual triggering phases is calculated by the microcontroller in the pump ECU.

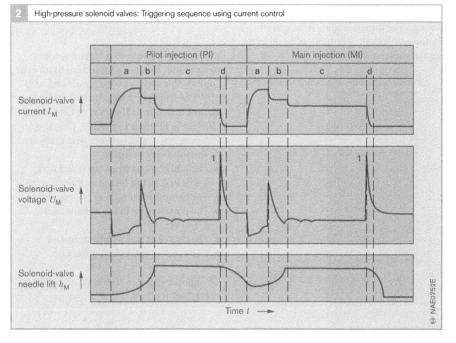

2 High-pressure solenoid valves: Triggering sequence using current control

Pilot injection (PI)   Main injection (MI)

a | b | c | d | a | b | c | d

Solenoid-valve current I_M

Solenoid-valve voltage U_M

Solenoid-valve needle lift h_M

Time t ⟶

NAE0752E

Fig. 2
a Pickup-current phase
b BIP detection,
c Holding-current phase
d High-speed quenching

1 Quenching voltage

Injection timing adjustment

Purpose

The point at which combustion commences relative to the position of the piston/crankshaft has a decisive impact on engine performance, exhaust emission levels and noise output. If the start of delivery remains constant without adjusting the injection timing as engine speed increases, the amount of crankshaft rotation between the start of delivery and the start of combustion would increase to such an extent that combustion would no longer take place at the correct time.

As with port-controlled distributor injection pumps, the timing device rotates the roller ring/cam ring so that the start of delivery occurs earlier or later relative to the position of the engine crankshaft. The interaction between the high-pressure solenoid valve and the timing device thus varies the start of injection and the injection pattern to suit the operating status of the engine.

Explanation of terms

For a proper understanding of injection timing adjustment, a number of basic terms require explanation.

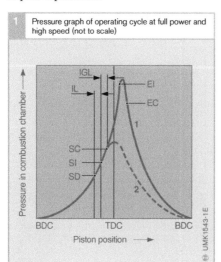

1 Pressure graph of operating cycle at full power and high speed (not to scale)

Fig. 1
1 Combustion
pressure
2 Compression
pressure

SD Start of delivery
TDC Engine-piston
Top Dead Center
SI Start of injection
EI End of injection
IL Injection lag
BDC Engine-piston
Bottom Dead
Center
SC Start of combustion
EC End of combustion
IGL Ignition lag

Injection lag

The Start of Delivery (SD, Fig. 1) occurs after the point at which the high-pressure solenoid valve closes. High pressure is generated inside the fuel line. The point at which that pressure reaches the nozzle opening pressure and opens the nozzle is the Start of Injection (SI). The time between the start of delivery and the start of injection is called the Injection Lag (IL).

The injection lag is largely independent of pump/engine speed. It is essentially determined by the propagation of the pressure wave along the high-pressure delivery line. The propagation time of the pressure wave is determined by the length of the delivery line and the speed of sound. In diesel fuel, the speed of sound is approx. 1,500 m/s.

If the engine speed increases, the amount of degrees of crankshaft rotation during the injection lag also increases. As a consequence, the nozzle opens later (relative to the position of the engine piston). This is undesirable. For this reason, the start of delivery must be advanced as engine speed increases.

Ignition lag

The diesel fuel requires a certain amount of time after the start of injection to form a combustible mixture with the air and to ignite. The length of time required from the start of injection to the Start of Combustion (SC) is the Ignition Lag (IGL). This, too, is independent of engine speed and is affected by the following variables:
- The ignition quality of the diesel fuel (indicated by the cetane number)
- The compression ratio of the engine
- The temperature in the combustion chamber
- The degree of fuel atomization and
- The exhaust-gas recirculation rate

The ignition lag is in the range of 2...9° of crankshaft rotation.

End of injection

When the high-pressure solenoid valve opens again, the high fuel pressure is released (End of Injection, EI) and the nozzle closes. This is followed by the End of Combustion (EC).

Other impacts

In order to limit pollutant emissions, the start of injection also has to be varied in response to engine load and temperature. The engine ECU calculates the required start of injection for each set of circumstances.

Design and method of operation

The timing device "advances" the position of the cams in the high-pressure pump relative to the diesel-engine crankshaft position as engine speed increases. This advances the start of injection. This compensates for the timing shift resulting from the injection lag and ignition lag. The impacts of engine load and temperature are also taken into account.

The injection timing adjustment is made up of the timing device itself, a timing-device solenoid valve and an angle-of-rotation sensor. Two types are used:

- The hydraulic timing device for axial-piston pumps and
- The hydraulically assisted timing device for radial-piston pumps

Hydraulic timing device

The hydraulic timing device (Fig. 2) is used on Type VP29 and VP30 axial-piston distributor injection pumps. Its design is the same as the version used on the electronically modulated Type VE..EDC port-controlled distributor injection pump.

The hydraulic timing device with its timing-device solenoid valve (5) and timing-device piston (3), which is located transversely to the pump axis, is positioned on the underside of the fuel-injection pump. The timing-device piston rotates the roller ring (1) according to load conditions and speed, so as to adjust the position of the rollers according to the required start of delivery.

As with the mechanical timing device, the pump intake-chamber pressure, which is proportional to pump speed, acts on the timing-device piston. That pressure on the

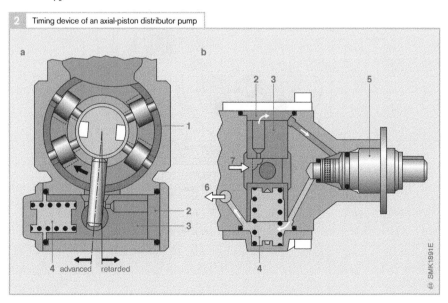

2   Timing device of an axial-piston distributor pump

a
b

2   3   5

1

7

6

2

3

4 advanced   retarded

4

SMK1891E

Fig. 2
a   Front view
b   Top view

1   Roller ring
2   Pump intake-
     chamber pressure
3   Timing-device piston
4   Pressure controlled
     by solenoid valve
5   Timing-device
     solenoid valve
6   Fuel return
7   Fuel inlet from pump
     intake chamber

pressure side of the timing device is regulated by the timing-device solenoid valve.

When the solenoid valve is open (reducing the pressure), the roller ring moves in the "retard" direction; when the valve is fully closed (increasing the pressure), it moves in the "advance" direction.

In between those two extremes, the solenoid valve can be "cycled", i.e. opened and closed in rapid succession by a Pulse-Width Modulation signal (PWM signal) from the pump control unit. This is a signal with a constant voltage and frequency in which the ratio of "on" time to "off" time is varied. The ratio between the "on" time and the "off"

time determines the pressure acting on the adjuster piston so that it can be held in any position.

Hydraulically assisted timing device
The hydraulically assisted timing device is used for radial-piston distributor injection pumps. It can produce greater adjustment forces. This is necessary to securely brace the cam ring with the greater drive power of the radial-piston pump. This type of timing device responds very quickly and regardless of the friction acting on the cam ring and the adjuster piston.

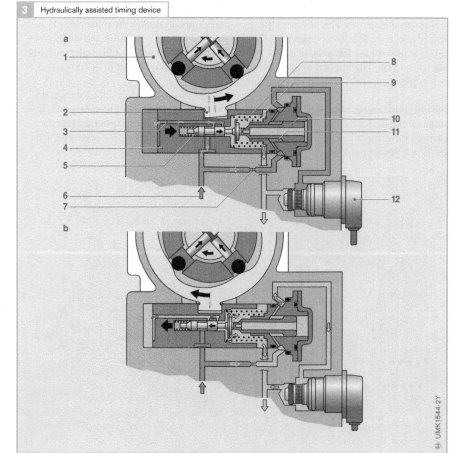

3   Hydraulically assisted timing device

Fig. 3
a Advance setting
b Retard setting

1   Cam ring
2   Ball pivot
3   Inlet channel/
    outlet channel
4   Timing-device piston
5   Control collar
6   Inlet from vane-type
    supply pump
7   Throttle
8   Control-plunger
    spring
9   Return spring
10 Control plunger
11 Annular chamber
    of hydraulic stop
12 Timing-device
    solenoid valve

UMK1544-2Y

The hydraulically assisted timing device, like the hydraulic timing device, is positioned on the underside of the injection pump (Fig. 3). It is also referred to by the Bosch type designation NLK.

The cam ring (1) engages in a cross-slot in the adjuster piston (4) by means of an adjuster lug (2) so that the axial movement of the adjuster piston causes the cam ring to rotate. In the center of the adjuster piston is a control sleeve (5) which opens and closes the control ports in the adjuster piston. In axial alignment with the adjuster piston is a spring-loaded hydraulic control piston (10) which defines the required position for the control sleeve.

At right-angles to the adjuster piston is the injection-timing solenoid valve (Pos. 12, shown schematically in Fig. 3 in the same plane as the timing device). Under the control of the pump control unit, the solenoid valve modulates the pressure acting on the control piston.

Advancing injection
When at rest, the adjuster piston (4) is held in the "retarded" position by a return spring (9). When in operation, the fuel supply pump pressure is regulated according to pump speed by means of the pressure control valve. That fuel pressure acts as the control pressure on the annular chamber (11) of the hydraulic stop via a restrictor bore (7) and when the solenoid valve (12) is closed moves the control piston (10) against the force of the control-piston spring (8) towards an "advanced" position (to the right in Fig. 3). As a result, the control sleeve (5) also moves in the "advance" direction so that the inlet channel (3) opens the way to the space behind the adjuster piston. Fuel can then flow through that channel and force the adjuster piston to the right in the "advance" direction.

The rotation of the cam ring relative to the resulting pump drive shaft causes the rollers to meet the cams sooner in the advanced position, thus advancing the start of injection. The degree of advance possible can be as much as 20° of camshaft rotation. On four-stroke engines, this corresponds to 40° of crankshaft rotation.

Retarding injection
The timing-device solenoid valve (12) opens when it receives the relevant PWM signal from the pump ECU. As a result, the control pressure in the annular chamber of the hydraulic stop (11) drops. The control plunger (10) moves in the "retard" direction (to the left in Fig. 3) by the action of the control-plunger spring (8).

The timing-device piston (4) remains stationary to begin with. Only when the control collar (5) opens the control bore to the outlet channel can the fuel escape from the space behind the timing-device piston. The force of the return spring (9) and the reactive torque on the cam ring then force the timing-device piston back in the "retard" direction and to its initial position.

Regulating the control pressure
The timing-device solenoid valve acts as a variable throttle. It can vary continuously the control pressure so that the control plunger can assume any position between the fully advanced and fully retarded positions. The hydraulically assisted timing device is more precise in this regard than the straightforward hydraulic timing device.

If, for example, the control plunger is to move more in the "advance" direction, the on/off ratio of the PWM signal from the pump ECU is altered so that the valve closes more (low ratio of "on" time to "off" time). Less fuel escapes through the timing-device solenoid valve and the control plunger moves to a more "advanced" position.

Timing-device solenoid valve

The timing-device solenoid valve (Fig. 4) is the same in terms of fitting and method of control as the one used on the port-controlled type VE..EDC electronically regulated distributor injection pump. It is also referred to as the Type DMV10 diesel solenoid valve.

The pump ECU controls the solenoid coil (6) by means of a PWM signal. The solenoid armature (5) is drawn back against the force of the valve spring (7) while the solenoid valve is switched on. The valve needle (3) connected to the solenoid armature opens the valve. The longer the "on" times of the PWM signal are, the longer the solenoid valve remains open. The on/off ratio of the signal thus determines the flow rate through the valve.

In order to avoid problems caused by resonance effects, the otherwise fixed timing frequency of the PWM signal (Fig. 5) does not remain constant over the entire speed range. It changes over to a different frequency (30...70 Hz) in specific speed bands.

Incremental angle-time system with angle-of-rotation sensor

The closed-loop position control of the timing device uses as its input variables the signal from the crankshaft speed sensor and the pump's internal incremental angle-time system signal from the angle-of-rotation sensor.

The incremental angle-time system is located in the fuel-injection pump between the vane-type supply pump and the roller ring/cam ring (Fig. 6). Its purpose is to measure the angular position of the engine camshaft and the roller ring/cam ring relative to one another. This information is used to calculate the current timing device setting. In addition, the angle-of-rotation sensor (2) supplies an accurate speed signal.

Design and method of operation

Attached to the pump drive shaft is the increment ring (4). This has a fixed number of 120 teeth (i.e. one tooth for every 3°). In addition, there are reference tooth spaces (3) according to the number of cylinders in the engine. The increment ring is also called the angle-sensor ring or the sensor ring.

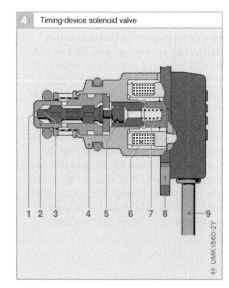

4   Timing-device solenoid valve

1 2 3   4 5   6   7 8   9

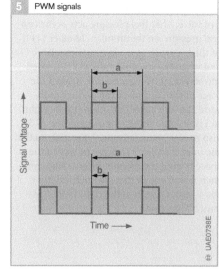

5   PWM signals

Signal voltage

Time

The speed sensor is connected to the bearing ring (5) of the timing device. As the increment ring rotates, the sensor produces an electrical signal by means of magnetically controllable semiconductor resistors and relative to the number of teeth that pass by the pickup. If the position of the timing device changes, the sensor moves along with the roller ring/cam ring. Consequently, the positions of the reference tooth spaces in the increment ring alter relative to the TDC signal from the crankshaft speed sensor.

The angular separation between the reference tooth spaces (or the synchronization signal produced by the tooth spaces) and the TDC signal is continuously detected by the pump ECU and compared with the stored reference figure. The difference between the two signals represents the actual position of the timing device.

If even greater accuracy in determining the start of injection is required, the start-of-delivery control system can be supplemented by a start-of-injection control system using a needle-motion sensor.

Start-of-injection control system

The start of delivery and start of injection are directly related to one another. This relationship is stored in the "wave-propagation time map" in the engine ECU. The engine ECU uses this data to calculate a start-of-injection setpoint according to the operating status of the engine (load, speed, temperature). It then sends the information to the pump ECU. The pump ECU calculates the necessary control signals for the high-pressure solenoid valve and the setpoint position for the timing-device piston.

The timing-device controller in the pump ECU continuously compares the actual position of the timing-device piston with the setpoint specified by the engine ECU and, if there is a difference, alters the on/off ratio of the signal which controls the timing-device solenoid valve. Information as to the actual start-of-injection setting is provided by the signal from an angle-of-rotation sensor or

alternatively, from a needle-motion sensor in the nozzle. This variable control method is referred to as "electronic" injection timing adjustment.

The benefits of a start-of-delivery control system are its rapid response characteristics, since all cylinders are taken into account. Another benefit is that it also functions when the engine is overrunning, i.e. when no fuel is injected. It means that the timing device can be preset for the next injection sequence.

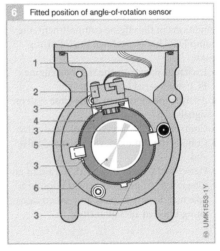

6   Fitted position of angle-of-rotation sensor

Fig. 6
1 Flexible conductive foil to ECU
2 Type DWS angle-of-rotation sensor
3 Tooth space
4 Increment ring
5 Bearing ring (connected to roller ring/cam ring)
6 Pump drive shaft

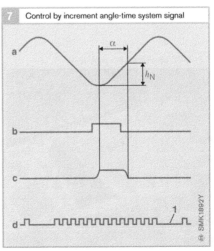

7   Control by increment angle-time system signal

Fig. 7
a Cam pitch
b Control pulse for high-pressure solenoid valve
c Valve lift of high-pressure solenoid valve
d Signal of angle-of-rotation sensor

1 Tooth space

h_N Effective stroke
α Delivery angle

Electronic control unit

Requirements

Screwed to the top of the fuel-injection pump and identifiable by its cooling fins is the pump ECU or combined pump/engine ECU. It is an LTCC (Low Temperature Cofired Ceramic) hybrid design. This gives it the capability to withstand temperatures of up to 125°C and vibration levels up to 100 g (acceleration due to gravity). These ECUs are available in 12-volt or 24-volt versions.

In addition to withstanding the external conditions in the engine compartment, the pump ECU has to perform the following tasks:
- Exchange data with the separate engine ECU via a serial bus system
- Analyze the increment angle-time system signals
- Control the high-pressure solenoid valve
- Control the timing-device solenoid valve and
- Detect fuel temperature with the aid of an integrated temperature sensor in order to take fuel density into account when calculating injected-fuel quantity

The control of fuel injection must operate very precisely so that the fuel-injection pump delivers precisely and consistently the required quantity of fuel at the required time in all engine operating conditions. Even minute discrepancies in start of injection and injection duration have a negative effect on the smooth running and noise levels in the diesel engine as well as its pollutant emissions.

The fuel-injection system also has to respond very quickly to changes. For this reason, the calculations taking place in the microcontroller and conversion of the control signals in the output stages are performed in real time (approx. 50 µs). In the case of a fuel-injection pump for a six-cylinder engine, the fuel-injection data is calculated up to 13,000 times a minute.

A Pre-Injection (PI) phase controlled by the high-pressure solenoid valve is also a viable option. This involves the injection of 1...2 mm³ of fuel before the Main Injection (MI) phase. This produces a more gradual increase in combustion pressure and therefore reduces combustion noise.

During the pre-injection phase, the fuel-quantity solenoid valve is operated ballistically, i.e. it is only partially opened. Consequently, it can be closed again more quickly. This keeps the injection gap as short as possible so that even at high speeds, there is sufficient cam lift remaining for the main injection phase. The entire injection sequence lasts approx. 1...2 ms.

The timing of the individual control phases is calculated by the microcontroller in the pump ECU. It also makes use of stored data maps. The maps contain settings for the specific vehicle application and certain engine characteristics along with data for checking the plausibility of the signals received. They also form the basis for determining various calculated variables.

1   LTCC circuit board of a pump ECU

In order to achieve specific and rapid opening of the solenoid valve, "fast extinction" of the energy stored in the solenoid valve takes place combined with a high extinction potential.

Two-ECU concept

Separate ECUs are used in diesel fuel-injection systems with solenoid-valve controlled axial-piston distributor injection pumps and first-generation radial-piston distributor injection pumps. These systems have a Type MSG engine ECU in the engine compartment and a Type PSG pump ECU mounted directly on the fuel-injection pump. There are two reasons for this division of functions: Firstly, it prevents the overheating of certain electronic components by removing them from the immediate vicinity of pump and engine. Secondly, it allows the use of short control leads for the high-pressure solenoid valve, so eliminating interference that may occur as a result of the very high currents (up to 20 A).

The pump ECU detects and analyzes the pump's internal sensor signals for angle of rotation and fuel temperature in order to adjust the start of injection. On the other hand, the engine ECU processes all engine and ambient data signals from external sensors and interfaces and uses them to calculate actuator adjustments on the fuel-injection pump.

The two control units communicate via a CAN interface.

Integrated engine and pump ECU on the fuel-injection pump

Increasing levels of integration using hybrid technology have made it possible to combine the engine-management control unit with the pump control unit on second-generation solenoid-valve controlled distributor injection pumps. The use of integrated ECUs allows a space-saving configuration. Other advantages are simpler installation and lower system costs due to fewer electrical interfaces.

The integrated engine/pump ECU is only used with radial-piston distributor injection pumps.

Summary

The overall system and the many modular assemblies are very similar on different types of solenoid-valve controlled distributor injection pump. Nevertheless, there are a number of differences. The main distinguishing features are detailed in Table 1.

1	Essential distinguishing features of solenoid-valve controlled distributor injection pumps		
Type	VP29 (VE..MV)	VP30 (VE..MV)	VP44 (VRV)
Application	IDI engines	DI engines	DI engines
Maximum injection pressure at nozzle	800 bar	1,400 bar	1,950 bar
High-pressure pump	Axial-piston	Axial-piston	Radial-piston
Delivery valve	Separate orifice check valve	Integrated orifice check valve	Integrated orifice check valve
Timing device	Hydraulic	Hydraulic	Hydraulically assisted
Vane-type delivery pump	Circular retaining ring Vanes without springs	Circular retaining ring Vanes without springs	Profiled retaining ring Vanes with springs
Engine ECU and pump ECU	Separate	Separate	Integrated or separate

Table 1

Overview of discrete cylinder systems

Diesel engines with discrete cylinder systems have a separate fuel-injection pump for each cylinder of the engine. Such individual fuel-injection pumps are easily adaptable to particular engines. The short high-pressure fuel lines enable the achievement of particularly good injection characteristics and extremely high injection pressures.

Continually increasing demands have led to the development of a variety of diesel fuel-injection systems, each of which is suited to different requirements. Modern diesel engines must offer low emissions, good fuel economy, high torque and power output while also being quiet-running.

There are basically three types of discrete cylinder system: the Type PF port-controlled discrete fuel-injection pump, and the solenoid-valve controlled unit injector and unit pump systems. Those systems differ not only in their design but also in their performance data and areas of application (Fig. 1).

Single-plunger fuel-injection pumps PF

Application
Type PF discrete injection pumps are particularly easy to maintain. They are used in the "off-highway" sector as
- Fuel-injection pumps for diesel engines with outputs of 4…75 kW/cylinder in small construction-industry machines, pumps, tractors and power generators and
- Fuel-injection pumps for large-scale engines with outputs of between 75 kW and 1,000 kW per cylinder. These versions are capable of working with high-viscosity diesel fuel and heavy oil

Design and method of operation
Type PF discrete fuel-injection pumps operate in the same way as Type PE in-line fuel-injection pumps. They have a single pump unit on which the injection quantity can be varied by means of a helix.

Each discrete fuel-injection pump is separately flanged-mounted to the engine and driven by the camshaft that controls the en-

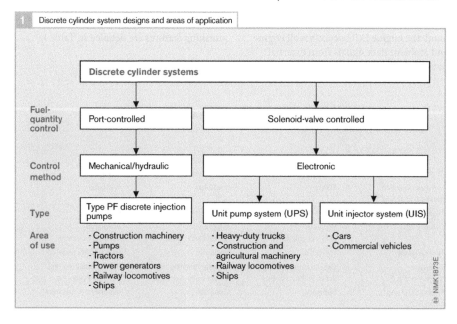

1 Discrete cylinder system designs and areas of application

	Discrete cylinder systems		
Fuel-quantity control	Port-controlled	Solenoid-valve controlled	
Control method	Mechanical/hydraulic	Electronic	
Type	Type PF discrete injection pumps	Unit pump system (UPS)	Unit injector system (UIS)
Area of use	- Construction machinery - Pumps - Tractors - Power generators - Railway locomotives - Ships	- Heavy-duty trucks - Construction and agricultural machinery - Railway locomotives - Ships	- Cars - Commercial vehicles

NMK1873E

gine valve timing. They can therefore be described as externally driven pumps. They may also be referred to as plug-in pumps.

Some of the smaller Type PF pumps come in 2, 3 and 4-cylinder versions. However, the majority of designs supply only a single cylinder and are therefore known as discrete or cylinder fuel-injection pumps.

Many discrete fuel-injection pumps have an integral roller tappet. In such cases they have the type designation PFR. With some designs for smaller engines, the roller tappet is mounted on the engine. Those versions have the type designation PFE.

Control

As with in-line fuel-injection pumps, a control rod incorporated in the engine acts on the fuel-injection pump units. A governor or control system moves the control rack, thereby varying the fuel delivery and injected-fuel quantity.

On large-scale engines, the governor is mounted directly on the engine block. Hydro-mechanical governors or electronic control systems may be used, or more rarely, purely mechanical governors.

Between the control rack for the discrete fuel-injection pumps and the actuating linkage from the governor, there is a sprung compensating link so that, in the event that the adjusting mechanism on one of the pumps jams, control of the other pumps is not compromised.

Fuel supply

Supply and filtering of the fuel and removal of air from the fuel-injection system is performed in the same way with Type PF dis-

crete fuel-injection pumps as for in-line fuel-injection pumps.

The fuel is fed to the individual fuel-injection pumps by a gear-type presupply pump. It delivers around 3...5 times as much fuel as the maximum full-load delivery of all individual fuel-injection pumps. The fuel pressure in this part of the system is around 3...10 bar.

The fuel is filtered by fine-pore filters with a pore size of 5...30 µm in order to keep suspended particles out of the fuel-injection system. Such particles would otherwise cause premature wear on the part of the high-precision fuel- injection components.

Heavy oil operation

Discrete fuel-injection pumps for engines with outputs of over 100 kW/cylinder are not only used to pump diesel fuel. They are also suitable for use with high-viscosity heavy oils with viscosities up to $700\,mm^2/s$ at 50°C. In order to do so, the heavy oil has to be pre-heated to temperatures as high as 150°C. This ensures that the required fuel-injection viscosity of $10...20\,mm^2/s$ is obtained.

---

2   Examples of Type PF discrete fuel-injection pumps

10 cm

UMK0455-1Y

Fig. 2
a   Type PFE 1 for small engines
b   Type PFR 1 for small engines
c   Type PFR 1 W for large-scale engines
d   Type PF 1 D for large-scale engines

Unit Injector System (UIS) and Unit Pump System (UPS)

The unit injector and unit pump fuel-injection systems achieve the highest injection pressures of all diesel fuel-injection systems currently available. They are capable of high-precision fuel injection that is infinitely variable in response to engine operating status. Diesel engines equipped with these systems produce low emission levels, are economical and quiet to run, and offer high performance and torque characteristics.

Areas of application

Unit Injector System (UIS)
The Unit Injector System (UIS) went into volume production for commercial vehicles in 1994 and for cars in 1998. It is a fuel-injection system with timer-controlled discrete fuel-injection pumps for diesel engines with direct injection (DI). This system offers a significantly greater degree of adaptability to individual engine designs than conventional port-controlled systems. It can be used on a wide range of modern diesel engines for cars and commercial vehicles extending to
- *Cars* and *light commercials* with engines ranging from three-cylinder 1.2 *l* units producing 45 kW (61 bhp) of power and 195 Nm of torque to 10-cylinder, 5 *l* engines with power outputs of 230 kW (bhp) and torque levels of 750 Nm.
- *Heavy-duty trucks* developing up to 80 kW/cylinder.

As it requires no high-pressure fuel lines, the unit injector system has excellent hydraulic characteristics. That is the reason why this system is capable of producing the highest injection pressures (up to 2,050 bar). The unit injector system for cars also offers the option of pre-injection.

Unit Pump System (UPS)
The Unit Pump System (UPS) is also referred to by the type designation PF..MV for large-scale engines.

Like the unit injector system, the unit pump system is a fuel-injection system with timer-controlled discrete fuel-injection pumps for Direct Injection (DI) diesel engines. There are three versions:
- The UPS12 for commercial-vehicle engines with up to 8 cylinders and power outputs of up to 35 kW/cylinder
- The UPS20 for heavy commercial-vehicle engines with up to 8 cylinders and power outputs of up to 80 kW/cylinder
- UPS for engines in construction and agricultural machinery, railway locomotives and ships with power outputs of up to 500 kW/cylinder and up to 20 cylinders

Design

System structure
The unit injector and unit pump systems are made up of four subsystems (Fig. 1):
- The *fuel supply system* (low-pressure system) provides suitably filtered fuel at the correct pressure.
- The *high-pressure system* generates the necessary injection pressure and injects the fuel into the combustion chamber.
- The *EDC electronic control system* consisting of the sensors, control unit and actuators performs all diesel engine management and control functions as well as providing all electrical and electronic interfaces.
- The *air-intake and exhaust-gas-systems* handle the supply of air for combustion, exhaust-gas recirculation and exhaust-gas treatment.

The modular design of the individual subsystems allows the entire fuel-injection system to be easily adapted to individual engine designs.

Differences

The essential difference between the unit injector system and the unit pump system lies in the way in which high pressure is generated (Fig. 2).

In the *Unit Injector System*, the high-pressure pump and the nozzle form a single unit – the "unit injector". There is a unit injector fitted in each cylinder of the engine. As there are no high-pressure fuel lines, extremely high injection pressures can be generated and precisely controlled injection patterns can be produced.

With the *Unit Pump System*, the high pressure pump – the "unit pump" – and the nozzle-and-holder assembly are separate units that are connected by a short length of high-pressure pipe. This arrangement has advantages in terms of use of space, pump-drive system, and servicing and maintenance.

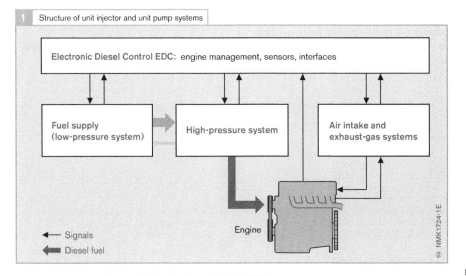

1 Structure of unit injector and unit pump systems

Electronic Diesel Control EDC: engine management, sensors, interfaces

Fuel supply (low-pressure system)

High-pressure system

Air intake and exhaust-gas systems

Engine

← Signals

← Diesel fuel

NMK1724-1E

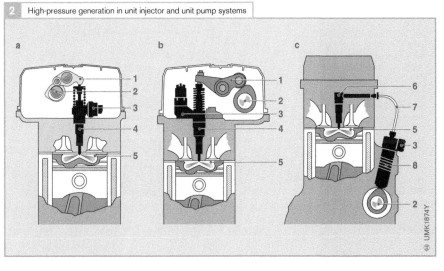

2 High-pressure generation in unit injector and unit pump systems

a   b   c

UMK1874Y

Fig. 2

a Unit Injector System for cars
b Unit Injector System for commercial vehicles
c Unit Pump System for commercial vehicles

1 Rocker arm
2 Camshaft
3 High-pressure solenoid valve
4 Unit injector
5 Engine combustion chamber
6 Nozzle-and-holder assembly
7 Short high-pressure line
8 Unit pump

System diagram of UIS for cars

Figure 1 shows all the components of a fully equipped unit injector system for an eight-cylinder diesel car engine. Depending on the type of vehicle and application, some of the components may not be used.

For the sake of clarity of the diagram, the sensors and desired-value generators (A) are not shown in their fitted positions. Exceptions to this are the components of the exhaust-gas treatment systems (F) as their proper fitted positions are necessary in order to understand the system.

The CAN bus in the interfaces section (B) enables exchange of data between a wide variety of systems and components including:
- The starter motor
- The alternator
- The electronic immobilizer
- The transmission control system
- The Traction Control System (TCS) and
- The Electronic Stability Program (ESP)

Even the instrument cluster (12) and the air-conditioning system (13) can be connected to the CAN bus.

For emission control, three alternative combination systems are shown (a, b and c).

Fig. 1
Engine, engine control unit and high-pressure
fuel-injection components
24 Fuel rail
25 Camshaft
26 Unit injector
27 Glow plug
28 Diesel engine (DI)
29 Engine control unit (master)
30 Engine control unit (slave)
M   Torque

A   Sensors and desired-value generators
1   Accelerator-pedal sensor
2   Clutch switch
3   Brake switches (2)
4   Operator unit for cruise control
5   Glow plug/starter switch ("ignition switch")
6   Vehicle-speed sensor
7   Crankshaft speed sensor (inductive)
8   Engine-temperature sensor (in coolant system)
9   Intake-air temperature sensor
10 Charge-air pressure sensor
11 Hot-film air-mass flow sensor (intake air)

B   Interfaces
12 Instrument cluster with signal output for fuel
    consumption, engine speed, etc.
13 Air-conditioning compressor with control
14 Diagnosis interface
15 Glow plug control unit
CAN Controller Area Network
    (vehicle's serial data bus)

C   Fuel supply system (low-pressure system)
16 Fuel filter with overflow valve
17 Fuel tank with filter and electric presupply pump
18 Fuel level sensor
19 Fuel cooler
20 Pressure limiting valve

D   Additive system
21 Additive metering unit
22 Additive control unit
23 Additive tank

E   Air-intake system
31 Exhaust-gas recirculation cooler
32 Charge-air pressure actuator
33 Charge-air (in this case with Variable
    Turbine Geometry VTG)
34 Intake manifold flap
35 Exhaust-gas recirculation actuator
36 Vacuum pump

F   Emission control systems
38 Broadband oxygen sensor Type LSU
39 Exhaust-gas temperature sensor
40 Oxidation-type catalytic converter
41 Particulate filter
42 Differential-pressure sensor
43 NO_x accumulator-type catalytic converter
44 Broadband oxygen sensor, optionally as NO_x sensor

1 Diesel fuel-injection system for cars using unit injector system

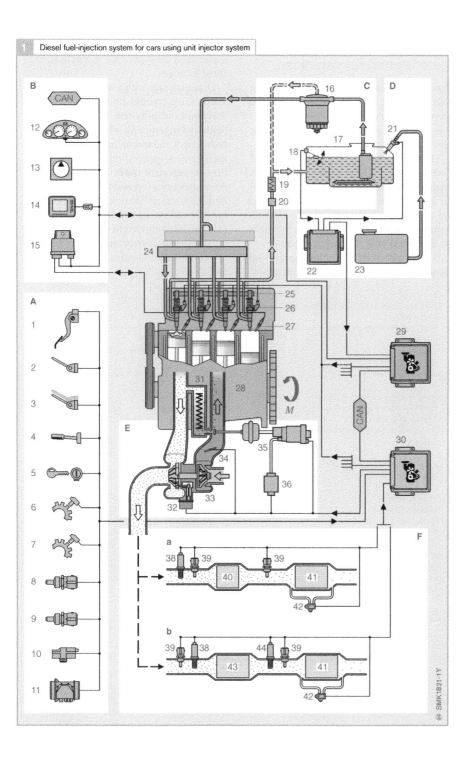

SMK1821-1Y

System diagram of UIS/UPS for commercial vehicles

Figure 2 shows all the components of a unit injector system for a six-cylinder diesel commercial-vehicle engine. Depending on the type of vehicle and application, some of the components may not be used.

The components of the Electronic Diesel Control system EDC (sensors, interfaces and engine control unit), the fuel-supply system, air-intake system and exhaust-gas treatment are very similar in the unit injector and unit pump systems. They differ only in the high-pressure section of the overall system.

For the sake of clarity of the diagram, only those sensors and desired-value generators

whose true position is necessary in order to understand the system are shown in their fitted locations.

Data exchange with a wide range of other systems (e.g. transmission control system, Traction Control System TCS, Electronic Stability Program ESP, oil quality sensor, tachograph, radar ranging sensor, vehicle management system, brake co-ordinator, fleet management system) involving up to 30 control units is possible via the CAN bus in the "Interfaces" section. Even the alternator (18) and the air-conditioning system (17) can be connected to the CAN bus.

For exhaust-gas treatment, three alternative combination systems are shown (a, b and c).

Fig. 2
Engine, engine control unit and high-pressure injection components
22 Unit pump and nozzle-and-holder assembly
23 Unit injector
24 Camshaft
25 Rocker arm
26 Engine control unit
27 Relay
28 Auxiliary equipment (e.g. retarder, exhaust flap for engine brake, starter motor, fan)
29 Diesel engine (DI)
30 Flame glow plug (alternatively grid heater)
M Torque

A Sensors and setpoint generators
1   Accelerator-pedal sensor
2   Clutch switch
3   Brake switches (2)
4   Engine brake switch
5   Parking brake switch
6   Control switch (e.g. cruise control, intermediate speed control, engine speed and torque reduction)
7   Starter switch ("ignition switch")
8   Charge-air speed sensor
9   Crankshaft speed sensor (inductive)
10 Camshaft speed sensor
11 Fuel temperature sensor
12 Engine-temperature sensor (in coolant system)
13 Charge-air temperature sensor
14 Charge-air pressure sensor
15 Fan speed sensor
16 Air-filter differential-pressure sensor

B Interfaces
17 Air-conditioning compressor with control
18 Alternator
19 Diagnosis interface
20 SCR control unit

21 Air compressor
CAN Controller Area Network (vehicle's serial data bus) (up to three data busses)

C Fuel supply system (low-pressure system)
31 Fuel pump
32 Fuel filter with water-level and pressure sensors
33 Control unit cooler
34 Fuel tank with filter
35 Fuel level sensor
36 Pressure limiting valve

D Air intake system
37 Exhaust-gas recirculation cooler
38 Control flap
39 Exhaust-gas recirculation actuator with exhaust-gas recirculation valve and position sensor
40 Intercooler with bypass for cold starting
41 Turbocharger (in this case with VTG) with position sensor
42 Charge-air pressure actuator

E Emission control systems
43 Exhaust-gas temperature sensor
44 Oxidation-type catalytic converter
45 Differential-pressure sensor
46 Particulate filter
47 Soot sensor
48 Fluid level sensor
49 Reducing agent tank
50 Reducing agent pump
51 Reducing agent injector
52 NO_x sensor
53 SCR catalytic converter
54 NH_3 sensor
55 Blocking catalytic converter
56 Catalyzed soot filter Type CSF
57 Hydrolyzing catalytic converter

2   Diesel fuel-injection system for commercial vehicles using Unit Injector or Unit Pump System

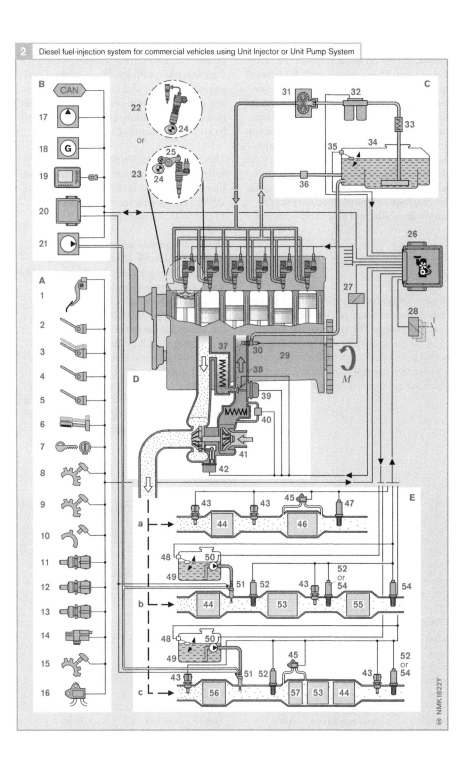

Single-plunger fuel-injection pumps PF

Single-plunger fuel-injection pumps PF are suitable for use on small, medium- and large-size engines. They are used for "off-highway" applications (agricultural and construction machinery, fixed installations such as pumps and power generators, railway locomotives, and small and large marine vessels). They are fitted to direct-injection and indirect-injection engines. On large-bore engines, they can also run on heavy oil. A single-plunger fuel-injection pump is fitted to each engine cylinder. Single-plunger fuel-injection pumps feature durability and ease of maintenance.

Design and method of operation

Single-plunger fuel-injection pumps PF operate in the same way as Type PE inline fuel-injection pumps. The layout of the timing edges on the pump plungers is identical on both types of pump. The delivery quantity is varied by a control rack and control sleeve which act as the pump plunger rotates.

In contrast with in-line fuel-injection pumps PE, however, single-plunger fuel-injection pumps PF are driven by the diesel engine's camshaft and not by a camshaft integrated in the pump housing. The designation code "PF" indicates that they are driven by an external camshaft.

This fuel-injection system has a particularly wide range of maximum injected-fuel quantities due to the wide variety of applications. Depending on version, it can range from 13,000 to 18,000 mm³ per stroke.

Single-plunger fuel-injection pumps are manufactured with aluminum or diecast housings. They are flanged to the engine. The standard design is a single-plunger variant, i.e. each engine cylinder is fitted with a fuel-injection pump. On multicylinder engines, this allows the use of very short fuel-injection tubing as each cylinder has a fuel-injection pump mounted close to it. The shorter the fuel-injection tubing, the better the fuel-injection system performance. Each type of diesel engine requires only one type of pump and fuel-injection tubing. This makes spare-parts inventory management much easier.

1　Technical data of Bosch single-plunger fuel-injection pumps PF					
Type designation	Area of application	Max. output per engine cylinder (kW)	Max. injection pressure at nozzle (bar)	Max. plunger lift (mm)	Piston diameter (mm)
Up to 75 kW per cylinder (light, medium and heavy duty)					
PFE 1Q..	Light duty	10	500	7	5...7
PFE 1A..	Medium duty	20	800	9	5...9
PFR 1K..	Medium duty	20	600	8	5...9
PFM[1] 1P	Heavy duty	50	1,150	12	9...10
75 kW per cylinder and above (large-bore engines)					
PF..Z	Large-bore engines	150	1,200	12	10...14
PF(R)..C	Large-bore engines	300	1,500	24	15...23
PF(R)..W	Large-bore engines	400	1,500	26	20...24
PF(R)..D	Large-bore engines	600	1,500	34	22...34
PF..E	Large-bore engines	700	1,200	45	25...36
PF(R)..H	Large-bore engines	1,000	1,500	48	32...46

Table 1
[1] M stands for monoblock

In exceptional cases, some of the smaller Type PF pumps come in 2-, 3- and 4-cylinder versions.

Single-plunger fuel-injection pumps are fitted a fixing device which maintains the injected-fuel quantity setting at maximum while the pump is being shipped. This prevents any accidental change in the setting. It is simplifies adjustment work when the pump is fitted to the engine.

There are basically two different types of single-plunger fuel-injection pump:
- Type PFR pumps have an integrated roller tappet. The roller is in direct contact with the cam of the engine camshaft.
- On Type PF and PFE pumps, the roller tappet is an integral component of the engine, and not part of the pump.

Injection timing

The drive cams for single-plunger fuel-injection pumps are located on the same camshaft that drives the engine valve timing. Therefore, it is not possible to change the injection timing by rotating the common camshaft relative to its drive gear.

If an intermediate device is adjusted – for example, an eccentric oscillating crank between the cam and the roller tappet – it is possible to obtain an advance angle of a few degrees of rotation (Fig. 1). This helps to optimize fuel consumption or exhaust-gas emissions, or adapt to the different ignition qualities of various types of fuel.

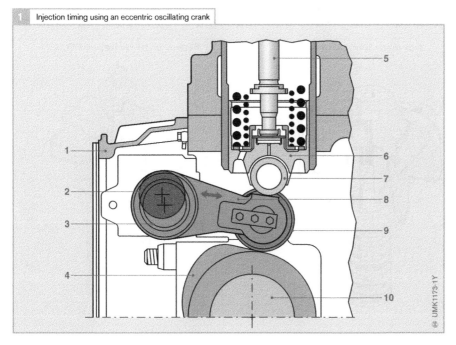

1   Injection timing using an eccentric oscillating crank

UMK1173-1Y

Fig. 1
1   Diesel engine
2   Profile of setting shaft
3   Oscillating crank bearing
4   Injection cam
5   Pump plunger
6   Roller tappet
7   Tappet roller
8   Oscillating crank
9   Camshaft roller
10   Engine camshaft

Sizes

Small fuel-injection pumps for power outputs of up to 50 kW/cylinder

These fuel-injection pumps are used for diesel engines on small construction machines, pumps, tractors and power generators.

Type *PFE 1A..* and *PFE 1Q..* pumps are single-plunger designs with no integrated roller tappet (Fig. 1). In these types of pump, the roller tappet runs in a guide bore inside the motor housing. With these type ranges, a control rack held in the motor housing engages in a control-sleeve lever of the control sleeve (6), which then rotates the pump plunger (5). This adjusts the delivery quantity.

The Type *PFR..K* pumps with integrated roller tappet (Fig. 2) are produced in 1, 2, 3, and 4-cylinder versions. They incorporate several plunger-and-barrel assemblies in a common pump housing, depending on the version. The plunger-and-barrel assemblies are driven by suitably positioned cams on the engine camshaft. With this type of pump, the delivery quantity is varied by means of a toothed control sleeve which engages in a control rack (5) guided inside the pump housing.

The maximum camshaft speed for driving small fuel-injection pumps is around 1,800 rpm. Depending on the plunger diameter, which can vary from 5 mm to 9 mm, the maximum WOT injected-fuel quantity may be as much as 95 mm^3 per stroke. The maximum peak injection pressure may be up to 600 bar on the pump side of the high-pressure fuel-injection tubing.

These types of pump are fitted as standard with constant-pressure valves with or without a return-flow restriction. Constant-pressure valves may be used if higher injection pressures or specific injected-fuel quantity stability are required.

Large fuel-injection pumps for power outputs greater than 50 kW/cylinder

These types of single-plunger fuel-injection pump are used for diesel engines with outputs of up to 1,000 kW per cylinder. They are capable of delivering high-viscosity diesel fu-

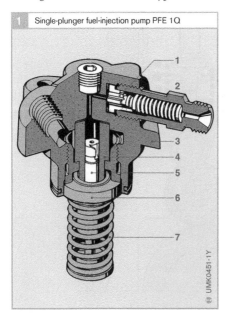

Fig. 1

1 Delivery valve
2 Delivery-valve holder
3 Pump housing
4 Pump barrel
5 Pump plunger
6 Control sleeve
7 Plunger return
   spring

Fig. 2

1 Delivery-valve holder
2 Delivery valve
3 Pump barrel
4 Pump plunger
5 Control rack
6 Control sleeve
7 Piston control arm
8 Roller tappet

1   Single-plunger fuel-injection pump PFE 1Q

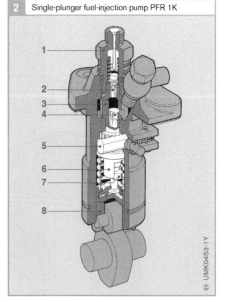

2   Single-plunger fuel-injection pump PFR 1K

els and heavy oil (Fig. 3). Pump barrels are designed with a plunger through-bore to achieve peak injection pressures at the pump of up to approx. 1,200 bar. For applications involving higher pressures, barrels are provided with a blind pocket to prevent the high fuel pressure from deforming the top of the plunger-and-barrel assembly (Fig. 4).

Pump plungers are symmetrical in design so that they can be guided centrally inside the pump barrels. Anti-erosion screws close to the spill ports in the pump barrel protect the pump housing from damage caused by the high-energy cutoff jet at the end of the spill.

The delivery valve is high-pressure-sealed against the pump barrel and the flange by lapped flat sealing faces. There is a rack-travel indicator attached to the control rack.

There may be several annular grooves machined into the pump barrel. They perform the following functions: The top groove closest to the pump interior acts as a leakage-return duct (5). It returns the leak fuel that escapes through the gap in the plunger-

and-barrel assembly back to the intake chamber via a hole in the pump barrel. Below that, there may be an oil-block groove. Blocking oil from the engine lube-oil circuit first passes through a fine filter and is then forced through a hole in this oil-block groove at a pressure of 3 to 5 bar. This pressure is higher than the pressure in the pump's fuel gallery at normal operating speeds. This prevents the engine lube oil from becoming diluted by the fuel. Between those two grooves, there may be another groove (13) to drain off mixed fuel and engine oil (emulsion). Emulsion goes into a separate collection tank.

For heavy-oil applications, the roller tappet on single-plunger fuel-injection pump PFR 1CY or, as the case may be, the guide bushing on Type PF pumps and the control sleeve, are lubricated by engine lube oil via a special connection.

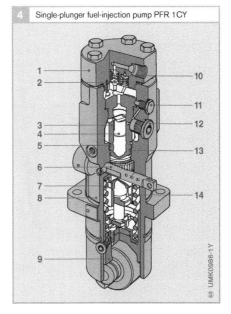

| 3 | Single-plunger fuel-injection pump PF 1D |
| 4 | Single-plunger fuel-injection pump PFR 1CY |

UMK0452-1Y

UMK0986-1Y

Fig. 3
1 Delivery valve
2 Vent screw
3 Pump barrel
4 Pump plunger
5 Control rack
6 Control sleeve
7 Guide bushing

Fig. 4
1 Flange
2 Forward-delivery valve
3 Pump barrel
4 Pump plunger
5 Leakage-return channel
6 Control rack
7 Pump spring
8 Pump housing
9 Roller tappet
10 Pressure-holding valve
11 Vent screw
12 Anti-erosion screw
13 Emulsion drain
14 Control sleeve

Unit Injector System (UIS)

In the Unit Injector System (UIS), the fuel-injection pump, high-pressure solenoid valve and nozzle form a single unit. The compact construction – with very short high-pressure lines integrated in the component between pump and nozzle – makes it easier to deliver higher injection pressures compared with other fuel-injection systems because the compression volume [1]) and thus the compression losses are lower. The peak injection pressure in the UIS currently varies, depending on the pump type, between 1,800 and 2,200 bar for commercial vehicles and up to 2,050 bar for cars.

Installation and drive

Each cylinder has its own Unit Injector (UI) which is installed directly in the cylinder head (Fig. 1). For passenger cars, there are two types of unit injector (UI-1, UI-2), which – while having an identical function –

differ in size. In a 2-valve engine, the UI-1 is secured using a clamping block at an angle of approx. 20° in the engine's cylinder head. In a 4-valve engine, the smaller injector (UI-2) is used on account of the smaller amount of space available; this injector is secured vertically in the cylinder head with anti-fatigue bolts.

The engine camshaft (2) has an actuating cam for each unit injector, the particular cam pitch being transferred to the pump plunger (6) by a rocker (1) The injection curve is influenced by the shape of the actuating cams. These are shaped so that the pump plunger moves more slowly when fuel is taken in (upward movement) than during injection (downward movement) in order, on the one hand, to prevent air from being accidentally drawn in and, on the other hand, to achieve a high delivery rate.

[1]) The compression volume is the fuel volume which is compressed

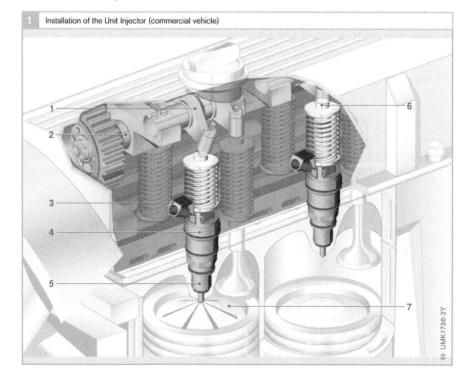

1   Installation of the Unit Injector (commercial vehicle)

Fig. 1
1   Rocker
2   Engine camshaft
3   Plug
4   Unit injector
5   Nozzle assembly
6   Pump plunger
7   Engine combustion
    chamber

UMK1736-2Y

Torsional vibration is induced in the camshaft by the forces applied to it during operation, and adversely affects injection characteristics and injected-fuel-quantity metering. It is therefore imperative that in order to reduce these vibrations the individual-pump drives are designed to be as rigid as possible (this applies to the camshaft drive, the camshaft itself, the rocker, and the rocker bearings).

The unit injector is installed in the engine's cylinder head and is therefore subject to very high temperatures. It is cooled by relatively cool fuel flowing back to the low-pressure stage.

Design and construction

Fuel is supplied to the UI for passenger cars through roughly 500 laser-drilled inlet passages in the steel sleeve of the injector. The fuel is filtered by the passages, which have a diameter of less than 0.1 mm.

The unit-injector body assembly serves as the pump barrel. The nozzle (Fig. 2, 7) is integrated in the stem of the unit injector. The stem and body assembly are connected to each other by means of a nozzle nut (13).

The follower spring (1) forces the pump plunger against the rocker (8) and the rocker against the actuating cam (9). This ensures that pump plunger, rocker and cam are always in contact during actual operation.

In the unit injector for commercial vehicles, the solenoid valve is integrated in the injector. In the UI for passenger cars, however, it is mounted externally on the pump body on account of the injector's smaller dimensions.

The design of the injector for passenger cars and commercial vehicles is shown on the following pages.

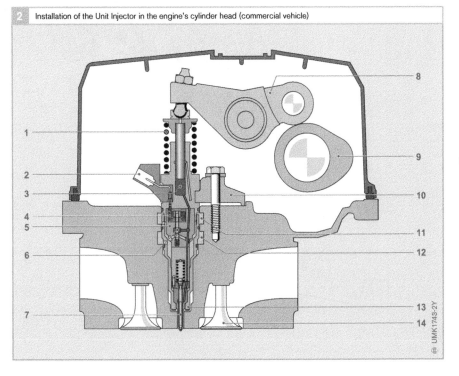

2   Installation of the Unit Injector in the engine's cylinder head (commercial vehicle)

Fig. 2

1 Follower spring
2 Plug
3 High-pressure chamber
4 Solenoid coil
5 Solenoid-valve body
6 Solenoid-valve needle
7 Nozzle assembly
8 Rocker
9 Actuating cam
10 Clamping element
11 Fuel return
12 Fuel inlet
13 Nozzle nut
14 Gas-exchange valve

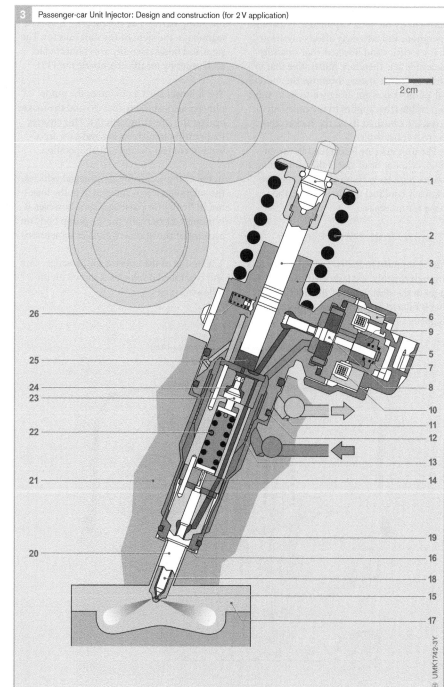

3   Passenger-car Unit Injector: Design and construction (for 2 V application)

2 cm

Fig. 3

1   Ball pin
2   Follower spring
3   Pump plunger
4   Pump-body
    assembly
5   Plug-in connection
6   Magnet core
7   Compensating
    spring
8   Solenoid-valve
    needle
9   Solenoid-valve coil
11   Fuel return
12   Seal
13   Inlet passages
    (laser-drilled holes
    acting as a filter)
14   Hydraulic stop
    (damping unit)
15   Needle seat
16   Sealing disc
17   Engine combustion
    chamber
18   Nozzle needle
19   Retaining nut
20   Integral nozzle
    assembly
21   Engine cylinder
    head
22   Needle-valve spring
23   Accumulator plunger
24   Accumulator
    chamber
25   High-pressure
    chamber
26   Solenoid-valve
    spring

UMK1742-3Y

4   Commercial-vehicle Unit Injector: Design and construction

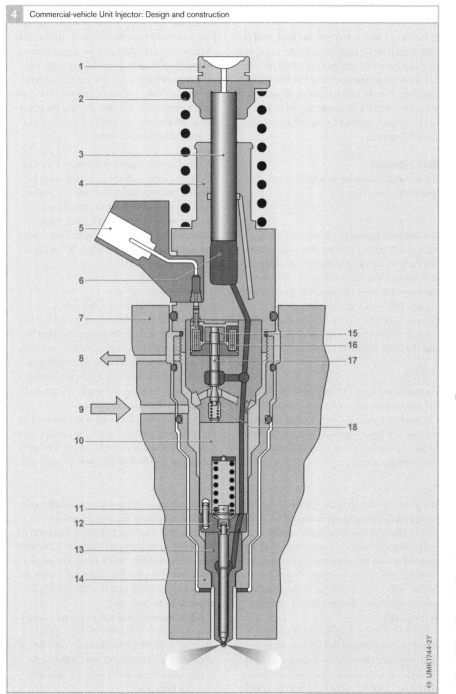

Fig. 4

1 Slide disc
2 Return spring
3 Pump plunger
4 Pump-body
   assembly
5 Plug-in connection
6 High-pressure
   chamber
7 Engine cylinder
   head
8 Fuel return
9 Fuel inlet
10 Spring retainer
11 Pressure pin
12 Shim
13 Integral nozzle
   assembly
14 Retaining nut
15 Armature
16 Solenoid-valve coil
17 Solenoid-valve
   needle
18 Solenoid-valve
   spring

UMK1744-2Y

Method of operation

In terms of main injection, the unit injector system essentially functions in the same way for passenger cars and for commercial vehicles. Pilot injection is effected in the UI for passenger cars by mechanical-hydraulic control and in the UI for commercial vehicles by electronic control (2 x activation of the solenoid valve).

Main injection

Main injection can be subdivided into four operating states (Fig. 1):

Suction stroke (a)

The follower spring (3) forces the pump plunger (2) upwards when the cam rotates. The fuel, which is permanently under pressure, flows from the fuel-supply system's low-pressure stage through the inlet passage (9) into the low-pressure passage (8). The solenoid valve is open. The fuel passes through the opened solenoid-valve seat (13) into the high-pressure chamber (5).

Initial stroke (b)

The actuating cam (1) continues to rotate and forces the pump plunger downwards. The solenoid valve is open so that the pump plunger can force the fuel through the low-pressure passage (8) into the fuel supply's low-pressure stage.

Delivery stroke and injection of fuel (c)

At a given instant in time, the ECU outputs the signal to energize the solenoid-valve coil (12) so that the solenoid-valve needle is pulled into the solenoid-valve seat (13) and the connection between the high-pressure chamber and the low-pressure stage is closed. This instant in time is the electric start of injection (Beginning of Injection Period, BIP).

Further volume displacement of the pump plunger causes the fuel pressure in the high-pressure chamber to increase, so that the fuel pressure in the injection nozzle also increases. Upon reaching the nozzle-needle

opening pressure of approx. 300 bar, the nozzle needle (11) is lifted from its seat and fuel is sprayed into the engine's combustion chamber (actual start of injection). Due to the pump plunger's high delivery rate, the pressure continues to increase throughout the whole of the injection process. The maximum pressure is reached during the transitional phase between delivery stroke and residual stroke (see below).

The real nozzle-opening pressure for main injection in the UIS for passenger cars is affected by both the initial tension of the nozzle spring and the hydraulic layout of the accumulator unit and spring retainer (accumulator-plunger lift, restrictor cross-sections) (see section entitled "Pilot injection (passenger cars)").

Residual stroke (d)

As soon as the solenoid-valve coil is switched off, the solenoid valve opens after a brief delay and opens the connection between the high-pressure chamber and the low-pressure stage. The pressure collapses abruptly, and when it drops below the nozzle-closing pressure, the nozzle closes and terminates the injection process.

The remaining fuel which is delivered by the pumping element until the cam's crown point is reached is forced into the low-pressure stage via the low-pressure passage (8).

Intrinsic safety

Single-pump systems are intrinsically safe. In other words, in the unlikely event of a malfunction, one uncontrolled injection of fuel is the most that can happen: If the solenoid valve remains open, no injection can take place since the fuel flows back into the low-pressure stage and it is impossible for pressure to be built up.

When the solenoid valve is permanently closed, no fuel can enter the high-pressure chamber since the chamber can only be filled via the opened solenoid-valve seat. In this case, at the most only a single injection can take place.

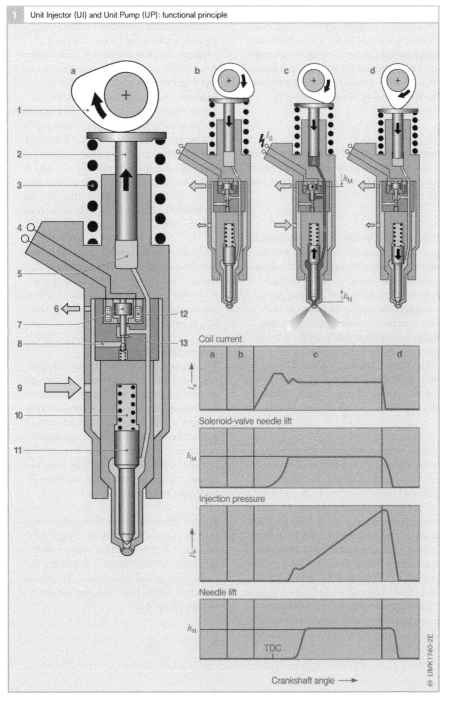

1   Unit Injector (UI) and Unit Pump (UP): functional principle

Coil current

Solenoid-valve needle lift

Injection pressure

Needle lift

TDC

Crankshaft angle ⟶

UMK1740-2E

Fig. 1
Operating states:
a Suction stroke
b Initial stroke
c Prestroke
d Residual stroke

1 Actuating cam
2 Pump plunger
3 Follower spring
4 Plug
5 High-pressure
   chamber
6 Fuel return
7 Solenoid-valve
   needle
8 Low-pressure
   passage
9 Fuel supply
10 Nozzle spring
11 Nozzle assembly
12 Solenoid-valve coil
13 Solenoid-valve seat

I_s Coil current
h_M Solenoid-valve
   needle stroke
p_o Injection pressure
h_N Nozzle-needle stroke

Pilot injection (passenger cars)

Mechanical-hydraulically controlled pilot injected is effected in the unit injector for passenger cars by an accumulator plunger and a damping unit. Pilot injection serves to reduce both noise and pollutant emissions. The functional principle is shown in Figure 2.

Initial position (a)

Nozzle needle (7) and accumulator plunger (3) are up against their seats. The solenoid valve is open, which means that no pressure can build up.

Start of pilot injection (b)

Pressure buildup starts as soon as the solenoid valve closes. At approx. 180 bar, the nozzle-opening pressure for pilot injection is significantly lower than that for main injection. When the nozzle opening pressure is reached, the needle lifts from its seat and pilot injection commences. During this phase, the nozzle needle's stroke is limited hydraulically by a damping unit.

The accumulator plunger initially remains on its seat because the nozzle needle opens first on account of its greater, hydraulically effective surface on which the pressure acts.

End of pilot injection (c)

Further pressure increase causes the accumulator plunger to be forced downwards and then lifted off its seat so that a connection is set up between the high-pressure (2) and the accumulator chambers (4). The resulting pressure drop in the high-pressure chamber, the pressure increase in the accumulator chamber, and the simultaneous increase in the initial tension of the compression spring (5) cause the nozzle needle to close. This marks the end of pilot injection.

For the most part, the pilot-injection quantity of approx. 1.5 mm³ is determined by the opening pressure and lift of the accumulator plunger.

Main injection (d)

The continuing movement of the pump plunger leads to the pressure in the high-pressure chamber continuing to increase. Because the compression spring is now subject to greater initial tension, main injection requires a higher opening pressure at the nozzle (approx. 300 bar) than pilot injection.

The accumulator plunger remains open during main injection. When it is open, it offers the fuel pressure a greater contact surface than the nozzle needle. That is why it is the nozzle needle which closes first when main injection is completed; after this, the accumulator plunger also returns to its initial position.

The time interval between pilot and main injection is chiefly determined by the accumulator-plunger lift (which for its parts is determined by the initial tension of the compression spring) and the engine speed. It is approx. 0.2...0.6 ms.

Future UI systems

Because mechanical-hydraulic pilot injection cannot be time-controlled, future-generation UI systems will feature electronically controlled pilot injection. A two-actuator concept will be adopted in the interests of a more flexible pilot-injection application (variable injection intervals). Nozzle opening will be controlled directly by an additional actuator instead of the mechanical-hydraulic accumulator.

Further developments will include increasing the injection pressure by using smaller nozzles. A higher injection pressure will significantly reduce emissions due to the improvement in fuel atomization. Future systems will deliver peak injection pressures of up to 2,400 bar.

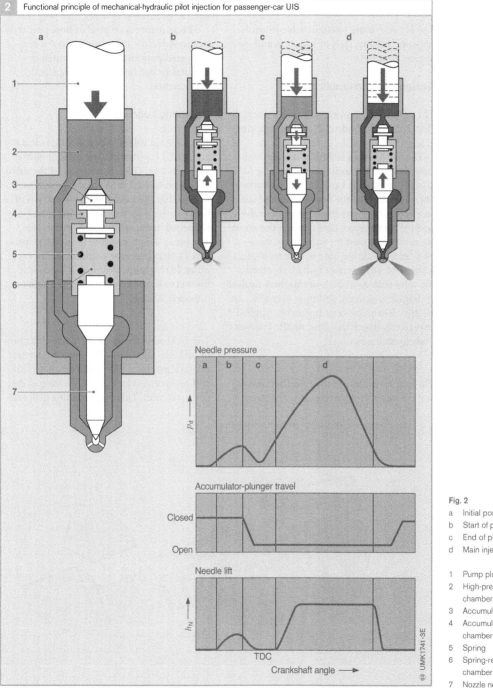

2 Functional principle of mechanical-hydraulic pilot injection for passenger-car UIS

Needle pressure

Accumulator-plunger travel

Closed

Open

Needle lift

TDC

Crankshaft angle →

Fig. 2
a   Initial position
b   Start of pilot injection
c   End of pilot injection
d   Main injection

1   Pump plunger
2   High-pressure
    chamber
3   Accumulator plunger
4   Accumulator
    chamber
5   Spring
6   Spring-retainer
    chamber
7   Nozzle needle

High-pressure solenoid valve

The high-pressure solenoid valve controls pressure buildup, start of injection and injection duration.

Design and construction

Valve

The valve itself comprises the valve needle (Fig. 1, 2), valve body (12) and valve spring (1).

The valve body's sealing surface is conically ground (10), and the valve needle is also provided with a conical sealing surface (11). The angle of the needle's ground surface is slightly larger than that of the valve body. With the valve closed, when the needle is forced up against the valve body, valve body and needle are only in contact along a line (and not a surface) which represents the valve seat. As a result of this dual-conical sealing arrangement, sealing is very efficient. High-precision processing must be applied to perfectly match the valve needle and valve body to each other.

Magnet

The magnet is comprised of the fixed stator and the movable armature (16). The stator it-self comprises the magnet core (15), a coil (6), and the electrical contacting with plug (8).

The armature is secured or non-positively connected to the valve needle. In the non-energized position, there is an initial or residual air gap between the stator and the armature.

Operating concept

Valve open

The solenoid valve is open as long as it is not actuated, i.e., when no current is applied across its coil. The force exerted by the valve spring pushes the valve needle up against the stop so that the valve flow cross-section (9) between the valve needle and the valve body is opened in the vicinity of the valve seat. The pump's high-pressure (3) and low-pressure (4) areas are now connected with each other. In this initial position, it is possible for fuel to flow into and out of the high-pressure chamber.

Valve closed

The ECU actuates the coil when an injection of fuel is to take place. The pickup current causes a magnetic flux in the magnetic-circuit components (magnet core, magnet disc and armature). This magnetic flux generates

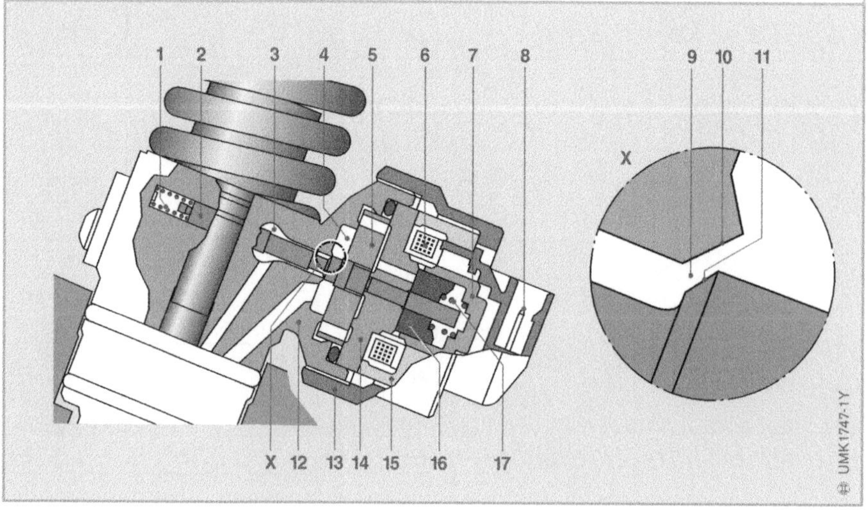

Fig. 1

1 Valve spring
2 Valve needle
3 High-pressure area
4 Low-pressure area
5 Shim
6 Solenoid-valve coil
7 Retainer
8 Plug
9 Valve flow cross-section
10 Valve-body sealing surface
11 Valve-needle sealing surface
12 Integral valve body
13 Union nut
14 Magnetic disc
15 Magnet core
16 Armature
17 Compensating spring

1   Passenger-car unit injector: high-pressure solenoid valve

a magnetic force which pulls in the armature towards the magnet disc (14) and with it moves the valve needle towards the valve body. The armature is pulled in until the valve needle and valve body come into contact at the seal seat and the valve closes. A residual air gap remains between the armature and the magnet disc.

The magnetic force is not only used to pull in the armature, but must at the same time overcome the force exerted by the valve spring and hold the armature against the spring force. In addition, the magnetic force must keep the sealing surfaces in contact with each other with a certain force in order also to withstand the pressure from the element chamber.

When the solenoid valve is closed, pressure is built up in the high-pressure chamber during the downward movement of the pump plunger to facilitate fuel injection. To stop the fuel-injection process, the current through the solenoid coil is switched off. As a result, the magnetic flux and the magnetic force collapse, and the spring forces the valve needle to its normal position against the stop. The valve seat is opened and the pressure in the high-pressure chamber is reduced.

Activation

In order for the high-pressure solenoid valve to close, it is activated with a relatively high pickup current (Fig. 2, a) with a steeply rising edge. This ensures short switching times for the solenoid valve and exact metering of the injected fuel quantity.

When the valve is closed, the pickup current can be reduced to a holding current (c) in order to keep the valve closed. This reduces the heat loss due to current flow. The holding current required is smaller, the near the armature is to the magnet disc because a small gap causes a greater magnetic force.

For a brief period between the pickup-current and holding-current phases, constant triggering current is applied to permit the detection of the solenoid-valve closing point ("BIP detection", phase b).

In order to ensure high-speed, defined opening of the solenoid valve at the end of the injection event, a high voltage is applied across the terminals for rapid quenching of the energy stored in the solenoid valve (phase d).

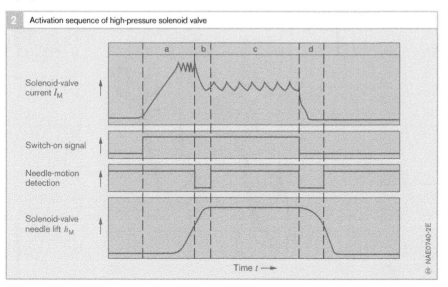

2   Activation sequence of high-pressure solenoid valve

Solenoid-valve current I_M

Switch-on signal

Needle-motion detection

Solenoid-valve needle lift h_M

Time $t \longrightarrow$

NAE0740-2E

Fig. 2
a   Pickup current (commercial-vehicle UIS/UPS: 12...20 A; passenger-car UIS: 20 A)
b   BIP detection
c   Holding current (commercial-vehicle UIS/UPS: 8...14 A; passenger-car UIS: 12 A)
d   Rapid quenching

Unit Pump System (UPS)

The Unit Pump System (UPS) is used in commercial vehicles and large engines. The Unit Pump (UP) works in the same way as the Unit Injector (UI) for commercial vehicles. In contrast to the UI, however, the nozzle and the injector are kept separate in the UP and connected to each other by a short line.

Installation and drive

The nozzle in the unit pump system is installed with a nozzle-holder assembly in the cylinder head whereas in the unit injector system it is integrated directly in the injector.

The pump is secured to the side of the engine block (Fig. 1) and driven directly by an injection cam (Fig. 2, 13) on the engine camshaft via a roller tappet (26). This offers the following advantages over the UI:

- New cylinder-head design is unnecessary,
- Rigid drive, since no rockers needed
- Simple handling for the workshop since the pumps are easy to remove

Design and construction

In contrast to the unit injector, high-pressure delivery lines are installed in the unit pump between the high-pressure pump and the nozzle. These lines must be able to permanently withstand the maximum pump pressure and the to some extent high-frequency pressure fluctuations which occur during the injection pauses. High-tensile, seamless steel tubes are therefore used. The lines are kept as short as possible and must be of identical length for the individual pumps of an engine.

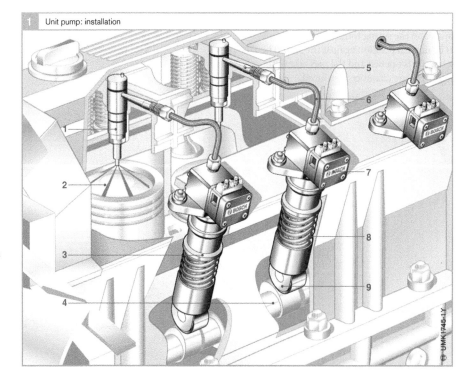

1   Unit pump: installation

Fig. 1
1 Stepped nozzle holder
2 Engine combustion chamber
3 Unit pump
4 Engine camshaft
5 Pressure fittings
6 High-pressure delivery line
7 Solenoid valve
8 Return spring
9 Roller tappet

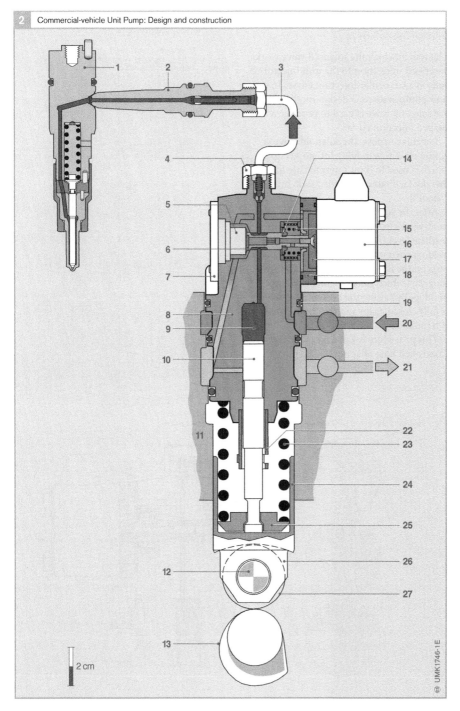

2   Commercial-vehicle Unit Pump: Design and construction

Fig. 2
1   Nozzle holder
2   Pressure fitting
3   High-pressure
    delivery line
4   Connection
5   Stroke stop
6   Solenoid-valve
    needle
7   Plate
8   Pump housing
9   High-pressure
    chamber
10 Pump plunger
11 Engine block
12 Roller-tappet pin
13 Cam
14 Spring seat
15 Solenoid-valve
    spring
16 Valve housing with
    coil and magnet
    core
17 Armature plate
18 Intermediate plate
19 Seal
20 Fuel inlet
21 Fuel return
22 Pump-plunger
    retention device
23 Tappet spring
24 Tappet body
25 Spring seat
26 Roller tappet
27 Tappet roller

2 cm

UMK1746-1E

Current-controlled rate shaping CCRS

The way in which the solenoid valve works described in relation to the unit injector results in a triangular injection curve. In some unit pump systems, a design modification of the solenoid valve is used to produce a boot-shaped injection curve.

For this purpose, the solenoid valve is equipped with a moving lift-stop (Fig. 2) which is used to limit intermediate lift and thereby facilitates a throttled switching state ("boot").

After the solenoid valve closes, the solenoid-valve current is returned to an intermediate level (Fig. 1, phase c_1) below the holding current (c_2) so that the valve needle rests on the lift-stop. This enables a throttling gap, which limits further pressure buildup. By raising the current, the valve is fully closed again and the "boot" phase is terminated.

This procedure is known as Current-Controlled Rate Shaping (CCRS).

Fig. 1
a   Pickup current
    (commercial-vehicle
    UPS: 12...20 A)
b   BIP detection
c₁ Holding current for
    boot-shaped fuel
    injection
c₂ Holding current
    (commercial-vehicle
    UPS: 8...14 A)
d   Rapid quenching

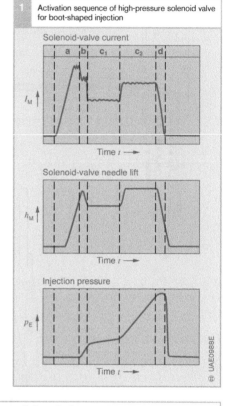

1   Activation sequence of high-pressure solenoid valve for boot-shaped injection

Solenoid-valve current

I_M

$a \quad b \quad c_1 \quad c_2 \quad d$

Time t ⟶

Solenoid-valve needle lift

h_M

Time t ⟶

Injection pressure

p_E

Time t ⟶

UAE0988E

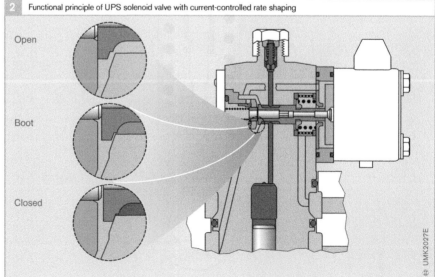

2   Functional principle of UPS solenoid valve with current-controlled rate shaping

Open

Boot

Closed

UMK2027E

1905

The idea behind the unit injector is practically as old as the diesel engine itself, and originated from Rudolf Diesel.

▼ Sketch from Rudolf Diesel's patent document from 1905

1999

It took the ideas, ingenuity, and hard work of countless engineers and technicians to turn Diesel's idea into a modern fuel-injection system.

But first of all, numerous difficulties and problems in the areas of materials technology, production engineering, control engineering, electronics, and fluid mechanics had to be overcome.

The Electronic Diesel Control (EDC) today places the Unit Injector System (UIS) in the position of being able to optimally control the various diesel-engine functions in a wide variety of operating states.

This system is capable of generating the highest injection pressures on today's market.

▼ Unit injector 1999

2000 onwards

Even considering today's high level of UIS development, there are still perspectives for the future. Further refinement of the electronic control, and even higher injection pressures, are only two of the possibilities on which the development departments are expending great effort.

In other words, the Unit Injector Systems (UIS) are well prepared for the future.

▼ Unit injector of the future

Overview of common-rail systems

The demands placed on diesel-engine fuel-injection systems are continuously increasing. Higher pressures, faster switching times, and a variable rate-of-discharge curve modified to the engine operating state have made the diesel engine economical, clean, and powerful. As a result, diesel engines have even entered the realm of luxury-performance sedans.

One of the advanced fuel-injection systems is the *common-rail (CR)* fuel-injection system. The main advantage of the common-rail system is its ability to vary injection pressure and timing over a broad scale. This was achieved by separating pressure generation (in the high-pressure pump) from the fuel-injection system (injectors). The rail here acts as a pressure accumulator.

Areas of application

The common-rail fuel-injection system for engines with diesel direct injection (Direct Injection, DI) is used in the following vehicles:

- *Passenger cars* ranging from high-economy 3-cylinder engines with displacements of 800 *cc*, power outputs of 30 kW (41 HP), torques of 100 Nm, and a fuel consumption of 3.5 l/100 km through to 8-cylinder engines in luxury-performance sedans with displacements of approx. 4 *l*, power outputs of 180 kW (245 HP), and torques of 560 Nm.
- *Light-duty trucks* with engines producing up to 30 kW/cylinder, and
- *Heavy-duty trucks, railway locomotives*, and *ships* with engines producing up to approx. 200 kW/cylinder

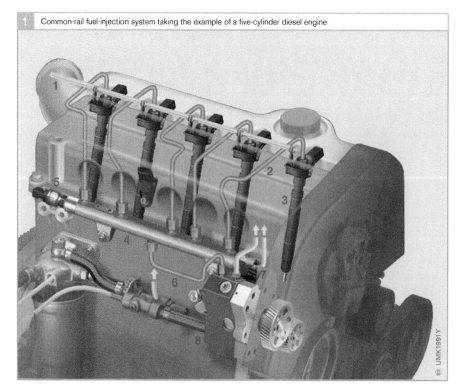

1   Common-rail fuel-injection system taking the example of a five-cylinder diesel engine

Fig. 1

1   Fuel return line
2   High-pressure fuel
    line to injector
3   Injector
4   Fuel rail
5   Rail-pressure sensor
6   High-pressure fuel
    line to rail
7   Fuel return line
8   High-pressure pump

The common-rail system is a highly flexible system for adapting fuel injection to the engine. This is achieved by:
- High injection pressure up to approx. 1,600 bar, in future up to 1,800 bar.
- Injection pressure adapted to the operating status (200...1,800 bar).
- Variable start of injection.
- Possibility of several pre-injection events and secondary injection events (even highly retarded secondary injection events).

In this way, the common-rail system helps to raise specific power output, lower fuel consumption, reduce noise emission, and decrease pollutant emission in diesel engines.

Today common rail has become the most commonly used fuel-injection system for modern, high-rev passenger-car direct-injection engines.

Design

The common-rail system consists of the following main component groups (Figs. 1 and 2):
- *The low-pressure stage,* comprising the fuel-supply system components.
- *The high-pressure system,* comprising components such as the high-pressure pump, fuel rail, injectors, and high-pressure fuel lines.
- *The electronic diesel control (EDC),* consisting of system modules, such as sensors, the electronic control unit, and actuators.

The key components of the common-rail system are the injectors. They are fitted with a rapid-action valve (solenoid valve or piezo-triggered actuator) which opens and closes the nozzle. This permits control of the injection process for each cylinder.

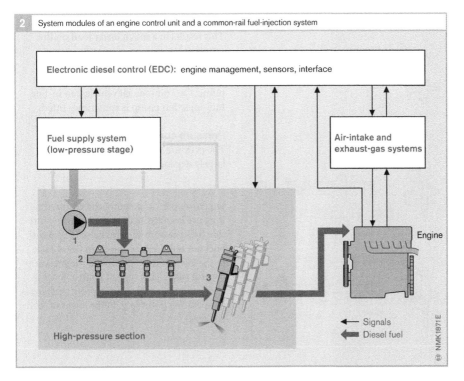

2   System modules of an engine control unit and a common-rail fuel-injection system

Electronic diesel control (EDC): engine management, sensors, interface

Fuel supply system (low-pressure stage)

Air-intake and exhaust-gas systems

Engine

High-pressure section

← Signals

← Diesel fuel

Fig. 2
1 High-pressure pump
2 Fuel rail
3 Injectors

All the injectors are fed by a common fuel rail, this being the origin of the term "common rail".

One of the main features of the common-rail system is that system pressure is variable dependent on the engine operating point. Pressure is adjusted by the pressure-control valve or the metering unit (Fig. 3).

The modular design of the common-rail system simplifies modification of the system to different engines.

Method of operation

In the common-rail fuel-injection system, the functions of pressure generation and fuel injection are separate. The injection pressure is generated independent of the engine speed and the injected fuel quantity. The Electronic Diesel Control (EDC) controls each of the components.

Pressure generation

Pressure generation and fuel injection are separated by means of an accumulator volume. Fuel under pressure is supplied to the accumulator volume of the common rail ready for injection.

A continuously operating high-pressure pump driven by the engine produces the desired injection pressure. Pressure in the fuel rail is maintained irrespective of engine speed or injected fuel quantity. Owing to the almost uniform injection pattern, the high-pressure pump design can be much smaller and its drive-system torque can be lower than conventional fuel-injection systems. This results in a much lower load on the pump drive.

The high-pressure pump is a radial-piston pump. On commercial vehicles, an in-line fuel-injection pump is sometimes fitted.

Pressure control

The pressure control method applied is largely dependent on the system.

Control on the high-pressure side

On passenger-car systems, the required rail pressure is controlled on the high-pressure side by a pressure-control valve (Fig. 3a, 4). Fuel not required for injection flows back to the low-pressure circuit via the pressure-control valve. This type of control loop allows rail pressure to react rapidly to changes in operating point (e.g. in the event of load changes).

Fig. 3

a   Pressure control on the high-pressure side by means of pressure-control valve for passenger-car applications

b   Pressure control on the suction side with a metering unit flanged to the high-pressure pump (for passenger cars and commercial vehicles)

c   Pressure control on the suction side with a metering unit and additional control with a pressure-control valve (for passenger cars)

1   High-pressure pump
2   Fuel inlet
3   Fuel return
4   Pressure-control valve
5   Fuel rail
6   Rail-pressure sensor
7   Injector connection
8   Return fuel connection
9   Pressure-relief valve
10   Metering unit
11   Pressure-control valve

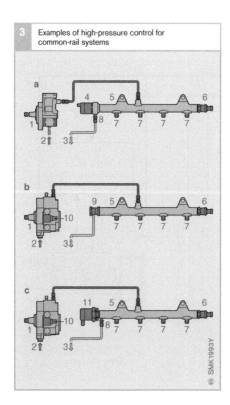

3   Examples of high-pressure control for common-rail systems

SMK1993Y

Control on the high-pressure side was adopted on the first common-rail systems. The pressure-control valve is mounted preferably on the fuel rail. In some applications, however, it is mounted directly on the high-pressure pump.

Fuel-delivery control on the suction side

Another way of controlling rail pressure is to control fuel delivery on the suction side (Fig. 3b). The metering unit (10) flanged on the high-pressure pump makes sure that the pump delivers exactly the right quantity of fuel to the fuel rail in order to maintain the injection pressure required by the system. In a fault situation, the pressure-relief valve (9) prevents rail pressure from exceeding a maximum.

Fuel-delivery control on the suction side reduces the quantity of fuel under high pressure and lowers the power input of the pump. This has a positive impact on fuel consumption. At the same time, the temperature of the fuel flowing back to the fuel tank is reduced in contrast to the control method on the high-pressure side.

Two-actuator system

The two-actuator system (Fig. 3c) combines pressure control on the suction side via the metering unit and control on the high-pressure side via the pressure-control valve, thus marrying the advantages of high-pressure-side control and suction-side fuel-delivery control (see the section on "Common-rail system for passenger cars").

Fuel injection

The injectors spray fuel directly into the engine's combustion chambers. They are supplied by short high-pressure fuel lines connected to the fuel rail. The engine control unit controls the switching valve integrated in the injector to open and close the injector nozzle.

The injector opening times and system pressure determine the quantity of fuel delivered. At a constant pressure, the fuel quantity delivered is proportional to the switching time of the solenoid valve. This is, therefore, independent of engine or pump speed (time-based fuel injection).

Potential hydraulic power

Separating the functions of *pressure generation* and *fuel injection* opens up further degrees of freedom in the combustion process compared with conventional fuel-injection systems; the injection pressure is more or less freely selectable within the program map. The maximum injection pressure at present is 1,600 bar; in future this will rise to 1,800 bar.

The common-rail system allows a further reduction in exhaust-gas emissions by introducing pre-injection events or multiple injection events and also attenuating combustion noise significantly. Multiple injection events of up to five per injection cycle can be generated by triggering the highly rapid-action switching valve several times. The nozzle-needle closing action is hydraulically assisted to ensure that the end of injection is rapid.

Control and regulation

Operating concept

The engine control unit detects the accelerator-pedal position and the current operating states of the engine and vehicle by means of sensors (see the section on "Electronic Diesel Control (EDC)"). The data collected includes:

- Crankshaft speed and angle
- Fuel-rail pressure
- Charge-air pressure
- Intake air, coolant temperature, and fuel temperature
- Air-mass intake
- Road speed, etc.

The electronic control unit evaluates the input signals. In sync with combustion, it calculates the triggering signals for the pressure-control valve or the metering unit, the injectors, and the other actuators (e.g. the EGR valve, exhaust-gas turbocharger actuators, etc.).

The injector switching times, which need to be short, are achievable using the optimized high-pressure switching valves and a special control system.

The angle/time system compares injection timing, based on data from the crankshaft and camshaft sensors, with the engine state (time control). The Electronic Diesel Control (EDC) permits a precise metering of the injected fuel quantity. In addition, EDC offers the potential for additional functions that can improve engine response and convenience.

Basic functions

The basic functions involve the precise control of diesel-fuel injection timing and fuel quantity at the reference pressure. In this way, they ensure that the diesel engine has low consumption and smooth running characteristics.

Correction functions for calculating fuel injection

A number of correction functions are available to compensate for tolerances between the fuel-injection system and the engine (see the section on "Electronic Diesel Control (EDC)"):

- Injector delivery compensation
- Zero delivery calibration
- Fuel-balancing control
- Average delivery adaption

Additional functions

Additional open- and closed-loop control functions perform the tasks of reducing exhaust-gas emissions and fuel consumption, or providing added safety and convenience. Some examples are:

- Control of exhaust-gas recirculation
- Boost-pressure control
- Cruise control
- Electronic immobilizer, etc.

Integrating EDC in an overall vehicle system opens up a number of new opportunities, e.g. data exchange with transmission control or air-conditioning system.

A diagnosis interface permits analysis of stored system data when the vehicle is serviced.

Control unit configuration

As the engine control unit normally has a maximum of only eight output stages for the injectors, engines with more than eight cylinders are fitted with two engine control units. They are coupled within the "master/slave" network via an internal, high-speed CAN interface. As a result, there is also a higher microcontroller processing capacity available. Some functions are permanently allocated to a specific control unit (e.g. fuel-balancing control). Others can be dynamically allocated to one or other of the control units as situations demand (e.g. to detect sensor signals).

▶ Diesel boom in Europe

Diesel engine applications

At the start of automobile history, the spark-ignition engine (Otto cycle) was the drive unit for road vehicles. The first time a diesel engine was mounted on a truck was 1927. Passenger cars had to wait until 1936.

The diesel engine made strong headway in the truck sector due to its fuel economy and long service life. By contrast, the diesel engine in the car sector was long relegated to a fringe existence. It was only with the introduction of supercharged direct-injection diesel engines – the principle of direct injection was already used in the first truck diesel engines – that the diesel engine changed its image. Meanwhile, the percentage of diesel-engined passenger cars among new registrations is fast approaching 50% in Europe.

Features of the diesel engine

What is the reason for the boom in diesel engines in Europe?

Fuel economy

Firstly, fuel consumption compared to gasoline engines is still lower – this is due to the greater efficiency of the diesel engine. Secondly, diesel fuel is subject to lower taxes in most European countries. For people who travel a lot, therefore, diesel is the more economical alternative despite the higher purchase price.

Driving pleasure

Almost all diesel engines on the market are supercharged. This produces a high cylinder charge at low revs. The metered fuel quantity can also be high, and this produces high engine torque. The result is a torque curve that permits driving at high torque and low revs.

It is torque, and not engine performance, that is the decisive factor for engine power. Compared to a gasoline engine without supercharging, a driver can experience more "driving pleasure" with a diesel engine of lower performance. The image of the "stinking slowcoach" is simply no longer true for diesel-engined cars of the latest generation.

Environmental compatibility

The clouds of smoke that diesel-engined cars produced when driven at high loads are a thing of the past. This was brought about by improved fuel-injection systems and Electronic Diesel Control (EDC). These systems can meter fuel quantity with high precision, adjusting it to the engine operating point and environmental conditions. This technology also meets prevailing exhaust-gas emission standards.

Oxidation-type catalytic converters, that remove carbon monoxide (CO) and hydrocarbons (HC) from exhaust gas, are standard equipment on diesel engines. Future, more stringent exhaust-gas emission standards, and even U.S. legislation, will be met by other exhaust-gas treatment systems, such as particulate filters and NO_x accumulator-type catalytic converters.

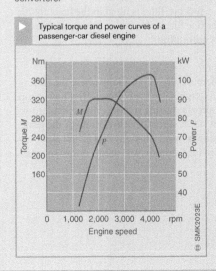

▶ Typical torque and power curves of a passenger-car diesel engine

Common-rail system for passenger cars

Fuel supply

In common-rail systems for passenger cars, electric fuel pumps or gear pumps are used to deliver fuel to the high-pressure pump.

Systems with electric fuel pump

The electric fuel pump is either part of the in-tank unit (in the fuel tank) or is fitted in the fuel line (in-line). It intakes fuel via a pre-filter and delivers it to the high-pressure pump at a pressure of 6 bar (Fig. 3). The maximum delivery rate is 190 *l*/h. To ensure fast engine starting, the pump switches on as soon as the driver turns the ignition key. This builds up the necessary pressure in the low-pressure circuit when the engine starts.

The fuel filter (fine filter) is fitted in the supply line to the high-pressure pump.

Systems with gear pump

The gear pump is flanged to the high-pressure pump and is driven by its input shaft (Figs. 1 and 2). In this way, the gear pump starts delivery only after the engine has started. Delivery rate is dependent on the engine speed and reaches rates up to 400 *l*/h at pressures up to 7 bar.

A fuel pre-filter is fitted in the fuel tank. The fine filter is located in the supply line to the gear pump.

Combination systems

There are also applications where the two pump types are used. The electric fuel pump improves starting response, in particular for hot starts, since the delivery rate of the gear pump is lower when the fuel is hot, and therefore, thinner, and at low pump speeds.

High-pressure control

On first-generation common-rail systems, rail pressure is controlled by the pressure-control valve. The high-pressure pump (type CP1) generates the maximum delivery quantity, irrespective of fuel demand. The pressure-control valve returns excess fuel to the fuel tank.

Second-generation common-rail systems control rail pressure on the low-pressure side by means of the metering unit (Figs. 1 and 2). The high-pressure pump (types CP3 and CP1H) need only deliver the fuel quantity that the engine actually requires. This lowers the energy demand of the high-pressure pump and reduces fuel consumption.

Third-generation common-rail systems feature piezo-inline injectors (Fig. 3).

If pressure is only adjustable on the low-pressure side, it takes too long to lower the pressure in the fuel rail when rapid negative load changes occur. Adapting pressure to dynamic changes in load conditions is then too slow. This is particularly the case with piezo-inline injectors due to their very low internal leakage. For this reason, some common-rail systems are equipped with an additional pressure-control valve (Fig. 3) besides the high-pressure pump and metering unit. This two-actuator system combines the advantages of control on the low-pressure side with the dynamic response of control on the high-pressure side.

Another advantage compared with control on the low-pressure side only is that the high-pressure side is also controllable when the engine is cold. The high-pressure pump then delivers more fuel than is injected and pressure is controlled by the pressure-control valve. Compression heats the fuel, thus eliminating the need for an additional fuel heater.

1 Example of a second-generation common-rail system for a 4-cylinder engine

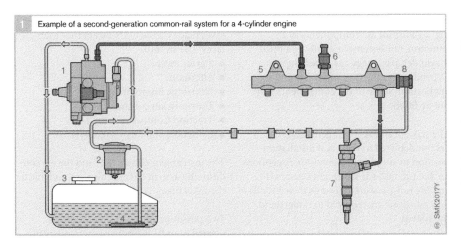

Fig. 1

1 High-pressure pump
   (CP3) with fitted
   geared presupply
   pump and metering
   unit
2 Fuel filter with water
   separator and heater
   (optional)
3 Fuel tank
4 Pre-filter
5 Fuel rail
6 Rail-pressure sensor
7 Solenoid-valve
   injector
8 Pressure-relief valve

2 Example of a second-generation common-rail system with two-actuator system for a V8 engine

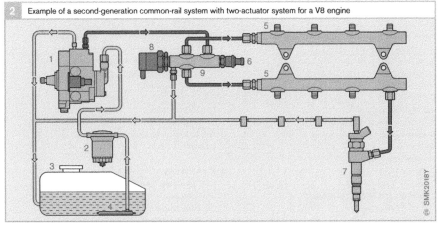

Fig. 2

1 High-pressure pump
   (CP3) with fitted
   geared presupply
   pump and metering
   unit
2 Fuel filter with water
   separator and heater
   (optional)
3 Fuel tank
4 Pre-filter
5 Fuel rail
6 Rail-pressure sensor
7 Solenoid-valve
   injector
8 Pressure-control valve
9 Function module
   (distributor)

3 Example of a third-generation common-rail system with two-actuator system for a 4-cylinder engine

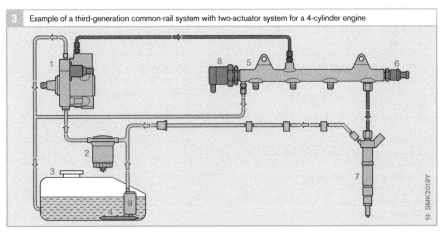

Fig. 3

1 High-pressure pump
   (CP1H) with
   metering unit
2 Fuel filter with water
   separator and heater
   (optional)
3 Fuel tank
4 Pre-filter
5 Fuel rail
6 Rail-pressure sensor
7 Piezo-inline injector
8 Pressure-control
   valve
9 Electric fuel pump

System diagram for passenger cars

Fig. 4 shows all the components in a common-rail system for a fully equipped, 4-cylinder, passenger-car diesel engine. Depending on the type of vehicle and its application, some of the components may not be fitted.

The sensors and setpoint generators (A) are not depicted in their real installation position to simplify presentation. Exceptions are the exhaust-gas treatment sensors (F) and the rail-pressure sensor as their installation positions are required to understand the system.

Data exchange between the various sections takes place via the CAN bus in the "Interfaces" (B) section:

● Starter motor
● Alternator
● Electronic immobilizer
● Transmission control
● Traction Control System (TSC)
● Electronic Stability Program (ESP)

The instrument cluster (13) and the air-conditioning system (14) are also connectable to the CAN bus.

Two possible combined systems are described (a or b) for exhaust-gas treatment.

Fig. 4

Engine, engine management, and high-pressure fuel-injection components
17 High-pressure pump
18 Metering unit
25 Engine ECU
26 Fuel rail
27 Rail-pressure sensor
28 Pressure-control valve (DRV 2)
29 Injector
30 Glow plug
31 Diesel engine (DI)
M Torque

A Sensors and setpoint generators
1   Pedal-travel sensor
2   Clutch switch
3   Brake contacts (2)
4   Operator unit for vehicle-speed controller (cruise control)
5   Glow-plug and starter switch ("ignition switch")
6   Road-speed sensor
7   Crankshaft-speed sensor (inductive)
8   Camshaft-speed sensor (inductive or Hall sensor)
9   Engine-temperature sensor (in coolant circuit)
10 Intake-air temperature sensor
11 Boost-pressure sensor
12 Hot-film air-mass meter (intake air)

B Interfaces
13 Instrument cluster with displays for fuel consumption, engine speed, etc.
14 Air-conditioner compressor with operator unit
15 Diagnosis interface
16 Glow control unit
CAN Controller Area Network
    (on-board serial data bus)

C Fuel-supply system (low-pressure stage)
19 Fuel filter with overflow valve
20 Fuel tank with pre-filter and Electric Fuel Pump, EFP (presupply pump)
21 Fuel-level sensor

D Additive system
22 Additive metering unit
23 Additive control unit
24 Additive tank

E Air supply
32 Exhaust-gas recirculation cooler
33 Boost-pressure actuator
34 Turbocharger (in this case with Variable Turbine Geometry (VTG))
35 Control flap
36 Exhaust-gas recirculation actuator
37 Vacuum pump

F Exhaust-gas treatment
38 Broadband lambda oxygen sensor, type LSU
39 Exhaust-gas temperature sensor
40 Oxidation-type catalytic converter
41 Particulate filter
42 Differential-pressure sensor
43 NO_x accumulator-type catalytic converter
44 Broadband lambda oxygen sensor, optional NO_x sensor

4   Common-rail diesel fuel-injection system for cars

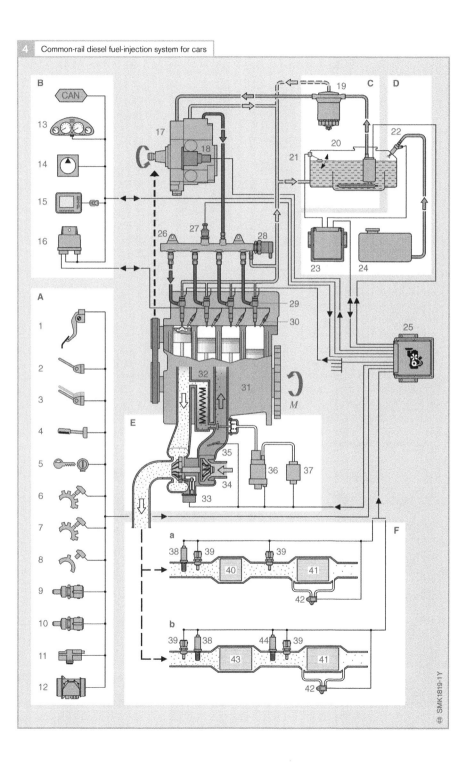

Areas of application

Diesel engines are characterized by high fuel economy. Since the first volume-production fuel-injection pump was introduced by Bosch in 1927, fuel-injection systems have experienced a process of continuous development.

Diesel engines are used in a wide variety of design for many different purposes (Fig. 1 and Table 1), for example

- To drive mobile power generators (up to approx. 10 kW/cylinder)
- As fast-running engines for cars and light-duty trucks (up to approx. 50 kW/cylinder)
- As engines for construction-industry and agricultural machinery (up to approx. 50 kW/cylinder)
- As engines for heavy trucks, omnibuses and tractor vehicles (up to approx. 80 kW/cylinder)
- To drive fixed installations such as emergency power generators (up to approx. 160 kW/cylinder)
- As engines for railway locomotives and ships (up to 1,000 kW/cylinder)

Requirements

Ever stricter statutory regulations on noise and exhaust-gas emissions and the desire for more economical fuel consumption continually place greater demands on the fuel-injection system of a diesel engine.

Basically, the fuel-injection system is required to inject a precisely metered amount of fuel at high pressure into the combustion chamber in such a way that it mixes effectively with the air in the cylinder as demanded by the type of engine (direct or indirect-injection) and its present operating status. The power output and speed of a diesel engine is controlled by means of the injected fuel quantity as it has no air intake throttle.

Mechanical control of diesel fuel-injection systems is being increasingly replaced by Electronic Diesel Control (EDC) systems. All new diesel-injection systems for cars and commercial vehicles are electronically controlled.

1   Applications for Bosch diesel fuel-injection systems

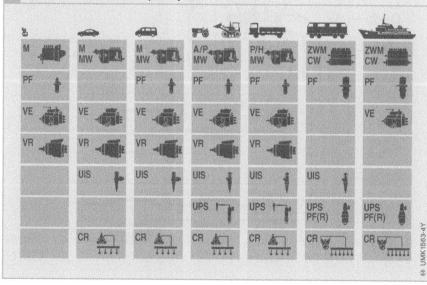

Fig. 1
M, MW,
A, P, H,
ZWM,
CW In-line fuel-injection
    pumps of increas-
    ing size
PF Discrete fuel-
    injection pumps
VE Axial-piston pumps
VR Radial-piston
    pumps
UIS Unit injector system
UPS Unit pump system
CR Common-rail
    system

Common-rail system for commercial vehicles

Fuel supply

Presupply

Common-rail systems for light-duty trucks differ very little from passenger-car systems. Electric fuel pumps or gear pumps are used for fuel presupply. On common-rail systems for heavy-duty trucks, only gear pumps are used to deliver fuel to the high-pressure pump (see the subsection "Gear-type fuel pump" in the section "Fuel supply to the low-pressure stage"). The presupply pump is normally flanged to the high-pressure pump (Figs. 1 and 2). In many applications, it is mounted on the engine.

Fuel filtering

As opposed to passenger-car systems, the fuel filter (fine filter) is fitted to the pressure side. For this reason, an exterior fuel inlet is required, in particular when the gear pump is flanged to the high-pressure pump.

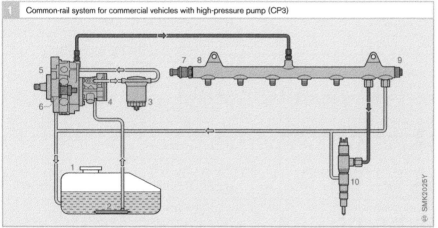

1 Common-rail system for commercial vehicles with high-pressure pump (CP3)

Fig. 1
1 Fuel tank
2 Pre-filter
3 Fuel filter
4 Gear presupply pump
5 High-pressure pump (CP3.4)
6 Metering unit
7 Rail-pressure sensor
8 Fuel rail
9 Pressure-relief valve
10 Injector

2 Common-rail system for commercial vehicles with high-pressure pump (CPN2)

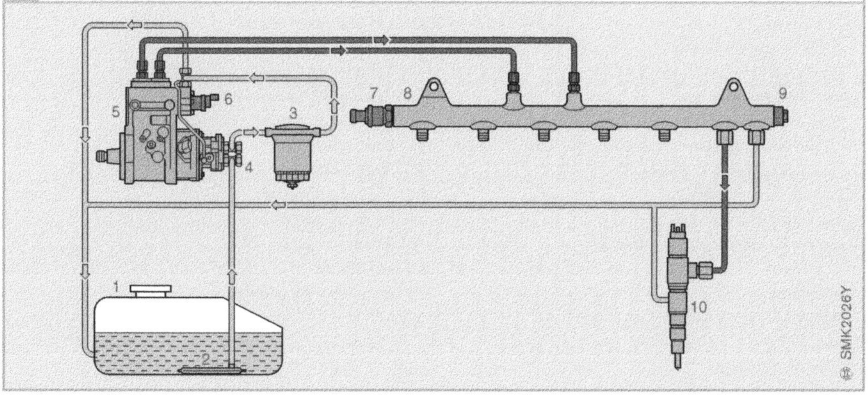

Fig. 2
1 Fuel tank
2 Pre-filter
3 Fuel filter
4 Gear presupply pump
5 High-pressure pump (CPN2.2)
6 Metering unit
7 Rail-pressure sensor
8 Fuel rail
9 Pressure-relief valve
10 Injector

System diagram for commercial vehicles

Fig. 3 shows all the components in a common-rail system for a 6-cylinder commercial-vehicle diesel engine. Depending on the type of vehicle and its application, some of the components may not be fitted.

Only the sensors and setpoint generators are depicted at their real position to simplify presentation, as their installation positions are required to understand the system.

Data exchange to the various sections takes place via the CAN bus in the "Interfaces" (B) section (e.g. transmission control,

Traction Control System (TCS), Electronic Stability Program (ESP), oil-grade sensor, trip recorder, **A**ctive **C**ruise **C**ontrol (ACC), brake coordinator – up to 30 ECUs). The alternator (18) and the air-conditioning system (17) are also connectable to the CAN bus.

Three systems are described for exhaust-gas treatment: a purely DPF system (a) mainly for the U.S. market, a purely SCR system (b) mainly for the EU market, and a combined system (c).

Fig. 3

Engine, engine management, and high-pressure fuel-injection components

22 High-pressure pump
29 Engine ECU
30 Fuel rail
31 Rail-pressure sensor
32 Injector
33 Relay
34 Auxiliary equipment (e.g. retarder, exhaust flap for engine brake, starter motor, fan)
35 Diesel engine (DI)
36 Flame glow plug (alternatively grid heater)
M Torque

A   Sensors and setpoint generators

1 Pedal-travel sensor
2 Clutch switch
3 Brake contacts (2)
4 Engine brake contact
5 Parking brake contact
6 Operating switch (e.g. vehicle-speed controller, intermediate-speed regulation, rpm- and torque reduction)
7 Starter switch ("ignition lock")
8 Turbocharger-speed sensor
9 Crankshaft-speed sensor (inductive)
10 Camshaft-speed sensor
11 Fuel-temperature sensor
12 Engine-temperature sensor (in coolant circuit)
13 Boost-air temperature sensor
14 Boost-pressure sensor
15 Fan-speed sensor
16 Air-filter differential-pressure sensor

B   Interfaces

17 Air-conditioner compressor with operator unit
18 Alternator
19 Diagnosis interface

20 SCR control unit
21 Air compressor
CAN **C**ontroller **A**rea **N**etwork (on-board serial data bus) (up to three data buses)

C   Fuel-supply system (low-pressure stage)

23 Fuel presupply pump
24 Fuel filter with water-level and pressure sensors
25 Control unit cooler
26 Fuel tank with pre-filter
27 Pressure-relief valve
28 Fuel-level sensor

D   Air intake

37 Exhaust-gas recirculation cooler
38 Control flap
39 Exhaust-gas recirculation positioner with exhaust-gas recirculation valve and position sensor
40 Intercooler with bypass for cold starting
41 Exhaust-gas turbocharger (in this case with variable turbine geometry) with position sensor
42 Boost-pressure actuator

E   Exhaust-gas treatment

43 Exhaust-gas temperature sensor
44 Oxidation-type catalytic converter
45 Differential-pressure sensor
46 Catalyst-coated particulate filter (CSF)
47 Soot sensor
48 Level sensor
49 Reducing-agent tank
50 Reducing-agent pump
51 Reducing-agent injector
52 NO_x sensor
53 SCR catalytic converter
54 NH_3 sensor

3 Common-rail diesel fuel-injection system for commercial vehicles

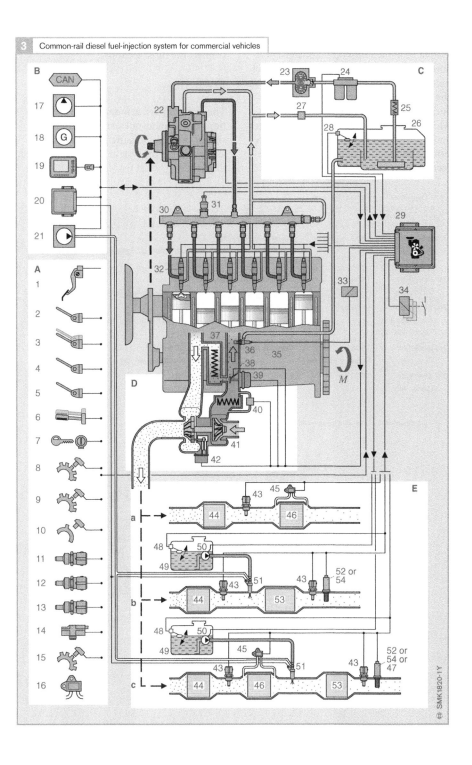

High-pressure components of common-rail system

The high-pressure stage of the common-rail system is divided into three sections: pressure generation, pressure storage, and fuel metering. The high-pressure pump assumes the function of pressure generation. Pressure storage takes place in the fuel rail to which the rail-pressure sensor and the pressure-control and pressure-relief valves are fitted. The function of the injectors is correct timing and metering the quantity of fuel injected. High-pressure fuel lines interconnect the three sections.

Overview

The main difference in the various generations of common-rail systems lie in the design of the high-pressure pump and the injectors, and in the system functions required (Table 1).

1   Overview of common-rail systems

CR generation	Maximum pressure	Injector	High-pressure pump
1st generation Pass. cars	1,350...1,450 bar	Solenoid-valve injector	CP1 Pressure control on high-pressure side by pressure-control valve
1st generation Comm. veh.	1,400 bar	Solenoid-valve injector	CP2 Suction-side fuel-delivery control by two solenoid valves
2nd generation Pass. cars and comm. veh.	1,600 bar	Solenoid-valve injector	CP3, CP1H Suction-side fuel-delivery control by metering unit
3rd generation Pass. cars	1,600 bar (in future 1,800 bar)	Piezo-inline injector	CP3, CP1H Suction-side fuel-delivery control by metering unit
3rd generation Comm. veh.	1,800 bar	Solenoid-valve injector	CP3.3NH Metering unit

Table 1

1   Common-rail fuel-injection system taking the example of a four-cylinder diesel engine

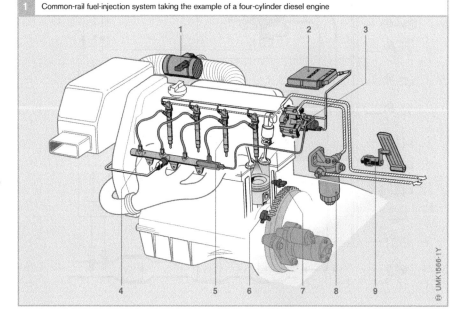

Fig. 1

1   Hot-film air-mass meter
2   Engine ECU
3   High-pressure pump
4   High-pressure accumulator (fuel rail)
5   Injector
6   Crankshaft-speed sensor
7   Engine-temperature sensor
8   Fuel filter
9   Pedal-travel sensor

UMK1566-1Y

The piezoelectric effect

In 1880 Pierre Curie and his brother Jacques discovered a phenomenon that is still very little known today, but is present in the everyday lives of millions of people: the piezoelectric effect. For example, it keeps the pointers of a crystal clock operating in time.

Certain crystals (e.g. quartz and turmaline) are piezoelectric: Electric charges are induced on the crystal surface by exerting a compression or elongation force along certain crystal axes. This electrical polarization arises by shifting positive and negative ions in the crystal relative to each other by exerting force (see Fig., b). The shifted centers of charge gravity within the crystal compensate automatically, but an electric field forms between the end faces of the crystal. Compressing and elongating the crystal create inverse field directions.

On the other hand, if an electrical voltage is applied to the end faces of the crystal, the effect reverses (inverse piezoelectric effect): The positive ions in the electric field migrate toward the negative electrode, and negative ions toward the positive electrode. The crystal then contracts or expands depending on the direction of the electric field strength (see Fig., c).

The following applies to piezoelectric field strength E_p:

$E_p = \delta \, \Delta x / x$

$\Delta x / x$: relative compression or elongation
δ: piezoelectric coefficient, numeric value 10^9 V/cm through 10^{11} V/cm

The change in length Δx results from the following when a voltage U is applied:
$U / \delta = \Delta x$ (using quartz as an example: deformation of about 10^{-9} cm at $U = 10$ V)

The piezoelectric effect is not only used in quartz clocks and piezo-inline injectors, it has many other industrial applications, either as a direct or inverse effect:

Piezoelectric sensors are used for knock control in gasoline engines. For example, they detect high-frequency engine vibrations as a feature of combustion knock. Converting mechanical vibration to electric voltage is also used in the crystal audio pickup of a record player or crystal microphones. The piezoelectric igniter (e.g. in a firelighter) causes mechanical pressure to produce the voltage to generate a spark.

On the other hand, if an alternating voltage is applied to a piezoelectrical crystal, it vibrates mechanically at the same frequency as the alternating voltage. Oscillating crystals are used as stabilizers in electrical oscillating circuits or as piezoelectric acoustic sources to generate ultrasound.

When used in clocks, the oscillating quartz is excited by an alternating voltage whose frequency is the same as the quartz's natural frequency. This is how an extremely time-constant resonant frequency is generated. In a calibrated quartz, it deviates by only approx. 1/1,000 second per year.

Principle of the piezoelectric effect (represented as a unit cell)

a   Quartz crystal SiO_2

b   Piezoelectric effect: When the crystal is compressed, negative O^{2-} ions shift upward, positive Si^{4+} ions shift downward: Electric charges are induced at the crystal surface.

c   Inverse piezoelectric effect: By applying an electrical voltage, O^{2-} ions shift upward, Si^{4+} ions shift downward: The crystal contracts.

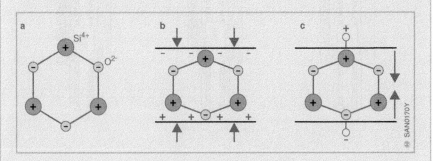

SAN0170Y

Injector

On a common-rail diesel injection system, the injectors are connected to the fuel rail by short high-pressure fuel lines. The injectors are sealed to the combustion chamber by a copper gasket. The injectors are fitted into the cylinder head by means of taper locks. Depending on the injection-nozzle design, common-rail injectors are intended for straight or inclined mounting in direct-injection diesel engines.

One of the system's features is that it generates an injection pressure irrespective of engine speed or injected fuel quantity. The start of injection and injected fuel quantity are controlled by the electrically triggered injector. The injection time is controlled by the angle/time system of the Electronic Diesel Control (EDC). This requires the use of sensors to detect the crankshaft position and the camshaft position (phase detection).

An optimum mixture formation is required to reduce exhaust-gas emissions and comply with continuous demands to reduce the noise of diesel engines. This calls for injectors with very small pre-injection quantities and multiple injection events.

There are presently three different injector types in serial production:
- Solenoid-valve injectors with one-part armature
- Solenoid-valve injectors with two-part armature
- Injector with piezo actuator

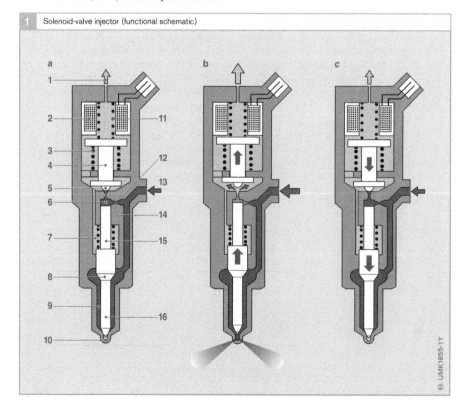

1   Solenoid-valve injector (functional schematic)

Fig. 1

a   Resting position
b   Injector opens
c   Injector closes

1   Fuel-return
2   Solenoid coil
3   Overstroke spring
4   Solenoid armature
5   Valve ball
6   Valve-control chamber
7   Nozzle spring
8   Pressure shoulder of nozzle needle
9   Chamber volume
10 Injection orifice
11 Solenoid-valve spring
12 Outlet restrictor
13 High-pressure connection
14 Inlet restrictor
15 Valve plunger (control plunger)
16 Nozzle needle

UMK1855-1Y

Solenoid-valve injector

Design

The injector can be subdivided into a number of function modules:

- The hole-type nozzle (see the section on "Injection nozzles")
- The hydraulic servo system
- The solenoid valve

Fuel is conveyed by the high-pressure connection (Fig. 1a, 13) via a supply passage to the injection nozzle and via an inlet restrictor (14) to the valve-control chamber (6). The valve-control chamber is connected to the fuel return (1) via the outlet restrictor (12) which can be opened by a solenoid valve.

Operating concept

The function of the injector can be subdivided into four operating states when the engine and the high-pressure pump are operating:

- Injector closed (with high pressure applied)
- Injector opens (start of injection)
- Injector fully open
- Injector closes (end of injection)

The operating states are caused by the balance of forces acting on the injector components. When the engine is not running and the fuel rail is not pressurized, the nozzle spring closes the injector.

Injector closed (resting position)

In its resting position, the injector is not triggered (Fig. 1a). The solenoid-valve spring (11) presses the valve ball (5) onto the seat of the outlet restrictor (12). Inside of the valve-control chamber, the pressure rises to the pressure in the fuel rail. The same pressure is also applied to the chamber volume (9) of the nozzle. The forces applied by the rail pressure to the end faces of the control plunger (15), and the force of the nozzle spring (7) retain the nozzle needle closed against the opening force applied to its pressure shoulders (8).

Injector opens (start of injection)

To begin with, the injector is in its resting position. The solenoid valve is triggered by the "pickup current". This makes the solenoid valve open very rapidly (Fig. 1b). The required rapid switching times are achieved by controlling solenoid-valve triggering in the ECU at high voltages and currents.

The magnetic force of the now triggered electromagnet exceeds the force of the valve spring. The armature raises the valve ball from the valve seat and opens the outlet restrictor. After a short time the increased pickup current is reduced to a lower holding current in the electromagnet. When the outlet restrictor opens, fuel flows from the valve-control chamber to the cavity above and then via the fuel-return line to the fuel tank. The inlet restrictor (14) prevents a complete pressure compensation. As a result, pressure in the valve-control chamber drops. Pressure in the valve-control chamber falls below the pressure in the nozzle chamber, which is still the same as the pressure in the fuel rail. The reduction in pressure in the valve-control chamber reduces the force acting on the control plunger and opens the nozzle needle. Fuel injection commences.

Injector fully open

The rate of movement of the nozzle needle is determined by the difference in the flow rates through the inlet and outlet restrictors. The control plunger reaches its upper stop and dwells there on a cushion of fuel (hydraulic stop). The cushion is created by the flow of fuel between the inlet and outlet restrictors. The injector nozzle is then fully open. Fuel is injected into the combustion chamber at a pressure approaching that in the fuel rail.

The balance of forces in the injector is similar to that during the opening phase. At a given system pressure, the fuel quantity injected is proportional to the length of time that the solenoid valve is open. This is entirely independent of the engine or pump speed (time-based injection system).

Injector closes (end of injection)
When the solenoid valve is no longer triggered, the valve spring presses the armature down and the valve ball closes the outlet restrictor (Fig. 1c). When the outlet restrictor closes, pressure in the control chamber rises again to that in the fuel rail via the inlet restrictor. The higher pressure exerts a greater force on the control plunger. The force on the valve-control chamber and the nozzle-spring force then exceed the force acting on the nozzle needle, and the nozzle needle closes. The flow rate of the inlet restrictor determines the closing speed of the nozzle needle. The fuel-injection cycle comes to an end when the nozzle needle is resting against its seat, thus closing off the injection orifices.

This indirect method is used to trigger the nozzle needle by means of a hydraulic servo system because the forces required to open the nozzle needle rapidly cannot be generated directly by the solenoid valve. The "control volume" required in addition to the injected fuel quantity reaches the fuel-return line via the restrictors in the control chamber.

In addition to the control volume, there are also leakage volumes through the nozzle-needle and valve-plunger guides. The control and leakage volumes are returned to the fuel tank via the fuel-return line and a collective line that comprises an overflow valve, high-pressure pump, and pressure-control valve.

Program-map variants
Program maps with fuel-quantity flat curve
With injectors, a distinction is made in the program map between ballistic and nonballistic modes. The valve plunger/nozzle needle unit reaches the hydraulic stop if the triggering period in vehicle operation is of sufficient length (Fig. 2a). The section until the nozzle needle reaches its maximum stroke is termed ballistic mode. The ballistic and nonballistic sections in the fuel-quantity map, where the injected fuel quantity is applied for the triggering period (Fig. 2b), is separated by a kink in the program map.

Another feature of the fuel-quantity map is the flat curve that occurs with small triggering periods. The flat curve is caused by the solenoid armature rebounding on opening. In this section, the injected fuel quantity is independent on the triggering period. This allows small injected fuel quantities to be represented as stable. Only after the armature has stopped rebounding does the injected fuel quantity curve continue to rise linearly as the triggering period becomes longer.

Injection events with small injected fuel quantities (short triggering periods) are used as pre-injection in order to suppress noise. Secondary injection events help to enhance soot oxidation in selected sections of the operating curve.

Program maps without fuel-quantity flat curve
The increasing stringency of emission-control legislation has lead to the use of the two system functions: *injector delivery compensation* (IMA) and *zero delivery calibration* (NMK), as well as to short intervals in injection between pre-injection, main injection, and secondary injection events. With injectors that have no flat-curve section, IMA allows a precise adjustment of the pre-injection fuel quantity when new. NMK corrects fuel-quantity drifts over time in the low-pressure section. The key condition for deploying these two system functions is a constant, linear rise in quantity, i.e. there is no flat curve in the fuel-quantity map (Fig. 2c). If the valve plunger/nozzle needle unit is operated in nominal mode without lift-stop at the same time, this represents a fully ballistic operating mode of the valve plunger and there is no kink in the fuel-quantity map.

Injector variants
A distinction is made between two different solenoid-valve concepts with solenoid-valve injectors:
- Injectors with one-part armature (1-spring system)
- Injectors with two-part armature (2-spring system)

The short intervals between injection events are ensured when the armature can return to its resting position very rapidly on closing. This is best achieved by a two-part armature with an overstroke stop. During the closing process, the armature plate moves down by positive locking. The bottoming-out of the armature plate is limited by an overstroke stop. As a result, the armature reaches its resting position faster. Armature rebound on closing can end faster by decoupling the armature masses and adapting the setting parameters. This helps to achieve shorter intervals between two injection events with the two-part armature concept.

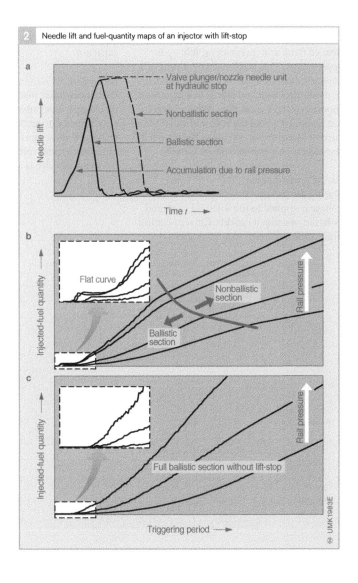

2   Needle lift and fuel-quantity maps of an injector with lift-stop

a

Needle lift

Valve plunger/nozzle needle unit at hydraulic stop

Nonballistic section

Ballistic section

Accumulation due to rail pressure

Time *t*

b

Injected-fuel quantity

Rail pressure

Flat curve

Nonballistic section

Ballistic section

c

Injected-fuel quantity

Rail pressure

Full ballistic section without lift-stop

Triggering period

UMK1983E

Triggering the solenoid-valve injector

In its resting position, the injector's high-pressure solenoid valve is not triggered and is therefore closed. The injector injects when the solenoid valve opens.

Triggering the solenoid valve is divided into five phases (Figs. 3 and 4).

Opening phase

Initially, in order to ensure tight tolerances and high levels of reproducibility for the injected fuel quantity, the current for opening the solenoid valve features a steep, precisely defined flank and increases rapidly up to approx. 20 A. This is achieved by means of a *booster voltage* of up to 50 V. It is generated in the control unit and stored in a capacitor (booster-voltage capacitor). When this voltage is applied across the solenoid valve, the current increases several times faster than it does when only battery voltage is used.

Pickup-current phase

During the pickup-current phase, battery voltage is applied to the solenoid valve and assists in opening it quickly. Current control limits pickup current to approx. 20 A.

Holding-current phase

In order to reduce power loss in the ECU and injector, the current is dropped to approx. 13 A in the holding-current phase. The energy which becomes available when pickup current and holding current are reduced is routed to the booster-voltage capacitor.

Switchoff

When the current is switched off in order to close the solenoid valve, the surplus energy is also routed to the booster-voltage capacitor.

Recharging the step-up chopper

Recharging takes place by means of a step-up chopper integrated in the ECU. The energy tapped during the opening phase is recharged at the start of the pickup phase until the original voltage required to open the solenoid valve is reached.

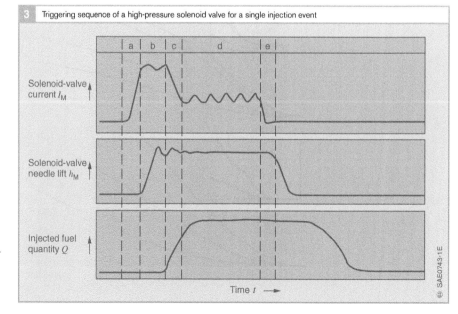

3 Triggering sequence of a high-pressure solenoid valve for a single injection event

Solenoid-valve current I_M

Solenoid-valve needle lift h_M

Injected fuel quantity Q

Time t ⟶

SAE0743-1E

Fig. 3
a　Opening phase
b　Pickup-current phase
c　Transition to holding-current phase
d　Holding-current phase
e　Switchoff

4   Common-rail system: Block diagram of the triggering phases for a cylinder group

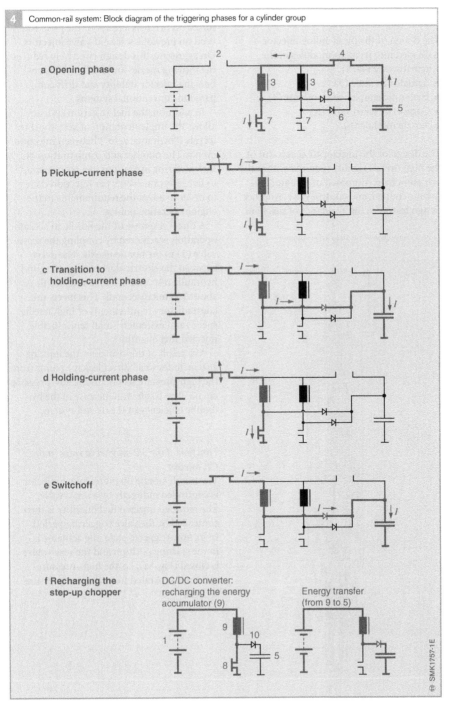

a Opening phase

b Pickup-current phase

c Transition to
   holding-current phase

d Holding-current phase

e Switchoff

f Recharging the
   step-up chopper

DC/DC converter:
recharging the energy
accumulator (9)

Energy transfer
(from 9 to 5)

SMK1757-1E

Fig. 4
1 Battery
2 Current control
3 Solenoid windings
   of the high-pressure
   solenoid valves
4 Booster switch
5 Booster-voltage
   capacitor
6 Free-wheeling
   diodes for energy
   recovery and high-
   speed quenching
7 Cylinder selector
   switch
8 DC/DC switch
9 DC/DC coil
10 DC/DC diode
I Current flow

Piezo-inline injector

Design and requirements

The design of the piezo-inline injector is divided into its main modules in the schematic (see Fig. 5):

- Actuator module (3)
- Hydraulic coupler or translator (4)
- Control or servo valve (5)
- Nozzle module (6)

The design of the injector took account of the high overall rigidity required within the actuator chain composed of actuator, hydraulic coupler, and control valve. Another design feature is the avoidance of mechanical forces acting on the nozzle needle. Such forces occurred as a result of the push rod used on previous solenoid-valve injectors. On aggregate, this design effectively reduces the moving masses and friction, thus enhancing injector stability and drift compared to conventional systems.

In addition, the fuel-injection system allows the implementation of very short intervals ("hydraulic zero") between injection events. The number and configuration of fuel-metering operations can represent up to five injection events per injection cycle in order to adapt the requirements to the engine operating points.

A direct response of the needle to actuator operation is achieved by coupling the servo valve (5) to the nozzle needle. The delay between the electric start of triggering and hydraulic response of the nozzle needle is about 150 microseconds. This meets the contradictory requirements of high needle speeds and extremely small reproducible injected fuel quantities.

As a result of this principle, the injector also includes small direct leakage points from the high-pressure section to the low-pressure circuit. The result is an increase in the hydraulic efficiency of the overall system.

Operating concept

Function of the 3/2-way servo valve in the CR injector

The nozzle needle on piezo-inline injector is controlled indirectly by a servo valve. The required injected fuel quantity is then controlled by the valve triggering period. In its non-triggered state, the actuator is in the starting position and the servo valve is closed (Fig. 6a), i.e. the high-pressure section is separated from the low-pressure section.

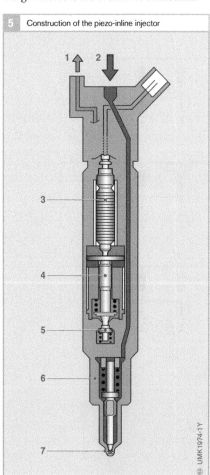

5 Construction of the piezo-inline injector

1 ↑ 2 ↓

3

4

5

6

7

UMK1974-1Y

Fig. 5
1 Fuel return
2 High-pressure connection
3 Piezo actuator module
4 Hydraulic coupler (translator)
5 Servo valve (control valve)
6 Nozzle module with nozzle needle
7 Injection orifice

The nozzle is kept closed by the rail pressure exerted in the control chamber (3). When the piezo actuator is triggered, the servo valve opens and closes the bypass passage (Fig. 6b). The flow-rate ratio between the outlet restrictor (2) and the inlet restrictor (4) lowers pressure in the control chamber and the nozzle (5) opens. The control volume flows via the servo valve to the low-pressure circuit of the overall system.

To start the closing process, the actuator is discharged and the servo valve releases the bypass passage. The control chamber is then refilled by reversing the inlet and outlet restrictors, and pressure in the control chamber is raised. As soon as the required pressure is attained, the nozzle needle starts to move and the injection process ends.

The valve design described above and the greater dynamic design of the actuator system result in much shorter injection times compared to injectors of conventional design, i.e. push rod and 2/2-way valve. Ultimately, this has a positive impact on exhaust-gas emissions and engine performance. Due to requirements regarding the engine in EU 4, the injector program maps were optimized to apply corrective functions (injector delivery compensation (IMA) and

zero delivery calibration (NMK). The pre-injection quantity can then be selected at will, and IMA can minimize the quantity spread in the program map using full ballistic mode (see Fig. 7).

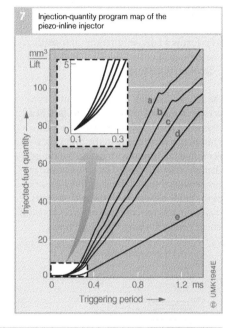

7 Injection-quantity program map of the piezo-inline injector

Fig. 7
Injected fuel quantities
at different injection
pressures
a   1,600 bar
b   1,200 bar
c   1,000 bar
d   800 bar
e   250 bar

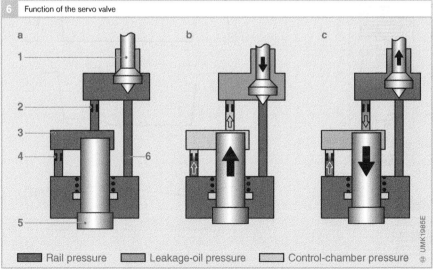

6 Function of the servo valve

Rail pressure   Leakage-oil pressure   Control-chamber pressure

Fig. 6
a   Start position
b   Nozzle needle opens
     (bypass closed,
     normal function
     with outlet and
     inlet restrictors)
c   Nozzle needle
     closes (bypass
     open, function with
     two inlet restrictors)

1   Servo valve
     (control valve)
2   Outlet restrictor
3   Control chamber
4   Inlet restrictor
5   Nozzle needle
6   Bypass

Function of the hydraulic coupler
Another key component in the piezo-inline injector is the hydraulic coupler (Fig. 8, 3) that implements the following functions:
- Translates and amplifies the actuator stroke.
- Compensates for any play between the actuator and the servo valve (e.g. caused by thermal expansion).
- Performs a failsafe function (automatic safety cutoff of fuel injection if electrical decontacting fails).

The actuator module and the hydraulic coupler are immersed in the diesel fuel flow at a pressure of about 10 bar. When the actuator is not triggered, pressure in the hydraulic coupler is in equilibrium with its surroundings. Changes in length caused by temperature are compensated by small leakage-fuel quantities via the guide clearances of the two plungers. This maintains the coupling of forces between actuator and switching valve at all times.

To generate an injection event, a voltage (110...150 V) is applied to the actuator until the equilibrium of forces between the switching valve and the actuator is exceeded. This increases the pressure in the coupler, and a small leakage volume flows out of the coupler via the piston guide clearances into the low-pressure circuit of the injector. The pressure drop caused in the coupler has no impact on injector function for a triggering period lasting several milliseconds.

At the end of the injection process, the quantity missing in the hydraulic coupler needs refilling. This takes place in the reverse direction via the guide clearances of the plungers as a result of the pressure difference between the hydraulic coupler and the low-pressure circuit of the injector. The guide clearances and the low-pressure level are matched to fill up the hydraulic coupler fully before the next injection cycle starts.

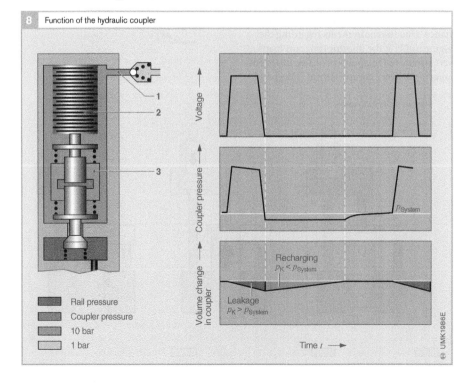

8   Function of the hydraulic coupler

Voltage →

Coupler pressure →

p_{System}

Volume change in coupler →

Recharging
$p_K < p_{System}$

Leakage
$p_K > p_{System}$

Time t →

Rail pressure
Coupler pressure
10 bar
1 bar

UMK1986E

Fig. 8
1   Low-pressure fuel rail with valve
2   Actuator
3   Hydraulic coupler (translator)

Triggering the common-rail piezo in-line injector

The injector is triggered by an engine control unit whose output stage was specially designed for these injectors. A reference triggering voltage is predetermined as a function of the rail pressure of the set operating point. The voltage signal is pulsed (Fig. 9) until there is a minimum deviation between the reference and the control voltage. The voltage rise is converted proportionally into a piezo-electric actuator stroke. The actuator stroke produces a pressure rise in the coupler by means of hydraulic translation until the equilibrium of forces is exceeded at the switching valve, and the valve opens. As soon as the switching valve reaches its end position, pressure in the control chamber starts to drop via the needle, and injection ends.

Benefits of the piezo-inline injector

- Multiple injection with flexible start of injection and intervals between individual injection events.
- Production of very small injected fuel quantities for pre-injection.
- Small size and low weight of injector (270 g compared to 490 g).
- Low noise (–3 dB [A]).
- Lower fuel consumption (–3%).
- Lower exhaust-gas emission (–20%).
- Increased engine performance (+7%).

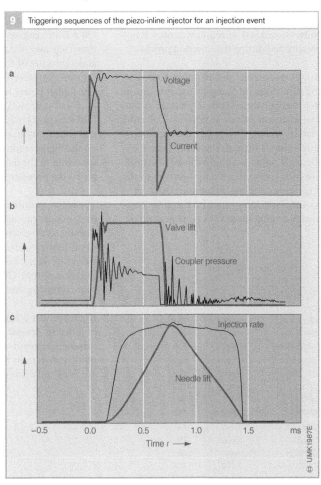

9 Triggering sequences of the piezo-inline injector for an injection event

a

Voltage

Current

b

Valve lift

Coupler pressure

c

Injection rate

Needle lift

-0.5   0.0   0.5   1.0   1.5   ms

Time t —→

UMK1987E

Fig. 9

a Current and voltage curves for triggering the injector

b Valve-lift curve and coupler pressure

c Valve-lift curve and injection rate

High-pressure pumps

Design and requirements

The high-pressure pump is the interface between the low-pressure and high-pressure stages. Its function is to make sure there is always sufficient fuel under pressure available in all engine operating conditions. At the same time it must operate for the entire service life of the vehicle. This includes providing a fuel reserve that is required for quick engine starting and rapid pressure rise in the fuel rail.

The high-pressure pump generates a constant system pressure for the high-pressure accumulator (fuel rail) independent of fuel injection. For this reason, fuel – compared to conventional fuel-injection systems – is not compressed during the injection process.

A 3-plunger radial-piston pump is used as the high-pressure pump to generate pressure in passenger-car systems. 2-plunger in-line fuel-injection pumps are also used on commercial vehicles. Preferably, the high-pressure pump is fitted to the diesel engine at the same point as a conventional distributor injection pump. The pump is driven by the engine via coupling, gearwheel, chain, or toothed belt. Pump speed is therefore coupled to engine speed via a fixed gear ratio.

The pump plunger inside of the high-pressure pump compresses the fuel. At three delivery strokes per revolution, the radial-piston pump produces overlapping delivery strokes (no interruption in delivery), low drive peak torques, and an even load on the pump drive.

On passenger-car systems, torque reaches 16 Nm, i.e. only 1/9th of the drive torque required for a comparable distributor injection pump. As a result, the common-rail system places fewer demands on the pump-drive system than conventional fuel-injection systems. The power required to drive the pump increases in proportion to the pressure in the fuel rail and the rotational speed of the pump (delivery quantity). On a 2-liter engine, the high-pressure pump draws a power of 3.8 kW at nominal speed and a pressure of 1,350 bar in the fuel rail (at a mechanical efficiency of approx. 90%). The higher power requirements of common-rail systems compared to conventional fuel-injection systems is caused by leakage and control volumes in the injector, and – on the high-pressure pump CP1 – the pressure drop to the required system pressure across the pressure-control valve.

The high-pressure radial-piston pumps used in passenger cars are lubricated by fuel. Commercial-vehicle systems may have fuel- or oil-lubricated radial-piston pumps, as well as oil-lubricated 2-plunger in-line fuel-injection pumps. Oil-lubricated pumps are more robust against poor fuel quality.

High-pressure pumps are used in a number of different designs in passenger cars and commercial vehicles. There are versions of pump generations that have different delivery rates and delivery pressures (Table 1).

1	Bosch high-pressure pumps for common-rail systems	
Pump	Pressure in bar	Lubrication
CP1	1,350	Fuel
CP1+	1,350	Fuel
CP1H	1,600	Fuel
CP1H-OHW	1,100	Fuel
CP3.2	1,600	Fuel
CP3.2+	1,600	Fuel
CP3.3	1,600	Fuel
CP3.4	1,600	Oil
CP3.4+	1,600	Fuel
CP2	1,400	Oil
CPN2.2	1,600	Oil
CPN2.2+	1,600	Oil
CPN2.4	1,600	Oil

Table 1

H   Increased pressure
    section
+   Higher delivery rate
OHW Off-Highway

Radial-piston pump (CP1)
Design
The drive shaft in the housing of CP1 is mounted in a central bearing (Fig. 1, 1). The pump elements (3) are arranged radially with respect to the central bearing and offset by 120°. The eccenter (2) fitted to the drive shaft forces the pump plunger to move up and down.

Force is transmitted between the eccentric shaft and the delivery plunger by means of a drive roller, a sliding ring mounted on the shaft eccenter, and a plunger base plate attached to the plunger base plate.

Operating concept
Fuel delivery and compression
The presupply pump – an electric fuel pump or a mechanically driven gear pump – delivers fuel via a filter and water separator to the inlet of the high-pressure pump (6). The inlet is located inside of the pump on passenger-car systems with a gear pump flanged to the high-pressure pump. A safety valve is fitted behind the inlet. If the delivery pressure of the presupply pump exceeds the opening pressure (0.5 to 1.5 bar) of the safety valve, the fuel is pressed through the restriction bore of the safety valve into the lubrication and cooling circuit of the high-pressure pump. The drive shaft with its eccenter moves the pump plunger up and down to mimic the eccentric lift. Fuel passed through the high-pressure pump's inlet valve (4) into the element chamber and the pump plunger moves downward (inlet stroke).

When the bottom-dead center of the pump plunger is exceeded, the inlet valve closes, and the fuel in the element chamber can no longer escape. It can then be pressurized beyond the delivery pressure of the presupply pump. The rising pressure opens the outlet valve (5) as soon as pressure reaches the level in the fuel rail. The pressurized fuel then passes to the high-pressure circuit.

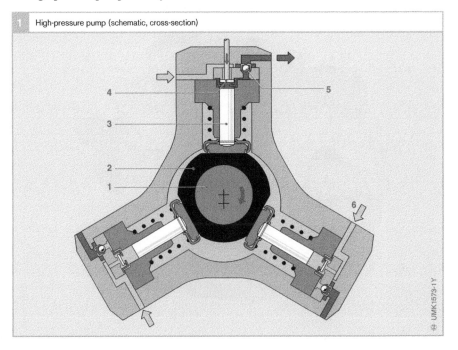

1 | High-pressure pump (schematic, cross-section)

UMK1573-1Y

Fig. 1
1 Drive shaft
2 Eccenter
3 Pump element with pump plunger
4 Inlet valve
5 Outlet valve
6 Fuel inlet

The pump plunger continues to deliver fuel until it reaches its top-dead center position (delivery stroke). The pressure then drops so that the outlet valve closes. The remaining fuel is depressurized and the pump plunger moves downward.

When the pressure in the element chamber exceeds the pre-delivery pressure, the inlet valve reopens, and the process starts over.

Transmission ratio

The delivery quantity of a high-pressure pump is proportional to its rotational speed. In turn, the pump speed is dependent on the engine speed. The transmission ratio between the engine and the pump is determined in the process of adapting the fuel-injection system to the engine so as to limit the volume of excess fuel delivered. At the same time it makes sure that the engine's fuel demand at WOT is covered to the full extent. Possible gear ratios are 1:2 or 2:3 relative to the crankshaft.

Delivery rate

As the high-pressure pump is designed for high delivery quantities, there is a surplus of pressurized fuel when the engine is idling or running in part-load range. On first-generation systems with a CP1, excess fuel delivered is returned to the fuel tank by the pressure-control valve on the fuel rail. As the compressed fuel expands, the energy imparted by compression is lost; overall efficiency drops. Compressing and then expanding the fuel also heats the fuel.

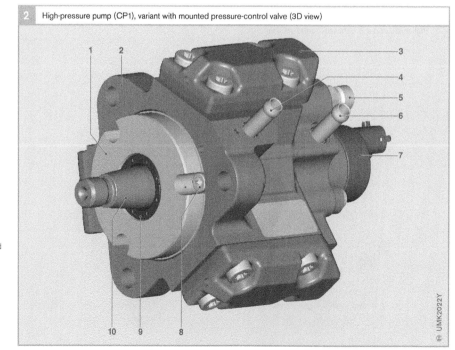

2   High-pressure pump (CP1), variant with mounted pressure-control valve (3D view)

Fig. 2

1 Flange
2 Pump housing
3 Engine cylinder head
4 Inlet connection
5 High-pressure inlet
6 Return connection
7 Pressure-control valve
8 Barrel bolt
9 Shaft seal
10 Eccentric shaft

Radial-piston pump (CP1H)

Modifications

An improvement in energetic efficiency is possible by controlling fuel delivery by the high-pressure pump on the fuel-delivery side (suction side). Fuel flowing into the pump element is metered by an infinitely variable solenoid valve (metering unit, ZME). This valve adapts the fuel quantity delivered to the rail to system demand. This fuel-delivery control not only drops the performance demand of the high-pressure pump, it also reduces the maximum fuel temperature. This system designed for the CP1H was taken over by the CP3.

Compared to the high-pressure pump CP1, the CP1H is designed for higher pressures up to 1,600 bar. This was achieved by reinforcing the drive mechanism, modifying the valve units, and introducing measures to increase the strength of the pump housing.

The metering unit is mounted on the high-pressure pump (Fig. 3, 13).

Design of the metering unit (ZME)

Fig. 4 shows the design of the metering unit. The plunger operated by solenoid force frees up a metering orifice depending on its position.

The solenoid valve is triggered by a PWM signal.

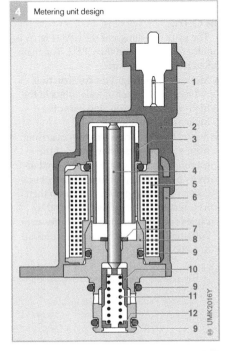

4 Metering unit design

Fig. 4
1 Plug with electrical interface
2 Magnet housing
3 Bearing
4 Armature with tappet
5 Winding with coil body
6 Cup
7 Residual air-gap washer
8 Magnetic core
9 O-ring
10 Plunger with control slots
11 Spring
12 Safety element

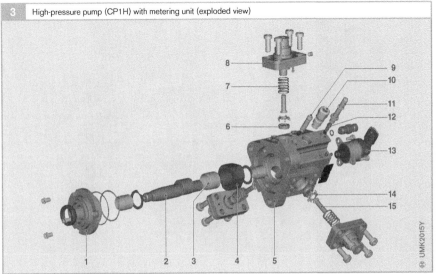

3 High-pressure pump (CP1H) with metering unit (exploded view)

Fig. 3
1 Flange
2 Eccentric shaft
3 Bushing
4 Drive roller
5 Pump housing
6 Plate
7 Spring
8 Engine cylinder head
9 Return-flow connection
10 Overflow valve
11 Inlet connection
12 Filter
13 Metering unit
14 Cage
15 Pump plunger

Radial-piston pump (CP3)

Modifications

The CP3 is a high-pressure pump with suction-side fuel-delivery control by means of a metering unit (ZME). This control was first used on the CP3 and was assumed later on the CP1H.

The principle design of the CP3 (Fig. 5) is similar to the CP1 and the CP1H. The main difference in features are:

- Monobloc housing: This construction reduces the number of leak points in the high-pressure section, and permits a higher delivery rate.
- Bucket tappets: Transverse forces arising from the transverse movement of the eccenter drive roller are not removed directly by the pump plungers but by buckets on the housing wall. The pump then has greater stability under load and is capable of withstanding higher pressures. Potentially, it can withstand pressures up to 1,800 bar.

Variants

Pumps of the CP3 family are used in both passenger cars and commercial vehicles. A number of different variants are used depending on the delivery rate required. The size, and thus the delivery rate, increases from the CP3.2 to the CP3.4. The oil-lubricated CP3.4 is only used on heavy-duty trucks. On light-duty trucks and vans, pumps primarily designed for passenger cars may also be used.

A special feature of systems for medium-duty and heavy-duty trucks is the fuel filter located on the pressure side. It is situated between the gear pump and the high-pressure pump, and permits a greater filter storage capacity before requiring a change. The high-pressure pump requires an external connection for the fuel inlet in any case, even if the gear pump is flanged onto the high-pressure pump.

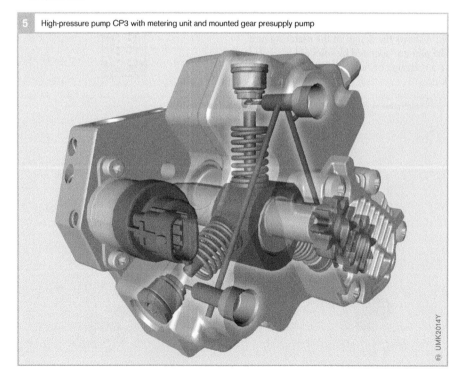

5   High-pressure pump CP3 with metering unit and mounted gear presupply pump

UMK2014Y

In-line piston pump (CP2)

Design

The oil-lubricated, quantity-controlled high-pressure pump (CP2) is only used on commercial vehicles. This is a 2-plunger pump with an in-line design, i.e. the two pump plungers are arranged adjacently (Fig. 6).

A gear pump with a high gear ratio is located on the camshaft extension. Its function is to draw fuel from the fuel tank and route it to the fine filter. From there, the fuel passes through another line to the metering unit located on the upper section of the high-pressure pump. The metering unit controls the fuel quantity delivered for compression dependent on actual demand in the same way as other common-rail high-pressure pumps of the recent generation.

Lube oil is supplied either directly via the mounting flange of the CP2 or a side-mounted inlet.

The drive gear ratio is 1:2. The CP2 is therefore mountable together with conventional in-line fuel-injection pumps.

Operating concept

Fuel enters the pump element and the compressed fuel is conveyed to the fuel rail via a combined inlet/outlet valve on the CP2.

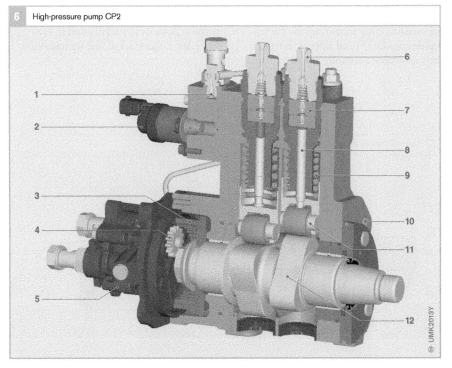

6   High-pressure pump CP2

Fig. 6
1 Zero delivery restrictor
2 Metering unit
3 Internal gear
4 Pinion
5 Gear presupply pump
6 High-pressure connection
7 Two-part inlet/outlet valve
8 C-coated plunger
9 Plunger return spring
10 Oil inlet
11 C-coated roller bolt
12 Concave cam

Fuel rail (high-pressure accumulator)

Function

The function of the high-pressure accumulator (fuel rail) is to maintain the fuel at high pressure. In so doing, the accumulator volume has to dampen pressure fluctuations caused by fuel pulses delivered by the pump and the fuel-injection cycles. This ensures that, when the injector opens, the injection pressure remains constant. On the one hand, the accumulator volume must be large enough to meet this requirement. On the other hand, it must be small enough to ensure a fast enough pressure rise on engine start. Simulation calculations are conducted during the design phase to optimize the performance features.

Besides acting as a fuel accumulator, the fuel rail also distributes fuel to the injectors.

Design

The tube-shaped fuel rail (Fig. 1, 1) can have as many designs as there are engine mounting variants. It has mountings for the rail-pressure sensor (5) and a pressure-relief valve or pressure-control valve (2).

Operating concept

The pressurized fuel delivered by the high-pressure pump passes via a high-pressure fuel line to the fuel-rail inlet (4). From there, it is distributed to the individual injectors (hence the term "common rail").

The fuel pressure is measured by the rail-pressure sensor and controlled to the required value by the pressure-control valve. The pressure-relief valve is used as an alternative to the pressure-control valve – depending on system requirements – and its function is to limit fuel pressure in the fuel rail to the maximum permissible pressure. The highly compressed fuel is routed from the fuel rail to the injectors via high-pressure delivery lines.

The cavity inside the fuel rail is permanently filled with pressurized fuel. The compressibility of the fuel under high pressure is utilized to achieve an accumulator effect. When fuel is released from the fuel rail for injection, the pressure in the high-pressure accumulator remains virtually constant, even when large quantities of fuel are released.

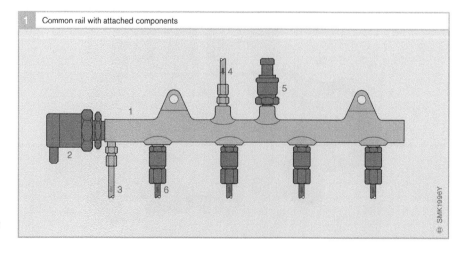

1   Common rail with attached components

Fig. 1

1   Fuel rail
2   Pressure-control
     valve
3   Return line from
     fuel rail to fuel tank
4   Inlet from high-
     pressure pump
5   Rail-pressure sensor
6   Fuel line to injector

SMK1996Y

Cleanliness quality

The sharp rise in the performance of new assemblies, e.g. the common-rail system for high-pressure diesel fuel injection, requires extreme precision in machining, and ever tighter tolerances and fits. Particle residue from the production process may lead to increased wear, or even the total failure of the assembly. This results in high requirements and tight tolerances for cleanliness quality, with a continuous reduction in the permitted particle size.

The cleanliness quality of components is currently determined in the production process by light-microscope image-analysis systems. They supply information on particle-size distribution. Additional information, such as the nature of the particles and their chemical composition, is required to develop innovative cleaning processes. This information is obtained by electron microscopes.

Particle-analysis system (SEM)

Bosch uses a particle-analysis system based on a Scanning Electron Microscope (SEM). This system performs an automated analysis of particles adhering to a product. The results of the analysis show particle-size distribution, the chemical composition of the particles, and images of the individual particles. Using this information, the source of the particles are identifiable. Action can then be taken to avoid, reduce, or wash off certain particle types. In this way, solutions are not based on the increased use of cleaning techniques, but on avoiding and reducing residual soiling during the production process.

The automated particle-analysis system (SEM) provides the cleanliness process with an analysis system that produces important information on the type of residual soiling. The precise identification of particles and their sources is vital to developing new cleaning techniques.

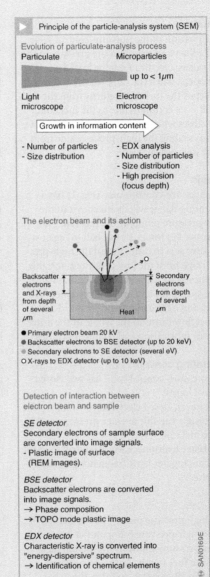

▶ Principle of the particle-analysis system (SEM)

Evolution of particulate-analysis process

Particulate	Microparticles

up to < 1μm

Light microscope	Electron microscope

Growth in information content ⟩

- Number of particles
- Size distribution

- EDX analysis
- Number of particles
- Size distribution
- High precision (focus depth)

The electron beam and its action

Backscatter electrons and X-rays from depth of several μm

Secondary electrons from depth of several μm

Heat

● Primary electron beam 20 kV
◉ Backscatter electrons to BSE detector (up to 20 keV)
◎ Secondary electrons to SE detector (several eV)
○ X-rays to EDX detector (up to 10 keV)

Detection of interaction between electron beam and sample

SE detector
Secondary electrons of sample surface are converted into image signals.
- Plastic image of surface (REM images).

BSE detector
Backscatter electrons are converted into image signals.
→ Phase composition
→ TOPO mode plastic image

EDX detector
Characteristic X-ray is converted into "energy-dispersive" spectrum.
→ Identification of chemical elements

SAN0169E

Pressure-control valve

Function
The function of the pressure-control valve is to adjust and maintain the pressure in the fuel rail as a factor of engine load, i.e.:
- It opens when the rail pressure is too high. Part of the fuel then returns from the fuel rail via a common line to the fuel tank.
- It closes when the rail pressure is too low, thus sealing the high-pressure side from the low-pressure side.

Design
The pressure-control valve (Fig. 1) has a mounting flange which attaches it to the high-pressure pump or the fuel rail. The armature (5) forces a valve ball (6) against the valve seat in order to seal the high-pressure stage from the low-pressure stage; this is achieved by the combined action of a valve spring (2) and an electromagnet (4) which force the armature downwards.

Fuel flows around the whole of the armature for lubrication and cooling purposes.

Operating concept
The pressure-control valve has two closed control loops:
- A slower, closed electrical control loop for setting a variable average pressure level in the fuel rail.
- A faster hydromechanical control loop for balancing out high-frequency pressure pulses.

Pressure-control valve not activated
The high pressure present in the fuel rail or at the high-pressure pump outlet is applied to the pressure-control valve via the high-pressure fuel supply. As the deenergized electromagnet exerts no force, the high-pressure force is greater than the spring force. The pressure-control valve opens to a greater or lesser extent depending on the delivery quantity. The spring is dimensioned to maintain a pressure of approx. 100 bar.

Pressure-control valve activated
When the pressure in the high-pressure circuit needs to be increased, the force of the electromagnet is added to that of the spring. The pressure-control valve is activated and closes until a state of equilibrium is reached between the high pressure and the combined force of the electromagnet and the spring. At this point, it remains in partly open position and maintains a constant pressure. Variations in the delivery quantity of the high-pressure pump and the withdrawal of fuel from the fuel rail by the injectors are compensated by varying the valve aperture. The magnetic force of the electromagnet is proportional to the control current. The control current is varied by pulse-width modulation. A pulse frequency of 1 kHz is sufficiently high to prevent adverse armature movement or pressure fluctuations in the fuel rail.

Designs
The pressure-control valve DRV1 is used in first-generation common-rail systems. Second- and third-generation CR systems operate using the two-actuator concept. Here, the rail pressure is adjusted by both a metering unit as well as a pressure-control valve.

Fig. 1
1 Electrical connections
2 Valve spring
3 Armature
4 Valve housing
5 Solenoid coil
6 Valve ball
7 Support ring
8 O-ring
9 Filter
10 High-pressure fuel supply
11 Valve body
12 Drain to low-pressure circuit

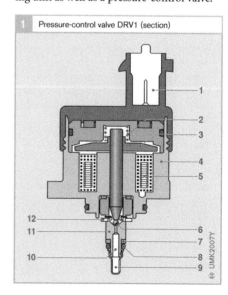

1 Pressure-control valve DRV1 (section)

In this case, either the pressure-control valve DRV2 is used or the DRV3 variant for higher pressures. This control strategy achieves lower fuel heating and eliminates the need for a fuel cooler.

The DRV2/3 (Fig. 2) differs from the DRV1 in the following features:
- Hard seal to the high-pressure interface (bite edge).
- Optimized magnetic circuit (lower power consumption).
- Flexible mounting concept (free plug orientation).

Pressure-relief valve

Function
The pressure-relief valve has the same function as a pressure limiter. The latest version of the internal pressure-relief valve now has an integrated limp-home function. The pressure-relief valve limits pressure in the fuel rail by releasing a drain hole when pressure exceeds a certain limit. The limp-home function ensures that a certain pressure is maintained in the fuel rail to permit the vehicle to continue running without any restriction.

Design and operating concept
The pressure-relief valve (Fig. 3) is a mechanical component. It consists of the following parts:
- A housing with an external thread for screwing to the fuel rail.
- A connection to the fuel-return line to the fuel tank (3).
- A movable plunger (2).
- A plunger return spring (5).

At the end which is screwed to the fuel rail, there is a hole in the valve housing which is sealed by the tapered end of the plunger resting against the valve seat inside the valve housing. At normal operating pressure, a spring presses the plunger against the valve seat so that the fuel rail remains sealed. Only if the pressure rises above the maximum system pressure is the plunger forced back against the action of the spring by the pressure in the fuel rail so that the high-pressure fuel can escape. The fuel is routed through passages into a central bore of the plunger and returned to the fuel tank via a common line. As the valve opens, fuel can escape from the fuel rail to produce a reduction in fuel-rail pressure.

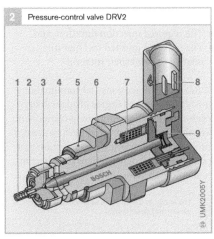

2 Pressure-control valve DRV2

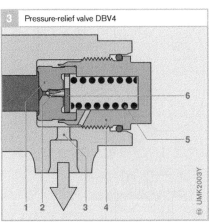

3 Pressure-relief valve DBV4

Fig. 2
1   Filter
2   Bite edge
3   Valve ball
4   O-ring
5   Union bolt with circlip
6   Armature
7   Solenoid coil
8   Electrical connection
9   Valve spring

Fig. 3
1   Valve insert
2   Valve plunger
3   Low-pressure section
4   Valve holder
5   Spring
6   Diaphragm disc

Injection nozzles

The injection nozzle injects the fuel into the combustion chamber of the diesel engine. It is a determining factor in the efficiency of mixture formation and combustion and, therefore has a fundamental effect on engine performance, exhaust-gas behavior, and noise. In order that injection nozzles can perform their function as effectively as possible, they have to be designed to match the fuel-injection system and engine in which they are used.

The injection nozzle is a central component of any fuel-injection system. It requires highly specialized technical knowledge on the part of its designers. The nozzle plays a major role in:
- Shaping the rate-of-discharge curve (precise progression of pressure and fuel distribution relative to crankshaft rotation)
- Optimum atomization and distribution of fuel in the combustion chamber, and
- Sealing off the fuel-injection system from the combustion chamber

Due to its exposed position in the combustion chamber, the nozzle is subjected to constant pulsating mechanical and thermal stresses from the engine and the fuel-injection system. The fuel flowing through the nozzle must also cool it. When the engine is overrunning, when no fuel is being injected, the nozzle temperature increases steeply. Therefore, it must have sufficient high-temperature resistance to cope with these conditions.

In fuel-injection systems based on in-line injection pumps (Type PE) and distributor injection pumps (Type VE/VR), and in Unit Pump (UP) systems, the nozzle is combined with the nozzle holder to form the nozzle-and-holder assembly (Fig. 1) and installed in the engine. In high-pressure fuel-injection systems, such as the Common Rail (CR) and Unit Injector (UI) systems the nozzle is a single integrated unit so that the nozzle holder is not required.

Indirect-Injection (IDI) engines use pintle nozzles, while direct-injection engines have hole-type nozzles.

The nozzles are opened by the fuel pressure. The nozzle opening, injection duration, and rate-of-discharge curve (injection pattern) are the essential determinants of injected fuel quantity. The nozzles must close rapidly and reliably when the fuel pressure drops. The closing pressure is at least 40 bar above the maximum combustion pressure in order to prevent unwanted post-injection or intrusion of combustion gases into the nozzle. The nozzle must be designed specifically for the type of engine in which it is used as determined by:
- The injection method (direct or indirect)
- The geometry of the combustion chamber
- The required injection-jet shape and direction
- The required penetration and atomization of the fuel jet
- The required injection duration, and
- The required injected fuel quantity relative to crankshaft rotation

Standardized dimensions and combinations provide the required degree of adaptability combined with the minimum of component diversity. Due to the superior performance combined with lower fuel consumption that it offers, all new engine designs use direct injection (and therefore hole-type nozzles).

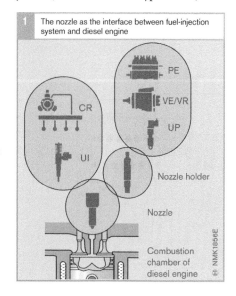

1 The nozzle as the interface between fuel-injection system and diesel engine

PE

VE/VR

CR

UP

UI

Nozzle holder

Nozzle

Combustion chamber of diesel engine

NMK1856E

The world of diesel fuel injection is a world of superlatives.

The valve needle of a commercial-vehicle nozzle will open and close the nozzle more than a billion times in the course of its service life. It provides a reliable seal at pressures as high as 2,050 bar as well as having to withstand many other stresses such as:
- The shocks caused by rapid opening and closing (on cars, this can take place as frequently as 10,000 times a minute if there are pre- and post-injection phases)
- The high flow-related stresses during fuel injection, and
- The pressure and temperature of the combustion chamber

The facts and figures below illustrate what modern nozzles are capable of:
- The pressure in the fuel-injection chamber can be as high as 2,050 bar. That is equivalent to the pressure produced by the weight of a large luxury sedan acting on an area the size of a fingernail.

- The injection duration is 1...2 milliseconds (ms). In one millisecond, the sound wave from a loudspeaker only travels about 33 cm.
- The injection durations on a car engine vary between 1 mm³ (pre-injection) and 50 mm³ (full-load delivery); on a commercial vehicle, between 3 mm³ (pre-injection) and 350 mm³ (full-load delivery). 1 mm³ is equivalent to half the size of a pinhead. 350 mm³ is about the same as 12 large raindrops (30 mm³ per raindrop). That amount of fuel is forced at a velocity of 2,000 km/h through an opening of less than 0.25 mm² in the space of only 2 ms.
- The valve-needle clearance is 0.002 mm (2 μm). A human hair is 30 times thicker (0.06 mm).

Such high-precision technology demands an enormous amount of expertise in development, materials, production, and measurement techniques.

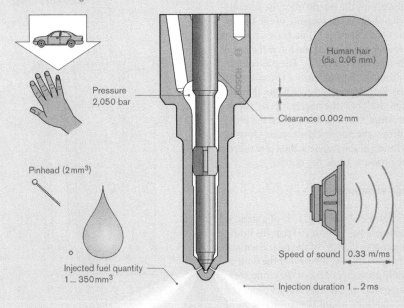

Pressure 2,050 bar

Human hair (dia. 0.06 mm)

Clearance 0.002 mm

Pinhead (2 mm³)

Speed of sound | 0.33 m/ms

Injected fuel quantity 1 ... 350 mm³

Injection duration 1 ... 2 ms

NMK1708-2E

Pintle nozzles

Usage

Pintle nozzles are used on Indirect Injection (IDI) engines, i.e. engines that have prechambers or whirl chambers. In this type of engine, the mixing of fuel and air is achieved primarily by the whirl effects created inside the cylinder. The shape of the injection jet can also assist the process. Pintle nozzles are not suitable for direct-injection engines as the peak pressures inside the combustion chamber would open the nozzle. The following types of pintle nozzle are available:

- Standard pintle nozzles
- Throttling pintle nozzles and
- Flatted-pintle nozzles

Design and method of operation

The fundamental design of all pintle nozzles is virtually identical. The differences between them are to be found in the geometry of the pintle (Fig. 1, 7). Inside the nozzle body is the nozzle needle (3) It is pressed downwards by the force F_F exerted by the spring and the pressure pin in the nozzle holder so that it seals off the nozzle from the combustion chamber. As the pressure of the fuel in the pressure chamber (5) increases, it acts on the pressure shoulder (6) and forces the nozzle needle upwards (force F_D). The pintle lifts away from the injector orifice (8) and opens the way for fuel to pass through into the combustion chamber (the nozzle "opens"; opening pressure 110...170 bar). When the pressure drops, the nozzle closes again. Opening and closing of the nozzle is thus controlled by the pressure inside the nozzle.

Design variations

Standard pintle nozzle

The nozzle needle of (Fig. 1, 3) of a standard pintle nozzle has a pintle (7) that fits into the injector orifice (8) of the nozzle with a small degree of play. By varying the dimensions and geometry of the of the pintle, the characteristics of the injection jet produced can be modified to suit the requirements of different engines.

Throttling pintle nozzle

One of the variations of the pintle nozzle is the throttling pintle nozzle. The profile of the pintle allows a specific rate-of-discharge curve to be produced. As the nozzle needle opens, at first only a very narrow annular orifice is provided which allows only a small amount of fuel to pass through (throttling effect).

As the pintle draws further back with increasing fuel pressure, the size of the gap through which fuel can flow increases. The greater proportion of the injected fuel quantity is only injected as the pintle approaches the limit of its upward travel. By modifying the rate-of-discharge curve in this way, "softer" combustion is produced because the pressure in the combustion chamber does not rise so quickly. As a result, combustion noise is reduced in the part-load range. This means that the shape of the pintle in combination with the throttling gap and the characteristic of the compression spring in the nozzle holder produces the desired rate-of-discharge curve.

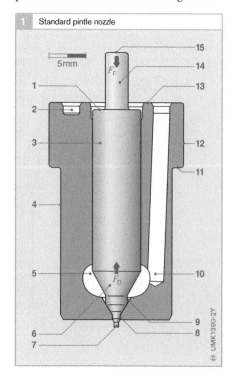

1   Standard pintle nozzle

5mm

Fig. 1

1 Stroke-limiting shoulder
2 Ring groove
3 Nozzle needle
4 Nozzle body
5 Pressure chamber
6 Pintle shoulder
7 Pintle
8 Injection orifice
9 Seat lead-in
10 Inlet port
11 Nozzle-body shoulder
12 Nozzle-body collar
13 Sealing face
14 Pressure pin
15 Pressure-pin contact face

F_F Spring force
F_D Force acting on pressure shoulder due to fuel pressure

Flatted-pintle nozzle

The flatted-pintle nozzle (Fig. 3) has a pintle with a flatted face on its tip which, as the nozzle opens (at the beginning of needle lift travel) produces a wider passage within the annular orifice. This helps to prevent deposits at that point by increasing the volumetric flow rate. As a result, flatted-pintle nozzles "coke" to a lesser degree and more evenly. The annular orifice between the jet orifice and the pintle is very narrow (< 10 µm). The flatted face is frequently parallel to the axis of the nozzle needle. By setting the flatted face at an angle, the volumetric flow rate, Q, can be increased in the flatter section of the rate-of-discharge curve (Fig. 4). In this way, a smoother transition between the initial phase and the fully-open phase of the rate-of-discharge curve can be obtained. Specially designed variations in pintle geometry allow the flow-rate pattern to be modified to suit particular engine requirements. As a result, engine noise in the part-load range is reduced and engine smoothness improved.

Heat shielding

Temperatures above 220°C also promote nozzle coking. Thermal-protection plates or sleeves (Fig. 2) help to overcome this problem by conducting heat from the combustion chamber into the cylinder head.

2 Thermal-protection sleeve

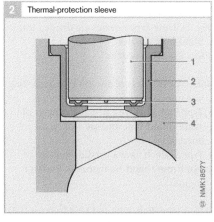

Fig. 2
1 Pintle nozzle
2 Thermal-protection sleeve
3 Protective disc
4 Cylinder head

3 Flatted-pintle nozzle

4 Volumetric flow rate as a function of pintle travel and nozzle design

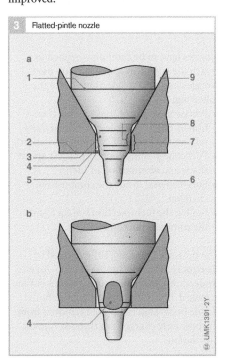

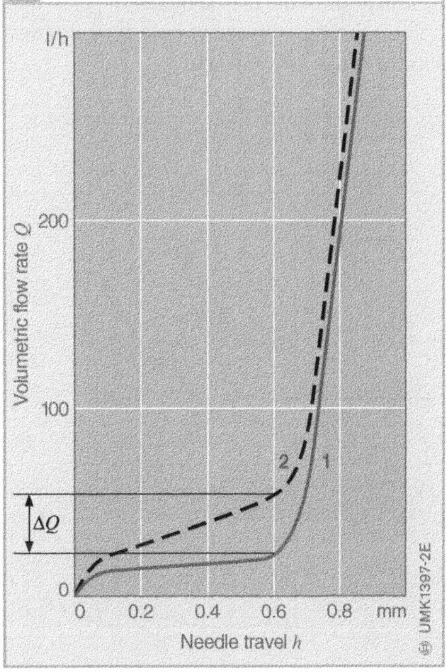

Fig. 3
a Side view
b Front view (rotation of 90° relative to side view)

1 Pintle seat face
2 Nozzle-body base
3 Throttling pintle
4 Flatted face
5 Injection orifice
6 Profiled pintle
7 Total contact ratio
8 Cylindrical overlap
9 Nozzle-body seat face

Fig. 4
1 Throttling pintle nozzle
2 Flatted-pintle nozzle (throttling pintle nozzle with flatted face)

ΔQ Difference in volumetric flow rate due to flatted face

Hole-type nozzles

Application

Hole-type nozzles are used on engines that operate according to the Direct-Injection process (DI). The position in which the nozzles are fitted is generally determined by the engine design. The injection orifices are set at a variety of angles according to the requirements of the combustion chamber (Fig. 1). Hole-type nozzles are divided into:
- Blind-hole nozzles
- Sac-less (vco) nozzles

Hole-type nozzles are also divided according to size into:
- *Type P* which have a needle diameter of 4 mm (blind-hole and sac-less (vco) nozzles).
- *Type S* which have a needle diameter of 5 or 6 mm (blind-hole nozzles for large engines).

In Common-Rail (CR) and Unit Injector (UI) fuel-injection systems, the hole-type nozzle is a single integrated unit. It combines, therefore, the functions of the nozzle holder.

The opening pressure of hole-type nozzles is in the range 150...350 bar.

Design

The injection orifices (Fig. 2, 6) are located on the sheath of the nozzle cone (7). The number and diameter are dependent on:
- The required injected fuel quantity
- The shape of the combustion chamber
- The air vortex (whirl) inside of the combustion chamber

The diameter of the injection orifices is slightly larger at the inner end than at the outer end. This difference is defined by the port taper factor. The leading edges of the injection orifices may be rounded by using the hydro-erosion (HE) process. This involves the use of an HE fluid that contains abrasive particles which smooth off the edges at points where high flow velocities occur (leading edges of injection orifices). Hydro-erosion can be used both on blind-hole and sac-less (vco) nozzles. Its purpose is to:
- optimize the flow-resistance coefficient
- pre-empt erosion of edges caused by particles in the fuel, and/or
- tighten flow-rate tolerances

Nozzles have to be carefully designed to match the engine in which they are used. Nozzle design plays a decisive role in the following:
- Precise metering of injected fuel (injection duration and injected fuel quantity relative to degrees of crankshaft rotation).
- Fuel conditioning (number of jets, spray shape and atomization of fuel).
- Fuel dispersal inside the combustion chamber.
- Sealing the fuel-injection system against the combustion chamber.

The pressure chamber (10) is formed by Electrochemical Machining (ECM). An electrode, through which an electrolyte solution is passed, is introduced into the pre-bored nozzle body. Material is then removed from the positively charged nozzle body (anodic dissolution).

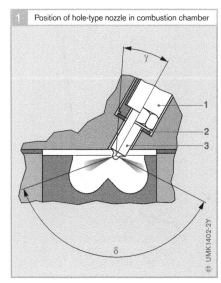

1   Position of hole-type nozzle in combustion chamber

γ

1

2

3

δ

UMK1402-2Y

Fig. 1
1   Nozzle holder or injector
2   Sealing washer
3   Hole-type nozzle

γ   Inclination
δ   Jet cone angle

Designs

Fuel in the volume below the nozzle-needle seat evaporates after combustion. This produces a large part of the engine's hydrocarbon emissions. For this reason, it is important to keep the dead volume, or "detrimental" volume, as small as possible.

In addition, the geometry of the needle seat and the shape of the nozzle cone have a decisive influence on the opening and closing characteristics of the nozzle. This, in turn, affects the soot and NO_x emissions produced by the engine.

The consideration of these various factors, in combination with the demands of the engine and the fuel-injection system, has resulted in a variety of nozzle designs.

There are two basic designs:
- Blind-hole nozzles
- Sac-less (vco) nozzles

Among the blind-hole nozzles, there are a number of variants.

Blind-hole nozzle

The injection orifices in the blind-hole nozzle (Fig. 2, 6) are arranged around a blind hole.

If the nozzle has a *rounded tip*, the injection orifices are drilled either mechanically or by electro-erosion, depending on the design.

In blind-hole nozzles with a *conical tip*, the injection orifices are generally created by electro-erosion.

Blind-hole nozzles may have a cylindrical or conical blind hole of varying dimensions.

Blind-hole nozzles with a cylindrical blind hole and rounded tip (Fig. 3), which consists of a cylindrical and a hemispherical section, offer a large amount of scope with regard to the number of holes, length of injection orifices, and spray-hole cone angle. The nozzle cone is hemispherical in shape, which – in combination with the shape of the blind hole – ensures that all the spray holes are of equal length.

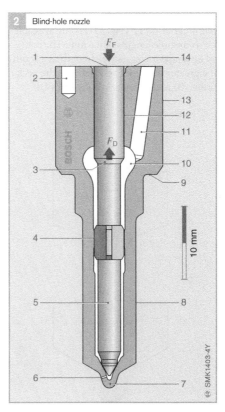

2 Blind-hole nozzle

F_F

1 · · · 14
2 · · ·
· · · 13
· · · 12
· · · 11
F_D
3 · · · 10
· · · 9

10 mm

4 · · ·

5 · · · 8

6 · · · 7

SMK1403-4Y

Fig. 2
1 Stroke-limiting
  shoulder
2 Fixing hole
3 Pressure shoulder
4 Secondary needle
  guide
5 Needle shaft
6 Injection orifice
7 Nozzle cone
8 Nozzle body
9 Nozzle-body shoulder
10 Pressure chamber
11 Inlet passage
12 Needle guide
13 Nozzle-body collar
14 Sealing face

F_F Spring force
F_D Force acting on
  pressure shoulder
  due to fuel pressure

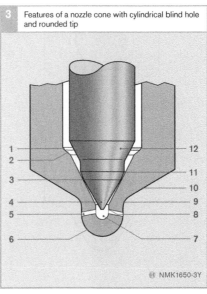

3 Features of a nozzle cone with cylindrical blind hole and rounded tip

1 · · · 12
2 · · ·
· · · 11
3 · · · 10
4 · · · 9
5 · · · 8
6 · · · 7

NMK1650-3Y

Fig. 3
1 Shoulder
2 Seat lead-in
3 Needle-seat face
4 Needle tip
5 Injection orifice
6 Rounded tip
7 Cylindrical blind hole
  (dead volume)
8 Injection orifice
  leading edge
9 Neck radius
10 Nozzle-cone taper
11 Nozzle-body seat
  face
12 Damping taper

The *blind-hole nozzle with a cylindrical blind hole and conical tip* (Fig. 4a) is only available for spray-hole lengths of 0.6 mm. The conical tip shape increases tip strength as a result of a greater wall thickness between the neck radius (3) and the nozzle body seat (4).

4   Nozzle cones

Blind-hole nozzles with conical blind holes and conical tip (Fig. 4b) have a smaller dead volume than nozzles with a cylindrical blind hole. The volume of the blind hole is between that of a sac-less (vco) nozzle and a blind-hole nozzle with a cylindrical blind hole. In order to obtain an even wall thickness throughout the tip, it is shaped conically to match the shape of the blind hole.

A further refinement of the blind-hole nozzle is the micro-blind-hole nozzle (Fig. 4c). Its blind-hole volume is around 30% smaller than that of a conventional blind-hole nozzle. This type of nozzle is particularly suited to use in common-rail systems, which operate with a relatively slow needle lift and, consequently, a comparatively long nozzle-seat restriction. The micro-blind-hole nozzle currently represents the best compromise between minimizing dead volume and even spray dispersal when the nozzle opens for common-rail systems.

Sac-less (vco) nozzles

In order to minimize dead volume – and therefore HC emissions – the injection orifice exits from the nozzle-body seat face. When the nozzle is closed, the nozzle needle more or less covers the injection orifice so that there is no direct connection between the blind hole and the combustion chamber (Fig. 4d). The blind-hole volume is considerably smaller than that of a blind-hole nozzle. Sac-less (vco) nozzles have a significantly lower stress capacity than blind-hole nozzles and can therefore only be produced with a spray-hole length of 1 mm. The nozzle tip has a conical shape. The injection orifices are generally produced by electro-erosion.

Special spray-hole geometries, secondary needle guides, and complex needle-tip geometries are used to further improve spray dispersal, and consequently mixture formation, on both blind-hole and sac-less (vco) nozzles.

Heat shield

The maximum temperature capacity of hole-type nozzles is around 300°C (heat resistance of material). Thermal-protection sleeves are available for operation in especially difficult conditions, and there are even cooled nozzles for large-scale engines.

Effect on emissions

Nozzle geometry has a direct effect on the engine's exhaust-gas emission characteristics.
● The *spray-hole geometry* (Fig. 5, 1) influences particulate and NO_x emissions.
● The *needle-seat geometry* (2) affects engine noise due to its effect on the pilot volume, i.e. the volume injected at the beginning of the injection process. The aim of optimizing spray-hole and seat geometry is to produce a durable nozzle capable of mass production to very tight dimensional tolerances.
● *Blind-hole geometry* (3) affects HC emissions, as previously mentioned. The designer can select and combine the various nozzle characteristics to obtain the optimum design for a particular engine and vehicle concerned.

For this reason, it is important that the nozzles are designed specifically for the vehicle, engine and fuel-injection system in which they are to be used. When servicing is required, it is equally important to use genuine OEM parts in order to ensure that engine performance is not impaired and exhaust-gas emissions are not increased.

Spray shapes

Basically, the shape of the injection jet for car engines is long and narrow because these engines produce a large degree of whirl inside of the combustion chamber. There is no whirl effect in commercial-vehicle engines. Therefore, the injection jet tends to be wider and shorter. Even where there is a large amount of whirl, the individual injection jets must not intermingle, otherwise fuel would be injected into areas where combustion has already taken place and, therefore, where there is a lack of air. This would result

in the production of large amounts of soot.

Hole-type nozzles have up to six injection orifices in passenger cars and up to ten in commercial vehicles. The aim of future development will be to further increase the number of injection orifices and to reduce their bore size (< 0.12 mm) in order to obtain even finer dispersal of fuel.

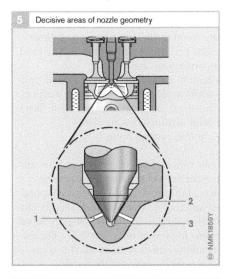

5 Decisive areas of nozzle geometry

Fig. 5
1 Injection-orifice geometry
2 Seat geometry
3 Blind-hole geometry

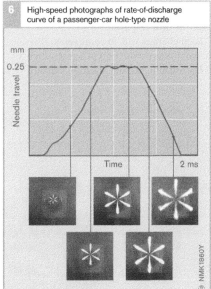

6 High-speed photographs of rate-of-discharge curve of a passenger-car hole-type nozzle

Future development of the nozzle

In view of the rapid development of new, high-performance engines and fuel-injection systems with sophisticated functionality (e.g. multiple injection phases), continuous development of the nozzle is a necessity. In addition, there are number of aspects of nozzle design which offer scope for innovation and further improvement of diesel engine performance in the future. The most important aims are:

- Minimize untreated emissions to reduce or totally avoid the outlay for an expensive exhaust-gas treatment (e.g. particulate filter).
- Minimize fuel consumption.
- Optimize engine noise.

There various different areas on which attention can be focused in the future development of the nozzle (Fig. 1) and a corresponding variety of development tools (Fig. 2). New materials are also constantly being developed to offer improvements in durability. The use of multiple-injection phases also has consequences for the design of the nozzle.

The use of other fuels (e.g. designer fuels) affects nozzle shape due to differences in viscosity or flow response.

Such changes will, in some cases, also demand new production processes, such as laser drilling for the injection orifices.

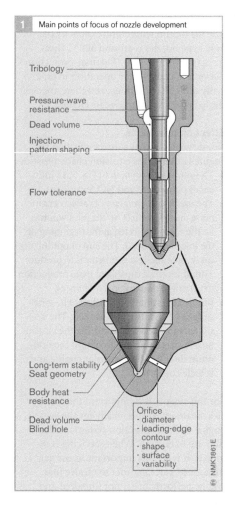

1   Main points of focus of nozzle development

Tribology

Pressure-wave resistance

Dead volume

Injection-pattern shaping

Flow tolerance

Long-term stability
Seat geometry

Body heat resistance

Dead volume
Blind hole

Orifice
- diameter
- leading-edge contour
- shape
- surface
- variability

NMK1861E

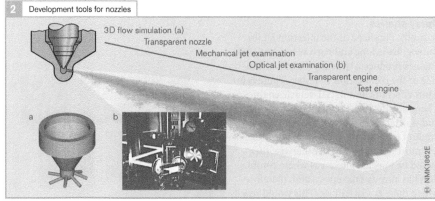

2   Development tools for nozzles

3D flow simulation (a)
Transparent nozzle
Mechanical jet examination
Optical jet examination (b)
Transparent engine
Test engine

a   b

NMK1862E

The image associated with diesel engines in many people's minds is more one of heavy-duty machinery than high-precision engineering. But modern diesel fuel-injection systems are made up of components that are manufactured to the highest degrees of accuracy and required to withstand enormous stresses.

The nozzle is the interface between the fuel-injection system and the engine. It has to open and close precisely and reliably for the entire life of the engine. When it is closed, it must not leak. This would increase fuel consumption, adversely affect exhaust-gas emissions, and might even cause engine damage.
  To ensure that the nozzles seal reliably at the high pressures generated in modern fuel-injection systems such as the VR (VP44), CR, UPS and UIS designs (up to 2,050 bar), they have to be specially designed and very precisely manufactured. By way of illustration, here are some examples:
• To ensure that the sealing face of the nozzle body (1) provides a reliable seal, its has a dimensional tolerance of 0.001 mm (1 µm). That means it must be accurate to within approximately 4,000 metal atom layers!
• The nozzle-needle guide clearance (2) is 0.002...0.004 mm (2...4 µm). The dimensional tolerances are similarly less than 0.001 mm (1 µm).

The injection orifices (3) in the nozzles are created by an electro-erosion machining process. This process erodes the metal by vaporization caused by the high temperature generated by the spark discharge between an electrode and the workpiece. Using high-precision electrodes and accurately configured parameters, extremely precise injection orifices with diameters of 0.12 mm can be produced. This means that the smallest injection orifice diameter is only twice the thickness of a human hair (0.06 mm). In order to obtain better injection characteristics, the leading edges of the noz-zle injection orifices are rounded off by special abrasive fluids (hydro-erosion machining).

The minute tolerances demand the use of highly specialized and ultra-accurate measuring equipment such as:
• Optical 3D coordinate measuring machine for measuring the injection orifices, or
• Laser interferometers for checking the smoothness of the nozzle sealing faces.

The manufacture of diesel fuel-injection components is thus "high-volume, high-technology".

▼ A matter of high-precision

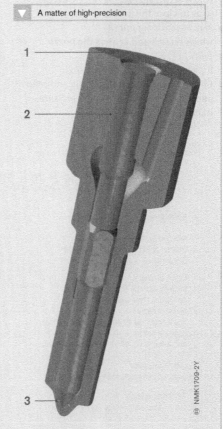

1 Nozzle-body sealing face
2 Guide clearance between nozzle needle and nozzle body
3 Injection orifice

⊕ NMK1709-2Y

Nozzle holders

A nozzle holder combines with the matching nozzle to form the nozzle-and-holder assembly. There is a nozzle-and-holder assembly fitted in the cylinder head for each engine cylinder (Fig. 1). These components form an important part of the fuel-injection system and help to shape engine performance, exhaust emissions and noise characteristics. In order that they are able to perform their function properly, they must be designed to suit the engine in which they are used.

The nozzle (4) in the nozzle holder sprays fuel into the diesel-engine combustion chamber (6). The nozzle holder contains the following essential components:

- *Valve spring(s)* (9)
  which act(s) against the nozzle needle so as to close the nozzle
- *Nozzle-retaining nut* (8)
  which retains and centers the nozzle
- *Filter* (11)
  for keeping dirt out of the nozzle
- *Connections* for the fuel supply and return lines which are linked via the *pressure channel* (10)

Depending on design, the nozzle holder may also contain seals and spacers. Standardized dimensions and combinations provide the required degree of adaptability combined with the minimum of component diversity.

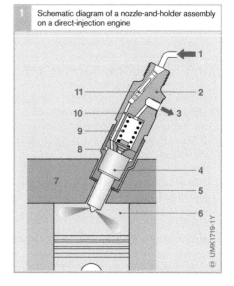

1   Schematic diagram of a nozzle-and-holder assembly on a direct-injection engine

Fig. 1
1   Fuel supply
2   Holder body
3   Fuel return
4   Nozzle
5   Sealing gasket
6   Combustion
    chamber of
    diesel engine
7   Cylinder head
8   Nozzle-retaining nut
9   Valve spring
10   Pressure channel
11   Filter

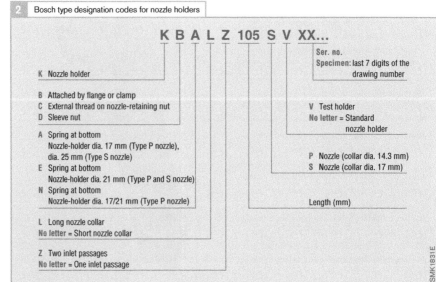

2   Bosch type designation codes for nozzle holders

$$\text{K B A L Z 105 S V XX...}$$

K   Nozzle holder

B   Attached by flange or clamp
C   External thread on nozzle-retaining nut
D   Sleeve nut

A   Spring at bottom
    Nozzle-holder dia. 17 mm (Type P nozzle),
    dia. 25 mm (Type S nozzle)
E   Spring at bottom
    Nozzle-holder dia. 21 mm (Type P and S nozzle)
N   Spring at bottom
    Nozzle-holder dia. 17/21 mm (Type P nozzle)

L   Long nozzle collar
No letter = Short nozzle collar

Z   Two inlet passages
No letter = One inlet passage

Ser. no.
Specimen: last 7 digits of the
    drawing number

V   Test holder
No letter = Standard
    nozzle holder

P   Nozzle (collar dia. 14.3 mm)
S   Nozzle (collar dia. 17 mm)

Length (mm)

The design of the nozzle holder for direct-injection (DI) and indirect-injection (IDI) engines is basically the same. But since modern diesel engines are almost exclusively direct-injection, the nozzle-and-holder assemblies illustrated here are mainly for DI engines. The descriptions, however, can be applied to IDI nozzles as well, but bearing in mind that the latter use pintle nozzles rather than the hole-type nozzles found in DI engines.

Nozzle holders can be combined with a range of nozzles. In addition, depending on the required injection pattern, there is a choice of
- *standard nozzle holder* (single-spring nozzle holder) or
- *two-spring nozzle holder* (not for unit pump systems).

A variation of those designs is the *stepped holder* which is particularly suited to situations where space is limited.

Depending on the fuel-injection system in which they are used, nozzle holders may or may not be fitted with *needle-motion sensors*.

The needle-motion sensor signals the precise start of injection to the engine control unit.

Nozzle holders may be attached to the cylinder block by flanges, clamps, sleeve nuts or external threads. The fuel-line connection is in the center or at the side.

The fuel that leaks past the nozzle needle acts as lubrication. In many nozzle-holder designs, it is returned to the fuel tank by a fuel-return line.

Some nozzle holders function without fuel leakage – i.e. without a fuel-return line. The fuel in the spring chamber has a damping effect on the needle stroke at high injection volumes and engine speeds so that a similar injection pattern to that of a two-spring nozzle holder is generated.

In the common-rail and unit-injector high-pressure fuel-injection systems, the nozzle is integral with the injector, so that a nozzle-and-holder assembly is unnecessary.

For large-scale engines with a per-cylinder output of more than 75 kW, there are application-specific fuel-injector assemblies which may also be cooled.

Fig. 3
a   Stepped nozzle holder for commercial vehicles
b   Standard nozzle holder for various engine types
c   Two-spring nozzle holder for cars
d   Standard nozzle holder for various engine types
e   Stepped nozzle holder without fuel-leakage connection for commercial vehicles
f   Stepped nozzle holder for commercial vehicles
g   Stepped nozzle holder for various engine types
h   Two-spring nozzle holder for cars
i   Stepped nozzle holder for various engine types
j   Standard nozzle holder with pintle nozzle for various types of IDI engine

3   Examples of nozzle-and-holder assemblies

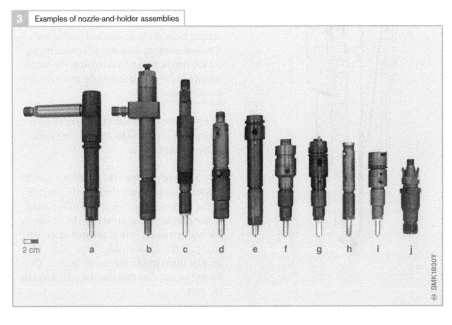

2 cm   a   b   c   d   e   f   g   h   i   j

SMK1830Y

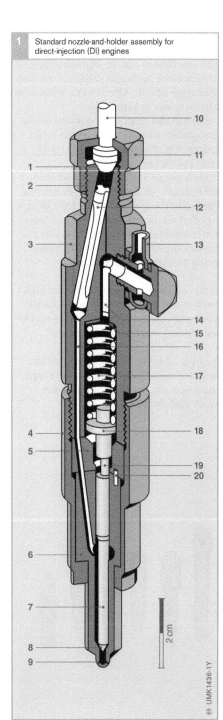

1   Standard nozzle-and-holder assembly for direct-injection (DI) engines

Fig. 1
1   Sealing cone
2   Screw thread for
    central pressure
    connection
3   Holder body
4   Nozzle-retaining nut
5   Intermediate disk
6   Nozzle body
7   Nozzle needle
8   Nozzle-body seat
    face
9   Injection orifice
10 Fuel inlet
11 Sleeve nut
12 Edge-type filter
13 Leak fuel connection
14 Leak fuel port
15 Shim
16 Pressure passage
17 Compression spring
18 Pressure pin
19 Pressure pin
20 Locating pin

Standard nozzle holders

Design and usage

The key features of standard nozzle holders are as follows:

- Cylindrical exterior with diameters of 17, 21, 25 and 26 mm
- Non-twist hole-type nozzles for engines with direct injection and
- Standardized individual components (springs, pressure pins, nozzle retaining nuts) that permit different combinations

The nozzle-and-holder assembly is made up of nozzle holder and nozzle (Fig. 1, with hole-type nozzle). The nozzle holder consists of the following components:

- Holder body (3)
- Intermediate disk (5)
- Nozzle-retaining nut (4)
- Pressure pin (18)
- Compression spring (17)
- Shim (15) and
- Locating pin (20)

The nozzle is attached centrally to the holder by the nozzle-retaining nut. When the retaining nut and holder body are screwed together, the intermediate disk is pressed against the sealing faces of the holder and nozzle body. The intermediate disk acts as a limiting stop for the needle lift and also centers the nozzle relative to the nozzle holder by means of the locating pins.

The pressure pin centers the compression spring and is guided by the nozzle-needle pressure pin (19).

The pressure passage (16) inside the nozzle holder body connects through the channel in the intermediate disk to the inlet passage of the nozzle, thus connecting the nozzle to the high-pressure line of the fuel-injection pump. If required, an edge-type filter (12) may be fitted inside the nozzle holder. This keeps out any dirt that may be contained in the fuel.

Method of operation

The compression spring inside the nozzle holder acts on the nozzle needle via the pressure pin. The spring tension is set by means of a shim. The force of the spring thus determines the opening pressure of the nozzle.

The fuel passes through the edge-type filter (12) to the pressure passage (16) in the holder body (3), through the intermediate disk (5) and finally through the nozzle body (6) to the space (8) surrounding the nozzle needle. During the injection process, the nozzle needle (7) is lifted upwards by the pressure of the fuel (110...170 bar for pintle nozzles and 150...350 bar for hole-type nozzles). The fuel passes through the injection orifices (9) into the combustion chamber. The injection process comes to an end when the fuel pressure drops to a point where the compression spring (17) is able to push the nozzle needle back against its seat. Start of injection is thus controlled by fuel pressure. The injected fuel quantity depends essentially on how long the nozzle remains open.

In order to limit needle lift for pre-injection, some designs have a nozzle-needle damper (Fig. 2).

Stepped nozzle holders

Design and usage

On multi-valve commercial-vehicle engines in particular, where the nozzle-and-holder assembly has to be fitted vertically because of space constraints, stepped nozzle-and-holder assemblies are used (Fig. 3). The reason for the name can be found in the graduated dimensions (1).

The design and method of operation are the same as for standard nozzle holders. The essential difference lies in the way in which the fuel line is connected. Whereas on a standard nozzle holder it is screwed centrally to the top end of the nozzle holder, on a stepped holder it is connected to the holder body (11) by means of a delivery connection (10). This type of arrangement is normally used to achieve very short injection fuel lines, and has a beneficial effect on the injection pressure because of the smaller dead volume in the fuel lines.

Stepped nozzle holders are produced with or without a leak fuel connection (9).

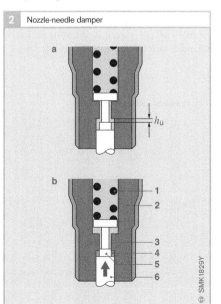

2   Nozzle-needle damper

a

h_u

b

1
2

3
4
5
6

SMK1829Y

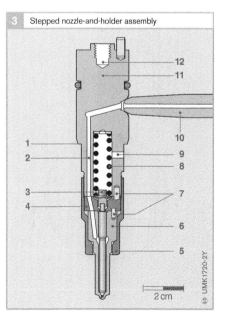

3   Stepped nozzle-and-holder assembly

12
11

10

1
2

9
8

3
4

7

6

5

2 cm

UMK1720-2Y

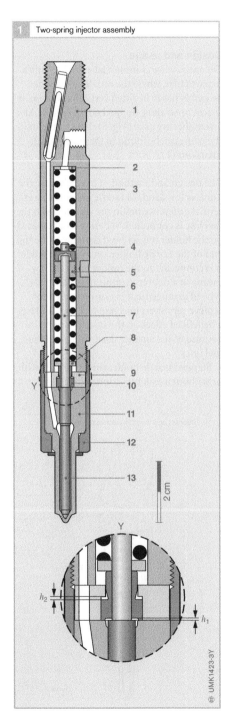

Fig. 1
1 Holder body
2 Shim
3 Compression
  spring 1
4 Pressure pin
5 Guide washer
6 Compression
  spring 2
7 Pressure pin
8 Spring seat
9 Intermediate disk
10 Stop sleeve
11 Nozzle body
12 Nozzle-retaining nut
13 Nozzle needle

h_1 Plunger lift to port
   closing
h_2 Main lift

Fig. 2
a Standard nozzle
  holder (single-spring)
b Two-spring nozzle
  holder

h_1 Plunger lift to port
   closing
h_2 Main lift

Two-spring nozzle holders

Usage
The two-spring nozzle holder is a refinement of the standard nozzle holder. It has the same external dimensions. Its graduated rate-of-discharge curve (Fig. 2) produces "softer" combustion and therefore a quieter engine, particularly at idle speed and part load. It is used primarily on direct-injection (DI) engines.

Design and method of operation
The two-spring nozzle holder (Fig. 1) has two compression springs positioned one behind the other. Initially, only one of the compression springs (3) is acting on the nozzle needle (13) and thus determines the opening pressure. The second compression spring (6) rests against a stop sleeve (10) which limits the plunger lift to port closing. During the injection process, the nozzle needle initially moves towards the plunger lift to port closing, h_1 (0.03...0.06 mm for DI engines, 0.1 mm for IDI engines). This allows only a small amount of fuel into the combustion chamber.

As the pressure inside the nozzle holder continues to increase, the stop sleeve overcomes the force of both compression springs (3 and 6). The nozzle needle then completes the main lift ($h_1 + h_2$, 0.2...0.4 mm) so that the main injected fuel quantity is injected.

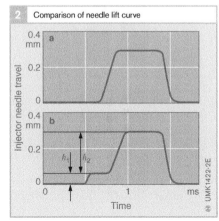

Comparison of needle lift curve

Nozzle holders with needle-motion sensors

Usage

Start of delivery is a key variable for optimizing diesel-engine performance. Detection of this variable allows the adjustment of start of delivery according to engine load and speed within a closed control loop. In systems with distributor and in-line fuel-injection pumps, this is achieved by means of a nozzle with a needle-motion sensor (Fig. 2) which transmits a signal when the nozzle needle starts to move upwards. It is sometimes also called a needle-motion sensor.

Design and method of operation

A current of approximately 30 mA is passed through the detector coil (Fig. 2, 11). This produces a magnetic field. The extended pressure pin (12) slides inside the guide pin (9). The penetration depth X determines the magnetic flux in the detector coil. By virtue of the change in magnetic flux in the coil, movement of the nozzle needle induces a velocity-dependent voltage signal (Fig. 1) in the coil which is processed by an analyzer circuit in the electronic control unit. When the signal level exceeds a threshold voltage, it is interpreted by the analyzer circuit to indicate the start of injection.

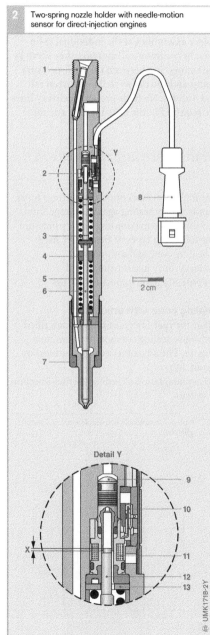

Two-spring nozzle holder with needle-motion sensor for direct-injection engines

2 cm

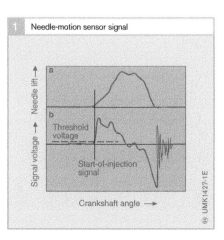

Needle-motion sensor signal

Detail Y

Fig. 1
a Needle-lift curve
b Corresponding coil
   signal voltage curve

Fig. 2
1 Holder body
2 Needle-motion
   sensor
3 Compression spring
4 Guide washer
5 Compression spring
6 Pressure pin
7 Nozzle-retaining nut
8 Connection to
   analyzer circuit
9 Guide pin
10 Contact tab
11 Detector coil
12 Pressure pin
13 Spring seat

X Penetration depth

High-pressure lines

Regardless of the basic system concept – in-line fuel-injection pump, distributor injection pump or unit pump systems – it is the high-pressure delivery lines and their connection fittings that furnish the links between the fuel-injection pump(s) and the nozzle-and-holder assemblies at the individual cylinders. In common-rail systems, they serve as the connection between the high-pressure pump and the rail as well as between rail and nozzles. No high-pressure delivery lines are required in the unit-injector system.

High-pressure connection fittings

The high-pressure connection fittings must supply secure sealing against leakage from fuel under the maximum primary pressure. The following types of fittings are used:
- Sealing cone and union nut
- Heavy-duty insert fittings, and
- Perpendicular connection fittings

Sealing cone with union nut
All of the fuel-injection systems described above use sealing cones with union nuts (Fig. 1). The advantages of this connection layout are:
- Easy adaptation to individual fuel-injection systems

- Fitting can be disconnected and reconnected numerous times
- The sealing cone can be shaped from the base material

At the end of the high-pressure line is the compressed pipe-sealing cone (3). The union nut (2) presses the cone into the high-pressure connection fitting (4) to form a seal. Some versions are equipped with a supplementary thrust washer (1). This provides a more consistent distribution of forces from the union nut to the sealing cone. The cone's open diameter should not be restricted, as this would obstruct fuel flow. Compressed sealing cones are generally manufactured in conformity with DIN 73 365 (Fig. 2).

Heavy-duty insert fittings
Heavy-duty insert fittings (Fig. 3) are used in unit-pump and common-rail systems as installed in heavy-duty commercial vehicles. With the insert fitting, it is not necessary to route the fuel line around the cylinder head to bring it to the nozzle holder or nozzle. This allows shorter fuel lines with associated benefits when it comes to space savings and ease of assembly.

The screw connection (8) presses the line insert (3) directly into the nozzle holder (1) or nozzle. The assembly also includes a mainte-

Fig. 1
1　Thrust washer
2　Union nut
3　Pipe sealing cone on high-pressure delivery line
4　Pressure connection on fuel-injection pump or nozzle holder

Fig. 2
1　Sealing surface

d　Outer line diameter
d_1　Inner line diameter
d_2　Inner cone diameter
d_3　Outer cone diameter
k　Length of cone
R_1, R_2　Radii

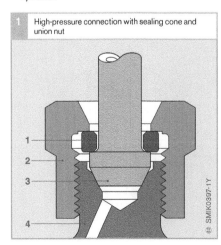

1 High-pressure connection with sealing cone and union nut

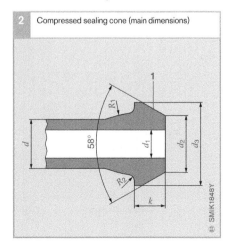

2 Compressed sealing cone (main dimensions)

nance-free edge-type filter (5) to remove coarse contamination from the fuel. At its other end, the line is attached to the high-pressure delivery line (7) with a sealing cone and union nut (6).

Perpendicular connection fittings

Perpendicular connection fittings (Fig. 4) are used in some passenger-car applications. They are suitable for installations in which there are severe space constraints. The fitting contains passages for fuel inlet and return (7, 9). A bolt (1) presses the perpendicular fitting onto the nozzle holder (5) to form a sealed connection.

High-pressure delivery lines

The high-pressure fuel lines must withstand the system's maximum pressure as well as pressure variations that can attain very high fluctuations. The lines are seamless precision-made steel tubing in killed cast steel which has a particularly consistent microstructure. Dimensions vary according to pump size (Table 1, next page).

All high-pressure delivery lines are routed to avoid sharp bends. The bend radius should not be less than 50 mm.

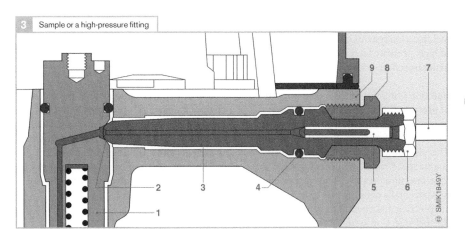

3   Sample or a high-pressure fitting

Fig. 3
1   Nozzle holder
2   Sealing cone
3   High-pressure fitting
4   Seal
5   Edge-type filter
6   Union nut
7   High-pressure delivery line
8   Screw connections
9   Cylinder head

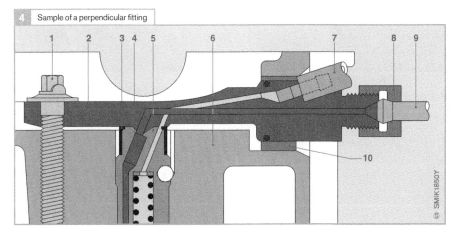

4   Sample of a perpendicular fitting

Fig. 4
1   Expansion bolt
2   Perpendicular fitting
3   Molded seal
4   Edge-type filter
5   Nozzle holder
6   Cylinder head
7   Fuel return line (leakage-fuel line)
8   Union nut
9   High-pressure delivery line
10 Clamp

Length, diameter and wall depth of the high-pressure lines all affect the injection process. To cite some examples: Line length influences the rate of discharge dependent on speed, while internal diameter is related to throttling loss and compression effects, which will be reflected in the injected-fuel quantity. These considerations lead to prescribed line dimensions that must be strictly observed. Tubing of other dimensions should never be installed during service and repairs. Defective high-pressure tubing should always be replaced by OEM lines. During servicing or maintenance, it is also important to observe precautions against fouling entering the system. This applies in any case to all service work on fuel-injection systems.

A general priority in the development of fuel-injection systems is to minimize the length of high-pressure lines. Shorter lines produce better injection-system performance.

Injection is accompanied by the formation of pressure waves. These are pulses that propagate at the speed of sound before finally being reflected on impact at the ends. This phenomenon increases in intensity as engine speed rises. Engineers exploit it to raise injection pressure. The engineering process entails defining line lengths that are precisely matched to the engine and the fuel-injection system.

All cylinders are fed by high-pressure delivery lines of a single, uniform length. More or less angled bends in the lines compensate for the different distances between the outlets from the fuel-injection pump or rail, and the individual engine cylinders.

The primary factor determining the high-pressure line's compression-pulsating fatigue strength is the surface quality of the inner walls of the lines, as defined by material and peak-to-valley height. Especially demanding performance requirements are satisfied by prestressed high-pressure delivery lines (for applications of 1,400 bar and over). Before installation on the engine, these customized lines are subjected to extremely high pressures (up to 3,800 bar). Then pressure is suddenly relieved. The process compresses the material on the inner walls of the lines to provide increased internal strength.

The high-pressure delivery lines for vehicle engines are normally mounted with clamp brackets located at specific intervals. This means that transfer of external vibration to the lines is either minimal or nonexistent.

The dimensions of high-pressure lines for test benches are subject to more precise tolerance specifications.

Table 1

d Outer line diameter
d_1 Inner line diameter

Wall thicknesses indicated in **bold** should be selected when possible.

Dimensions for high-pressure lines are usually indicated as follows:
$d \times s \times l$
l Line length

1 Main dimensions of major high-pressure delivery lines in mm

d \ d_1	1.4	1.5	1.6	1.8	2.0	2.2	2.5	2.8	3.0	3.6	4.0	4.5	5.0	6.0	7.0	8.0	9.0
			Wall thickness s														
4	1.3	1.25	**1.2**														
5	1.8	1.75	1.7	1.6													
6		**2.25**	**2.2**	2.1	2	1.9	1.75	1.6	**1.5**								
8				3	2.9	**2.75**	**2.6**	2.5	2.2	**2**							
10					3.75	3.6	**3.5**	**3.2**	**3**	2.75	2.5						
12							**4.5**	4.2	**4**	3.75	3.5						
14									**5**	4.75	**4.5**	**4**		**3**			
17											6	**5.5**	5	**4.5**			
19																	5
22														7			

▶ Cavitation in the high-pressure system

Cavitation can damage fuel-injection systems (Fig. 1). The process takes place as follows:

Local pressure variations occur at restrictions and in bends when a fluid enters an enclosed area at extremely high speeds (for instance, in a pump housing or in a high-pressure line). If the flow characteristics are less than optimum, low-pressure sectors can form at these locations for limited periods of time, in turn promoting the formation of vapor bubbles.

These gas bubbles implode in the subsequent high-pressure phase. If a wall is located immediately adjacent to the affected sector, the concentrated high energy can create a cavity in the surface over time (erosion effect). This is called cavitation damage.

As the vapor bubbles are transported by the fluid's flow, cavitation damage will not necessarily occur at the location where the bubble forms. Indeed, cavitation damage is frequently found in eddy zones.

The causes behind these temporary localized low-pressure areas are numerous and varied. Typical factors include:

- Discharge processes
- Closing valves
- Pumping between moving gaps and
- Vacuum waves in passages and lines

Attempts to deal with cavitation problems by improving material quality and surface-hardening processes cannot produce anything other than very modest gains. The ultimate objective is and remains to prevent the vapor bubbles from forming, and, should complete prevention prove impossible, to improve flow behavior to limit the negative impacts of the bubbles.

1 Cavitation damage in the distributor head of a VE pump

SMK1851Y

Fig. 1
1   Cavitation

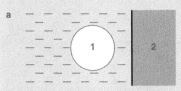

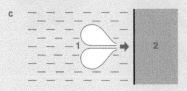

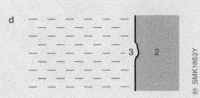

2 Implosion of a cavitation bubble

SMK1852Y

Fig. 2
a   A vapor bubble is formed
b   The vapor bubble collapses
c   The collapsed sections form a sharp edge with extremely high energy
d   The imploding vapor bubble leaves a recess on the surface

1   Vapor bubble
2   Wall
3   Recess

Start-assist systems

The colder the diesel engine, the more reluctant it is to start. Leakage and thermal losses reduce the compression pressure in the cold cylinders and with it the temperature of the compressed air. Cold engines have an outside-temperature limit, under which starting without the assistance of auxiliary start-assist devices is no longer possible.

Compared with gasoline, diesel fuel is very easily combustible. That is why warm precombustion-chamber and whirl-chamber diesel engines and direct-injection (DI) engines will start spontaneously at low outside temperatures down to ≥0°C. Here, the spontaneous ignition temperature for diesel fuel of 250°C is achieved with the engine turning at starting speed. Cold precombustion-chamber and whirl-chamber diesel engines require start assistance at ambient temperatures of < 40°C and < 20°C respectively, while DI engines only need such intervention below 0°C.

Overview

Systems for passenger cars and light commercial vehicles

Preheating systems are used for passenger cars and light commercial vehicles. These systems increase starting comfort and help the engine to run smoothly and with minimal emissions after starting and in the warm-up period.

Preheating systems consist of sheathed-element glow plugs, a switch a preheating software in the engine-management system. Conventional preheating systems use glow plugs with a nominal voltage of 11 V which are activated by the vehicle system voltage. New low-voltage preheating systems require glow plugs with nominal voltages below 11 V whose heat output is adapted to the engine's requirements by an electronic glow control unit.

In precombustion-chamber and whirl-chamber diesel engines (IDI), the glow plug extends into the secondary combustion chamber, while in DI engines, it extends into the main combustion chamber of the engine cylinder.

1   Components of a preheating system

SMK2028Y

The air/fuel mixture is directed past the hot tip of the glow plug and heated. The ignition temperature is reached in combination with the heating of the intake air during the compression cycle.

Preheating systems are usually replaced by flame starting systems in diesel engines with a displacement of more than 1 *l*/cylinder (commercial vehicles).

Requirements

The more exacting comfort requirements of today's diesel drivers have decisively shaped the development of modern preheating systems. Nowadays, drivers will no longer accept a cold start together with the "diesel minute's silence".

More stringent emission limits and the desire for higher specific engine power have resulted in the development of low-compression engines. The cold-starting and cold-running performance of these engines is problematic, but this can be controlled by higher preheating temperatures and longer preheating times.

Due to the dramatic increase in the number of electrical loads/consumers, low power consumption by electrical components will become increasingly important in the future.

To summarize, a preheating system meet satisfy the following requirements:
- Fastest possible preheating rate (1,000°C/s) even in the event of a dip in the vehicle system voltage
- High preheating-system service life (commensurate with the engine service life)
- Extended post- and intermediate-heating times in minutes
- Ideal adaptation of heating output to engine requirements
- Continuous heating temperature up to 1,150°C for low-compression engines
- Reduced load on the vehicle electrical system
- EURO IV- and US 07-compatible
- On-Board Diagnosis according to OBD II and EOBD

Preheating systems

Preheating phases
The preheating process consists of five phases.
- During preheating, the glow plug is heated to operating temperature.
- During standby heating, the preheating system maintains a glow-plug temperature necessary for starting for a defined period.
- Glow-plug start assist is used while the engine runs up.
- The post-heating phase starts after starter release.
- The glow plugs are subjected to intermediate heating after engine cooling by overrunning or to support particulate-filter regeneration.

Conventional preheating system
Design and construction
Conventional preheating systems consist of
- a metal glow plug with a nominal voltage of 11 V
- a relay glow control unit and
- a software module for the preheating function integrated in the engine control unit (Electronic Diesel Control, EDC)

Method of operation
The preheating software in EDC starts and ends the preheating process in accordance with operation of the glow-plug starter switch and parameters stored in the software. The glow control unit activates the glow plugs with vehicle system voltage via a relay during the preheating, standby, start and post-heating phases. The nominal voltage of the glow plugs is 11 V. The heating output is therefore dependent on the current vehicle system voltage and the glow-plug thermistor (PTC). The glow plug thus has a self-regulating function. In conjunction with an engine-load-dependent cutout function in the engine-management preheating software, it is possible to safely prevent the glow plug from suffering temperature overload. Adaptation of the post-heating time to the

engine requirements extends the glow plug's service life and enhances its cold-running properties.

Duraterm sheathed-element glow plug
Design and properties
The glow element consists of a tubular heating element which is sealed inside the gastight plug body (Fig. 1, 3). The tubular heating element consists of a hot-gas and corrosion-resistant element sheath (4) which encloses a filament surrounded by compressed magnesium oxide powder (6). That filament is made up of two resistors connected in series – the heating filament (7) located in the tip of the sheath, and the control filament (5).

Whereas the heating filament has an electrical impedance that is independent of temperature, the control filament has a positive temperature coefficient (PTC). In the latest generation of glow plugs (Type GSK2), its impedance increases even more steeply as the temperature rises than with the older designs (Type S-RSK). The Type GSK2 glow plugs are thus faster at reaching the temperature required for igniting the diesel fuel (850°C in 4 s) and also have a lower steady-state temperature. This means that the temperature is kept below the critical level for the glow plug. Consequently, it can remain in operation for up to three minutes after the engine has started. This post-heating function results in a more effective engine cold-idle phase with substantially lower noise and emission output.

The heating filament is welded into the cap of the element sheath for grounding. The control filament is contacted at the terminal stud, which establishes the connection to the vehicle electrical system.

Function
When voltage is applied to the glow plug, most of the electrical energy of the heating filament is initially converted into heat; the temperature at the tip of the glow plug increases sharply. The temperature of the control filament – and with it also the impedance – increase with a time delay. The current draw and thus the total heating output of the glow plug decrease and the temperature approaches a steady-state condition. The heating characteristics shown in Figure 2 are produced.

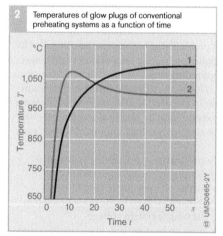

2   Temperatures of glow plugs of conventional preheating systems as a function of time

Temperature T (°C), Time t (s)

1,050, 950, 850, 750, 650

0   10   20   30   40   50

UMS0665-2Y

Fig. 2
1 S-RSK
2 GSK2

Fig. 1
1 Connector
2 Insulating washer
3 Body
4 Element sheath
5 Control filament
6 Magnesium oxide powder
7 Heating filament
8 Element seal
9 Double seal
10 Knurled nut

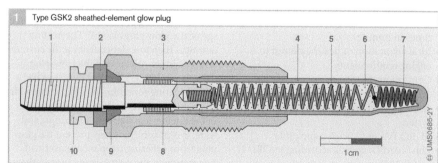

1   Type GSK2 sheathed-element glow plug

1 cm

UMS0685-2Y

Low-voltage preheating system

Design and construction

Depending on the application, the low-volt-age preheating system contains

- ceramic Rapiterm sheathed-element glow plugs or HighSpeed metal sheathed-element glow plugs of low-voltage configuration < 11 V,
- an electronic glow control unit, and
- a software module for the preheating function integrated in the engine control unit (Electronic Diesel Control, EDC).

Operating concept

The glow control unit activates the glow plugs in such a way that the heating temperature is adapted to the engine requirements in the preheating, standby, start, post-heating and intermediate-heating phases. In order during preheating to achieve as quickly as possible the heating temperature required for engine starting, the glow plugs are briefly operated in this phase with push voltage, which is above the glow-plug nominal voltage. The activating voltage is then reduced to the glow-plug nominal voltage during start-standby heating.

During glow-plug start assist, the activation voltage is increased again in order to compensate for the cooling of the glow plug by the cold intake air. This is also possible in the post- and intermediate-heating phases. The required voltage is generated from the vehicle system voltage by Pulse-Width Modulation (PWM). Here, the associated PWM value is taken from a program map, which is adapted to the relevant engine within an application. The program map is stored in the preheating module of the EDC software and contains the following parameters:

- Speed
- Injected fuel quantity (i.e., load)
- Time after starter release (currently three post-heating phases are defined, within which the temperature of the glow plug can be adapted)
- Coolant temperature

Map-controlled activation reliably prevents thermal overloading of the glow plug in all engine operating states.

The heating function implemented in EDC contains an overheating-protection facility in the event of repeat heating. This is brought about by an energy integration model. During heating, the energy introduced into the glow plug is integrated. After deactivation, the amount of energy dissipated from the glow plug by radiation and heat discharge is subtracted from this amount of energy. This enables the current temperature of the glow plug to be estimated. If the temperature drops below a threshold stored in EDC, the glow plug can be heated up again with push voltage.

The heating temperature which can be adjusted as a function of the coolant temperature prolongs the service life of the glow plug while maintaining its excellent cold-starting and cold-running properties. This is achieved by lowering the glow-plug temperature when the coolant is "hot" – e.g. in TDI engines from approx. –10°C – and shortening the post-heating time. The application of the preheating system can therefore be matched to the requirements of the vehicle manufacturer.

These preheating systems facilitate a rapid start when HighSpeed metal glow plugs are used and an immediate start when Rapiterm glow plugs are used similar to a gasoline engine up to –28°C.

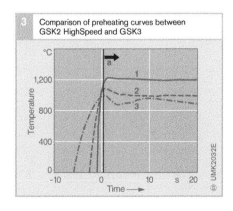

3   Comparison of preheating curves between GSK2 HighSpeed and GSK3

Fig. 3
a   From $t = 0$ s flow velocity of 11 m/s

1   Rapiterm glow plug (7 V)
2   HighSpeed metal glow plug (5 V)
3   Metal glow plug (11 V)

HighSpeed metal sheathed-element glow plug

Figure 4 shows a HighSpeed metal sheathed-element glow plug with a nominal voltage of 4.4 V (push voltage of 11 V when heating for 1.8 s, then reduction to nominal voltage) with an M8 body.

The basic design and operation of the HighSpeed glow plug is the same as that of the Duraterm. The heating and control filaments are designed here for a lower nominal voltage and a high preheating rate.

The slender shape is designed to suit the restricted space in four-valve engines. The glow element (dia. 4/3.3 mm) is tapered at the front in order to accommodate the heating filament closer to the element sheath. This allows preheating rates of up to 1,000°C/3 s with the push mode used here. The maximum heating temperature is in excess of 1,000°C. The temperature during start-standby heating and in post-heating mode is approx. 980°C. These functional properties are adapted to the requirements of diesel engines with a compression ratio of $\varepsilon \geq 18$.

Rapiterm sheathed-element glow plug

Rapiterm sheathed-element glow plugs (Fig. 5) have glow elements made from a new, high-temperature-resistant, ceramic composite material with adjustable electric conductivity. On account of their very high oxidation stability and thermal-shock resistance, they allow an immediate start, maximum heating temperatures of 1,300°C, and post- and intermediate heating lasting minutes at 1,150°C. Their low power consumption and long service life make them superior to other sheathed-element glow plugs. This is achieved by

- the special properties of the composite material
- the configuration of the low-voltage glow plug
- the heating zone situated on the surface and
- optimized activation by the combination of glow control unit and EDC

Bosch has developed this Rapiterm glow plug for the special requirements of engines with a low compression ratio of $\varepsilon \leq 16$.

Reduced emissions in diesel engines with a low compression ratio

By lowering the compression ratio in modern diesel engines from $\varepsilon = 18$ to $\varepsilon = 16$, it is possible to reduce the NO_X and soot emissions while simultaneously increasing the specific power. However, the cold-starting and cold-running performance is problematic in these engines. In order to obtain minimal exhaust-gas opacity values and height-

| 4 | HighSpeed metal sheathed-element glow plug |

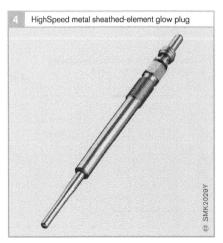

SMK2029Y

| 5 | Rapiterm sheathed-element glow plug |

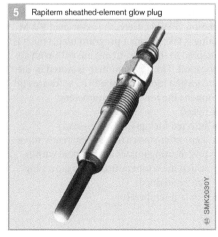

SMK2030Y

ened smooth running during cold starting and cold running, temperatures at the glow plug of over 1,150°C are required – 850°C is sufficient for conventional engines. During the cold-starting phase, these low emission values – blue-smoke and soot emissions – can only be maintained by post-heating lasting several minutes. Compared with standard preheating systems, the Rapiterm preheating system from Bosch reduces the exhaust-gas opacity values by up to 60%.

Glow control unit

The glow control unit controls the glow plugs via a power relay or power transistors. It receives its starting signal from the engine management module or a temperature sensor.

Autarkic glow control units assume all the control and display functions. The preheating process is controlled by temperature sensors in these systems. If the injected fuel quantity exceeds a critical level, a load switch interrupts the post-heating process. This prevents the glow plugs from overheating. These systems have in the meantime been superseded by EDC-controlled glow control units.

EDC-controlled relay glow control unit for 11 V glow plugs

The glow control unit activates the 11 V glow plugs with vehicle system voltage via a relay in accordance with the EDC specifications. The heating output of the preheating system is therefore dependent on the current vehicle system voltage and the glow-plug thermistor (PTC characteristic). Preheating systems with relay glow control units are characterized by low application expenditure. Malfunctioning glow plugs or relay faults are detected and signaled to EDC by diagnostic flag.

EDC-controlled transistor glow control unit for low-voltage glow plugs

The new electronic glow control units allow specific voltage control of the low-voltage glow plugs. The required effective voltage is generated from the vehicle system voltage by Pulse-Width Modulation (PWM). Here, the associated PWM value is taken from an engine-specific program map which is stored in the preheating module of the EDC software. In this way, the heating output of the preheating system can be perfectly adapted to the engine requirements. Staggered activation of the glow plugs reduces the maximum load on the vehicle electrical system during the cold-starting and post-heating phases to a minimum.

The glow control unit incorporates self-diagnosis and glow-plug monitoring facilities. Faults occurring in the preheating system are reported to and stored in the EDC ECU. This facilitates On-Board Diagnosis in accordance with OBD II (USA) and EOBD (Europe). The error codes stored in EDC enable the service personnel to identify quickly and clearly the failure cause – a glow plug, the glow control unit or the main fuse.

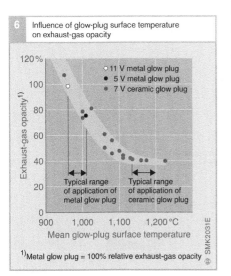

6   Influence of glow-plug surface temperature on exhaust-gas opacity

○ 11 V metal glow plug
● 5 V metal glow plug
● 7 V ceramic glow plug

Exhaust-gas opacity[1]

Typical range of application of metal glow plug

Typical range of application of ceramic glow plug

Mean glow-plug surface temperature

SMK2031E

[1]Metal glow plug = 100% relative exhaust-gas opacity

Fig. 6
Starting temperature: –20°C
Compression ratio: 16:1

Minimizing emissions inside of the engine

When the air/fuel mixture is burned, the main byproducts include pollutants such as NO_x, soot, CO, and HC. The amount of these pollutants contained in the untreated exhaust gas (exhaust gas after combustion before exhaust-gas treatment) mainly depends on engine operating conditions. Besides combustion-chamber shape and airflow path (supercharging/turbocharging, exhaust-gas recirculation, whirl control), the fuel-injection system plays a key role in minimizing emissions.

The introduction of new emission standards in Europe (Euro 3 since the year 2000) has placed much more stringent requirements on combustion processes in passenger-car diesel engines. To obtain the best tradeoff between conflicting factors, such as NO_x emissions and low combustion noise, pre-injection and main injection must occur at precisely the right time and in the exact quantities. This can only be achieved by electronic control of fuel-injection systems. Electronic Diesel Control (EDC) allows enhanced fuel-delivery control, more precise adjustment of start of injection, optimized combustion processes dependent on the operating point, reduced fuel consumption, and lower pollutant emissions.

In future more severe standards and greater customer demands related to convenience and drivability will only be achievable for diesel engines through the use of modern fuel-injection systems, such as the Unit Injector System/Unit Pump System, or common rail.

It will no longer be possible to meet continuing reductions in emission limits by making internal modifications to the engine alone. There will also be a need for additional exhaust-gas treatment methods. By the time Euro 5 is introduced in Europe, it will be absolutely necessary to fit a particulate filter to comply with the very low particulate limits.

New, even more flexible high-pressure fuel-injection systems are under development to comply with the ultra-low U.S. NO_x limits of Tier 2 (valid since 2004) in order to dispense with complex systems required to extract NO_x from the exhaust gas.

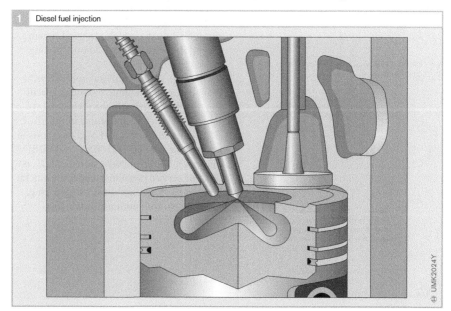

1   Diesel fuel injection

Fig. 1
In order to atomize fuel very finely, it is injected through the nozzle holes into the combustion chamber at high pressure.

UMK2024Y

Combustion process

The combustion process and its adjustment play a key role in the diesel engine when it comes to achievable performance, fuel consumption, and emissions.

Engine performance is limited by the black-smoke emission value (maximum permitted exhaust-gas opacity at full load) and the maximum permitted exhaust-gas temperature. The material properties of the exhaust-gas turbocharger define the limit value of the exhaust-gas temperature at the turbine inlet.

Combustion in the diesel engine can be divided into three phases:
- Ignition lag, i.e. the time between start of injection and start of ignition
- Premixed combustion
- Diffusion flame (mixture-controlled combustion)

Ignition lag, and thus a small quantity of injected fuel, is required during the first phase to limit combustion noise. After combustion starts, a good mixture formation is needed to achieve low soot and NO_x emissions.

The following factors have a decisive influence on the combustion phases:
- Pressure and temperature states within the combustion chamber
- Mass, composition, and movement of the charge
- Injection pressure process

These parameters are adjustable firstly by engine-specific parameters, and secondly by variable operating parameters.

The following fixed, engine-specific parameters are important for a given cylinder displacement:
- Compression ratio
- Stroke/bore ratio
- Shape of piston recess
- Intake port geometry
- Intake and exhaust valve timing

The fuel-injection system plays a key role in the combustion process since it defines the point of 50% mass fraction burnt and mixture formation by determining the injection point and rate-of-discharge curve. In turn, the last two parameters are key factors controlling emissions and efficiency.

Besides the fuel-injection system, development focus is increasing on the air-flow system since compliance with ever more stringent NO_x emission limits requires very high exhaust-gas recirculation rates.

Fig. 2 (on the next page) shows the main engine-specific and operation-dependent influencing variables that affect the combustion process.

Fuel-injection system

On the air-intake side, mixture formation is influenced by movement of the charge inside of the cylinder. This, in turn, depends on intake-duct geometry and combustion-chamber shape. As injection pressures have risen, the function of mixture formation has gradually shifted to the fuel-injection system. As a result, this has led to the development of the low-whirl combustion process.

On the fuel-injection side, extremely small nozzle holes with flow-optimized geometries promote good mixture formation as the injected fuel is then well prepared. At the same time this shortens ignition lag, and only small quantities of fuel are injected. During diffusion combustion that follows, optimized atomization results in high EGR compatibility, and this produces less NO_x and soot.

Air-flow system

Besides the fuel-injection system, more attention is also focusing on the air-flow system, since compliance with ever more stringent NO_x emission limits requires very high EGR compatibility of the combustion process. This minimizes the formation of NO_x so that the particulate filter (now fitted in ever greater numbers) can cope with the quantity of particulate emissions produced.

It requires a system that is capable of combining comparatively high charge-air pressures at high, precise EGR rates identical for all cylinders, and at the lowest possible intake temperatures.

Cylinder charge

Other measures carried out on the engine have an impact on cylinder charge from peripheral systems, and ultimately on the concentration of pollutants in the exhaust gas.

The most important measure for minimizing pollutants here is exhaust-gas recirculation. Exhaust gas recirculated to the intake manifold raises the proportion of inert gas, thus causing a drop in peak combustion temperature. It also reduces the production of nitrogen oxides (see the section on "Exhaust-gas recirculation").

Exhaust-gas turbocharging

Although turbocharging is primarily intended to increase specific performance, it also increases EGR compatibility in the program map by producing a greater charge mass in the combustion chamber. This makes for a more favorable NO_x/soot trade-off. It also makes Variable-Turbine-Geometry (VTG) turbochargers indispensable since they vary charge-air pressure by means of variable turbine blades. This variability makes it possible to use a larger turbine that has a lower exhaust-gas backpressure than a wastegate turbocharger. Decreasing the compression ratio, i.e. making the piston recess bigger, means that the free jet length at full load is larger, and air efficiency increases. At the same time the final compression temperature is lower, peak temperatures during combustion drop, and there is less formation of NO_x.

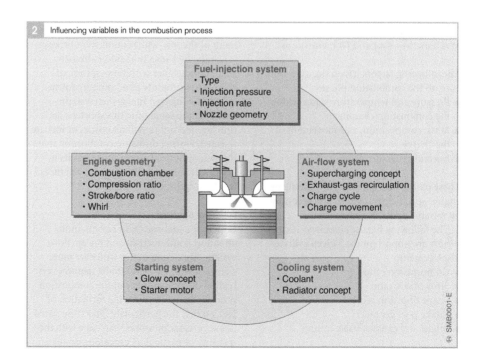

2   Influencing variables in the combustion process

Fuel-injection system
- Type
- Injection pressure
- Injection rate
- Nozzle geometry

Engine geometry
- Combustion chamber
- Compression ratio
- Stroke/bore ratio
- Whirl

Air-flow system
- Supercharging concept
- Exhaust-gas recirculation
- Charge cycle
- Charge movement

Starting system
- Glow concept
- Starter motor

Cooling system
- Coolant
- Radiator concept

SMB0001-E

Combustion temperature

Together with the excess-air factor, the combustion temperature has a significant influence on the formation of NO_x. High temperatures and excess air ($\lambda > 1$) promote the formation of nitrogen oxides. In heterogeneous diffusion combustion, local, lean zones are inevitable, thus increasing the formation of nitrogen oxides. The aim of optimizing the combustion process, therefore, is to lower peak temperatures in the combustion chamber by raising the inert-gas component (EGR), and optimizing mixture formation at the same time in order to lessen the slight increase in soot production. In poor combustion conditions and at low temperatures, the flame front tends to extinguish prematurely. This occurs mainly when the engine is cold and load is low, causing a strong rise in CO and HC, which are the products of incomplete combustion. To counteract this, EGR coolers are bypassed when the engine is running cold. These coolers normally have a high cooling capacity required to reduce NO_x emissions when the engine is operating at normal running temperatures.

Nitrogen oxides form at high temperatures with excess air. Localized peak temperatures and localized, high excess-air factors must then be lowered. This is only achievable by retarding the start of injection at high injection rates during diffusion combustion. Combustion starts shortly before top dead center. This avoids almost any compression of combustion products that could increase the temperature. The high injection rate results in rapid turnover with 50% mass fraction burnt and high EGR compatibility. High combustion-chamber temperatures promote the formation of NO_x.

Other impacts on pollutant emissions

Engine speed

A higher engine speed means greater friction losses in the engine and a higher power input by the ancillary assemblies (e.g. water pump). Engine efficiency, therefore, drops as engine speed increases.

If a specific performance is produced at high engine speed, it requires a greater fuel quantity than if the same performance is produced at low engine speed. It also produces more pollutant emissions.

Nitrogen oxides (NO_x)

As the time available to form NO_x in the combustion chamber is shorter at higher engine speeds, NO_x emissions decrease as engine speed increases. In addition, the residual-gas content in the combustion chamber must be considered since it causes lower peak temperatures. As the residual-gas content normally drops with rising engine speed, this effect runs counter to the interdependence described above.

Hydrocarbons (HC) and carbon monoxide (CO)

As engine speed rises, HC and CO emissions rise as the time for mixture preparation and combustion shortens. As piston speed rises, combustion-chamber pressure drops faster in the expansion phase. This results in poorer combustion conditions, especially at low loads, and combustion efficiency suffers. On the other hand, charge movement and turbulence increase the rate of combustion as engine speed rises. Combustion time becomes shorter, and this compensates at least partly for the poorer marginal conditions.

Soot

Normally, soot becomes less as engine speed increases, since charge movement is more intense, thus resulting in better mixture formation.

Torque

As torque increases, so does the temperature in the combustion chamber, and this improves the combustion conditions. The untreated NO_x emissions, therefore, increase, whereas the products of incomplete combustion, such as CO and HC emissions, initially decrease. As full load approaches, and thus low excess-air factors ($\lambda < 1.4$), soot and CO emissions increase due to oxygen deficiency.

Soot

Soot occurs due to localized oxygen deficiency due to thermal cracking of the hydrocarbon molecules at local temperatures above approx. 1,500 K. Consequently, enhanced air efficiency leads to the formation of less soot, or it allows the injection of larger quantities of fuel, and thus increased performance for the same soot factors.

Fuel

Another decisive factor for enhanced exhaust-gas values are quality improvements in the fuel. For example, sulfur-dioxide emissions have dropped to negligible values in road traffic since the introduction of low-sulfur or sulfur-free fuel.

With respect to conventional combustion, diesel fuel should have the highest possible cetane number, i.e. optimized ignition quality. This shortens ignition lag, and has a favorable impact on combustion noise. The fuel should also possess good lubricity, and a low water and impurity content to ensure proper functioning of the fuel-injection system throughout its service life.

The demands placed on fuel quality have also risen due to constantly rising engine performance. Various additives increase the cetane number, enhance fuel lubricity and flowability, and protect the fuel system from corrosion.

Fuel consumption

The quantity of emitted CO_2 is proportional to fuel consumption – therefore, any reduction in CO_2 is only obtainable by lowering fuel consumption.

Measures for diminishing NO_x emissions to comply with more stringent limits, for example increasing the rate of exhaust-gas recirculation, result in a lower rate of combustion. In turn, this retards combustion into the expansion phase. The point of 50% mass fraction burnt also shifts towards the expansion phase. The generally poorer combustion conditions lead to a drop in engine efficiency. Without measures to lower fuel consumption, e.g. optimizing friction, this will hike fuel consumption, as was the case when Euro 3 was introduced.

Development of homogeneous combustion processes

New homogeneous combustion processes are presently under development with a view to complying with future NO_x limits (in Europe, Euro 4/Euro 5, in the U.S., Tier 2, Bin 5). They have enormous potential for minimizing NO_x emissions compared to standard combustion processes.

The aim is to inject the largest possible fuel quantity, or the total quantity, during the ignition lag to reduce or totally avoid the diffusive phase. Striving for homogenization in the cylinder charge (air, fuel, exhaust gas from exhaust-gas recirculation) minimizes localized differences in the excess-air factor. This will almost stop the formation of NO_x and soot.

In a first stage (partly Homogeneous Compressed Combustion Ignition (pHCCI)), partial homogenization is implementable on conventional diesel engines within a limited engine-speed and load range. This is mainly achieved by adapting the fuel-injection strategy with high exhaust-gas recirculation rates to control ignition lag and the rate of combustion. This will be followed by further development phases that aim to achieve totally premixed combustion (Homogeneous Compressed

Combustion Ignition (HCCI)) in extended program-map areas. This requires an optimization of systems and components, such as the shape of the combustion chamber and injection nozzles.

The disadvantage of these processes is much higher HC and CO emissions compared to standard combustion processes since the air/fuel mixture punches through to the combustion-chamber wall as a result of premixing. This produces wall-quenching effects [1] similar to gasoline engines. The high exhaust-gas recirculation rates also lead to a reduction in bulk quenching, and thus to a rise in the products of incomplete combustion.

Homogenization becomes gradually more problematic as increases occur in load, injected fuel quantity, combustion-chamber temperatures, and pressures. Under these conditions, a change to standard combustion is avoidable, with the result that it must always be possible to control both operating modes. Homogeneous combustion requires more development work and control concepts to cope with operating-mode switch-over, sensitivity to the slightest fluctuations in the exhaust-gas recirculation rate with regard to combustion noise and stability, and increased HC and CO emissions.

[1] Flame front extinguished by low temperatures at the cylinder wall

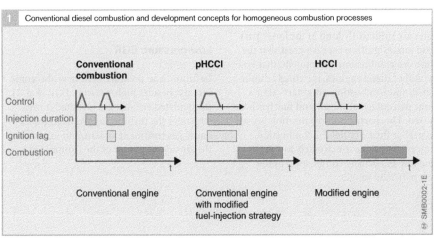

1 Conventional diesel combustion and development concepts for homogeneous combustion processes

Conventional combustion

pHCCI

HCCI

Control
Injection duration
Ignition lag
Combustion

Conventional engine

Conventional engine with modified fuel-injection strategy

Modified engine

SMB0002-1E

Fig. 1
Conventional combustion:
Influencing ignition lag by means of mixture formation and the state of the combustion chamber; premixed and diffusive combustion

pHCCI:
Influencing ignition lag mainly by means of the EGR rate; largely premixed combustion

HCCI:
Influencing ignition lag mainly by means of the EGR rate; premixed combustion only

Exhaust-gas recirculation

Concept

Exhaust-Gas Recirculation (EGR) is a highly effective internal engine measure to lower NO_x emissions on diesel engines. A distinction is made between:

- Internal EGR, which is determined by valve timing and residual gas.
- External EGR, which is routed to the combustion chamber through additional lines and a control valve.

The NO_x-reducing effect is mainly due to the following causes:

- Reduction in exhaust-gas mass flow.
- Drop in the rate of combustion, and thus local peak temperatures due to an increase in the inert-gas component in the combustion chamber.
- Reduction in partial oxygen pressure or local excess-air factor.

Since high local temperatures ($> 2,000\,K$) and a sufficiently high partial oxygen pressure are required to form NO_x, the measures listed above result in a drastic reduction in the formation of NO_x as the EGR rate rises. Reducing the reactive components in the combustion chamber also leads to a rise in black smoke, which limits the quantity of recirculated exhaust gas.

The quantity of recirculated exhaust gas also affects the period of ignition lag. If EGR rates are sufficiently long in the lower part-load range, ignition lag is so great that the diffusive combustion component, that is so typical of diesel engines, is strongly diminished, and combustion only starts after a large percentage of the air and fuel has been mixed. This partial homogenization is used in new or future (p)HCCI combustion processes to achieve extremely low-NO_x and low-particulate combustion in the low part-load ranges.

High-pressure EGR

Operating concept

EGR systems that are presently in production are high-pressure EGRs (Fig. 1). This means that exhaust gas is tapped upstream of the exhaust-gas turbocharger turbine and is routed through a mixer to the engine upstream of the intake plenum. The EGR volume depends on the pressure difference between the exhaust-gas backpressure upstream of the turbine, intake-manifold pressure, and the position of the pneumatically or electrically operated EGR valve.

On car engines, the driving pressure drop in the emission-related program-map areas is sufficient for the most part. Only in the lowest load points is it frequently necessary to restrict gas flow on the intake-manifold side to achieve sufficiently high EGR rates.

On truck engines, suitable measures, e.g. VTG superchargers, venturi mixers, or flutter valves, are always required to implement EGR. This is because of the extended emission-related load range up to full-load, and enhanced turbocharger efficiencies.

EGR control

Standard EGR-rate control on passenger cars is presently performed by measuring the air mass and can be made more precise by combining this with a lambda closed-loop control. On trucks, control is by means of a differential-pressure signal sent to a measuring venturi.

Low-pressure EGR

Operating concept

In future, low-pressure EGR may also come into use besides high-pressure EGR (Fig. 2). Recirculated exhaust gas is routed downstream of the turbine, tapped from the exhaust-gas treatment system, and injected on the air side upstream of the compressor.

The advantages of this are:
- Optimized EGR uniform distribution between individual cylinders.
- More intensive cooling of the homogeneous mixture of exhaust gas and fresh air after passing through the compressor and the intercooler.
- Increase and complete decoupling of possible charge-air pressure from EGR rate since the full exhaust-gas mass flow is routed through the turbine.

On the other hand, low-pressure EGR is less favorable than high-pressure EGR because of the larger volume contaminated with exhaust gas in dynamic operation.

Exhaust-gas cooling

In order to enhance the effects of EGR, the recirculated exhaust-gas quantity is cooled in a heat exchanger cooled by engine coolant. This raises gas density in the intake manifold and causes a lower final compression temperature. In general, the effects of higher localized excess-air factors cancel each other out as a result of increased charge density and reduced peak temperatures due to the lower final compression temperature. At the same time, however, EGR compatibility rises to produce possibly higher exhaust-gas recirculation rates at much lower NO_x emissions.

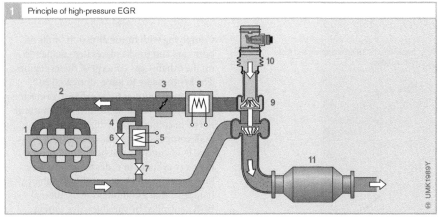

1   Principle of high-pressure EGR

Fig. 1
1   Engine
2   Intake manifold
3   Throttle
4   Bypass
5   EGR cooler
6   Bypass valve
7   EGR valve
8   Intercooler
9   Turbocharger
10   Air-mass meter
11   Oxidation-type
     catalytic converter

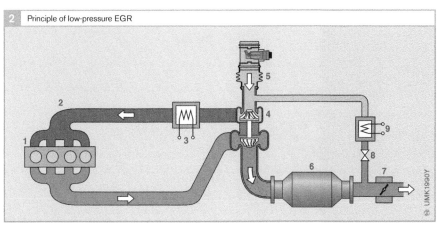

2   Principle of low-pressure EGR

Fig. 2
1   Engine
2   Intake manifold
3   Intercooler
4   Turbocharger
5   Air-mass meter
6   Oxidation-type
     catalytic converter
7   Throttle
8   EGR valve
9   EGR cooler

Since diesel-engine exhaust gas already has a low temperature at very low load points anyway, cooling the recirculated exhaust gas at the high EGR rates required to reduce NO_x emissions leads to unstable combustion. This then results in a significant rise in HC and CO emissions. A switchable EGR cooler is very effective to increase combustion-chamber temperature, stabilize combustion, reduce untreated HC and CO emissions, and raise exhaust-gas temperature. In particular, this occurs in the cold start phase of the car emission test, during which the oxidation-type catalytic converter has not reached its lightoff temperature. It also helps the oxidation-type catalytic converter to reach its operating temperature much faster.

Outlook

EGR with variable valve timing
An internal EGR achieved using variable valve timing would be suitable to obtain the best possible dynamic operating behavior. Conceivable scenarios could be, for example, increasing the amount of residual exhaust gas in the cylinder by advancing the "Close exhaust port" signal, or opening the exhaust valves during the intake phase, or opening the intake valves during the exhaust stroke. This would allow an adjustment of the exhaust-gas recirculation rate from one working cycle to the other by controlling the valve gear. The tradeoff, however, would be a high temperature of the recirculated exhaust gas, which would considerably limit the possible EGR rates.

NO_x emission minimizing concepts
Complying with future emission limits on both cars and trucks places high demands on the exhaust-gas concept of diesel engines. The key measure to lower untreated NO_x emissions continues to be exhaust-gas recirculation. As a result, EGR rates will continue to rise in conjunction with an increase in EGR compatibility of the combustion process. EGR control must be fast and precise for every cylinder in order to achieve very low emissions and best possible drivability. It appears that a combination of internal EGR controlled by variable valve timing and low-pressure EGR is an ideal solution.

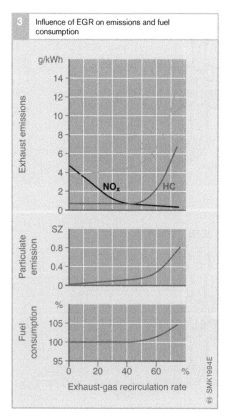

3 Influence of EGR on emissions and fuel consumption

Positive crankcase ventilation

Blowby gas

Crankcase ventilation gas (blowby gas) is produced as a result of combustion processes in an internal-combustion engine. Gas flows out of the combustion chamber and into the crankcase through design-related gaps between the cylinder walls and pistons, pistons and piston rings, through the ring gaps of the piston rings, and through valve seals. Crankcase blowby gases, related to engine exhaust gas, may contain a multiple of hydrocarbon concentrations. In addition to products arising from complete and incomplete combustion, water (vapor), soot, and carbon residue, this gas contains engine oil in the form of minute droplets.

Particularly in connection with turbocharged diesel engines and spark-ignition engines with direct fuel injection, engine oil and soot contained in the blowby gas can cause deposits that form on turbochargers, in the intercooler, on valves, and in the downstream soot or particulate filter (ash deposits from inorganic additives in the engine oil). Consequently, this may impair operation.

One way to minimize oil consumption caused by engine oil escaping through crankcase ventilation is to return the oil via an oil separator and only ventilate the gas.

Ventilation

In a closed-circuit ventilation system, the untreated gas flow from the crankcase is routed through a ventilation system comprising additional components (e.g. oil separator, pressure-control devices, non-return valves) to the combustion-air intake, and from there it is returned to the combustion chamber. In open-loop ventilation systems, the treated gas is blown off directly to the atmosphere. However, legislation now restricts the use of open systems to only a few exceptional cases.

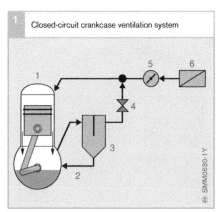

1   Closed-circuit crankcase ventilation system

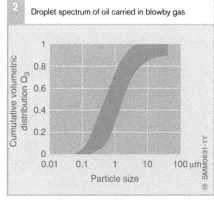

2   Droplet spectrum of oil carried in blowby gas

Fig. 1
1   Engine
2   Oil return
3   Oil separator
4   Vacuum-limit valve
5   Throttle valve
6   Intake filter

Fig. 2
Aerodynamic diameter, determined in various types of engine.

Exhaust-gas treatment

In the past, diesel-engine emissions were minimized mainly by measures implemented inside of the engine. However, the untreated emissions produced by many diesel-engined cars will exceed future emission limits in Europe, the U.S.A., and Japan. The high minimization rates required will presumably only be achievable by an efficient combination of measures inside and downstream of the engine. For this reason, intense development work is also ongoing on exhaust-gas treatment systems for diesel engines in analogy to tried and tested processes for gasoline engines.

In the 1980s the three-way catalytic converter was introduced on gasoline engines to reduce nitrogen oxides (NO_x), hydrocarbons (HC), and carbon monoxide (CO) to form nitrogen. Three-way catalytic converters operate at a λ value of 1.

In diesel engines operating with excess air, the three-way catalytic converter can be used not only for NO_x reduction. This is because, in the lean diesel exhaust gas, HC and CO emissions in the catalytic converter prefer not to react with NO_x but with exhaust-gas oxygen.

It is relatively simple to remove HC and CO emissions from diesel exhaust gas by means of an oxidation catalyst. However, it is more complicated to remove nitrogen oxides when oxygen is present in the exhaust gas. Basically, it is possible to remove nitrogen in a NO_x storage catalyst, or an SCR catalytic converter (Selective Catalytic Reduction).

Mixture formation inside of the diesel engine produces much larger quantities of soot emissions than in the gasoline engine. The present trend on passenger cars is to remove soot by means of a particulate filter downstream of the engine, and concentrate on internal engine measures for NO_x reduction and noise suppression. On trucks, NO_x emissions are normally reduced by an SCR system downstream of the engine.

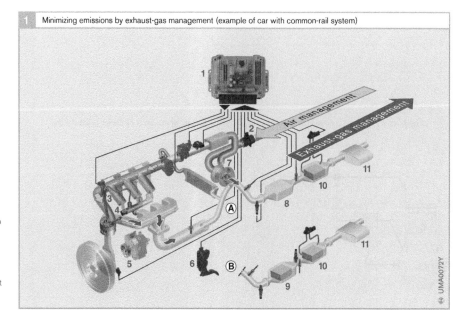

1 Minimizing emissions by exhaust-gas management (example of car with common-rail system)

NOₓ storage catalyst

The NOₓ Storage Catalyst (NSC) reduces nitrogen oxides in two stages:
- Loading phase: Continuous NOₓ storage in storage components of the catalyst in the lean exhaust gas.
- Regeneration: Periodic NOₓ removal and conversion in the rich exhaust gas.

Depending on the operating point, the loading phase lasts from 30 to 300 s. Accumulator regeneration takes from 2 to 10 s.

NOₓ storage

The NOₓ storage catalyst is coated with chemical compounds that have a strong tendency to bind strongly with NO_2. However, this is a chemically reversible bond. Examples here are the oxides and carbonates of alkalines and alkaline metals. Due to the temperature response, barium nitrate is the main chemical used.

As only NO_2 can be stored directly, but not NO, the NO components in the exhaust gas are oxidized on the surface of a platinum coating in an upstream or integrated oxidation catalyst to form NO_2. This reaction takes place in several stages since the concentration of free NO_2 in the exhaust gas is reduced during storage, and additional NO is then oxidized to form NO_2.

In the NOₓ storage catalyst, NO_2 reacts with the compounds on the catalyst surface (e.g. barium carbonate, $BaCO_3$, as storage material), and oxygen (O_2) from the lean diesel gas to form nitrates:

$$BaCO_3 + 2\,NO_2 + {}^1/_2\,O_2 \rightleftharpoons BA(NO_3)_2 + CO_2.$$

The NOₓ storage catalyst stores nitrogen oxides in this way. Storage is only optimized in a material-dependent temperature interval of the exhaust gas ranging from 250 to 450°C. Below this temperature, NO oxidizes very slowly to form NO_2; above this temperature, the NO_2 is non-stable. Accumulator-type catalytic converters also have a certain storage capability (surface storage) at low

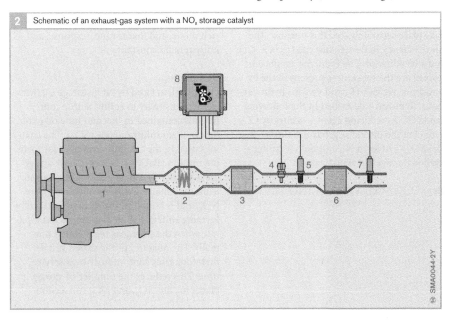

2   Schematic of an exhaust-gas system with a NOₓ storage catalyst

Fig. 2
1   Diesel engine
2   Exhaust heater (optional)
3   Oxidation catalyst
4   Temperature sensor
5   Broadband lambda oxygen sensor
6   NOₓ storage catalyst
7   NOₓ sensor
8   Engine control unit

SMA0044-2Y

temperatures. This is sufficient to store nitrogen oxides produced at low temperatures during the start process.

As the amount of stored nitrogen oxides (saturation) increases, the ability to continue to bind additional nitrogen oxides decreases. The volume of nitrogen oxides passing through the catalytic converter then increases with time. There are two ways of detecting when the catalytic converter is charged to such a degree that the storage phase needs to be terminated.
- A model-based process calculates the volume of stored nitrogen oxides, taking account of the converter state, and derives the remaining storage capacity.
- A NO$_x$ sensor downstream of the NO$_x$ storage catalyst measures the concentration of nitrogen oxide in the exhaust gas, thus determining the present storage level.

NO$_x$ removal and conversion
At the end of the storage phase, the catalytic converter must be regenerated, i.e. the stored nitrogen oxides must be removed from the storage components and converted into nitrogen (N$_2$) and carbon dioxide (CO$_2$). The processes for NO$_x$ removal and conversion take place separately. For this purpose, the air deficiency in the exhaust gas (rich, $\lambda < 1$) must be adjusted. The reducing agents employed are the substances present in the exhaust gas, i.e. CO, H$_2$, and various hydrocarbons. Removal – described in the following using CO as reducing agent – occurs by CO reducing the nitrate (e.g. barium nitrate, Ba(NO$_3$)$_2$) to form N$_2$, and then forming a carbonate together with barium:

$$Ba(NO_3)_2 + 3\,CO \rightarrow BaCO_3 + 2\,NO + 2\,CO_2$$

This results in CO$_2$ and NO. A rhodium coating then reduces the nitrogen oxides into N$_2$ and CO$_2$ using CO in the familiar way known from the three-way catalytic converter:

$$2\,NO + 2\,CO \rightarrow N_2 + 2\,CO_2$$

There are two methods of detecting when the removal phase is complete:
- The model-based process calculates the quantity of nitrogen oxides remaining in the NO$_x$ storage catalyst.
- A lambda oxygen sensor downstream of the catalytic converter measures the oxygen excess in the exhaust gas, and indicates a change in voltage from "lean" to "rich" when removal has ended.

On diesel engines, rich operating conditions ($\lambda < 1$) can be adjusted by retarding the injection point and throttling the intake air. During this phase, the engine operates at poorer efficiency. In order to minimize any additional fuel consumption, the regeneration phase should be as short as possible compared to the storage phase. It must be guaranteed that the vehicle remains fully drivable during the switchover from lean to rich mode, and that torque, response, and noise remain constant.

Desulfating
One problem faced by NO$_x$ storage catalysts is their sensitivity to sulfur. Sulfur compounds contained in fuel and lube oil oxidize to form sulfur dioxide (SO$_2$). The coatings used in the catalytic converter for forming nitrates (BaCO$_3$), however, have a very great affinity to sulfate, i.e. SO$_2$ is removed from the exhaust gas more effectively than NO$_x$, and is bound in the storage material by forming sulfate. Sulfate bonding is not separate when the storage is regenerated normally. The quantity of the stored sulfate, therefore, rises continuously over service time. This reduces the number of storage places for the NO$_x$, and NO$_x$ conversion

decreases. To ensure sufficient NO$_x$ storage capacity, the catalytic converter must be desulfated (sulfur regeneration) at regular intervals. If fuel contains 10 mg/kg of sulfur ("sulfur-free fuel"), regeneration must take place at intervals of about 5,000 km.

During the desulfating process, the catalytic converter is heated to a temperature of over 650°C for a period of over 5 minutes, and it is purged with rich exhaust gas ($\lambda < 1$). To increase the temperature, the same measures are used to regenerate the Diesel Particulate Filter (DPF). As opposed to DPF regeneration, however, O$_2$ is completely removed from the exhaust gas by controlling the combustion process. Under these conditions, barium sulfate is converted back to barium carbonate.

The choice of a suitable desulfating process control (e.g. oscillating λ about 1) must make sure that the SO$_2$ removed is not reduced to hydrogen sulfide (H$_2$S) by a continuous deficiency of exhaust-gas oxygen, O$_2$. H$_2$S is already highly toxic in very small concentrations and is perceptible by its intensive odor.

The controlled conditions for desulfating must also avoid excessive aging of the catalyst. Although high temperatures (>750°C) accelerate the desulfating process, they also speed up catalyst aging. For this reason, optimized catalyst desulfating must take place within a limited temperature and excess-air-factor window, and may only have a negligible impact on drivability.

A high sulfur content in fuel will speed up catalyst aging, since desulfating takes place more frequently, and will also increase fuel consumption. Ultimately, the use of storage catalysts is dependent on general availability of sulfur-free fuel at the filling station.

Selective catalytic reduction of nitrogen oxides

Overview

As opposed to the NSC process (NO$_x$ Storage Catalyst), Selective Catalytic Reduction (SCR process) operates continuously, and does not intervene in engine operation. The process is now being launched on production truck models. It offers the possibility of minimizing NO$_x$ emissions and reducing fuel consumption at the same time. On the other hand, NO$_x$ removal and conversion causes a higher fuel consumption in the NSC process.

In large furnaces, selective catalytic reduction has become a tried and tested method of waste-gas denitrification. It is based on reducing certain nitrogen oxides (NO$_x$) in the presence of oxygen using selected reducing agents. Here, "selective" means that the reducing agent prefers to oxidize selectively with the oxygen contained in the nitrogen oxides instead of with the molecular oxygen present in much greater quantities in the exhaust gas. Ammonia (NH$_3$) has proven to be a highly selective reducing agent in this case.

In a car environment, the quantities of NH$_3$ required would raise safety issues due to the toxicity of the chemical. However, NH$_3$ can be produced from nontoxic carrier substances, such as urea or ammonium carbamate. Urea has proved to be a good catalyst carrier. Urea, (NH$_2$)$_2$CO, is produced on an industrial scale as fertilizer and feedstuff. It is biologically compatible with groundwater, and chemically stable for the environment. Urea is highly soluble in water, and can therefore be added to the exhaust gas as an easy-to-meter urea/water solution.

At a mass concentration of 32.5% urea in water, the freezing point has a localized minimum at −11°C: A eutectic solution forms, but does not separate when frozen.

The DENOXTRONIC 1 system was developed to meter the reducing agent in the exhaust gas precisely. This system is resistant to freezing. The main components can be heated to ensure the metering function starts shortly after a cold start.

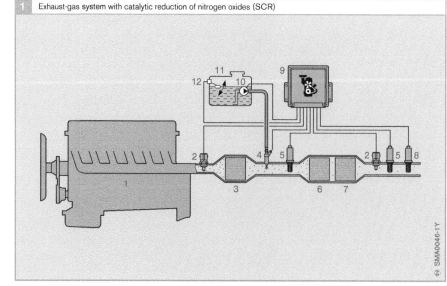

1   Exhaust-gas system with catalytic reduction of nitrogen oxides (SCR)

Fig. 1
1 Diesel engine
2 Temperature sensor
3 Oxidation catalyst
4 Injector for reducing
    agent
5 NO$_x$ sensor
6 SCR catalytic
    converter
7 NH$_3$ blocking
    catalytic converter
8 NH$_3$ sensor
9 Engine control unit
10 Reducing-agent
    pump
11 Reducing-agent tank
12 Fluid-level sensor

The urea/water solution will be obtainable under the brand name of AdBlue – firstly at depots, and then at all highway filling stations. The first official AdBlue pump opened in Stuttgart (Germany) at the end of 2003.

AdBlue complies with the draft standard DIN 70 070 which defines the solution properties.

Chemical reactions

Urea first forms ammonia before the actual SCR reaction starts. This takes place in two reaction steps, which are together termed a hydrolysis reaction. Firstly, NH_3 and isocyanic acid are formed in a thermolysis reaction:

$(NH_2)_2CO \rightarrow NH_3 + HNCO$ (thermolysis)

Then isocyanic acid is converted with water in a hydrolysis reaction to form ammonia and carbon dioxide.

$HNCO + H_2O \rightarrow NH_3 + CO_2$ (hydrolysis)

To prevent the precipitation of solids, the second reaction must take place rapidly by selecting suitable catalysts and temperatures that are sufficiently high (starting at 250°C). Modern SCR reactors also assume the function of a hydrolyzing catalyst, thus dispensing with an upstream hydrolyzing catalyst as previously necessary. Ammonia produced by thermohydrolysis reacts in the SCR catalytic converter according to the following equations:

- $4 NO + 4 NH_3 + O_2 \rightarrow 4 N_2 + 6H_2O$ (Eq. 1)
- $NO + NO_2 + 2 NH_3 \rightarrow 2 N_2 + 3 H_2O$ (Eq. 2)
- $6 NO_2 + 8 NH_3 \rightarrow 7 N_2 + 12 H_2O$ (Eq. 3)

At low temperatures ($< 300°C$), conversion mainly takes place using reaction 2. For this reason, it is necessary to set a $NO_2 : NO$ ratio of about $1 : 1$ to achieve good conversion at low temperatures. Under these circumstances, reaction 2 can take place at temperatures starting at 170 to 200°C.

Oxidizing NO to form NO_x occurs in an upstream oxidation catalyst, and this is necessary to achieve optimized efficiency.

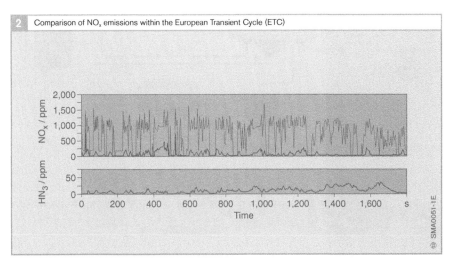

2   Comparison of NO_x emissions within the European Transient Cycle (ETC)

Fig. 2
— With no admixture of urea/water solution: 10.9 g/kWh
— With admixture of 32.5% urea/water solution: 1.0 g/kWh

If more reducing agent is dispensed than is converted in reduction with NO_x, it may result in NH_3 leakage. NH_3 is a gas and has a very low odor threshold (15 ppm). This may cause a nuisance to the environment, but it is avoidable. NH_3 is removable by placing an additional oxidation catalyst downstream of the SCR catalytic converter. This blocking catalytic converter oxidizes any ammonia that may occur to form N_2 and H_2O. In addition, a careful application of metered AdBlue is essential.

One key parameter for the application is the feed ratio α, which is defined as the molar ratio of metered NH_3 as a factor of NO_x present in the exhaust gas. Under ideal operating conditions (no NH_3 leakage, no secondary reactions, no NH_3 oxidation), α is directly proportional to the NO_x reduction rate: At $\alpha = 1$, NO_x reduction of 100% is achievable in theory. In practice, however, a NO_x reduction of 90% in fixed and mobile operation is achievable at an NH_3 leakage of

< 20 ppm. The quantity of AdBlue required for this is approximately equivalent to 5% of the quantity of diesel fuel used.

The reducing-agent requirement depends on the specific NO_x emission (g_{NO_x}/kg_{diesel}). The SCR process can compensated for higher NO_x emissions in the untreated exhaust gas occurring in the efficiency-optimized combustion process by adding AdBlue.

By arranging the hydrolysis reaction upstream, modern SCR catalytic converters achieve a NO_x conversion rate of > 50% only at temperatures above approx. 250°C. Optimized conversion rates are attained within a temperature window of 250 to 450°C. Present catalytic-converter research is concentrating on extending the work temperature window, and, in particular, on optimizing activity at low temperatures.

3   Modular design of DENOXTRONIC 1 system

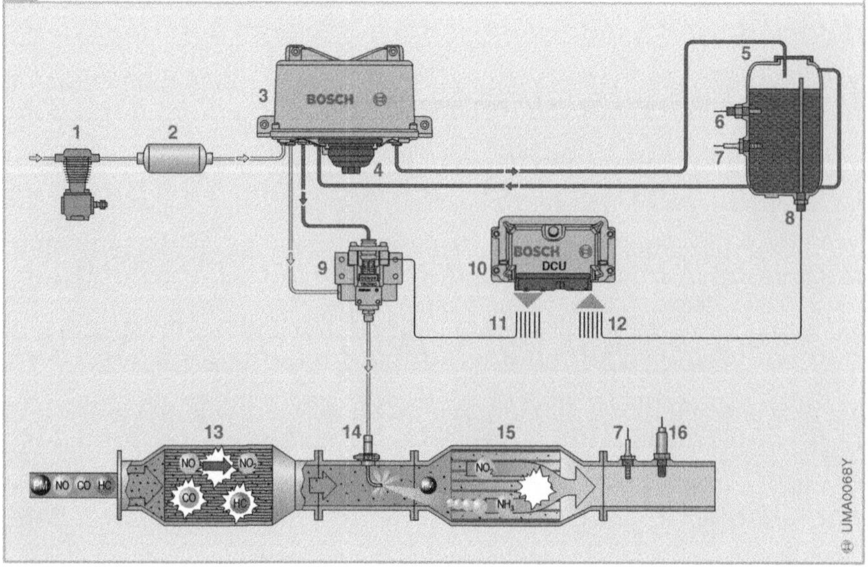

Fig. 3
1 Compressor
2 Air tank
3 Delivery module
4 Filter
5 AdBlue tank
6 AdBlue quality sensor
7 Temperature sensor
8 Level sensor
9 Dosing module
10 ECU
11 Actuators
12 Sensors
13 Diesel oxidation catalyst
14 Dosing tube
15 SCR catalytic converter
16 Exhaust-gas sensor

DENOXTRONIC 1

System overview

The modular DENOXTRONIC 1 system meters the reducing agent and comprises the following modules:

- Delivery module, to deliver the urea/water solution at the required pressure to the dosing module.
- Dosing module, to meter the precise quantity of urea/water solution and add compressed air.
- Dosing tube, to atomize and distribute the urea/water solution in the exhaust pipe.
- Control unit, to exchange information with the engine ECU via CAN. This unit is fitted in the delivery module.

Compressed air is provided from an onboard supply to enhance atomization of the reducing-agent solution in the exhaust-gas system.

Delivery module

In the delivery module, reducing agent flows through a prefilter to a diaphragm pump which has an integrated pressure attenuator and overflow valve. The solution then runs through the main filter to the dosing module at a max. pressure of 3.5 bar. Pressure and temperature are monitored constantly. As required, a bleeder valve opens. It is electrically switched and connected to the supply reservoir. After conditioning in this way, the reducing agent flows finally to the dosing module.

Compressed air from the onboard air tank enters the delivery module via a separate line. An air-control valve and air-pressure sensor ensure a constant admission pressure upstream of the central choke valve. Central choke-valve operation is supercritical. Pressure control ensures a constant air-mass flow from the delivery module to the dosing module. This is a key factor for precision dosing. A second air-pressure sensor located downstream of the central choke valve provides additional safety.

The delivery module is mounted as usual on the vehicle chassis.

Dosing module

The urea/water solution is metered precisely in the dosing module by a clocked solenoid valve. The clock rate is normally 4 Hz. Precise metering is achieved in a different control range if system dynamics are still sufficient. The metered urea/water solution then flows into the venturi. Compressed air also enters the venturi through a non-return valve and forms a carrier air flow that transports the urea/water solution in the form of an aerosol and a film to the dosing tube.

The dosing module is mounted as close to the dosing point as possible to achieve high system dynamics.

Dosing tube

The dosing tube represents the interface between the dosing module and the exhaust-gas system. It provides optimized mixture formation and homogeneous mixture distribution in the exhaust pipe. There are normally eight holes with a diameter of 0.5 mm arranged symmetrically around the dosing-tube circumference. The carrier-air flow passes through these holes at high velocity, atomizing the entrained urea/water solution into small drops that quickly evaporate. This accelerates the drops to high velocities to reach all parts of the exhaust pipe.

Control unit, DCU 15

The control unit is located in the delivery module. It reads out signals from internal and external sensors, triggers internal and external actuators, and performs control and monitoring functions. Components are monitored, and all inputs are checked for plausibility. The main function is to calculate the required dosing quantity based on the preset dosing strategy. The control unit also monitors the temperatures of system components and secures operation at low temperatures by actively triggering a heating system.

Besides internal sensors, other parameters detected are the fluid level in the tank, tank temperature, and the exhaust-gas temperature in the catalytic converter. Later construction phases planned include other exhaust-gas components and exhaust-gas monitoring.

Communication with the engine control unit runs over a CAN bus. Diagnostics take place over an ISO-K or CAN interface.

DENOXTRONIC dosing strategy

A model-based calculation for the optimum dosing quantity is required to optimize NO_x reduction while minimizing NH_3 leakage – i.e. permeation of NH_3 through the catalyst system. The quantity measured – e.g. on the engine test bench – is corrected as a factor of the catalytic-converter temperature and the quantity of NH_3 stored in the catalytic converter.

Standard model

The dosing quantity for the reducing agent is stored as a function of injected fuel quantity and engine speed In program map A, measured either on the test bench, or calculated by *a priori* assumptions. The system feeds in correction parameters, such as engine temperature (to consider the effects of operating temperature on NO_x production), and the number of system operating hours (to consider aging).

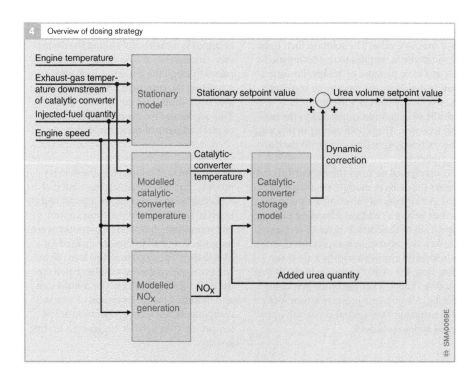

4   Overview of dosing strategy

Engine temperature
Exhaust-gas temperature downstream of catalytic converter
Injected-fuel quantity
Engine speed

Stationary model → Stationary setpoint value

Urea volume setpoint value

Modelled catalytic-converter temperature

Catalytic-converter temperature

Catalytic-converter storage model

Dynamic correction

Added urea quantity

Modelled NO_x generation → NO_x

SMA0069E

The difference between fixed catalytic-converter temperature (stored in program map B) and the exhaust-gas temperature measured downstream of the catalytic converter is used to determine a correction factor for reducing-agent dosing in a third program map C when switching between two fixed operating points. This correction factor minimizes NH_3 leakage.

Expansion with storage block

On catalytic converters with high NH_3 storage capacity, in particular, it is recommended to model the transient processes and the quantity of NH_3 actually stored. Since NH_3 storability in SCR catalytic converters drops as temperature rises, there may otherwise be some undesirable occurrence of NH_3 leakage in transient mode and, in particular, as exhaust-gas temperature rises.

To avoid this effect, the catalytic-converter temperature and the quantity of NO_x produced are estimated by program maps and time-lag devices. Catalytic-converter efficiency is saved to a program map as a function of temperature and the quantity of NH_3 stored. The product of the catalyst coefficient and the NO_x present in the catalyst is equivalent to the converted quantity of reducing agent. The difference between added and converted reducing agent results in a (positive or negative) factor added to the quantity of ammonia stored in the catalyst. This is calculated continuously. If the figure for the quantity of NH_3 stored exceeds a fixed temperature-dependent threshold, the dosing quantity is reduced to avoid NH_3 leakage. If the quantity of NH_3 stored drops below the threshold, the dosing quantity is increased to optimize NO_x conversion.

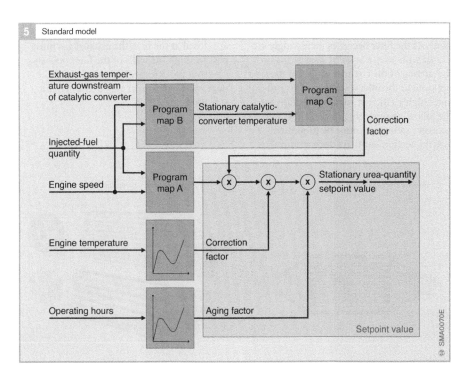

5   Standard model

Exhaust-gas temperature downstream of catalytic converter

Injected-fuel quantity

Engine speed

Engine temperature

Operating hours

Program map B

Program map A

Stationary catalytic-converter temperature

Program map C

Correction factor

Correction factor

Aging factor

Stationary urea-quantity setpoint value

Setpoint value

SMA0070E

Diesel Particulate Filter (DPF)

Soot particles emitted from a diesel engine can be efficiently removed from the exhaust gas by Diesel Particulate Filters (DPF). Particulate filters used so far in passenger cars consist of porous ceramics. Particulate filters made of sintered metal are now under development.

Ceramic particulate filter

Ceramic particulate filters consist of a honeycomb structure made of silicon carbide or Cordierite, which has a large number of parallel, mostly square channels. The thickness of the channel walls is typically 300 to 400 µm. Channel size is specified by their cell density (channels per square inch (cpsi); typical value: 100...300 cpsi).

Adjacent channels are closed off at each end by ceramic plugs to force the exhaust gas to penetrate through the porous ceramic walls. As soot particles pass through the walls, they are transported into the pore walls by diffusion (inside of the ceramic walls) where they adhere (deep-bed filtration). As the filter becomes increasingly saturated with soot, a layer of soot forms on the surface of the channel walls (on the side opposite to the inlet channels). This provides highly efficient surface filtration for the following operating phase. However, excessive saturation must be prevented (see the section entitled "Regeneration").

As opposed to deep-bed filters, wall-flow filters store the particles on the surface of the ceramic walls (surface filtration).

Besides filters with a symmetrical arrangement of square inlet and outlet channels, ceramic "octosquare substrates" are now on offer (Fig. 2). They have larger octagonal inlet channels and smaller square outlet channels. The large inlet channels considerably increase the storability of the particulate filter for ash, non-combustible residue from burned engine oil, and additive ash (see the section entitled "Additive system"). Octosquare filters will shortly be launched onto the market.

Particulate filters made of sintered metal

In the sintered metal filter, the filter surfaces comprise a metallic carrier structure composed of mesh filled with sintered metal powder. The filter's design has a specific geometry: The filter surfaces form concentric, wedge-shaped filter pockets through which the exhaust gas flows. Since the slats are closed at the rear, the exhaust gas must pass through the walls of the filter pockets. The soot particles are deposited on the pore walls in a similar way to the ceramic substrate.

Fig. 1

1   Inflowing exhaust
    gas
2   Housing
3   Ceramic plug
4   Honeycomb ceramic
5   Outflowing exhaust
    gas

Fig. 2

a   Square channel
    cross-section
b   Octosquare design

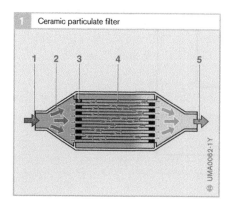

1   Ceramic particulate filter

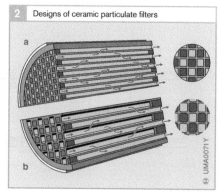

2   Designs of ceramic particulate filters

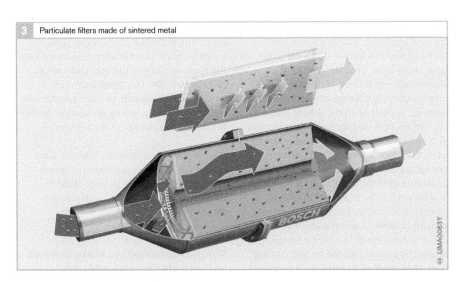

Both particulate filters made of sintered metal and ceramic filters achieve a retention efficiency of more than 95% for particles across the entire spectrum range in question (10 nm to 1 µm).

Regeneration

Irrespective of the particulate-filter material – ceramics or sintered metal – the material must be freed from time to time from the deposited particles, i.e. it must be regenerated. The growing amount of soot deposited in the filter gradually increases the exhaust-gas backpressure. This impairs engine efficiency and acceleration power.

Regeneration must be carried out approximately every 500 kilometers; this figure is subject to strong fluctuation dependent on the untreated soot emission and filter size (approx. 300 to 800 kilometers). Regeneration takes about 10 to 15 minutes. The additive system takes slightly less time. It is also dependent on engine operating conditions.

The filter is regenerated by burning off the soot that has collected in the filter. The particle carbon component can be oxidized (burned) using the oxygen constantly present in the exhaust gas above a temperature of approx. 600°C to form nontoxic CO_2. Such high temperatures occur only when the engine is operating at rated output. It is highly rare in normal vehicle operation. For this reason, measures must be taken to lower the soot burnoff temperature and/or raise the exhaust-gas temperature.

Soot oxidizes at temperatures as low as 300...450°C using NO_2 as oxidizer. This method is used industrially in the continuously regenerating trap (CRT®) system.

Compared with the ceramic filter, the sintered metal filter has the advantage that its thermal conductivity is better. After the soot is ignited in part of the filter, the reaction heat occurring there is transported more easily to adjacent areas. The soot layer is burned off evenly. This avoids any non-regenerated soot areas which may occur in ceramic filters under worst-case scenarios.

Additive system

The soot oxidation temperature of 600°C can be lowered to approx. 450...500°C by using an additive – usually cerium or iron compounds – in diesel fuel. But even this temperature is not always reached in the exhaust-gas system when the vehicle is operating. The result is that the soot is not burned off continuously. Active regeneration is triggered, therefore, above a specific level of soot saturation in the particulate filter (see the section entitled "Saturation detection"). To achieve this, the combustion control in the engine is modified so that the exhaust-gas temperature rises to the soot burnoff temperature. This is possible by retarding the injection point (see the section entitled "Measures inside of the engine to raise the exhaust-gas temperature").

The additive in the fuel is retained in the filter as a residue (ash) after regeneration. This ash, as well as ash from engine-oil or fuel residue, gradually clogs the filter, thus raising the exhaust-gas backpressure. The pressure loss across the sintered metal filter is lower than for the ceramic filter with the same ash saturation.

To reduce the pressure rise, the ash storability of sintered metal filters and ceramic octosquare filters must be increased by making the cross-sections of the inlet channels as large as possible. This provides the filters with sufficient capacity to accommodate all the ash residue occurring on burnoff during the normal service life of the vehicle.

With conventional ceramic filters, it is assumed that the filter is removed approximately every 120,000 km when additive-based regeneration is used, and cleaned mechanically.

Catalyzed Diesel Particulate Filter (CDPF)

Soot-particle burnoff can also be improved by coating the filter with noble metals (mainly platinum). However, the effect is less than when using an additive.

The CDPF requires further measures for regeneration to raise the exhaust-gas temperature, similar to the measures taken with the additive system. Compared with the additive system, however, the catalyzed coating has the advantage that no additive ash occurs in the filter.

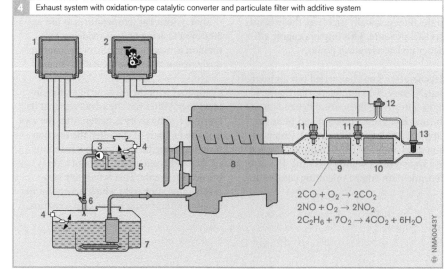

4   Exhaust system with oxidation-type catalytic converter and particulate filter with additive system

$$2CO + O_2 \rightarrow 2CO_2$$
$$2NO + O_2 \rightarrow 2NO_2$$
$$2C_2H_6 + 7O_2 \rightarrow 4CO_2 + 6H_2O$$

Fig. 4

1 Additive control unit
2 Engine control unit
3 Additive pump
4 Fluid-level sensor
5 Additive tank
6 Additive metering
    unit
7 Fuel tank
8 Diesel engine
9 Oxidation catalyst
10 Particulate filter
11 Temperature sensor
12 Differential-pressure
    sensor
13 Soot sensor

The catalyzed coating has several functions:
- It oxidizes CO and HC
- It oxidizes NO to form NO_2
- It oxidizes CO to form CO_2

Oxidizing CO and HC
It also oxidizes CO and HC, as in the oxidation-type catalytic converter. At high CO and HC emissions, the energy released can be significant (see the section entitled "Diesel oxidation catalyst"). The resulting temperature hike acts directly on the point where high temperatures are required to ignite the soot. This avoids heat losses that may occur with an upstream catalytic burner.

Oxidizing NO to form NO_2
NO is oxidized on the catalyzed coating to form NO_2. NO_2 is a more active oxidizer than O_2 and, therefore, oxidizes soot at lower temperatures (CRT® effect). In this reaction, NO_2 is again reduced to NO. Since exhaust-gas flow velocity through the filter wall is slow, any NO occurring can diffuse contrary to the flow direction and oxidize soot in further oxidation reducing cycles.

Oxidizing CO to form CO_2
Another phenomenon is oxidizing CO produced during soot oxidation at low regeneration temperatures to form CO_2. Soot burnoff is enhanced by localized heat generation.

CRT® system
Truck engines run close to maximum torque more frequently than car engines, i.e. causing comparatively high NO_x emissions. On trucks, therefore, it is possible to perform continuous regeneration of the particulate filter based on the CRT® principle (Continuously Regenerating Trap).

According to this principle, soot combusts with NO_2 at temperatures as low as 300 to 450°C. The process is reliable at these temperatures if the mass ratio of NO_2/soot is greater than 8 : 1. To apply the process, an oxidation catalyst is arranged upstream of the particulate filter to oxidize NO into NO_2. In most cases, this provides ideal conditions for regeneration using the CRT® system on trucks at normal operation. The method is also termed "passive regeneration", since soot is burned continuously without the need for active measures.

The efficiency of this process was demonstrated in truck fleet trials. However, other regeneration processes are provided for trucks.

On cars, which normally run in the low-load range, complete regeneration of the particulate filter is not possible using the CRT® effect.

System configuration
Irrespective of the applied particulate filter process, the regeneration process requires a system for control and monitoring. The system senses the state of the filter (state functions), i.e. it detects saturation level, defines regeneration strategy, and monitors the filter. In addition, it controls regeneration by intervening in the fuel-injection and air-intake systems. When operating an additive system, there are also functions for tank-level detection and additive dosing.

The basic configuration is almost identical for all systems.

Besides the particulate filter (DPF), the DPF system comprises other components and sensors:

- *Diesel Oxidation Catalyst (DOC)*
  The main function of the DOC is to minimize HC and CO emissions. In DPF applications, it also acts as a "catalytic burner": By oxidizing specific hydrocarbons entrained (retarded secondary injection) in the DOC, the required regeneration temperature is reached in the exhaust gas. The DOC is also required for oxidizing NO into NO_2 in the CRT® system.
- *Differential-pressure sensor*
  The differential-pressure sensor measures the pressure drop across the particulate filter; this figure is then used to calculate filter saturation. The differential pressure is also used to calculate the exhaust-gas backpressure in the engine so that it can be limited to the maximum permitted level. Optionally, an absolute-pressure sensor can be fitted upstream of the DPF instead of a differential-pressure sensor.
- *Temperature sensor upstream of the DPF*
  In regeneration mode, the temperature upstream of the DPF is the key parameter to determine soot burnoff in the filter.
- *Temperature sensor upstream of the DOC*
  The temperature upstream of the DOC helps to determine HC convertibility (*lightoff*) in the DOC.
- *Lambda oxygen sensor*
  The lambda oxygen sensor is not directly a DPF system component, but it enhances system response even for the DPF, since a specific emission behavior is achieved by precise exhaust-gas recirculation.

Control-unit functions

Saturation detection

Two processes are used in parallel for saturation detection. Flow resistance in the particulate filter is calculated from the pressure drop across the filter and volumetric flow. This is a measure of filter permeability, and thus soot mass.

In addition, a model is used to calculate the soot mass stored in the DPF. The (untreated) soot-mass flow of the engine is integrated in this model. Corrections for system dynamics, such as the residual oxygen part in the exhaust gas, etc. are taken into consideration, as well as continuous particulate oxidation by NO_2. During thermal regeneration, soot burnoff is calculated in the control unit as a factor of DPF temperature and oxygen mass flow.

Soot mass is calculated by a coordinator using the soot masses determined in both processes, and this becomes the key factor in the regeneration strategy.

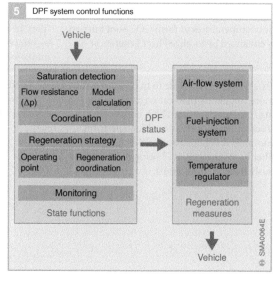

5   DPF system control functions

Vehicle

Saturation detection
Flow resistance (Δp)   Model calculation

Coordination

Regeneration strategy
Operating point   Regeneration coordination

Monitoring
State functions

DPF status

Air-flow system

Fuel-injection system

Temperature regulator

Regeneration measures

Vehicle

SMA0064E

Regeneration strategy

As the quantity of soot mass rises in the filter, regeneration must be triggered in good time. As filter saturation increases, the quantity of heat released during soot burnoff increases, as well as peak temperatures in the filter. To prevent these temperatures from destroying the filter, regeneration must be triggered before a critical saturation state is reached. Depending on the filter material, 5 to 10 g of soot per liter of filter volume is cited as the critical saturation mass.

A good strategy is to bring forward regeneration when conditions are favorable (e.g. when driving on the highway), or delay regeneration when the conditions are poor.

The regeneration strategy defines when and what regeneration measures are performed, dependent on the soot mass in the filter, and the engine and vehicle operating states. These parameters are transmitted as status values to all other engine-control functions.

Monitoring

The differential-pressure sensor helps to monitor whether the filter is possibly blocked, broken, or removed. The DPF-system sensors are also monitored. Besides standard monitoring, the plausibility of values of the downstream differential-pressure sensor is also monitored. In dynamic operation, the supply line between the exhaust-gas system and the pressure sensor is monitored by a signal-curve evaluation circuit. The temperature sensors upstream of the DOC and the DPF are checked at cold start for plausibility with other EDC temperature sensors.

Regeneration measures in the fuel-injection and air-intake systems

When a regeneration request is received, the fuel-injection and air-intake systems are switched over to other setpoint parameters via ramp signals. The driver does not perceive any change in torque or noise. The interventions to achieve the required regeneration temperature in the exhaust gas depend on the operating point (see the section entitled "Measures inside of the engine to raise the exhaust-gas temperature").

Exhaust-gas temperature controller

The exhaust-gas temperature is controlled during unfavorable ambient conditions and throughout the entire filter service life to ensure reliable regeneration. The controller design is cascaded to mirror the split in regeneration measures (see the section entitled "Measures inside of the engine to raise the exhaust-gas temperature").

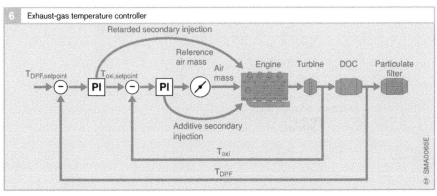

6   Exhaust-gas temperature controller

SMA0065E

Measures inside of the engine to raise the exhaust-gas temperature

The temperature level of 550 to 650°C required for regeneration during standard diesel-engine operation is only reached at high engine speeds and at full load.

The main measures taken inside of the engine *(engine burner)* to increase exhaust-gas temperature are advanced, "burnoff" or "additive" secondary injection, retarded main injection, and intake-air throttling. Depending on the engine operating point, one or several of these measures are triggered during regeneration. In some operating points, these measures are supplemented by retarding secondary injection. This leads to a further increase in exhaust-gas temperature due to oxidation of fuel in the DOC no longer converted in the combustion chamber *(cat burner)*.

Figures 7 and 8 shows typical exhaust-gas temperatures and engine measures required for regeneration as a factor of engine speed and load. Using the described combination of measures, only one temperature of 600°C is set downstream of the DOC at a residual oxygen content of > 5%. The residual oxygen content is important since soot burnoff is too slow at lower O_2 concentrations.

Across the entire program map, exhaust-gas recirculation is shut down during regeneration to avoid high components of unburned hydrocarbons in the combustion air. At the same time this provides a stable single-controller concept for air-mass control.

The program map is roughly divided into six areas which feature different measures for raising temperature.

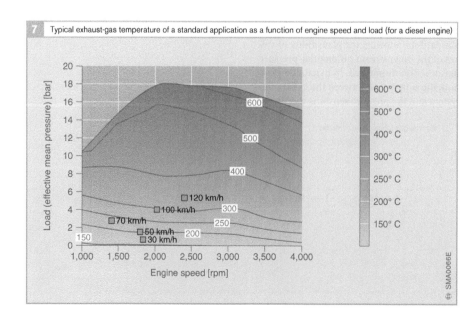

7   Typical exhaust-gas temperature of a standard application as a function of engine speed and load (for a diesel engine)

Area 1:
No engine measures are required since the exhaust-gas temperature in the basic application is over 600°C.

Area 2:
Firstly, the start of injection for main injection is retarded; secondly, a secondary injection event is added. Here, secondary injection is still part of the combustion process and contributes to the torque produced.

Area 3:
Due to the low supercharge and the large quantity of fuel, the excess-air factor in this area is $\lambda < 1.4$. An added, i.e. advanced, secondary injection event would lead to localized, extremely low excess-air factors, and thus to a drastic increase in black-smoke emissions; for this reason, a delayed, i.e. retarded secondary injection event is applied instead.

Area 4:
The required temperature rise is achieved by a combination of lowering charge-air pressure, triggering secondary injection, and by retarding main injection. The parts of the individual measures must be optimized with respect to emissions, fuel consumption, and noise. In most cases, not all of these measures are required at the same time.

Area 5:
This area requires a large temperature hike compared with normal operation. For this reason, the air mass must be reduced by the throttle valve in addition to the measures described above. Further measures for stabilizing the combustion process are needed, e.g. increasing the quantity of fuel for pre-injection, and adapting the time interval between pre-injection and main injection.

Area 6:
Only in this small area is it not possible to achieve stable regeneration at temperatures > 600°C downstream of the catalytic converter.

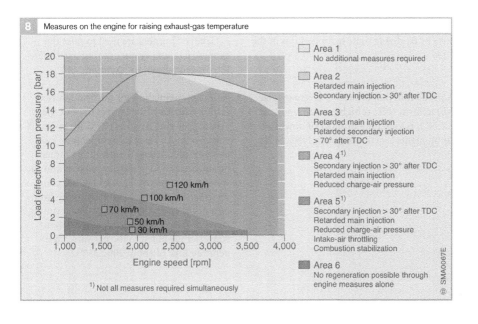

8   Measures on the engine for raising exhaust-gas temperature

Load (effective mean pressure) [bar]

Engine speed [rpm]

□ 120 km/h
□ 100 km/h
□ 70 km/h
□ 50 km/h
□ 30 km/h

Area 1
No additional measures required

Area 2
Retarded main injection
Secondary injection > 30° after TDC

Area 3
Retarded main injection
Retarded secondary injection
> 70° after TDC

Area 4[1]
Secondary injection > 30° after TDC
Retarded main injection
Reduced charge-air pressure

Area 5[1]
Secondary injection > 30° after TDC
Retarded main injection
Reduced charge-air pressure
Intake-air throttling
Combustion stabilization

Area 6
No regeneration possible through
engine measures alone

[1] Not all measures required simultaneously

SMA0067E

Diesel oxidation catalyst

Operation

The Diesel Oxidation Catalyst (DOC) fulfils a variety of functions for exhaust-gas treatment:

- Reduction in CO and HC emissions.
- Reduction in particle mass.
- Oxidation of NO to form NO_2.
- Use as catalytic burner.

Reduction in CO and HC emissions

Carbon monoxide (CO) and hydrocarbons (HC) are oxidized to form carbon dioxide (CO_2) and water vapor (H_2O) in the DOC. Oxidation in the DOC is almost complete, starting from a specific limit temperature, i.e. the lightoff temperature. Depending on exhaust-gas composition, flow velocity, and catalyst composition, the lightoff temperature takes place at about 170 to 200°C. Starting with this temperature, conversion rises up to over 90% within a temperature interval of 20...30°C.

Reduction in particle mass

The particles emitted by the diesel engine consist partly of hydrocarbons which desorb from the particle core as temperature rises. Particle Mass (PM) can be reduced by 15 to 30% by oxidizing these hydrocarbons in the DOC.

Oxidation of NO to form NO_2

A prime function of the DOC is to oxidize NO to form NO_2. A high NO_2 component in the NO_x is vital for a number of downstream components (particulate filter, NSC, SCR).

In the untreated engine exhaust gas, the NO_2 component in the NO_x is only about 1 : 10 at most operating points. NO_2 is in temperature-dependent equilibrium with NO in the presence of oxygen (O_2). This equilibrium is on the part of NO_2 at low temperatures ($< 250°C$). Above about 450°C, however, NO becomes the thermodynamically preferred component. The function of the DOC is to raise the $NO_2 : NO$ ratio at low temperatures by inducing thermodynamic equilibrium. Depending on the catalyst coating and composition of the exhaust gas, this is achieved above a temperature of 180 to 230°C, when the concentration of NO_2 rises sharply within this temperature range. In compliance with thermodynamic equilibrium, the NO_2 concentration continues to drop as the temperature rises.

Catalytic burner

The oxidation catalyst can also be used as a catalytic heater (catalytic burner, cat burner). Reaction heat released when CO and HC are oxidized is used to raise the exhaust-gas temperature downstream of the DOC. CO and HC emissions are raised specifically for this purposes by means of an engine secondary injection, or a fuel injector downstream of the engine.

Catalytic burners are used to raise the exhaust-gas temperature during particulate-filter regeneration, for example.

As an approximation for the heat released during oxidation, the temperature of the exhaust gas rises by about 90°C for every 1% volume of CO. Since the temperature rise is very rapid, a steep temperature gradient becomes set in the catalytic converter. In the worst-case scenario, CO and HC are converted and heat is released only in the front area of the catalytic converter. The resulting stress in the ceramic carrier and catalytic converter is limited to the permitted temperature hike of about 200...250°C.

Design

Structural design

Oxidation catalysts consist of a carrier structure made of ceramics or metal, an oxide mixture (washcoat) composed of aluminum oxide (Al_2O_3), cerium (IV) oxide (CeO_2), zirconium oxide (ZrO_2), and active catalytic noble metals, such as platinum (Pt), palladium (Pd), and rhodium (Rh).

The prime function of the washcoat is to provide a large surface area for the noble metal, and to slow down catalyst sintering that occurs at high temperatures, leading to an irreversible drop in catalyst activity. The highly porous structure of the washcoat must be stable enough to resist against sintering processes.

The quantity of noble metals used for the coating, often referred to as the loading, is specified in g/ft^3. The loading is approximately 50...90 g/ft^3 (1.8...3.2 g/l). Since only surface atoms are chemically active, development aims at producing and stabilizing noble-metal particles that are as small as possible (of the order of magnitude of a few nm) in order to minimize the use of noble metals.

Differences in the structural design of the catalytic converter and choice of catalyst composition change major properties, such as lightoff temperature, conversion, temperature stability, tolerance to poisoning, as well as manufacturing costs.

Internal structure

The main parameters of a catalytic converter are channel density (specified in cpsi (channels per square inch)), wall thickness of the individual channels, and the external dimensions of the catalytic converter (cross-sectional area and length). Channel density and wall thickness determine heatup response, exhaust-gas backpressure, and mechanical stability of the catalytic converter.

Design

The catalyst volume V_{cat} is defined as a factor of exhaust-gas volumetric flow, which is itself proportional to the swept volume V_{stroke} of the engine. Typical design figures for an oxidation catalyst are $V_{cat}/V_{stroke} =$ 0.6...0.8.

The ratio of exhaust-gas volumetric flow to catalyst volume is termed flow velocity (unit: h^{-1}). Typical figures for an oxidation catalyst are 150,000...250,000 h^{-1}.

Operating conditions

Besides use of the correct catalyst, the main factors governing the efficiency of exhaust-gas treatment are the correct operating conditions. They are adjustable by the engine management system within a wide range.

If the operating temperatures are excessively high, sintering processes will occur, i.e. several small noble-metal particles will clump together to form a larger particle with a correspondingly smaller surface area, and thus reduced activity. The function of exhaust-gas temperature management is, therefore, to enhance the service life of the catalytic converter by avoiding excessive temperatures.

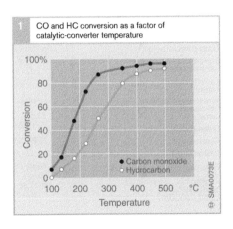

1   CO and HC conversion as a factor of catalytic-converter temperature

Electronic Diesel Control (EDC)

Electronic control of a diesel engine allows fuel-injection parameters to be varied precisely for different conditions. This is the only means by which a modern diesel engine is able to satisfy the many demands placed upon it. The EDC (Electronic Diesel Control) system is subdivided into three areas, "Sensors and desired-value generators", "Control unit", and "Actuators".

System overview

Requirements

Present-day development in the field of diesel technology is focused on lowering fuel consumption and exhaust-gas emissions (NO_x, CO, HC, particulate), while increasing engine performance and torque. In recent years this has led to an increase in the popularity of the direct-injection (DI) diesel engine, which uses much higher fuel-injection pressures than indirect-injection (IDI) engines with whirl or prechamber systems. Due to the more efficient mixture formation and the absence of flow-related losses between the whirl chamber/prechamber and the main combustion chamber, the fuel consumption of direct-injection engines is 10...20% lower than that achieved by indirect-injection designs.

In addition, diesel engine development has been influenced by the high levels of comfort and convenience demanded in modern cars. Noise levels, too, are subject to more and more stringent requirements.

As a result, the performance demanded of fuel-injection and engine-management systems has also increased, specifically with regard to:

- High fuel-injection pressures
- Rate-of-discharge curve control
- Pre-injection and, where applicable, secondary injection
- Variation of injected fuel quantity, charge-air pressure, and start of injection to suit operating conditions
- Temperature-dependent excess-fuel quantity for starting
- Control of idle speed independent of engine load
- Controlled exhaust-gas recirculation (cars)
- Cruise control
- Tight tolerances for injection duration and injected fuel quantity, and maintenance of high precision over the service life of the system (long-term performance)

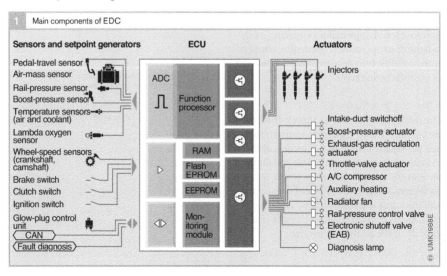

1   Main components of EDC

Sensors and setpoint generators	ECU	Actuators

Sensors and setpoint generators:
Pedal-travel sensor
Air-mass sensor
Rail-pressure sensor
Boost-pressure sensor
Temperature sensors (air and coolant)
Lambda oxygen sensor
Wheel-speed sensors (crankshaft, camshaft)
Brake switch
Clutch switch
Ignition switch
Glow-plug control unit
CAN
Fault diagnosis

ECU:
ADC
Function processor
RAM
Flash EPROM
EEPROM
Monitoring module

Actuators:
Injectors
Intake-duct switchoff
Boost-pressure actuator
Exhaust-gas recirculation actuator
Throttle-valve actuator
A/C compressor
Auxiliary heating
Radiator fan
Rail-pressure control valve
Electronic shutoff valve (EAB)
Diagnosis lamp

UMK1988E

Conventional mechanical governing of engine speed uses a number of adjusting mechanisms to adapt to different engine operating conditions and ensures a high mixture formation quality. Nevertheless, it is restricted to a simple engine-based control loop and there are a number of important influencing variables that it cannot take account of or cannot respond quickly enough to.

As demands have increased, what was originally a straightforward system using electric actuator shafts has developed into the present-day EDC, a complex electronic control system capable of processing large amounts of data in real time. It can form part of an overall electronic vehicle control system ("drive-by-wire"). And a result of the increasing integration of electronic components, the control-system circuitry can be accommodated in a very small space.

Operating concept

Electronic Diesel Control (EDC) is capable of meeting the requirements listed above as a result of microcontroller performance that has risen considerably in the last few years.

In contrast with diesel-engine vehicles with conventional mechanically controlled fuel-injection pumps, the driver of a vehicle equipped with EDC has no direct control over the injected fuel quantity via the accelerator pedal and cable. The injected fuel quantity is actually determined by a number of different influencing variables. They include:

- The vehicle response desired by the driver (accelerator-pedal position)
- The engine operating status
- The engine temperature
- Interventions by other systems (e.g. TCS)
- The effect on exhaust-gas emission levels, etc.

The control unit calculates the injected fuel quantity on the basis of all these influencing variables. Start of delivery can also be varied. This demands a comprehensive monitoring concept that detects inconsistencies and initiates appropriate actions in accordance with the effects (e.g. torque limitation or limp-home mode in the idle-speed range). EDC, therefore incorporates a number of control loops.

Electronic diesel control allows data exchange with other electronic systems, such as the Traction Control System (TCS), Electronic Transmission Control (ETC), or Electronic Stability Program (ESP). As a result, the engine management system can be integrated in the vehicle's overall control system network, thereby enabling functions such as reduction of engine torque when the automatic transmission changes gear, regulation of engine torque to compensate for wheel spin, disabling of fuel injection by the engine immobilizer, etc.

The EDC system is fully integrated in the vehicle's diagnostic system. It meets all OBD (On-Board Diagnosis) and EOBD (European OBD) requirements.

System modules

Electronic Diesel Control (EDC) is divided into three system modules (Fig. 1):

1. *Sensors and setpoint generators* detect operating conditions (e.g. engine speed) and setpoint values (e.g. switch position). They convert physical variables into electrical signals.

2. The *electronic control unit* processes data from the sensors and setpoint generators based on specific open- and closed-loop control algorithms. It controls the actuators by means of electrical output signals. In addition, the control unit acts as an interface to other systems and to the vehicle diagnostic system.

3. *Actuators* convert electrical output signals from the control unit into mechanical parameters (e.g. the solenoid valve for the fuel-injection system).

▶ Where does the word "electronics" come from?

This term actually goes back to the ancient Greeks. For them, the word "electron" meant amber. Its force of attraction on woollen threads or similar was known to Thales von Milet over 2,500 years ago.

Electrons, and therefore electronics as such, are extremely fast due to their very small mass and electric charge. The term "electronics" comes directly from the word "electron".

The mass of an electron has as little effect on a gram of any given substance as a 5 gram weight has on the total mass of our earth.

The word "electronics" was born in the 20th century. There is no evidence available as to when the word was used for the first time. It could be Sir John Ambrose Fleming, one of the inventors of the electron tube in about 1902.

Even the first "Electronic Engineer" already existed in the 19th century. Fleming was listed in the 1888 edition of "Who's Who", published during the reign of Queen Victoria. The official title was "Kelly's Handbook of Titled, Landed and Official Classes". The Electronic Engineer can be found under the title "Royal Warrant Holders", that is the list of persons who had been awarded a Royal Warrant.

What was this Electronic Engineer's job? He was responsible for the correct functioning and cleanliness of the gas lamps at court. And why did he have such a splendid title? Because he knew that "electrons" in ancient Greece stood for glitter, shine, and sparkle.

Source:
"Basic Electronic Terms" ("Grundbegriffe der Elektronik") – Bosch publication (reprint from the "Bosch Zünder" (Bosch Company Newspaper)), 1988.

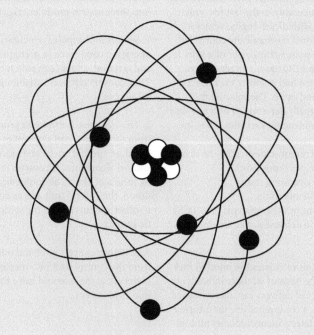

LAE0047Y

In-line fuel-injection pumps

1   Overview of the EDC components for inline fuel-injection pumps

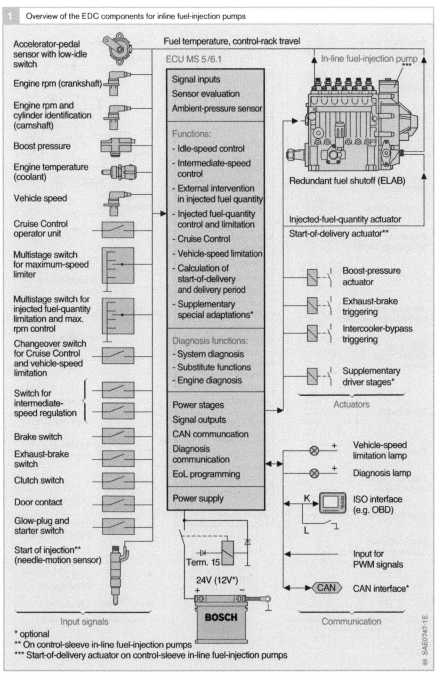

Accelerator-pedal sensor with low-idle switch

Engine rpm (crankshaft)

Engine rpm and cylinder identification (camshaft)

Boost pressure

Engine temperature (coolant)

Vehicle speed

Cruise Control operator unit

Multistage switch for maximum-speed limiter

Multistage switch for injected fuel-quantity limitation and max. rpm control

Changeover switch for Cruise Control and vehicle-speed limitation

Switch for intermediate-speed regulation

Brake switch

Exhaust-brake switch

Clutch switch

Door contact

Glow-plug and starter switch

Start of injection** (needle-motion sensor)

Fuel temperature, control-rack travel

ECU MS 5/6.1

Signal inputs
Sensor evaluation
Ambient-pressure sensor

Functions:
- Idle-speed control
- Intermediate-speed control
- External intervention in injected fuel quantity
- Injected fuel-quantity control and limitation
- Cruise Control
- Vehicle-speed limitation
- Calculation of start-of-delivery and delivery period
- Supplementary special adaptations*

Diagnosis functions:
- System diagnosis
- Substitute functions
- Engine diagnosis

Power stages
Signal outputs
CAN communcation
Diagnosis communication
EoL programming

Power supply

Term. 15

24V (12V*)

In-line fuel-injection pump***

Redundant fuel shutoff (ELAB)

Injected-fuel-quantity actuator

Start-of-delivery actuator**

Boost-pressure actuator

Exhaust-brake triggering

Intercooler-bypass triggering

Supplementary driver stages*

Actuators

Vehicle-speed limitation lamp

Diagnosis lamp

K   ISO interface (e.g. OBD)

L

Input for PWM signals

CAN   CAN interface*

Input signals

BOSCH

Communication

SAE0747-1E

* optional
** On control-sleeve in-line fuel-injection pumps
*** Start-of-delivery actuator on control-sleeve in-line fuel-injection pumps

Helix and port-controlled axial-piston distributor pumps

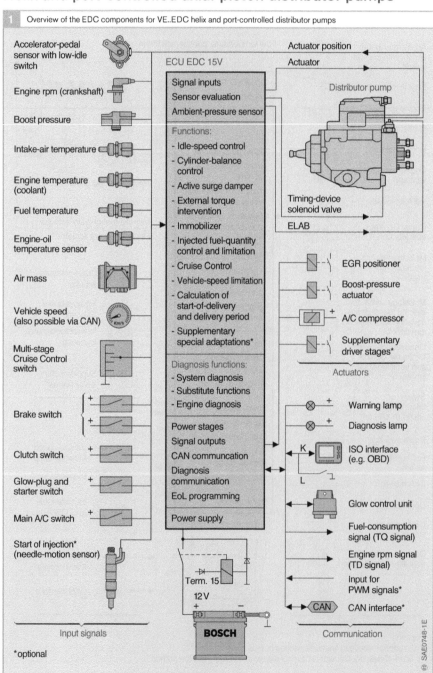

1   Overview of the EDC components for VE..EDC helix and port-controlled distributor pumps

Accelerator-pedal sensor with low-idle switch

Engine rpm (crankshaft)

Boost pressure

Intake-air temperature

Engine temperature (coolant)

Fuel temperature

Engine-oil temperature sensor

Air mass

Vehicle speed (also possible via CAN)

Multi-stage Cruise Control switch

Brake switch

Clutch switch

Glow-plug and starter switch

Main A/C switch

Start of injection* (needle-motion sensor)

ECU EDC 15V

Signal inputs
Sensor evaluation
Ambient-pressure sensor

Functions:
- Idle-speed control
- Cylinder-balance control
- Active surge damper
- External torque intervention
- Immobilizer
- Injected fuel-quantity control and limitation
- Cruise Control
- Vehicle-speed limitation
- Calculation of start-of-delivery and delivery period
- Supplementary special adaptations*

Diagnosis functions:
- System diagnosis
- Substitute functions
- Engine diagnosis

Power stages
Signal outputs
CAN communcation
Diagnosis communication
EoL programming
Power supply

Actuator position
Actuator

Distributor pump

Timing-device solenoid valve
ELAB

EGR positioner

Boost-pressure actuator

A/C compressor

Supplementary driver stages*

Actuators

Warning lamp

Diagnosis lamp

ISO interface (e.g. OBD)

Glow control unit

Fuel-consumption signal (TQ signal)

Engine rpm signal (TD signal)

Input for PWM signals*

CAN interface*

Term. 15
12 V

BOSCH

Input signals

Communication

*optional

SAE0748-1E

Solenoid-valve-controlled axial-piston and radial-piston distributor pumps

2   Overview of the EDC components for VE..MV, VR solenoid-valve-controlled distributor pumps

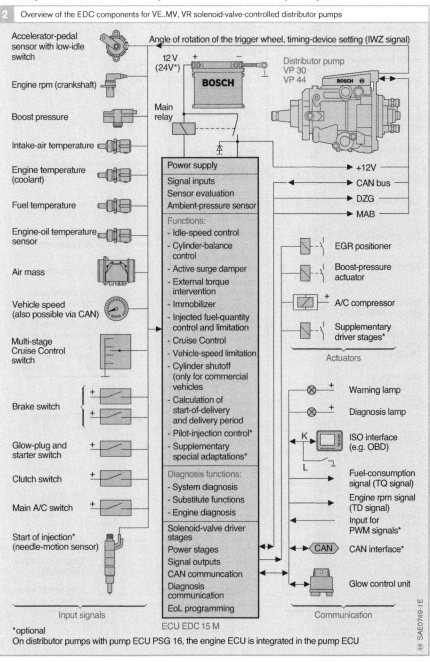

Accelerator-pedal sensor with low-idle switch

Angle of rotation of the trigger wheel, timing-device setting (IWZ signal)

12 V (24V*)

BOSCH

Distributor pump VP 30 VP 44

BOSCH

Engine rpm (crankshaft)

Boost pressure

Main relay

Intake-air temperature

Engine temperature (coolant)

Fuel temperature

Engine-oil temperature sensor

Air mass

Vehicle speed (also possible via CAN)

Multi-stage Cruise Control switch

Brake switch

Glow-plug and starter switch

Clutch switch

Main A/C switch

Start of injection* (needle-motion sensor)

Power supply

Signal inputs
Sensor evaluation
Ambient-pressure sensor

Functions:
- Idle-speed control
- Cylinder-balance control
- Active surge damper
- External torque intervention
- Immobilizer
- Injected fuel-quantity control and limitation
- Cruise Control
- Vehicle-speed limitation
- Cylinder shutoff (only for commercial vehicles)
- Calculation of start-of-delivery and delivery period
- Pilot-injection control*
- Supplementary special adaptations*

Diagnosis functions:
- System diagnosis
- Substitute functions
- Engine diagnosis

Solenoid-valve driver stages
Power stages
Signal outputs
CAN communcation
Diagnosis communication
EoL programming

ECU EDC 15 M

+12V

CAN bus

DZG

MAB

EGR positioner

Boost-pressure actuator

A/C compressor

Supplementary driver stages*

Actuators

Warning lamp

Diagnosis lamp

ISO interface (e.g. OBD)

Fuel-consumption signal (TQ signal)

Engine rpm signal (TD signal)

Input for PWM signals*

CAN interface*

Glow control unit

Input signals

Communication

*optional
On distributor pumps with pump ECU PSG 16, the engine ECU is integrated in the pump ECU

SAE0749-1E

Unit Injector System (UIS) for passenger cars

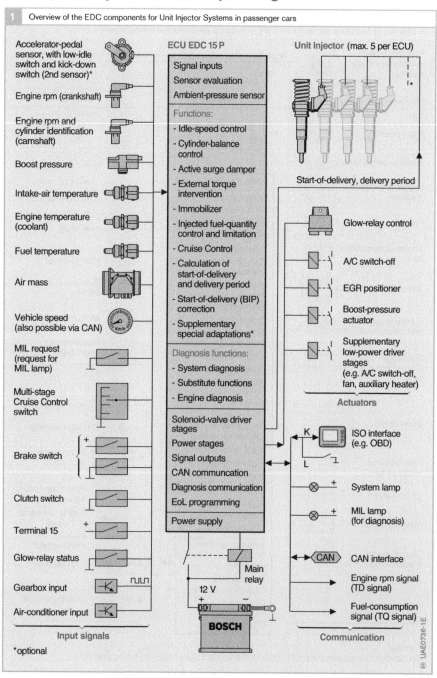

1   Overview of the EDC components for Unit Injector Systems in passenger cars

Input signals

- Accelerator-pedal sensor, with low-idle switch and kick-down switch (2nd sensor)*
- Engine rpm (crankshaft)
- Engine rpm and cylinder identification (camshaft)
- Boost pressure
- Intake-air temperature
- Engine temperature (coolant)
- Fuel temperature
- Air mass
- Vehicle speed (also possible via CAN)
- MIL request (request for MIL lamp)
- Multi-stage Cruise Control switch
- Brake switch
- Clutch switch
- Terminal 15
- Glow-relay status
- Gearbox input
- Air-conditioner input

*optional

ECU EDC 15 P

Signal inputs
Sensor evaluation
Ambient-pressure sensor

Functions:
- Idle-speed control
- Cylinder-balance control
- Active surge damper
- External torque intervention
- Immobilizer
- Injected fuel-quantity control and limitation
- Cruise Control
- Calculation of start-of-delivery and delivery period
- Start-of-delivery (BIP) correction
- Supplementary special adaptations*

Diagnosis functions:
- System diagnosis
- Substitute functions
- Engine diagnosis

Solenoid-valve driver stages
Power stages
Signal outputs
CAN communcation
Diagnosis communication
EoL programming
Power supply

Main relay

12 V

BOSCH

Unit Injector (max. 5 per ECU)

*

Start-of-delivery, delivery period

Glow-relay control

A/C switch-off

EGR positioner

Boost-pressure actuator

Supplementary low-power driver stages (e.g. A/C switch-off, fan, auxiliary heater)

Actuators

ISO interface (e.g. OBD)

System lamp

MIL lamp (for diagnosis)

CAN interface

Engine rpm signal (TD signal)

Fuel-consumption signal (TQ signal)

Communication

UAE0736-1E

Unit Injector System (UIS)
and Unit Pump System (UPS) for commercial vehicles

2 Overview of the EDC components for Unit Injector Systems (UIS) and Unit Pump Systems (UPS) in commercial vehicles

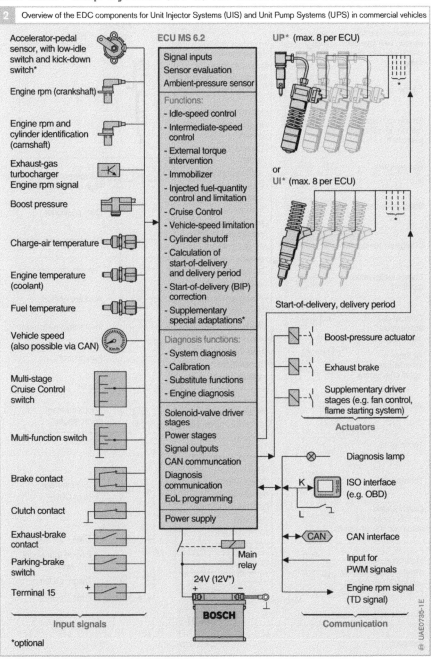

Accelerator-pedal sensor, with low-idle switch and kick-down switch*

Engine rpm (crankshaft)

Engine rpm and cylinder identification (camshaft)

Exhaust-gas turbocharger Engine rpm signal

Boost pressure

Charge-air temperature

Engine temperature (coolant)

Fuel temperature

Vehicle speed (also possible via CAN)

Multi-stage Cruise Control switch

Multi-function switch

Brake contact

Clutch contact

Exhaust-brake contact

Parking-brake switch

Terminal 15

Input signals

*optional

ECU MS 6.2

Signal inputs
Sensor evaluation
Ambient-pressure sensor

Functions:
- Idle-speed control
- Intermediate-speed control
- External torque intervention
- Immobilizer
- Injected fuel-quantity control and limitation
- Cruise Control
- Vehicle-speed limitation
- Cylinder shutoff
- Calculation of start-of-delivery and delivery period
- Start-of-delivery (BIP) correction
- Supplementary special adaptations*

Diagnosis functions:
- System diagnosis
- Calibration
- Substitute functions
- Engine diagnosis

Solenoid-valve driver stages
Power stages
Signal outputs
CAN communcation
Diagnosis communication
EoL programming

Power supply

UP* (max. 8 per ECU)

or

UI* (max. 8 per ECU)

Start-of-delivery, delivery period

Boost-pressure actuator
Exhaust brake
Supplementary driver stages (e.g. fan control, flame starting system)

Actuators

Diagnosis lamp
ISO interface (e.g. OBD)
CAN interface
Input for PWM signals
Engine rpm signal (TD signal)

Main relay

24V (12V*)

BOSCH

Communication

UAE0735-1E

Common Rail System (CRS) for passenger cars

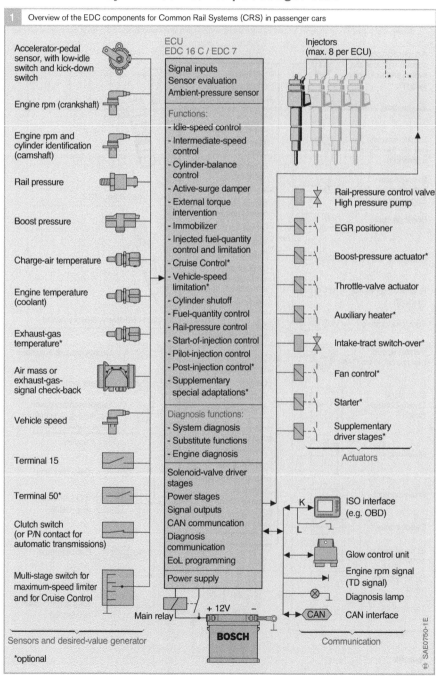

1 Overview of the EDC components for Common Rail Systems (CRS) in passenger cars

Sensors and desired-value generator

- Accelerator-pedal sensor, with low-idle switch and kick-down switch
- Engine rpm (crankshaft)
- Engine rpm and cylinder identification (camshaft)
- Rail pressure
- Boost pressure
- Charge-air temperature
- Engine temperature (coolant)
- Exhaust-gas temperature*
- Air mass or exhaust-gas-signal check-back
- Vehicle speed
- Terminal 15
- Terminal 50*
- Clutch switch (or P/N contact for automatic transmissions)
- Multi-stage switch for maximum-speed limiter and for Cruise Control

ECU EDC 16 C / EDC 7

Signal inputs
Sensor evaluation
Ambient-pressure sensor

Functions:
- Idle-speed control
- Intermediate-speed control
- Cylinder-balance control
- Active-surge damper
- External torque intervention
- Immobilizer
- Injected fuel-quantity control and limitation
- Cruise Control*
- Vehicle-speed limitation*
- Cylinder shutoff
- Fuel-quantity control
- Rail-pressure control
- Start-of-injection control
- Pilot-injection control
- Post-injection control*
- Supplementary special adaptations*

Diagnosis functions:
- System diagnosis
- Substitute functions
- Engine diagnosis

Solenoid-valve driver stages
Power stages
Signal outputs
CAN communcation
Diagnosis communication
EoL programming

Power supply

Main relay

+ 12V −

BOSCH

Injectors (max. 8 per ECU)

Actuators
- Rail-pressure control valve High pressure pump
- EGR positioner
- Boost-pressure actuator*
- Throttle-valve actuator
- Auxiliary heater*
- Intake-tract switch-over*
- Fan control*
- Starter*
- Supplementary driver stages*

Communication
- K — ISO interface (e.g. OBD)
- L
- Glow control unit
- Engine rpm signal (TD signal)
- Diagnosis lamp
- CAN — CAN interface

*optional

SAE0750-1E

Common Rail System (CRS) for commercial vehicles

2 Overview of the EDC components for Common Rail Systems (CRS) in commercial vehicles

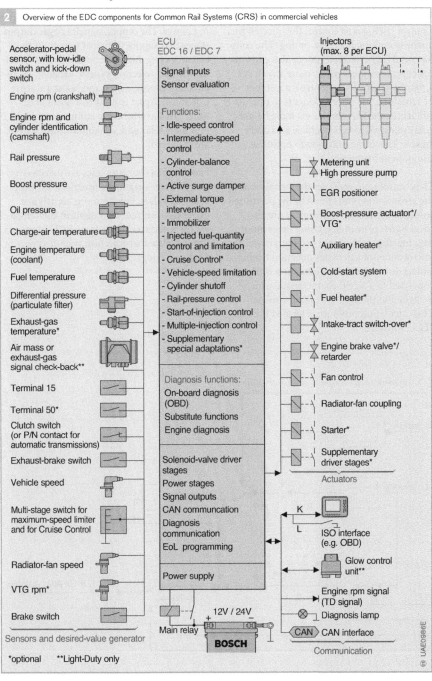

Accelerator-pedal sensor, with low-idle switch and kick-down switch

Engine rpm (crankshaft)

Engine rpm and cylinder identification (camshaft)

Rail pressure

Boost pressure

Oil pressure

Charge-air temperature

Engine temperature (coolant)

Fuel temperature

Differential pressure (particulate filter)

Exhaust-gas temperature*

Air mass or exhaust-gas signal check-back**

Terminal 15

Terminal 50*

Clutch switch (or P/N contact for automatic transmissions)

Exhaust-brake switch

Vehicle speed

Multi-stage switch for maximum-speed limiter and for Cruise Control

Radiator-fan speed

VTG rpm*

Brake switch

Sensors and desired-value generator

ECU
EDC 16 / EDC 7

Signal inputs
Sensor evaluation

Functions:
- Idle-speed control
- Intermediate-speed control
- Cylinder-balance control
- Active surge damper
- External torque intervention
- Immobilizer
- Injected fuel-quantity control and limitation
- Cruise Control*
- Vehicle-speed limitation
- Cylinder shutoff
- Rail-pressure control
- Start-of-injection control
- Multiple-injection control
- Supplementary special adaptations*

Diagnosis functions:
On-board diagnosis (OBD)
Substitute functions
Engine diagnosis

Solenoid-valve driver stages
Power stages
Signal outputs
CAN communcation
Diagnosis communication
EoL programming

Power supply

12V / 24V
Main relay

BOSCH

Injectors
(max. 8 per ECU)

Metering unit
High pressure pump

EGR positioner

Boost-pressure actuator*/ VTG*

Auxiliary heater*

Cold-start system

Fuel heater*

Intake-tract switch-over*

Engine brake valve*/ retarder

Fan control

Radiator-fan coupling

Starter*

Supplementary driver stages*

Actuators

K
L
ISO interface (e.g. OBD)

Glow control unit**

Engine rpm signal (TD signal)

Diagnosis lamp

CAN CAN interface

Communication

UAE0986E

*optional   **Light-Duty only

Data processing

The main function of the Electronic Diesel Control (EDC) is to control the injected fuel quantity and the injection timing. The common-rail fuel-injection system also controls injection pressure. Furthermore, on all systems, the engine ECU controls a number of actuators. For all components to operate efficiently, the EDC functions must be precisely matched to every vehicle and every engine. This is the only way to optimize component interaction (Fig. 2).

The control unit evaluates the signals sent by the sensors and limits them to the permitted voltage level. Some input signals are also checked for plausibility. Using these input data together with stored program maps, the microprocessor calculates injection timing and its duration. This information is then converted to a signal characteristic which is aligned to the engine's piston strokes. This calculation program is termed the "ECU software".

The required degree of accuracy together with the diesel engine's outstanding dynamic response requires high-level computing power. The output signals trigger output stages that supply sufficient power for the actuators (e.g. high-pressure solenoid valves for the fuel-injection system, exhaust-gas recirculation positioners, and boost-pressure actuators). Apart from this, a number of other auxiliary-function components (e.g. glow relay and air-conditioning system) are triggered.

Faulty signal characteristics are detected by output-stage diagnostic functions for the solenoid valves. Furthermore, signals are exchanged with other systems in the vehicle via the interfaces. The engine ECU monitors the complete fuel-injection system as part of a safety strategy.

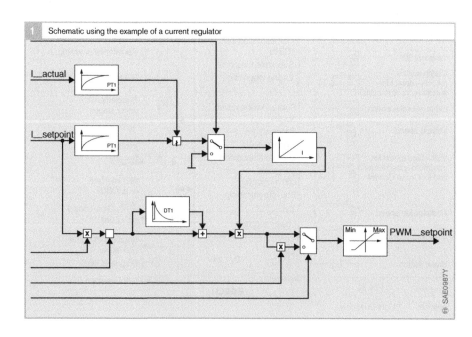

1　Schematic using the example of a current regulator

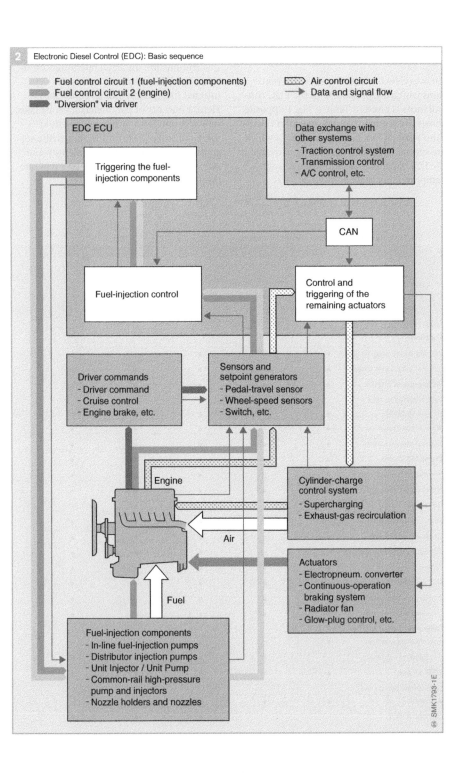

2 Electronic Diesel Control (EDC): Basic sequence

Fuel control circuit 1 (fuel-injection components)
Fuel control circuit 2 (engine)
"Diversion" via driver

Air control circuit
Data and signal flow

EDC ECU

Triggering the fuel-injection components

Data exchange with other systems
- Traction control system
- Transmission control
- A/C control, etc.

CAN

Fuel-injection control

Control and triggering of the remaining actuators

Driver commands
- Driver command
- Cruise control
- Engine brake, etc.

Sensors and setpoint generators
- Pedal-travel sensor
- Wheel-speed sensors
- Switch, etc.

Engine

Cylinder-charge control system
- Supercharging
- Exhaust-gas recirculation

Air

Actuators
- Electropneum. converter
- Continuous-operation braking system
- Radiator fan
- Glow-plug control, etc.

Fuel

Fuel-injection components
- In-line fuel-injection pumps
- Distributor injection pumps
- Unit Injector / Unit Pump
- Common-rail high-pressure pump and injectors
- Nozzle holders and nozzles

SMK1793-1E

Fuel-injection control

An overview of the various control functions which are possible with the EDC control units is given in Table 1. Fig. 1 opposite shows the sequence of fuel-injection calculations with all functions, a number of which are special options. These can be activated in the ECU by the workshop when retrofit equipment is installed.

In order that the engine can run with optimal combustion under all operating conditions, the ECU calculates exactly the right injected fuel quantity for all conditions. Here, a number of parameters must be taken into account. On a number of solenoid-valve-controlled distributor pumps, the solenoid valves for injected fuel quantity and start of injection are triggered by a separate pump ECU (PSG).

1   EDC variants for road vehicles: Overview of functions

Fuel-injection system	In-line injection pumps	Helix-controlled distributor injection pumps	Solenoid-valve-controlled distributor injection pumps	Unit Injector System and Unit Pump System	Common Rail System
	PE	VE-EDC	VE-M, VR-M	UIS, UPS	CR
Function					
Injected-fuel-quantity limitation	•	•	•	•	•
External torque intervention	•[3]	•	•	•	•
Vehicle-speed limitation	•[3]	•	•	•	•
Vehicle-speed control (Cruise Control)	•	•	•	•	•
Altitude compensation	•	•	•	•	•
Boost-pressure control	•	•	•	•	•
Idle-speed control	•	•	•	•	•
Intermediate-speed control	•[3]	•	•	•	•
Active surge damping	•[2]	•	•	•	•
BIP control	–	–	•	•	–
Intake-tract switch-off	–	–	•	•[2]	•
Electronic immobilizer	•[2]	•	•	•	•
Controlled pilot injection	–	–	•	•[2]	•
Glow control	•[2]	•	•	•[2]	•
A/C switch-off	•[2]	•	•	•	•
Auxiliary coolant heating	•[2]	•	•	–	•
Cylinder-balance control	•[2]	•	•	•	•
Control of injected fuel quantity compensation	•[2]	–	•	•	•
Fan (blower) triggering	–	•	•	•	•
EGR control	•[2]	•	•	•[2]	•
Start-of-injection control with sensor	•[1,3]	•	•	–	–
Cylinder shutoff	–	–	•[3]	•[3]	•[3]

Table 1

1   Only control-sleeve in-line injection pumps

2   Passenger cars only

3   Commercial vehicles only

1   Calculation of fuel-injection process in the ECU

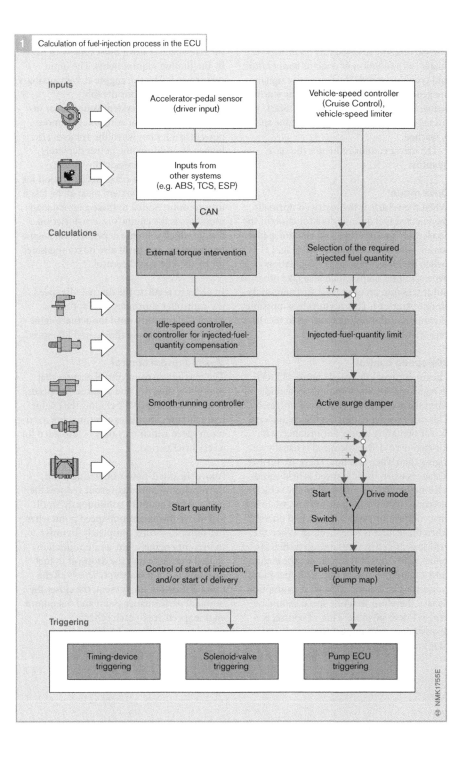

Start quantity

For starting, the injected fuel quantity is calculated as a function of coolant temperature and cranking speed. Start-quantity signals are generated from the moment the starting switch is turned (Fig. 1, switch in "Start" position) until a given minimum engine speed is reached.

The driver cannot influence the start quantity.

Drive mode

When the vehicle is being driven normally, the injected fuel quantity is a function of the accelerator-pedal setting (accelerator-pedal sensor) and of the engine speed (Fig. 1, switch in "Drive" position). Calculation depends upon maps which also take other influences into account (e. g. fuel and intake-air temperature). This permits best-possible alignment of the engine's output to the driver's wishes.

Idle-speed control

The function of idle speed control (LLR) is to regulate a specific setpoint speed at idle when the accelerator pedal is not operated. This can vary depending on the engine's particular operating mode. For instance, with the engine cold, the idle speed is usually set higher than when it is hot. There are further instances when the idle speed is held somewhat higher. For instance, when the vehicle's electrical-system voltage is too low, when the air-conditioning system is switched on, or when the vehicle is freewheeling. When the vehicle is driven in stop-and-go traffic, together with stops at traffic lights, the engine runs a lot of the time at idle. Considerations concerning emissions and fuel consumption dictate, therefore, that idle speed should be kept as low as possible. This, of course, is a disadvantage with respect to smooth-running and pulling away.

When adjusting the stipulated idle speed, the idle-speed control must cope with heavily fluctuating requirements. The input power needed by the engine-driven auxiliary equipment varies considerably.

At low electrical-system voltages, for instance, the alternator consumes far more power than it does when the voltages are higher. In addition, the power demands from the A/C compressor, the steering pump, and the high-pressure generation for the diesel injection system must all be taken into account. Added to these external load moments is the engine's internal friction torque which is highly dependent on engine temperature, and must also be compensated for by the idle-speed control.

In order to regulate the desired idle speed, the controller continues to adapt the injected fuel quantity until the actual engine speed corresponds to the desired idle speed.

Maximum-rpm control

The maximum-rpm control ensures that the engine does not run at excessive speeds. To avoid damage to the engine, the engine manufacturer stipulates a permissible maximum speed which may only be exceeded for a very brief period.

Above the rated-power operating point, the maximum-speed governor reduces the injected fuel quantity continuously, until just above the maximum-speed point when fuel-injection stops completely. In order to prevent engine surge, a ramp function is used to ensure that the drop-off in fuel injection is not too abrupt. This is all the more difficult to implement, the closer the nominal performance point and maximum engine speed are to each other.

Intermediate-speed control

Intermediate-speed control (ZDR) is used on commercial vehicles and light-duty trucks with power take-offs, e.g. crane), or for special vehicles (e.g. ambulances with a power generator). With the control in operation, the engine is regulated to a load-independent intermediate speed.

With the vehicle stationary, the intermediate-speed control is activated via the cruise-control operator unit. A fixed rotational speed can be called up from the data store at the push of a button. In addition, this operator unit can be used for preselecting specific engine speeds. The intermediate-speed control is also applied on passenger cars with automated transmissions (e.g. Tiptronic) to control the engine speed during gearshifts.

Vehicle-speed controller (cruise control)

Cruise control allows the vehicle to be driven at a constant speed. It controls the vehicle speed to the speed selected by the driver without him/her needing to press the accelerator pedal. The driver can set the required speed either by operating a lever or by pressing buttons on the steering wheel. The injected fuel quantity is either increased or decreased until the desired (set) speed is reached.

On some cruise-control applications, the vehicle can be accelerated beyond the current set speed by pressing the accelerator pedal. As soon as the accelerator pedal is released, cruise control regulates the speed back down to the previously set speed.

If the driver depresses the clutch or brake pedal while cruise control is activated, control is terminated. On some applications, the control can be switched off by the accelerator pedal.

If cruise control has been switched off, the driver only needs to shift the lever to the restore position to reselect the last speed setting.

The operator controls can also be used for a step-by-step change of the selected speed.

Vehicle-speed limiter
Variable limitation
Vehicle-speed limitation (FGB, also called the limiter) limits the maximum speed to a set value, even if the driver continues to depress the accelerator pedal. On very quiet vehicles, where the engine can hardly be heard, this is a particular help for the driver who can no longer exceed speed limits inadvertently.

The vehicle-speed limiter keeps the injected fuel quantity down to a limit corresponding to the selected maximum speed. It can be deactivated by pressing the lever or depressing the kickdown switch. In order to reselect the last speed setting, the driver only needs to press the lever to the restore position. The operator controls can also be used for a step-by-step change of the selected speed.

Fixed limitation
In a number of countries, fixed maximum speeds are mandatory for certain classes of vehicles (for instance, for heavy trucks). Vehicle manufacturers also limit the maximum speeds of their heavy vehicles by installing a fixed speed limit which cannot be deactivated.

In the case of special vehicles, the driver can also select from a range of fixed, programmed speed limits (for instance, when workers are standing on the platform of a garbage truck).

Active-surge damping

Sudden engine-torque changes excite the vehicle's drivetrain, which, as a result, goes into bucking oscillation. These oscillations are perceived by the vehicle's occupants as unpleasant periodic changes in acceleration (Fig. 2, a). The function of the active-surge damper (ARD) is to reduce these changes in acceleration (b).

Two different methods are used:
- In case of sudden changes in the torque required by the driver (through the accelerator pedal), a precisely matched filter function reduces drivetrain excitation (1).
- The speed signals are used to detect drivetrain oscillations which are then damped by an active control. In order to counteract the drivetrain oscillations (2), the active control reduces the injected fuel quantity when rotational speed increases, and increases it when speed drops.

Smooth-running control (SRC)/ Control of injected-fuel-quantity compensation (MAR)

Presuming a constant injection duration, not all of the engine's cylinders generate the same torque. This can be due to differences in cylinder-head sealing, as well as differences in cylinder friction, and hydraulic-injection components. These differences in torque output lead to rough engine running and an increase in exhaust-gas emissions.

Smooth-running control (LRR) or fuel-balancing control (MAR) have the function of detecting these differences based on the resulting fluctuations in engine speed, and to compensate by adjusting the injected fuel quantity in the cylinder affected. Here, the rotational speed at a given cylinder after injection is compared to a mean speed. If the particular cylinder's speed is too low, the injected fuel quantity is increased; if it is too high, the fuel quantity is reduced (Fig. 3).

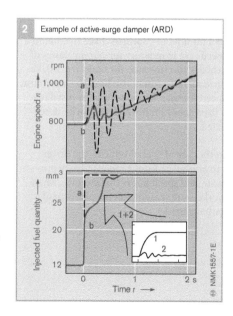

Fig. 2
a   Without active-surge damper
b   With active-surge damper
1   Filter function
2   Active correction

2   Example of active-surge damper (ARD)

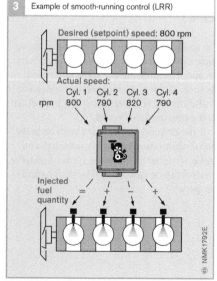

3   Example of smooth-running control (LRR)

Smooth-running control is a convenience feature. Its primary object is to ensure that the engine runs smoothly at near-idle. The injected-fuel-quantity compensation function is aimed at not only improving comfort at idle, but also at reducing exhaust-gas emissions in the medium-speed ranges by ensuring identical injected fuel quantities for all cylinders.

On commercial vehicles, smooth-running control is also known as the AZG (adaptive cylinder equalization).

Injected-fuel-quantity limit
There are a number of reasons why the fuel quantity actually required by the driver, or that which is physically possible, should not always be injected. The injection of such fuel quantities could have the following effects:
- Excessive exhaust-gas emissions
- Excessive soot
- Mechanical overloading due to high torque or excessive engine speed
- Thermal overloading due to excessive temperatures of the exhaust gas, coolant, oil, or turbocharger
- Thermal overloading of the solenoid valves if they are triggered too long

To avoid these negative effects, a number of input variables (for instance, intake-air quantity, engine speed, and coolant temperature) are used to generate this limitation figure. The result is that the maximum injected fuel quantity is limited and with it the maximum torque.

Engine-brake function
When a truck's engine brake is applied, the injected fuel quantity is either reduced to zero, or the idle fuel quantity is injected. For this purpose, the ECU detects the position of the engine-brake switch.

Altitude compensation
Atmospheric pressure drops as altitude increases so that the cylinder is charged with less combustion air. This means that the injected fuel quantity must be reduced accordingly, otherwise excessive soot will be emitted. In order that the injected fuel quantity can be reduced at high altitudes, the atmospheric pressure is measured by the ambient-pressure sensor in the ECU. This reduces the injected fuel quantity at higher elevations. Atmospheric pressure also has an effect on boost-pressure control and torque limitation.

Cylinder shutoff
If less torque is required at high engine speeds, very little fuel needs to be injected. As an alternative, cylinder shutoff can be applied to reduce torque. Here, half of the injectors are switched off (commercial-vehicle UIS, UPS, and CRS). The remaining injectors then inject correspondingly more fuel which can be metered with even higher precision.

When the injectors are switched on and off, special software algorithms ensure smooth transitions without noticeable torque changes.

Injector delivery compensation

New functions are added to common-rail (CR) and UIS/UPS systems to enhance the high precision of the fuel-injection system further, and ensure them for the service life of the vehicle.

With injector delivery compensation (IMA), a mass of measuring data is detected for each injector during the injector manufacturing process. The data is then affixed to the injector in the form of a data-matrix code. With piezo-inline injectors, data on lift response is included. This data is transferred to the ECU during vehicle production. While the engine is running, these values are used to compensate for deviations in metering and switching response.

Zero delivery calibration

The reliable mastery of small pre-injection events for the service life of the vehicle is vitally important to achieve the required level of comfort (through reduced noise) and exhaust-gas emission targets. There must be some form of compensation for fuel-quantity drifts in the injectors. For this reason, a small quantity of fuel is injected in one cylinder in overrun conditions in second- and third-generation CR systems. The wheel-speed sensor detects the resulting torque increase as a minor dynamic change in engine speed. This increase in torque, which remains imperceptible to the driver, is clearly linked to the injected fuel quantity. The process is then repeated for all cylinders and at various operating points. A teach-in algorithm detects minor changes in pre-injection quantity and corrects the injector triggering period accordingly for all pre-injection events.

Average delivery adaption

The deviation of the actually injected fuel quantity from the setpoint value is required to adapt exhaust-gas recirculation and charge-air pressure correctly. The average delivery adaption (MMA), therefore determines the average value of the injected fuel quantity for all cylinders from the signals received from the lambda oxygen sensor and the air-mass sensor. Correction values are then calculated from the setpoint and actual values (see "Lambda closed-loop control for passenger-car diesel engines").

The MMA teach-in function ensures a constant level of favorable exhaust-gas emission values in the lower part-load range for the service life of the vehicle.

Pressure-wave correction

Injection events trigger pressure waves in the line between the nozzle and the fuel rail in all CR systems. These pressure pulses affect systematically the injected fuel quantity of later injection events (pre-injection/main injection/secondary injection) within a combustion cycle. The deviations of later injection events are dependent on the fuel quantity previously injected, the time interval between injection events, rail pressure, and fuel temperature. The control unit can calculate a correction factor by including these parameters in suitable compensation algorithms.

However, extremely high application resources are required for this correction function. The benefit is the possibility of flexibly adjusting the interval between pre-injection and main injection, for example, in order to optimize combustion.

Injector delivery compensation

Functional description

Injector delivery compensation (IMA) is a software function to make fuel quantity metering more precise and increase injector efficiency on the engine. The feature has the function of correcting injected fuel quantity to the setpoint value over the entire program map individually for every injector in a CR system. This reduces system tolerances and exhaust-gas emission spread. The compensation values required for IMA represent the difference from the setpoint value of each factory test point, and are inscribed on each injector in encoded form.

The entire engine environment is corrected by means of a correction program map that uses compensation values to calculate a correction quantity. At the end of the line of the car assembly plant, the EDC compensation values belonging to the injectors fitted and their cylinder assignment are programmed in the electronic control unit using EOL programming. The compensation values are also reprogrammed when an injector is replaced at the customer service workshop.

Necessity for this function

The technical resources required for a further restriction of the manufacturing tolerances for injectors rise exponentially and appear to be financially unprofitable. IMA is a viable solution to increase efficiency, enhance the metering precision of fuel quantity injected in the engine, and reduce exhaust-gas emissions.

Measured values in testing

The end-of-line test measures every injector at several points that are representative for the spread of the particular injector type. Deviations from setpoint values at these points (compensation values) are calculated and then inscribed on the injector head.

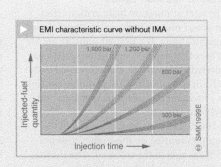

EMI characteristic curve without IMA

Fig. 1
Curves of various injectors as a function of rail pressure.
IMA reduces curve spread.
EMI Injected-fuel-quantity indicator

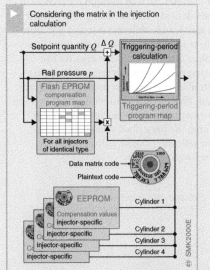

Considering the matrix in the injection calculation

Fig. 2
Calculation of injector triggering period based on setpoint quantity, rail pressure, and correction values

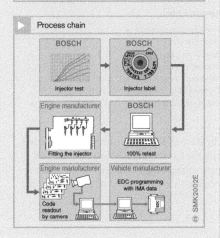

Process chain

Fig. 3
Schematic of process chain from injector delivery compensation at Bosch through to end-of-line programming at the vehicle manufacturer's plant

Start-of-injection control

The start of injection has a critical effect on power output, fuel consumption, noise, and emissions. The desired value for start of injection depends on engine speed and injected fuel quantity, and it is stored in the ECU in special maps. Adaptation is possible as a function of coolant temperature and ambient pressure.

Tolerances in manufacture and in the pump mounting on the engine, together with changes in the solenoid valve during its lifetime, can lead to slight differences in the solenoid-valve switching times which in turn lead to different starts of injection. The response behaviour of the nozzle-and-holder assembly also changes over the course of time. Fuel density and temperature also have an effect upon start of injection. This must be compensated for by some form of control strategy in order to stay within the prescribed emissions limits. The following closed-loop controls are employed (Table 2):

Closed-loop control using the needle-motion sensor

The inductive needle-motion sensor is fitted in an injection nozzle (reference nozzle, usually cylinder 1). When the needle opens (and closes) the sensor transmits a pulse (Fig. 4). The needle-opening signal is used by the ECU as confirmation of the start of injection. This means that inside a closed control loop the start of injection can be precisely aligned to the desired value for the particular operating point.

The needle-motion sensor's untreated signal is amplified and interference-suppressed before being converted to precision square-wave pulses which can be used to mark the start of injection for a reference cylinder.

The ECU controls the actuator mechanism for the start of injection (for in-line pumps the solenoid actuator, and for distributor pumps the timing-device solenoid valve) so that the actual start of injection always corresponds to the desired/setpoint start of injection.

Table 2

2 Start-of-injection control			
Closed-loop control	Control using needle-motion sensor	Start-of-delivery control	BIP control
Injection system			
In-line injection pumps	●	–	–
Helix-controlled distributor pumps	●	–	–
Solenoid-valve-controlled distributor pumps	●	●	–
Common Rail	–	–	–
Unit Injector/Unit Pump	–	–	●

The start-of-injection control is not needed with the Common Rail System, since the high-voltage triggering used in the CRS permits highly reproducible starts of injection.

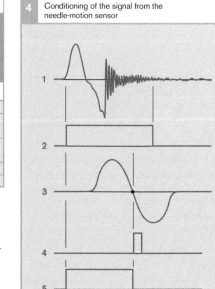

4   Conditioning of the signal from the needle-motion sensor

Fig. 4

1   Untreated signal from the needle-motion sensor (NBF),

2   Signal derived from the NBF signal,

3   Untreated signal from the inductive engine-speed sensor

4   Signal derived from untreated engine-speed signal,

5   Evaluated start-of-injection signal

The start-of-injection signal can only be evaluated when fuel is being injected and when the engine speed is stable. During starting and overrun (no fuel injection), the needle-motion sensor cannot provide a signal which is good enough for evaluation. This means that the start-of-injection control loop cannot be closed because there is no signal available confirming the start-of-injection.

In-line fuel-injection pumps
On in-line pumps, a special digital current controller improves the control's accuracy and dynamic response by aligning the current to the start-of-injection controller's setpoint value practically without any delay at all.

In order to ensure start-of-injection accuracy in open-loop-controlled operation too, the start-of-delivery solenoid in the control-sleeve actuator mechanism is calibrated to compensate for the effects of tolerances. The current controller compensates for the effects of the temperature-dependent solenoid-winding resistance. All these measures ensure that the setpoint value for current as derived from the start map leads to the correct stroke of the start-of-delivery solenoid and to the correct start of injection.

Start-of-delivery control using the incremental angle/time signal (IWZ)
On the solenoid-valve-controlled distributor pumps (VP30, VP44), the start of injection is also very accurate even without the help of a needle-motion sensor. This high level of accuracy was achieved by applying positioning control to the timing device inside the distributor pump. This form of closed-loop control serves to control the start of delivery and is referred to as start-of-delivery control. Start of delivery and start of injection have a certain relationship to each other and this is stored in the so-called *wave-propagation-time map* in the engine ECU.

The signal from the crankshaft-speed sensor and the signal from the incremental angle/time system (IWZ signal) inside the pump, are used as the input variables for the timing-device positioning control.

The IWZ signal is generated inside the pump by the rotational-speed or angle-of-rotation sensor (1) on the trigger wheel (2) attached to the driveshaft. The sensor shifts along with the timing device (4) which, when it changes position, also changes the position of the tooth gap (3) relative to the TDC pulse of the crankshaft-speed sensor. The angle between the tooth gap, or the synchronization pulse generated by the tooth gap, and the TDC pulse is continually registered by the pump ECU and compared with the stored reference value. The difference between the two angles represents the timing device's actual position, and this is continually compared with its setpoint/desired position. If the timing-device position deviates, the triggering signal for the timing-device solenoid valve is changed until actual and setpoint position coincide with each other.

Since all cylinders are taken into account, the advantage of this form of start-of-delivery control lies in the system's rapid response. It has a further advantage in that it also functions during overrun when no fuel

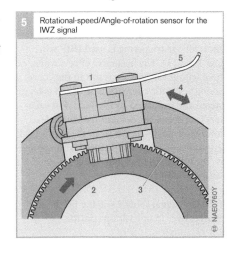

5 Rotational-speed/Angle-of-rotation sensor for the IWZ signal

Fig. 5
1 Rotational-speed/
   angle-of-rotation
   sensor inside the
   injection pump
2 Trigger wheel
3 Trigger-wheel tooth
   gap
4 Shift due to timing
   device
5 Electrical plug-in
   connection

injection takes place which means that the timing device can be preset for when the next injection event occurs.

In case even more severe demands are made on the accuracy of the start of injection, the start-of-delivery control can have an optional start-of-injection control with needle-motion sensor superimposed upon it.

BIP control

BIP control is used with the solenoid-valve-controlled Unit Injector System (UIS) and Unit Pump System (UPS). The start of delivery – or BIP (Begin of Injection Period) – is defined as the instant in time in which the solenoid closes. As from this point, pressure buildup starts in the pump high-pressure chamber. The nozzle opens as soon as the nozzle-opening pressure is exceeded, and injection can commence (start of injection). Fuel metering takes place between start of delivery and end of solenoid-valve triggering. This period is termed the delivery period.

Since there is a direct connection between the start of delivery and the start of injection, all that is needed for the precise control of the start of injection is information on the instant of the start of delivery.

So as to avoid having to apply additional sensor technology (for instance, a needle-motion sensor), electronic evaluation of the solenoid-valve current is used in detecting the start of delivery. Around the expected instant of closing of the solenoid valve, constant-voltage triggering is used (BIP window, Fig. 6, 1). The inductive effects when the solenoid valve closes result in the curve having a specific characteristic which is registered and evaluated by the ECU. For each injection event, the deviation of the solenoid-valve closing point from the theoretical setpoint is registered and stored, and applied for the following injection sequence as a compensation value.

If the BIP signal should fail, the ECU changes over to open-loop control.

Shutoff

The *auto-ignition* principle of operation means that in order to stop the diesel engine it is only necessary to cut off its supply of fuel.

With EDC (Electronic Diesel Control), the engine is switched off due to the ECU outputting the signal "Fuel quantity zero" (that is, the solenoid valves are no longer triggered, or the control rack is moved back to the zero-delivery setting).

There are also a number of redundant (supplementary) shutoff paths (for instance, the electrical shutoff valve (ELAB) on the port-and-helix controlled distributor pumps).

The UIS and UPS are intrinsically safe, and the worst thing that can happen is that one single unwanted injection takes place. Here, therefore, supplementary shutoff paths are not needed.

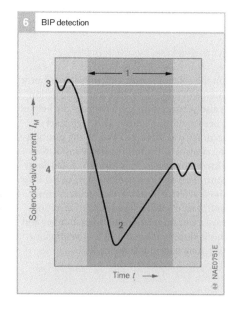

6   BIP detection

Solenoid-valve current I_M

Time t

NAE0751E

Fig. 6
1   BIP window
2   BIP signal
3   Level of pickup
     current
4   Holding-current level

Further special adaptations

In addition to those described here, EDC permits a wide range of other functions. For instance, these include:

Drive recorder
On commercial vehicles, the Drive Recorder is used to record the engine's operating conditions (for instance, how long was the vehicle driven, under what temperatures and loads, and at what engine speeds). This data is used in drawing up an overview of operational conditions from which, for instance, individual service intervals can be calculated.

Special application engineering for competition trucks
On race trucks, the 160 km/h maximum speed may be exceeded by no more than 2 km/h. On the other hand, this speed must be reached as soon as possible. This necessitates special adaptation of the ramp function for the vehicle-speed limiter.

Adaptations for off-highway vehicles
Such vehicles include diesel locomotives, rail cars, construction machinery, agricultural machinery, boats and ships. In such applications, the diesel engine(s) is/are far more often run in the full-load range than is the case with road vehicles (90% full-load operation compared with 30%). The power output of such engines must therefore be reduced in order to ensure an adequate service life.

The mileage figures which are often used as the basis for the service interval on road vehicles are not available for such equipment as agricultural or construction machinery, and in any case if they were available they would have no useful significance. Instead, the Drive Recorder data is used here.

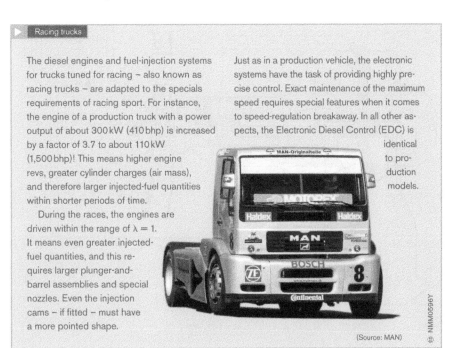

Racing trucks

The diesel engines and fuel-injection systems for trucks tuned for racing – also known as racing trucks – are adapted to the specials requirements of racing sport. For instance, the engine of a production truck with a power output of about 300 kW (410 bhp) is increased by a factor of 3.7 to about 110 kW (1,500 bhp)! This means higher engine revs, greater cylinder charges (air mass), and therefore larger injected-fuel quantities within shorter periods of time.

During the races, the engines are driven within the range of λ = 1. It means even greater injected-fuel quantities, and this requires larger plunger-and-barrel assemblies and special nozzles. Even the injection cams – if fitted – must have a more pointed shape.

Just as in a production vehicle, the electronic systems have the task of providing highly precise control. Exact maintenance of the maximum speed requires special features when it comes to speed-regulation breakaway. In all other aspects, the Electronic Diesel Control (EDC) is identical to production models.

(Source: MAN)

NMM0596Y

Lambda closed-loop control for passenger-car diesel engines

Application

The lawmakers are continually increasing the severity of legislation governing exhaust-gas emission limits for cars powered by diesel engines. Apart from the measures taken to optimize the engine's internal combustion, the open and closed-loop control of functions related to exhaust-gas emissions are continuing to gain in importance. Introduction of lambda closed-loop control offers major potential for reducing emission-value spread in diesel engines.

A broadband lambda oxygen sensor in the exhaust pipe (Fig. 1, 7) measures the residual oxygen content in the exhaust gas. This is an indicator of the A/F ratio (excess-air-factor lambda λ). The lambda oxygen-sensor signal is adapted while the engine is running. This ensures a high level of signal accuracy throughout the sensor's service life. The lambda oxygen-sensor signal is used as the basis for a number of lambda functions, which will be described in more detail in the following.

Lambda closed-loop control circuits are used to regenerate NO_X accumulator-type catalytic converters.

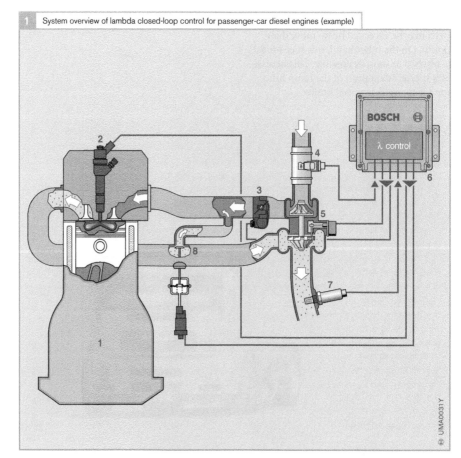

1 System overview of lambda closed-loop control for passenger-car diesel engines (example)

Fig. 1
1   Diesel engine
2   Diesel injection component (here, common-rail injector)
3   Control flap
4   Hot-film air-mass meter
5   Exhaust-gas turbocharger (here, VTG version)
6   Engine ECU for EDC
7   Broadband lambda oxygen sensor
8   EGR valve

Lambda closed-loop control is designed for all passenger-car fuel-injection systems with engine control units dating dating from the EDC16 generation.

Basic functions

Pressure compensation

The unprocessed lambda oxygen-sensor signal is dependent on the oxygen concentration in the exhaust gas and the exhaust-gas pressure at the sensor installation point. The influence of pressure on the sensor signal must, therefore, be compensated.

The *pressure-compensation* function incorporates two program maps, one for exhaust-gas pressure, and one for pressure dependence of the lambda oxygen-sensor output signal. These two maps are used to correct the sensor output signal with reference to the particular operating point.

Adaption

In overrun mode (trailing throttle), lambda oxygen-sensor adaption takes into account the deviation of the measured oxygen concentration from the fresh-air oxygen concentration (approx. 21%). As a result, the system "learns" a correction value which is used at every engine operating point to correct the measured oxygen concentration. This leads to a precise, drift-compensated lambda output signal for the service life of the lambda oxygen sensor.

Lambda-based EGR control

Compared with air-mass-based exhaust-gas recirculation, detecting oxygen concentration in the exhaust gas allows tighter emission tolerance bands for an automotive manufacturer's entire vehicle fleet. For future limits, an emission advantage of approx. 10...20% can be gained in this way for the exhaust-gas test.

---

2   Operating concept of average delivery adaption in "indirect control" mode

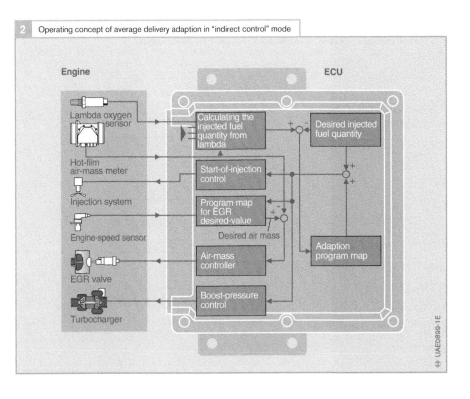

Average delivery adaption

Average delivery adaption supplies a precise injection quantity signal to form the setpoint for the exhaust-gas-related closed control loop. Correction of exhaust-gas recirculation plays a major role in emissions here. Average delivery adaption operates in the lower part-load range and determines the average deviation in the injected fuel quantity of all cylinders.

Fig. 2 (previous page) shows the basic structure of average delivery adaption and its influence on the exhaust-gas-related closed control loops.

The lambda oxygen-sensor signal and the air-mass signal are used to calculate the actually injected fuel mass, which is then compared to the desired injected fuel mass. Differences are stored in an adaption map in defined "learning points". This procedure ensures that, when the operating point requires an injected fuel quantity correction, it can be implemented without delay even during dynamic changes of state.

These correction quantities are stored in the EEPROM of the ECU and are available immediately the engine is started.

Basically speaking, there are two average-delivery adaption operating modes. They differ in the way they apply detected deviations in injected fuel quantity:

Operating mode: Indirect Control
In *Indirect Control* mode (Fig. 2), a precise injection quantity setpoint is used as the input variable in various exhaust-gas-related reference program maps. The injected fuel quantity is not corrected during the fuel-metering process.

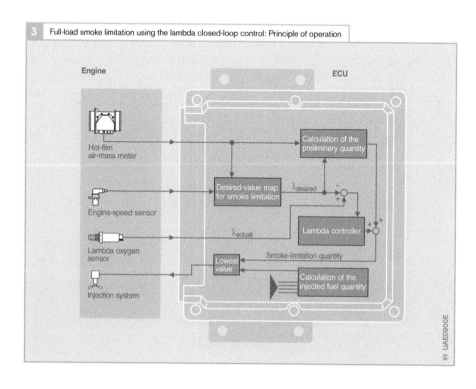

3   Full-load smoke limitation using the lambda closed-loop control: Principle of operation

Engine

ECU

Hot-film air-mass meter

Engine-speed sensor

Lambda oxygen sensor

Injection system

Calculation of the preliminary quantity

Desired-value map for smoke limitation $\lambda_{desired}$

λ_{actual}

Lambda controller

Smoke-limitation quantity

Lowest value

Calculation of the injected fuel quantity

UAE0900E

Operating mode: Direct Control

In *Direct Control* mode, the quantity deviation is used in the metering process to correct the injected fuel quantity so that the actual fuel quantity injected coincides more precisely with the reference injected fuel quantity. In this case, this is (more or less) a closed quantity control loop.

Full-load smoke limitation

Fig. 3 shows the block diagram of the control structure for full-load smoke limitation using a lambda oxygen sensor. The objective here is to determine the maximum fuel quantity which may be injected without exceeding a given smoke-emission value.

The signals from the air-mass meter and the engine-speed sensor are applied together with a smoke-limitation map to determine the desired air/fuel ratio value $\lambda_{desired}$. This, in turn, is applied together with the air mass to calculate the precontrol value for the maximum permissible injected fuel quantity.

This form of control is already in serial production, and has a lambda closed-loop control imposed on it. The lambda controller calculates a correction fuel quantity from the difference between the desired air/fuel ratio $\lambda_{desired}$ and the actual air/fuel ratio value λ_{actual}. The maximum full-load injected fuel quantity is the total of the pilot-control quantity and the correction quantity.

This control architecture permits a high level of dynamic response due to pilot control, and improved precision due to the superimposed lambda control loop.

Detection of undesirable combustion

The lambda oxygen sensor signal helps to detect the occurrence of undesirable combustion in overrun mode. It is detected if the lambda oxygen-sensor signal drops below a calculated threshold. In this case, the engine can be switched off by closing a control flap and the EGR valve. The detection of undesirable combustion represents an additional engine safeguard function.

Summary

A lambda-based exhaust-gas recirculation system can substantially reduce emission-value spread over a manufacturer's vehicle fleet due to production tolerances or aging drift. This is achieved by using average delivery adaption.

Average delivery adaption supplies a precise injection quantity signal to form the setpoint for the exhaust-gas-related closed control loop. The precision of these control loops is increased as a result. Correction of exhaust-gas recirculation plays the major role on emissions here.

In addition, the application of lambda closed-loop control permits the precise metering of the full-load smoke quantity and detection of undesirable combustion in overrun (trailing throttle) mode.

Furthermore, the lambda oxygen sensor's high-precision signal can be used in a lambda closed control loop to regenerate NO_X catalytic converters.

Application

The *closed-loop* and *open-loop* control applications are of vital importance for various on-board systems.

The term *(open-loop) control* is used in many cases, not only for the process of controlling, but also for the entire system in which control takes place (for this reason, the general term *"control unit"* is used, although it may perform a closed-loop control function). Accordingly, arithmetic processes run in control units to calculate both closed-loop and open-loop functions.

Closed-loop control

Closed-loop control is a process in which a parameter (controlled variable x) is detected continuously, compared to another parameter (reference variable w_1), and adapted to the reference variable in an adjustment process depending on the result of the comparison. The resulting action takes place in a closed circuit (closed control loop).

Closed-loop control has the function of adjusting the value of controlled variables to a value specified by a reference variable, despite any disturbance influences that may occur.

The *closed control loop* (Fig. 1a) is a closed-loop control circuit with a discrete action. Controlled variable x acts within a loop configuration in a form of negative feedback. Contrary to open-loop control, closed-loop control considers the impact of all disturbance values (z_1, z_2) occurring within the control loop. Examples of closed-loop systems in a vehicle:

- Lambda closed-loop control
- Idle-speed control
- ABS/TSC/ESP control
- Air conditioning (interior temperature)

Open-loop control

Open-loop control is the process within a system in which one or several parameters act as input variables affecting other parameters due to intrinsic laws governing the system. A feature of open-loop control is the open action sequence across an individual transfer element or the open control loop.

An *open control loop* (Fig. 1b) is an arrangement of elements that interact on each other in a loop structure. It may interact in any possible way with other systems as an entity within a higher-level system. The open control loop can only counter the impact of a disturbance value measured by the control unit (e.g. z_1); other disturbance values (e.g. z_2) may act unimpaired. Examples of open-loop systems in a vehicle:

- Electronic Transmission Control (ETC)
- Injector delivery compensation and pressure-wave correction for calculating injected fuel quantity

Fig. 1
a   Closed control loop
b   Open control loop
c   Block diagram of a digital closed-control loop

w Reference variable
x Controlled variable (closed loop)
x_A Controlled variable (open loop)
y Manipulated variable
z_1, z_2 Disturbance values

T Sampling time
*   Digital signal values
A   Analog
D   Digital

1 Closed-loop and open-loop control applications

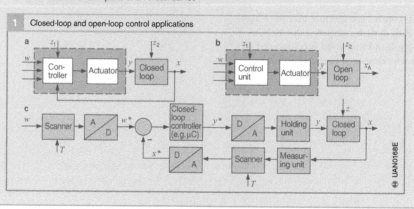

Torque-controlled EDC systems

The engine-management system is continually being integrated more closely into the overall vehicle system. Through the CAN bus, vehicle dynamics systems such as TCS, and comfort and convenience systems such as cruise control, have a direct influence on the Electronic Diesel Control (EDC). Apart from this, much of the information registered and/or calculated in or by the engine management system must be passed on to other ECUs through the CAN bus.

In order to be able to incorporate the EDC even more efficiently in a functional alliance with other ECUs, and implement other changes rapidly and effectively, it was necessary to make far-reaching changes to the newest-generation controls. These changes resulted in the torque-controlled EDC which was introduced with the EDC16. The main feature is the changeover of the module interfaces to the parameters, as commonly encountered in practice in the vehicle.

Engine parameters

Essentially, an IC engine's output can be defined using the three parameters: power P, engine speed n, and torque M.

For 2 diesel engines. Fig. 1 compares typical curves of torque and power as a function of engine speed. Basically speaking, the following equation applies:

$$P = 2 \cdot \pi \cdot n \cdot M$$

In other words, it suffices to use the torque as the reference (command) variable. Engine power then results from the above equation. Since power output cannot be measured directly, torque has turned out to be a suitable reference (command) variable for engine management.

Torque control

When accelerating, the driver uses the accelerator pedal (sensor) to directly demand a given torque from the engine. At the same time, but independent of the driver's requirements, via the interfaces other vehicle systems submit torque demands resulting from the power requirements of the particular component (e.g. air conditioner, alternator). Using these torque-requirement inputs, the engine management calculates the output torque to be generated by the engine and controls the fuel-injection and air-system actuators accordingly. This method has the following advantages:

- No single system (for instance, boost pressure, fuel injection, pre-glow) has a direct effect on engine management. This enables the engine management to also take into account higher-level optimization criteria (such as exhaust-gas emissions and fuel consumption) when processing external requirements, and thus control the engine in the most efficient manner,
- Many of the functions which do not directly concern the engine management can be designed to function identically for diesel and gasoline engines.
- Extensions to the system can be implemented quickly.

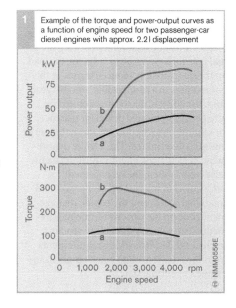

1 Example of the torque and power-output curves as a function of engine speed for two passenger-car diesel engines with approx. 2.2 l displacement

Power output (kW): 75, 50, 25, 0
b
a

Torque (N·m): 300, 200, 100, 0
b
a

Engine speed: 0, 1,000, 2,000, 3,000, 4,000 rpm

NMM0556E

Fig. 1
a Year of manufacture 1968
b Year of manufacture 1998

Engine-management sequence

Fig. 2 shows (schematically) the processing of the setpoint inputs in the engine ECU. In order to be able to fulfill their assignments efficiently, the engine management's control functions all require a wide range of sensor signals and information from other ECUs in the vehicle.

Propulsion torque

The driver's input (that is, the signal from the accelerator-pedal sensor) is interpreted by the engine management as the request for a propulsive torque. The inputs from the cruise control and the vehicle speed limiter are processed in exactly the same manner.

Following this selection of the desired propulsive torque, should the situation arise, the vehicle-dynamics system (TCS, ESP) increases the desired torque value when there is the danger of wheel lockup and decreases it when the wheels show a tendency to spin.

Further external torque demands

The drivetrain's torque adaptation must be taken into account (drivetrain transmission ratio). This is defined for the most part by the ratio of the particular gear, or by the torque-converter efficiency in the case of automatic transmissions. On vehicles with an automatic-gearbox, the transmission control stipulates the torque requirement during the actual gear shift. Apart from reducing the load on the transmission, reduced torque at this point results in a comfortable, smooth gear shift. In addition, the torque required by other engine-powered units (for instance, air-conditioner compressor, alternator, servo pump) is determined. This torque requirement is calculated either by the units themselves or by the engine management.

Calculation is based on unit power and rotational speed, and the engine management adds up the various torque requirements. The vehicle's drivability remains unchanged despite varying requirements from the auxiliary units and changes in the engine's operating state.

Internal torque demands

At this stage, the idle-speed control and the active surge damper intervene.

For instance, if demanded by the situation, in order to prevent mechanical damage, or excessive smoke due to the injection of too much fuel, the torque limitation reduces the internal torque requirement. In contrast to the previous engine-management systems, limitations are no longer only applied to the injected fuel-quantity, but instead, depending upon the required effects, also to the particular physical quantity involved.

The engine's losses are also taken into account (e.g. friction, drive for the high-pressure pump). The torque represents the engine's measurable effects to the outside. The engine management, though, can only generate these effects in conjunction with the correct fuel injection together with the correct injection point, and the necessary marginal conditions as apply to the air-intake system (e.g. boost pressure and EGR rate). The required injected fuel quantity is determined using the current combustion efficiency. The calculated fuel quantity is limited by a protective function (for instance, protection against overheating), and if necessary can be varied by Smooth-Running Control (SRC). During engine start, the injected fuel quantity is not determined by external inputs such as those from the driver, but rather by the separate "start-quantity control" function.

Actuator triggering

Finally, the desired values for the injected fuel quantity are used to generate the triggering data for the injection pump and/or the injectors, and for defining the optimum operating point for the intake-air system.

2 Engine-management sequence for torque-controlled diesel injection

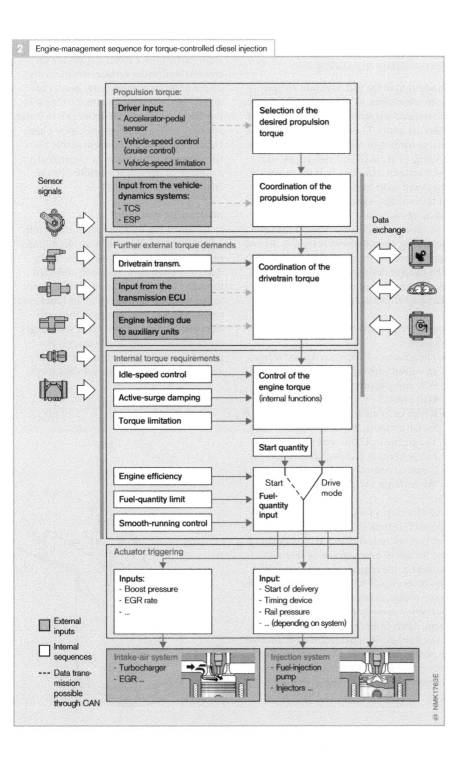

Sensor signals

Data exchange

Propulsion torque:

Driver input:
- Accelerator-pedal sensor
- Vehicle-speed control (cruise control)
- Vehicle-speed limitation

Selection of the desired propulsion torque

Input from the vehicle-dynamics systems:
- TCS
- ESP

Coordination of the propulsion torque

Further external torque demands

Drivetrain transm.

Input from the transmission ECU

Engine loading due to auxiliary units

Coordination of the drivetrain torque

Internal torque requirements

Idle-speed control

Active-surge damping

Torque limitation

Control of the engine torque (internal functions)

Start quantity

Engine efficiency

Fuel-quantity limit

Smooth-running control

Start Drive mode

Fuel-quantity input

Actuator triggering

Inputs:
- Boost pressure
- EGR rate
- ...

Input:
- Start of delivery
- Timing device
- Rail pressure
- ... (depending on system)

☐ External inputs

☐ Internal sequences

--- Data transmission possible through CAN

Intake-air system
- Turbocharger
- EGR ...

Injection system
- Fuel-injection pump
- Injectors ...

NMK1763E

Control and triggering of the remaining actuators

In addition to the fuel-injection components themselves, EDC is responsible for the control and triggering of a large number of other actuators. These are used for cylinder-charge control, or for the control of engine cooling, or are used in diesel-engine start-assist systems. Here too, as is the case with the closed-loop control of injection, the inputs from other systems (such as TCS) are taken into account.

A variety of different actuators are used, depending upon the vehicle type, its area of application and the type of fuel injection. This chapter deals with a number of examples, and further actuators are covered in the Chapter "Actuators".

A variety of different methods are used for triggering:
- The actuators are triggered directly from an output (driver) stage in the engine ECU using appropriate signals (e.g. the EGR valve).
- If high currents are involved (for instance for fan control), the ECU triggers a relay.
- The engine ECU transfers signals to an independent ECU, which is then used to trigger or control the remaining actuators (for instance, for glow control).

The advantage of incorporating all engine-control functions in the EDC ECU lies in the fact that not only the injected fuel quantity and instant of injection can be taken into account in the engine control concept, but also other engine functions such as EGR and boost-pressure control. This leads to a considerable improvement in engine management. Apart from this, the engine ECU has a vast amount of information at its disposal as needed for other functions (for instance, engine and intake-air temperature as used for glow control on the diesel engine).

Auxiliary coolant heating

High-performance diesel engines are very efficient, and under certain circumstances do not generate enough waste heat to adequately heat the vehicle's interior. One solution for overcoming this problem is to install auxiliary coolant heating using glow plugs. Depending upon the power available from the alternator, this system is triggered in a number of steps. It is controlled by the engine ECU as used for EDC.

Intake-duct switch-off

In the lower engine-rpm ranges and at idle, a flap (Fig. 1, 6) operated by an electropneumatic transducer closes one of the intake ducts (5). Fresh air in then only inducted through the turbulence duct (2). This leads to improved air turbulence in the lower rpm ranges which in turn results in more efficient combustion. In the higher rpm ranges, the engine's volumetric efficiency is improved thanks to the open intake duct (5) and the power output increases as a result.

Fig. 1
1   Intake valve
2   Turbulence duct
3   Cylinder
4   Piston
5   Intake duct
6   Flap

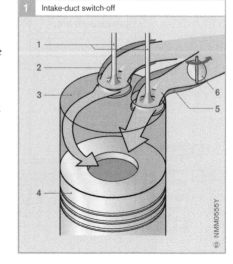

1   Intake-duct switch-off

Boost-pressure control

Boost-pressure control applied to the exhaust-gas turbocharger improves the engine's torque curve in full-load operation, and its exhaust and refill cycle in the part-load range. The optimum (desired) boost pressure is a function of engine speed, injected fuel quantity, coolant and fuel temperature, and the surrounding air pressure. This optimum (desired) boost pressure is compared with the actual value registered by the boost-pressure sensor and, in the case of deviation, the ECU either operates the bypass valve's electropneumatic transducer or the guide blades of the VTG (Variable Turbine Geometry) exhaust-gas turbocharger (refer also to the Chapter "Actuators").

Fan triggering

When a given engine temperature is exceeded, the engine ECU triggers the engine cooling fan, which continues to rotate for a brief period after the engine is switched off. This run-on period is a function of the coolant temperature and the load imposed on the engine during the preceding driving cycle.

Exhaust-gas recirculation (EGR)

In order to decrease the NO_X emissions, exhaust gas is directed into the engine's intake duct through a channel, the cross section of which can be varied by an EGR valve. The EGR valve is triggered by an electropneumatic transducer or by an electric actuator.

Due to the high temperature of the exhaust gas and its high proportion of contamination, it is difficult to precisely measure the exhaust-gas flow which is recirculated back into the engine. Control, therefore, takes place indirectly through an air-mass meter located in the flow of fresh intake air. The meter's output signal is then compared in the ECU with the engine's theoretical air requirement which has been calculated from a variety of data (e.g. engine rpm). The lower the measured mass of the incoming fresh air compared to the theoretical air requirement, the higher is the proportion of recirculated exhaust gas.

Substitute functions

If individual input signals should fail, the ECU is without the important information it needs for calculations. In such cases, substitute functions are used. Two examples are given below:

Example 1: The fuel temperature is needed for calculation of the injected fuel quantity. If the fuel-temperature sensor fails, the ECU uses a substitute value for its calculations. This must be selected so that excessive soot formation is avoided, although this can lead to a reduction of engine power in certain operating ranges.

Example 2: Should the camshaft sensor fail, the ECU applies the crankshaft-sensor signal as a subsitute. Depending on the vehicle manufacturer, there are a variety of different concepts for using the crankshaft signal to determine when cylinder 1 is in the compression cycle. The use of substitute functions leads to engine restart taking slightly longer.

Substitute functions differ according to vehicle manufacturer, so that many vehicle-specific functions are possible.

The diagnosis function stores data on all malfunctions that occur. This data can then be accessed in the workshop (refer also to the Chapter "Electronic Diagnosis (OBD)").

Data exchange with other systems

Fuel-consumption signal

The engine ECU (Fig. 1, 3) determines fuel consumption and sends this signal via CAN to the instrument cluster or a separate on-board computer (6), where the driver is informed of current fuel consumption and/or the range that can be covered with the remaining fuel in the tank. Older systems used Pulse-Width Modulation (PWM) for the fuel-consumption signal.

Starter control

The starter motor (8) can be triggered from the engine ECU. This ensures that the driver cannot operate the starter motor with the engine already running. The starter motor only turns long enough to allow the engine to reach a self-sustaining speed reliably. This function leads to a lighter, and thus lower-priced, starter motor.

Glow control unit

The glow control unit (GZS, 5) receives information from the engine ECU to control glow start and duration. It then triggers the glow plugs accordingly and monitors the glow process, and reports back to the engine ECU on any faults (diagnostic function). The pre-glow indicator lamp is usually triggered from the engine ECU.

Electronic immobilizer

To prevent unauthorized starting and drive-off, the engine cannot be started before a special immobilizer (7) ECU removes the block from the engine ECU.

The driver can signal the immobilizer ECU that he/she is authorized to use the vehicle, either by remote control or by means of the glow-plug and starter switch ("Ignition" key). The immobilizer ECU then removes the block on the engine ECU to allow engine start and normal operation.

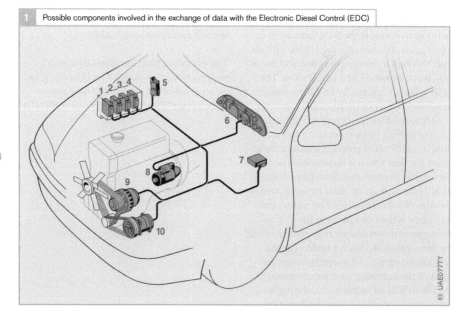

1 Possible components involved in the exchange of data with the Electronic Diesel Control (EDC)

Fig. 1

1 ESP ECU (with ABS and TCS)
2 ECU for transmission-shift control
3 Engine ECU (EDC)
4 A/C ECU
5 Glow control unit
6 Instrument cluster with onboard computer
7 Immobilizer ECU
8 Starter motor
9 Alternator
10 A/C compressor

External torque intervention

In the case of external torque intervention, the injected fuel quantity is influenced by another (external) ECU (for instance, for transmission-shift control, or TCS). This informs the engine ECU whether the engine torque is to be changed, and if so, by how much (this defines the injected fuel quantity).

Alternator control

By means of a standard serial interface, the EDC can control and monitor the alternator (9) remotely. The regulator voltage can be controlled, just the same as the complete alternator assembly can be switched off. In case of low battery power, for instance, the alternator's charging curve can be improved by increasing the idle speed. It is also possible to perform simple alternator diagnosis through this interface.

Air conditioner

In order to maintain comfortable temperatures inside the vehicle when the ambient temperature is high, the air conditioner (A/C) cools down cabin air with the help of an A/C compressor (10). Depending on the engine and operating conditions, the A/C compressor may draw as much as 30% of the engine's output power.

Immediately the driver hits the accelerator pedal (in other words he/she wishes maximum torque), the compressor can be switched off briefly by the engine ECU to concentrate all of the engine's power to the wheels. Since the compressor is only switched off very briefly, this has no noticeable effect on interior temperature.

Serial data transmission (CAN)

Modern-day vehicles are equipped with a constantly increasing number of electronic systems. Along with their need for extensive exchange of data and information in order to operate efficiently, the data volumes and speeds are also increasing at a rapid rate.

Although CAN (Controller Area Network) is a linear bus system (Fig. 1) specifically designed for automotive applications, it has already been introduced in other sectors (for instance, in building automation).

Data is relayed in serial form, that is, one after another on a common bus line. All CAN stations have access to this bus, and via a CAN interface in the ECUs, they can receive and transmit data over the CAN bus line. Since a considerable amount of data can be exchanged and repeatedly accessed on a single bus line, this network results in far fewer lines required. On conventional systems, data exchange takes place point to point over individually assigned data lines.

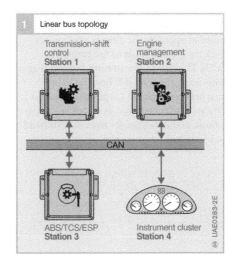

1   Linear bus topology

Transmission-shift control
Station 1

Engine management
Station 2

CAN

ABS/TCS/ESP
Station 3

Instrument cluster
Station 4

UAE0283-2E

Applications in the vehicle

For CAN in the vehicle there are four areas of application each of which has different requirements. These are as follows:

Multiplex applications

Multiplex is suitable for use with applications controlling the open and closed-loop control of components in the sectors of body electronics, and comfort and convenience. These include climate control, central locking, and seat adjustment. Transfer rates are typically between 10 kbaud and 125 kbaud (1 kbaud = 1 kbit/s) (low-speed CAN).

Mobile communications applications

In the area of mobile communications, CAN networks such components as navigation system, telephone, and audio installations with the vehicle's central display and operating units. Networking here is aimed at standardizing operational sequences as far as possible, and at concentrating status information at one point so that driver distraction is reduced to a minimum. With this application, large quantities of data are transmitted and data transfer rates are in the 125 kbaud range. It is impossible to directly transmit audio or video data here.

Diagnosis applications

The diagnosis applications using CAN are aimed at applying the already existing network for the diagnosis of the connected ECUs. The presently common form of diagnosis using the special K line (ISO 9141) then becomes invalid. Large quantities of data are also transferred in diagnostic applications, and data transfer rates of 250 kbaud and 500 kbaud are planned.

Real-time applications

Real-time applications serve for the open- and closed-loop control of the vehicle's movements. Here, such electronic systems as engine management, transmission-shift control and Electronic Stability Program (ESP) are networked with each other via the CAN bus. Commonly, data transfer rates of between 125 kbaud and 1 Mbaud (high-speed CAN) are needed to guarantee the required real-time response.

Bus configuration

Configuration is understood to be the layout and interaction between the components in a given system. The CAN bus has a linear bus topology, which in comparison with other logical structures (ring bus and/or star bus) features a lower failure probability. If one of the stations fails, the bus still remains fully accessible to all the other stations. The stations connected to the bus can be either ECUs, display devices, sensors, or actuators. They operate using the Multi-Master principle, whereby the stations concerned all have equal priority regarding their access to the bus. It is not necessary to have a higher-order administration.

Content-based addressing

The CAN bus system does not address each station individually according to its features, but rather according to its message contents. It allocates each "message" a fixed "identifier" (message name) which identifies the contents of the message in question (e.g., engine speed). This identifier has a length of 11 bits (standard format) or 29 bits (extended format).

With content-based addressing each station must itself decide whether it is interested in the message or not ("message filtering" Fig. 2). This function can be performed by a special CAN module (Full-CAN), so that less load is placed on the ECU's central microcontroller. Basic CAN modules "read" all messages. Using content-based addressing, instead of allocating station addresses, makes the complete system highly flexible so that equipment variants are easier to install and operate. If one of the ECUs requires new information which is already on the bus, all it needs to do is call it up from the bus. Similarly, provided they are receivers, new stations can be connected (implemented) without it being necessary to modify the already existing stations.

Bus arbitration

The identifier not only indicates the data content, but also defines the message's priority rating. An identifier corresponding to a low binary number has high priority and vice versa. Message priorities are a function for instance of the speed at which their contents change, or their significance with respect to safety. There are never two (or more) messages of identical priority in the bus.

Each station can begin message transmission as soon as the bus is unoccupied. Conflict regarding bus access is avoided by applying bit-by-bit identifier arbitration (Fig. 3), whereby the message with the highest priority is granted first access without delay and without loss of data bits (nondestructive protocol).

The CAN protocol is based on the logical states "dominant" (logical 0) and "recessive" (logical 1). The "Wired And" arbitration principle permits the dominant bits transmitted by a given station to overwrite the recessive bits of the other stations. The station with the lowest identifier (that is, with the highest priority) is granted first access to the bus. The transmitters with low-priority messages automatically become receivers, and repeat their transmission attempt as soon as the bus is vacant again.

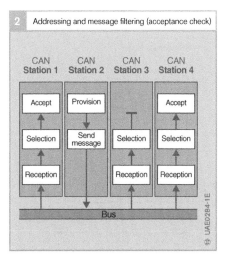

2   Addressing and message filtering (acceptance check)

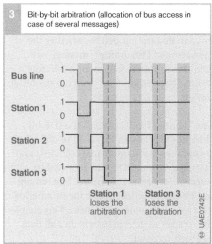

3   Bit-by-bit arbitration (allocation of bus access in case of several messages)

Fig. 2
Station 2 transmits, Station 1 and 4 accept the data.

Fig. 3
Station 2 gains first access (Signal on the bus = signal from Station 2)

0   Dominant level
1   Recessive level

In order that all messages have a chance of entering the bus, the bus speed must be appropriate to the number of stations participating in the bus. A cycle time is defined for those signals which fluctuate permanently (e.g. engine speed).

Message format
CAN permits two different formats which only differ with respect to the length of their identifiers. The standard-format identifier is 11 bits long, and the extended-format identifier 29 bits. Both formats are compatible with each other and can be used together in a network. The data frame comprises seven consecutive fields (Fig. 4) and is a maximum of 130 bits long (standard format) or 150 bits (extended format).

The bus is recessive at idle. With its dominant bit, the *"Start of frame"* indicates the beginning of a message and synchronizes all stations.

The *"Arbitration Field"* consists of the message's identifier (as described above) and an additional control bit. While this field is being transmitted, the transmitter accompanies the transmission of each bit with a check to ensure that it is still authorized to transmit or whether another station with a higher-priority message has accessed the Bus. The control bit following the identifier is designated the RTR-bit (Remote Transmission Request). It defines whether the message is a "Data frame" (message with data) for a receiver station, or a "Remote frame" (request for data) from a transmitter station.

The *"Control Field"* contains the IDE bit (Identifier Extension **Bit**) used to differentiate between standard format (IDE = 0) and extended format (IDE = 1), followed by a bit reserved for future extensions. The remaining 4 bits in this field define the number of data bytes in the next data field. This enables the receiver to determine whether all data has been received.

The *"Data Field"* contains the actual message information comprised of between 0 and 8 bytes. A message with data length = 0 is used to synchronize distributed processes. A number of signals can be transmitted in a single message (e.g., engine temperature and engine speed).

The *"CRC Field"* (Cyclic Redundancy Check) contains the frame check word for detecting possible transmission interference.

The *"ACK Field"* contains the **acknowledgement** signals used by the receiver stations to confirm receipt of the message in non-corrupted form. This field comprises the ACK slot and the recessive ACK delimiter. The ACK slot is also transmitted recessively and overwritten "dominantly" by the receivers upon the message being correctly received. Here, it is irrelevant whether the message is of significance or not for the particular receiver in the sense of the message filtering or acceptance check. Only correct reception is confirmed.

The *"End of Frame"* marks the end of the message and comprises 7 recessive bits.

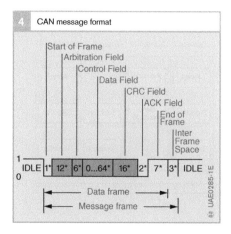

4   CAN message format

Start of Frame
Arbitration Field
Control Field
Data Field
CRC Field
ACK Field
End of Frame
Inter Frame Space

IDLE | 1* | 12* | 6* | 0...64* | 16* | 2* | 7* | 3* | IDLE

Data frame

Message frame

UAE0285-1E

Fig. 4
0   Dominant level
1   Recessive level
*   Number of bits

The *"Inter-Frame Space"* comprises three bits which serve to separate successive messages. This means that the bus remains in the recessive IDLE mode until a station starts a bus access.

As a rule, a sending station initiates data transmission by sending a "data frame". It is also possible for a receiving station to call in data from a sending station by transmitting a "remote frame".

Detecting errors

A number of control mechanisms for detecting errors are integrated in the CAN protocol.

In the *"CRC Field"*, the receiving station compares the received CRC sequence with the sequence calculated from the message.

With the *"Frame Check"*, frame errors are recognized by checking the frame structure.
     The CAN protocol contains a number of fixed-format bit fields which are checked by all stations.

The *"ACK Check"* is the receiving stations' confirmation that a message frame has been received. Its absence signifies for instance that a transmission error has been detected.

"Monitoring" indicates that the sender observes (monitors) the bus level and compares the differences between the bit that has been sent and the bit that has been checked.

Compliance with *"Bitstuffing"* is checked by means of the "Code check". The stuffing convention stipulates that in every *"Data Frame"* or *"Remote Frame"* a maximum of five successive equal-priority bits may be sent between the *"Start of Frame"* and the end of the *"CRC Field"*. As soon as five identical bits have been transmitted in succession, the sender inserts an opposite-priority bit. The receiving station erases these opposite-priority bits after receiving the message. Line errors can be detected using the "bitstuffing" principle.

If one of the stations detects an error, it interrupts the actual transmission by sending an "Error frame" comprising six successive dominant bits. Its effect is based on the intended violation of the stuffing rule, and the object is to prevent other stations accepting the faulty message.

Defective stations could have a derogatory effect upon the bus system by sending an "error frame" and interrupting faultless messages. To prevent this, CAN is provided with a function which differentiates between sporadic errors and those which are permanent, and which is capable of identifying the faulty station. This takes place using statistical evaluation of the error situations.

Standardization

The International Organization for Standardization (ISO) and SAE (Society of Automotive Engineers) have issued CAN standards for data exchange in automotive applications:
- For low-speed applications up to 125 kbit/s: ISO 11 519-2, and
- For high-speed applications above > 125 kBit/s: ISO 11 898 and SAE J 22 584 (passenger cars) and SAE J 1939 (trucks and buses).
- Furthermore, an ISO Standard on CAN Diagnosis (ISO 15 765 – Draft) is being prepared.

Application-related adaptation[1] of car engines

Application-related adaptation means modification of an engine to suit a particular type of vehicle intended for a specific type of use. Adaptation of the fuel-injection system – and specifically of electronic diesel control EDC – is a major part of that process.

All new diesel engines for cars are now direct-injection (DI) engines. And they all have to comply with the Euro III emission control standards that have been in force since 2000, or other comparable standards. These emission standards – combined with the higher expectations in the area of vehicle user-friendliness – can only be met by the use of sophisticated electronic control systems. Such systems have the capability – and reflect the necessity – of controlling thousands of parameters (approx. 6,000 in the case of the present EDC generation). Those parameters are subdivided into:

- Individual parameter values (e.g. temperature thresholds at which specific functions are activated) and
- Ranges of parameter values in the form of two-dimensional or multi-dimensional data maps (e.g. injection point t_E as a function of engine speed n, injected-fuel quantity m_e and start of delivery FB)

The optimization potential of EDC systems has become so great that it is now limited only by the constraints of time available and the cost of the personnel and the work involved in adapting and testing the various functions and their interaction.

Adaptation phases

Application-related adaptation of car engines is subdivided into the three stages described below.

Hardware adaptation

In the context of application-related adaptation of car engines, items such as the combustion chamber, the injection pump and the injectors are referred to as hardware. That hardware is primarily adapted in such a way that the performance and emission figures demanded are obtained. Hardware adaptation is performed initially on an engine test bench under static conditions. If dynamic tests are possible on the test bench, they are used to further optimize the engine and the fuel-injection system.

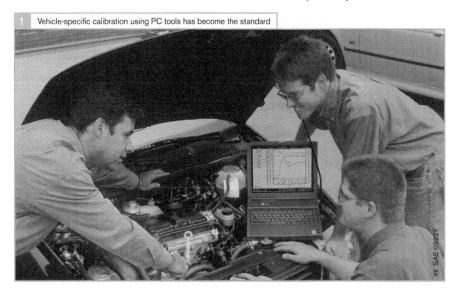

1   Vehicle-specific calibration using PC tools has become the standard

SAE 0922Y

Software adaptation

Once the hardware adaptation is complete, the control-unit software is accordingly configured and adapted for optimum mixture preparation and combustion control. For example, this includes calculating and programming the engine data maps for start of injection, exhaust-gas recirculation and charge-air pressure. As with hardware adaptation, this work is carried out on the test bench.

Vehicle-related adaptation

When the basis for the initial vehicle trials has been established, adaptation of all parameters that affect engine response and dynamic characteristics takes place. This third stage involves the essential adaptation to the particular vehicle concerned. The work is for the greater part performed with the engine in situ (Fig. 1).

Interaction between the three phases

As there are reciprocal effects between the adaptation phases, recursions (repeated procedures) are required. As soon as possible, it is also necessary to run all three phases simultaneously with the engine in the vehicle and on the test bench.

For example, at low engine loads a very high exhaust-gas recirculation rate is aimed at in order to reduce the NO_X emissions. Under dynamic conditions, this can lead to poor "accelerator response" on the part of the engine. In order to obtain good acceleration characteristics, the static emissions settings programmed in the software adaptation phase must be re-adjusted. In turn, this may result in negative effects on emissions under certain engine operating conditions which have to be compensated for under other conditions.

In the example outlined, there is a fundamental conflict between the various objectives: on the one hand, strict requirements have to be met (e.g. statutory limits for exhaust emission levels), while on the other hand there are "optional" demands that are more attributable to the desire for comfort and performance (engine response, noise, etc.). The latter can result in opposing conclusions. A compromise between the different objectives offers the vehicle manufacturer the opportunity to imbue the vehicle with some of the features that make up its characteristic brand identity.

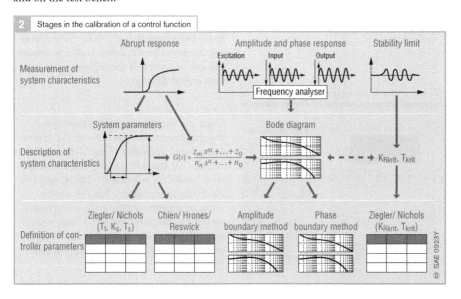

2   Stages in the calibration of a control function

Adaptation to differing ambient conditions

The various controllers and adjustment parameters must be configured for a wide variety of different ambient conditions. To control idle speed, for example, there are several parameter sets for each individual gear which are further differentiated according to whether

- The vehicle is stationary or moving
- The engine is warm or cold
- The clutch is engaged or disengaged

That means that for this function alone, there are as many as 50 parameter sets.

The EDC also provides adaptation functions for extreme ambient conditions. These generally have to be verified by specifically targeted special trials involving

- Cold-weather testing in temperatures down to –25°C (e.g. winter trials in Sweden)
- Hot-weather testing in temperatures over 40°C (e.g. summer trials in Arizona)
- High-altitude/low atmospheric pressure testing (e.g. in the Alps) and
- Combined hot-weather and altitude or cold-weather and altitude testing, e.g. towing a heavy trailer over mountain passes (e.g. in Spain's Sierra Nevada or in the Alps)

For cold starting, very specific adjustments have to be made to the injected fuel quantity and the start of delivery based on engine coolant temperature. In addition, the glow plugs have to be switched on. At high altitudes with a cold engine, the effectively available pull-away torque is very low. For some applications, EDC suspends turbocharger operation for that short period because it would otherwise "use up" a large proportion of the engine's torque output. Particularly in the case of vehicles with automatic transmission, this would prevent the vehicle from pulling away at all, as the torque available at the driving wheels would be insufficient.

Altitude compensation for turbocharged engines demands limitation of the required turbocharger pressure in response to atmospheric pressure, as otherwise the turbocharger would be destroyed by over-revving.

Other adjustments

Safety functions

As well as the functions that determine emission levels, power output and user-friendliness, there are also numerous safety functions that require adaptation (e.g. response to failure of a sensor or actuator).

Such safety functions are primarily intended to restore the vehicle to a safe operating condition for the driver and/or to ensure the safe operation of the engine (e.g. to prevent engine damage).

Communication

There are also numerous functions which require communication between the engine control unit and other control units on the vehicle (e.g. traction control, ESP, transmission control for automatic transmission and electronic immobilizer). For this reason, a special communication code is employed (input and output variables). Where necessary, additional measured data has to be calculated and encoded in the appropriate form.

3 Screen of an engine test-bench monitor (example)

Examples of adaptation

Since the arrival of the EDC system in 1986, the possibilities for optimization, especially with regard to the convenience features, have considerably expanded. A wide variety of software functions (e.g. control functions) are used, all of which have to be specifically adapted to each individual vehicle. Some examples are outlined below.

Idle-speed control

This function controls the speed at which the engine runs when the accelerator pedal is not depressed. Idle-speed control must operate with absolute reliability under all possible engine operating conditions. Therefore, extensive adaptation work is required. Adjustment of the coasting response in all gears, for example, is highly involved, especially with regard to the interplay with the twin-mass flywheels generally used. This type of flywheel produces highly complex rotational vibration effects throughout the drivetrain.

The first stage of the process is an analytical definition (i.e. recording of the controlled system response, description of the controlled system by algorithms and definition of the control parameters).

This is followed by a comprehensive road test. A circular track (test track) provides the possibility for virtually unlimited flat-road driving. Particularly with active surge damping, conflict between objectives can arise as this function may prevent rapid compensation in response to abrupt changes in engine speed or load.

Apart from the drivetrain, the engine mountings also play an important part. In order to diminish the various conflicts in objectives, therefore, some applications employ variable-characteristic engine mountings which are controlled by the EDC. These can be set to a softer setting when the engine is idling and to a harder response when the engine is under load.

Smooth-running control

The engine smooth-running function ensures that the injection volumes are the same for all cylinders and in so doing improves engine smoothness and emission levels. Under certain circumstances, a malfunction can occur at very high or very low ambient temperatures if the vibration damping characteristics of the belt drive systems for auxiliary units (e.g. alternator, power-steering pump, air-conditioning compressor) significantly alter. Depending on the frequencies generated as a result of periodic speed fluctuations, the engine smooth-running function may attempt to even them out by alteration of the injected-fuel quantity volume for individual cylinders. Under unfavourable conditions, this may then result in higher exhaust-emission levels or make the engine run even more unevenly. For that reason, this function must be thoroughly tested under all operating conditions.

Pressure-charging controller

Almost all existing DI car diesel engines are fitted with turbochargers. On most of those engines, the charge-air pressure is controlled by the EDC system. The aim is to obtain optimum response characteristics (rapid generation of charge-air pressure) while ensuring reliable protection of the engine against excessive charge-air pressure and consequent excessively high cylinder pressure.

Exhaust-Gas Recirculation EGR

Exhaust-gas recirculation EGR is now a standard feature of DI car diesel engines. As previously indicated, together with the control of turbocharger pressure it is a determining factor in the amount of air that enters the engine. In order to ensure smokeless and low-NO_X combustion, the air-fuel mixture must conform to precisely defined parameters for all engine operating conditions. Those parameters are initially optimized under static conditions on the engine test bench. The control function then has the task of maintaining those parameters under dynamic operating conditions without adversely affecting the response characteristics of the engine.

[1]) Some parts of the
adaptation process
are also referred to
as calibration.

Application-related adaptation[1]) of commercial-vehicle engines

Particularly because of its economy and durability, the diesel has established itself as the engine of choice for commercial vehicles. Today all new engines are direct-injection (DI) designs.

Optimization objectives

For commercial-vehicle engines, the following attributes are optimized.

Torque

The aim is to obtain the maximum possible torque under all operating conditions in order to be able to move heavy loads in even the most difficult situations (e.g. when negotiating steep gradients or using PTO drives). When pursuing that objective, the engine's limits (e.g. maximum permissible cylinder pressure and exhaust temperature) as well as the smoke emission limit have to be taken into account.

Fuel consumption

For commercial vehicles, economy is a decisive factor. For that reason fuel consumption occupies a position of greater importance for commercial vehicles than is the case with cars. Minimizing fuel consumption (or CO_2 emissions) is therefore of prime significance in engine adaptation.

Durability

Modern commercial-vehicle engines are expected to be able to complete over a million kilometers of service.

Pollutant emissions

Since October 2000, new commercial vehicles registered in the European Union have been required to conform to the Euro III emission-control standard. Engine adaptation must ensure that the limits for NO_X, particulate, HC and CO emission and exhaust opacity are reliably complied with.

Comfort/convenience

The demands relating to such aspects as engine response, quietness, smoothness and starting characteristics must also be taken into account.

Adaptation phases

The aim of adaptation is to ensure that the objectives outlined above are achieved as fully as possible, i.e. that the best possible compromise is reached between competing demands. This involves adaptation of engine and fuel-injection hardware components as

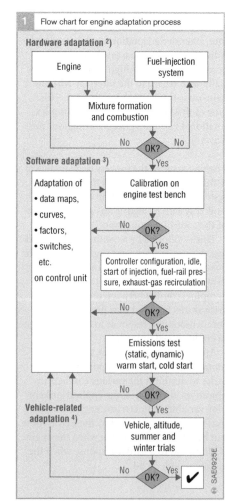

1   Flow chart for engine adaptation process

Fig. 1

[2]) Criteria:
- Full-load response
- Emissions
- Fuel consumption

[3]) Additional criterion:
- Dynamic adaptation

[4]) Other criteria:
- Starting characteristics
- Smoothness, etc.

well as software functions performed by the engine-management module.

As with car engines, the phases of hardware, software and vehicle-related adaptation can be distinguished (Fig. 1).

Hardware adaptation

Hardware adaptation involves making modifications to all significant "components" of the engine and fuel-injection system. Significant engine-hardware components include the combustion chamber, the turbocharger, the air-intake system (e.g. swirl-imparting

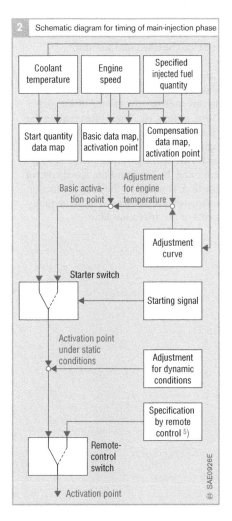

2 Schematic diagram for timing of main-injection phase

Coolant temperature | Engine speed | Specified injected fuel quantity

Start quantity data map | Basic data map, activation point | Compensation data map, activation point

Basic activation point | Adjustment for engine temperature

Adjustment curve

Starter switch

Starting signal

Activation point under static conditions

Adjustment for dynamic conditions

Specification by remote control [5]

Remote-control switch

Activation point

SAE0926E

components) and, if necessary, the exhaust-recirculation system. Significant components of the fuel-injection system are the injection pump, the high-pressure fuel lines if applicable, and the injectors. Hardware adaptation is carried out on the engine test bench.

Software adaptation

Once the hardware adaptation is complete, the control-unit software is configured accordingly. Stored in the software are the relationships between a vast number of engine and fuel-injection parameters (for examples, see Fig. 2). This work too is carried out on the engine test bench. An application control unit, which – as with the adaptation of car engines – is linked to a PC with operator software, provides access to the software to be adapted.

The following tasks are performed in the course of software adaptation:
- Calibration of the basic engine-data maps under static operating conditions
- Control function configuration
- Calibration of compensation data maps
- Optimisation of engine-data maps under dynamic conditions

First of all, adjustments to the system-specific parameters – such as start of injection, injection pressure, exhaust-gas recirculation, charge-air pressure and, if applicable, pre- and post-injection – are carried out under static operating conditions on the engine test bench. The test results are assessed with reference to the target criteria (emission levels, fuel consumption, etc.). Based on those results, the appropriate parameter values, data curves and data maps are then calculated and programmed (Fig. 3 overleaf). Because of the ever increasing number of such parameters, automation of parameter configuration is a continuing aim.

Following adaptation of the basic data maps, the effect of such variables as ambient temperature, atmospheric pressure, engine-coolant temperature and fuel temperature

Fig. 2
[5] Specification of set values in order to bypass data maps during calibration

on the major parameters is factored into so-called compensation data maps. In addition, existing control functions are adapted (e.g. fuel-rail pressure control for common-rail injection systems, charge-air pressure control). The data established under static operating conditions is finally optimized under dynamic conditions.

Vehicle-related adaptation

The process of vehicle-related adaptation involves modifying the basic design of the engine arrived at on the test bench to the specifics of the vehicle in which it is to be used, and testing conformity with requirements under as wide a range as possible of real operating and ambient conditions.

The adaptation/testing of the basic functions such as idle-speed control, engine response and starting characteristics is essentially performed in the same way as for cars, though the assessment criteria may differ according to the particular type of application. When adapting an engine for use in a bus, for example, more emphasis is placed on comfort aspects or low noise output, whereas a truck engine for long-distance operation would be designed more for reliable and economical transportation of heavy loads.

Examples of adaptation

Idle-speed control

When adapting the idle-speed control function for a commercial-vehicle engine, major emphasis is generally placed on good load response and minimal undershoot. This ensures good pulling away and manoeuvring capabilities even when carrying heavy loads.

The behavior of the drivetrain as a controlled system depends heavily on temperature and transmission ratio. For that reason the engine-management module has multiple parameter sets for idle-speed control. When defining those parameters, changes in the drivetrain response over its service life must also be taken into account.

Power take-off (PTO) drives

Many commercial vehicles have PTO drives that are used to drive cranes, lifting platforms, pumps, etc. These often require the diesel engine to run at a virtually constant, higher operating speed that is unaffected by load. This can be governed by the EDC system using the "intermediate-speed control" function. Once again, the control function parameters can be adapted to the requirements of the driven machine.

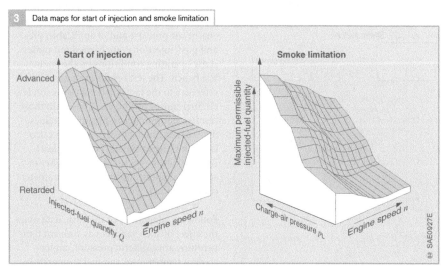

3 Data maps for start of injection and smoke limitation

Start of injection

Advanced

Retarded

Injected-fuel quantity Q

Engine speed n

Smoke limitation

Maximum permissible injected-fuel quantity

Charge-air pressure p_L

Engine speed n

SAE0927E

Engine response characteristics

In the process of adaptation, engine response characteristics, i.e. the way in which accelerator-pedal position is translated into injected-fuel quantity and engine torque output, are to a large extent infinitely variable through control-unit configuration. It ultimately depends on the application as to whether an "RQ characteristic [6]" or "RQV characteristic [7]" engine response is programmed, or a mixture of the two.

Communication

The EDC control unit on a commercial vehicle is normally part of a network of multiple electronic control units. The exchange of data between vehicle, transmission, brake and engine control units takes place via an electronic data bus (usually a CAN). Correct interaction between the various control units involved cannot be fully tested and optimized until they are installed in the vehicle, as the process of basic configuration on the engine test bench usually involves only the engine-management module on its own.

A typical example of the interaction between two vehicle control units is the process of changing gear with an automatic transmission. The transmission control unit sends a request via the data bus for a reduction in injection quantity at the optimum point in the gear-shifting operation. The engine control unit then makes the requested reduction – without input from the driver – thus enabling the transmission control unit to disengage the current gear. If necessary, the transmission control unit may request an increase in engine speed at the appropriate point to facilitate engagement of the new gear. Once the operation is complete, control over the injected fuel quantity is passed back to the driver.

Electromagnetic compatibility

The large number of electronic vehicle systems and the wide use of other electronic communications equipment (e.g. radio telephones, two-way radios, GPS navigation systems) in commercial vehicles make it necessary to optimize the Electromagnetic Compatibility (EMC) of the engine-management module and all its connecting leads in terms both of immunity to external interference and of emission of interference signals. Of course, a large proportion of this optimization work is carried out during the development of the control units and sensors concerned. Since, however, the dimensioning (e.g. length of cable runs, type of shielding) and routing of the wiring looms in the actual vehicle has a major influence on immunity to and creation of interference, testing and, if necessary, optimization of the complete vehicle inside an EMC room is absolutely essential.

Fault diagnosis

The diagnostic capabilities demanded of commercial-vehicle systems are also very extensive. Reliable diagnosis of faults ensures maximum possible vehicle availability.

The engine control unit constantly checks that the signals from all connected sensors and actuators are within the specified limits and also tests for loose contacts, short circuits to ground or to battery voltage, and for plausibility with other signals. The signal range limits and plausibility criteria must be defined by the application developer. As with car engines, those limits must on the one hand be sufficiently broad to ensure that extreme conditions (e.g. hot or cold weather, high altitudes) do not produce false diagnoses, and on the other, sufficiently narrow to provide adequate sensitivity to real faults. In addition, fault response procedures must be defined which specify whether and in what way the engine may continue to be operated if a specific fault is detected. Finally, detected faults have to be stored in a fault memory in order that service technicians can quickly locate and remedy the problem.

[6] Control function for minimum and maximum speed or maximum speed only
[7] Variable-speed or incremental control function

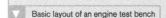

Engine test bench

1 Intake air
2 Filter
3 Cold-water inlet
4 Hot-water inlet
5 Fuel
6 Coolant
7 Heater
8 Quick-change system
9 Transfer modules for supply fluids
10 Engine control unit (EDC)
11 Intercooler
12 Fuel-injection system
13 Engine
14 Control and sensor signals
15 Catalytic converter
16 Power supply
17 Measuring-data interface
18 Electric dynamometer
19 Accelerator positioner
20 Test-bench computer
21 Indexing system (rapid synchronized measured-data acquisition)
22 Exhaust-gas analyzing equipment (e.g. analyzers for gaseous emissions, opacimeter, Fourier Transformed Infra-Red (FTIR) spectroscope, mass spectrometer, particle counter)
23 Dilution tunnel
24 Dilution air
25 Mixing section
26 Volume meter
27 Fan
28 Particle sampling system
29 CVS bag system
30 Changeover valve

A fuel-injection system is tested on an engine test bench as part of its development process. Engine test benches are designed to allow easy access to the various parts of the engine.

By conditioning the supply fluids such as intake air, fuel and engine coolant, (i.e. controlling their temperature and/or pressure) reproducible results can be obtained.

In addition to measurements under static operating conditions, dynamic tests with rapid load and engine-speed changes are increasingly demanded. For such purposes there are test benches with electric dynamometers (18). They can not only retard but also drive the test vehicle (e.g. in order to simulate overrun when traveling downhill). Using appropriate simulation software, the statutory emission control tests can then be run on the test bench rather than on a vehicle tester with the engine in situ.

The test-bench computer (20) is responsible for controlling and monitoring the engine and the testing equipment. It also takes care of data recording and storage. With the aid of automation software, calibration operations (e.g. data-map measurements) can be carried out very efficiently.

Using a suitable quick-change system (8), the pallets with the engines to be tested can be changed over within about twenty minutes. This increases test-bench capacity utilization.

Basic layout of an engine test bench

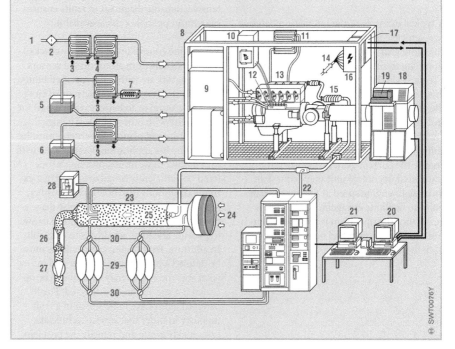

SWT0076Y

Calibration tools

The traditional calibration tools (for car and commercial-vehicle applications) include
- The "transparent" engine (usually a single-cylinder engine which has small windows and mirrors that allow the combustion process to be observed)
- The engine test bench
- The EMC room and
- A wide variety of special devices such as microphones for measuring sound levels or strain gauges for measuring mechanical stress

Computer simulation of hardware and software components is also becoming increasingly important. A large part of the adaptation work, however, is carried out using PC-based calibration tools. Such programs allow developers to modify the engine-management software. One such calibration tool is the INCA (**In**tegrated **C**alibration and **A**cquisition System) program, compromising a number of different tools. It is made up of the following components:
- The *Core System* incorporates all measurement and adjustment functions.
- The *Offline Tools (standard specification)* comprise the software for analysis of measured data and management of adjustment data, and the programming tool for the Flash EPROM.

The use and function of the calibration tools can be illustrated by the description below of a typical calibration process.

1   Hardware for use with INCA calibration tool

Fig. 1
a *Thermo-Scan*
  Interface module for temperature sensors
b *Dual-Scan*
  Interface module for analog signals and temperature sensors
c *Lambda Meter*
  Interface module for broadband oxygen sensor
d *Baro-Scan*
  Testing module for pressures
e *AD-Scan*
  Interface module for analog signals
f *CAN-link card*
g *KIC 2*
  Calibration module for diagnostic interface

SAE 0928Y

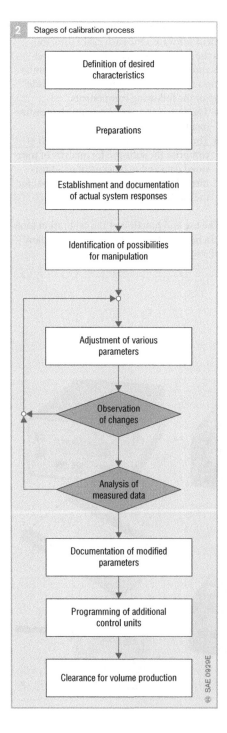

2 Stages of calibration process

- Definition of desired characteristics
- Preparations
- Establishment and documentation of actual system responses
- Identification of possibilities for manipulation
- Adjustment of various parameters
- Observation of changes
- Analysis of measured data
- Documentation of modified parameters
- Programming of additional control units
- Clearance for volume production

SAE 0929E

Software calibration process

Defining the desired characteristics

The desired characteristics (e.g. dynamic response, noise output, exhaust composition) are defined by the engine manufacturer and the (exhaust emissions) legislation. The aim of calibration is to alter the characteristics of the engine so that those requirements are met. This necessitates testing on the engine test bench and in the vehicle.

Preparations

Special electronic engine control units are used for calibration. Compared with the control units used on the production models, they allow the alteration of parameters that are fixed for normal operation. An important aspect of the preparations is choosing and setting up the appropriate hardware and/or software interface.

Additional measuring equipment (e.g. temperature sensors, flow meters) enables the recording of other physical variables for special tests.

Establishing and documenting the actual system responses

The recording of specific measured data is carried out using the INCA core system. The information concerned can be displayed on the screen and analyzed in the form of numerical values or graphs.

The measured data can not only be viewed after the measurements have been taken but while measurement is still in progress. In that way, the response of the engine to changes (e.g. in the exhaust-gas recirculation rate) can be investigated. The data can also be recorded for subsequent analysis of transient processes (e.g. engine starting).

Identifying possibilities for manipulation

With the help of the control-unit software documentation (data framework) it is possible to identify which parameters are best suited to altering system behavior in the manner desired.

Alteration of selected parameters

The parameters stored in the control-unit software can be displayed as numerical values (in tables) or as graphs (curves) on the PC and altered. Each time an alteration is made, the system response is observed.

All parameters can be altered while the engine is running so that the effects are immediately observable and measurable.

In the case of short-lived or transient processes (e.g. engine starting) it is effectively impossible to alter the parameters while the process is in progress. In such cases, therefore, the process has to be recorded during the course of a test, the measured data saved in a file and then the parameters that are to be altered identified by analyzing the recorded data.

Further tests are performed in order to evaluate the success of the adjustments made or to learn more about the process.

Analyzing measured data

Analysis and documentation of the measured data is performed with the aid of the offline tool MDA (Measured Data Analyzer). This stage of the calibration process involves comparing and documenting the system behavior before and after alteration of parameters. Such documentation encompasses improvements as well as problems and malfunctions.

Documentation is important because several people will be involved in the process of engine optimization at different times.

Documenting the modified parameters

The changes to the parameters are also compared and documented. This is done with the offline tool ADM (Application Data Manager), sometimes also called CDM (Calibration Data Manager).

The calibration data obtained by various technicians is compared and merged into a single data record.

Programming additional control units

The new parameter settings arrived at can also be used on other engine control units for further calibration. This necessitates re-programming of the Flash EPROMs of those control units. This is carried out using the INCA core system tool PROF (Programming of Flash EPROM).

Depending on the extent of the calibration and the design innovations, multiple looping of the steps described above may take place.

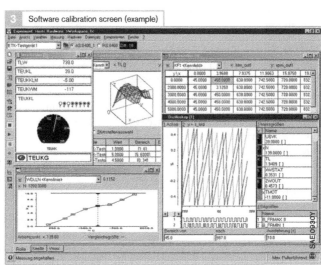

3   Software calibration screen (example)

Electronic Control Unit (ECU)

Digital technology furnishes an extensive array of options for open and closed-loop control of automotive electronic systems. A large number of parameters can be included in the process to support optimal operation of various systems. After receiving the electric signals transmitted by the sensors, the control unit processes these data in order to generate control signals for the actuators. The control program, the "software", is stored in a special memory and implemented by a microcontroller. The control unit and its components are referred to as "hardware". The EDC control unit contains all of the algorithms for open and closed-loop control needed to govern the engine-management processes.

Operating conditions

The ECU is subjected to very high demands with respect to
- extreme ambient temperatures (during normal operation from −40°C to +60...+125°C),
- violent temperature fluctuations,
- resistance to the effects of such materials as oil and fuel, etc.,
- surrounding dampness, and
- mechanical stresses such as engine vibrations.

The ECU must continue to perform flawlessly during starts with a weak battery (cold starts, etc.) as well as at high charge voltages (surge in onboard electrical system).

Other requirements arise from the need for EMC (Electromagnetic Compatibility). The standards of electromagnetic interference immunity and the limitations on emission of high-frequency interference signals are extremely strict.

Design and construction

The pcb (printed-circuit board) with the electronic components (Fig. 1) is installed in a metal case, and connected to the sensors, actuators, and power supply through a multi-pole plug-in connector (4). The high-power driver stages (6) for the direct triggering of the actuators are integrated in the ECU case in such a manner that excellent heat dissipation to the case is ensured.

When the ECU is mounted directly on the engine, an integrated heat sink is used to dissipate the heat from the ECU case to the fuel which permanently flushes the ECU. This ECU cooler is only used on commercial vehicles. Compact, engine-mounted hybrid-technology ECUs are available for even higher levels of temperature loading.

The majority of the electronic components use SMD technology (Surface-Mounted Device). so that a particularly space-saving and weight-saving design can be used. Conventional wiring is only applied at some of the power-electronics components and at the plug-in connections,

Data processing

Input signals
In their role as peripheral components, the actuators and the sensors represent the interface between the vehicle and the ECU in its role as the processing unit. The electrical signals from the sensors travel through the wiring harness and the plug to reach the control unit. These signals can be of the following type:

Analog input signals
Within a given range, analog input signals can assume practically any voltage value. Examples of physical quantities which are available as analog measured values are intake-air mass, battery voltage, intake-manifold and boost pressure, coolant and intake-air temperature. An Analog/Digital Converter (ADC) within the ECU's microcon-

troller transforms the signal data in the digital form required by the microcontroller's central processing unit. The maximum resolution of these analog signals is 5 mV. This translates into roughly 1,000 incremental graduations based on an overall monitoring range of 0...5 V.

Digital input signals
Digital input signals only have two states. They are either "high" or "low" (logical 1 and logical 0 respectively). Examples of digital input signals are on/off switching signals, or digital sensor signals such as the rotational-speed pulses from a Hall generator or a magnetoresistive sensor. Such signals are processed directly by the microcontroller.

Pulse-shaped input signals
The pulse-shaped signals from inductive sensors containing information on rotational speed and reference mark are conditioned in their own ECU stage. Here, spurious pulses are suppressed and the pulse-shaped signals converted into digital rectangular signals.

Signal conditioning
Protective circuitry is used to limit the input signals to a permissible maximum voltage. By applying filtering techniques, the superimposed interference signals are to a great extent removed from the useful signal which, if necessary, is then amplified to the permissible input-signal level for the microcontroller (0...5 V).

Signal conditioning can take place completely or partially in the sensor depending upon the sensor's level of integration.

Signal processing
The control unit is the switching center governing all of the functions and sequences regulated by the engine-management system. The closed and open-loop control functions are executed in the microcontroller. The input signals from sensors and interfaces linking other systems (e.g., CAN bus) serve as the input parameters and are subjected to a further plausibility check in the computer. The ECU program supports generation of the output signals used to control the actuators.

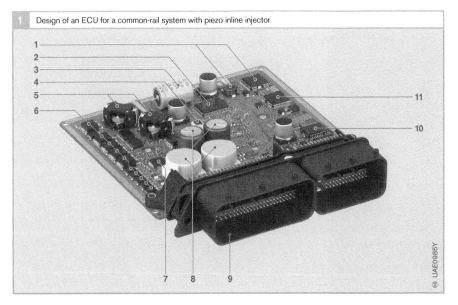

1   Design of an ECU for a common-rail system with piezo inline injector

1
2
3
4
5
6
7   8   9
11
10

UAE0985Y

Fig. 1
1   Switched-mode power supply with voltage stabilization
2   Flash-EPROM
3   Battery backup capacitor (for high-voltage generation)
4   Atmospheric-pressure sensor
5   High-voltage power supply
6   High-power driver stages
7   ASIC for driver-stage triggering
8   High-voltage store (high-voltage charge carrier)
9   Connector
10   Bridge driver stage
11   Multiple switching driver stage

Further components (e.g. of microcontroller) are mounted on the underside.

Microcontroller

The microcontroller is the ECU's central component (Fig. 2) and controls its operative sequence. Apart from the CPU (Central Processing Unit), the microcontroller contains not only the input and output channels, but also timer units, RAMs, ROMs, serial interfaces, and further peripheral assemblies, all of which are integrated on a single microchip. Quartz-controlled timing is used for the microcontroller.

Program and data memory

In order to carry out the computations, the microcontroller needs a program – the "software". This is in the form of binary numerical values arranged in data records and stored in a program memory. These binary values are accessed by the CPU which interprets them as commands which it implements one after the other (refer also to the Chapter "Electronic open and closed-loop control").

This program is stored in a Read-Only Memory (ROM, EPROM, or Flash-EPROM) which also contains variant-specific data (individual data, characteristic curves, and maps). This is non-variable data which cannot be changed during vehicle operation. It is used to regulate the program's open and closed-loop control processes.

The program memory can be integrated in the microcontroller and, depending upon the particular application, expanded by the addition of a separate component (e.g., by an external EPROM or a Flash-EPROM).

ROM

Program memories can be in the form of a ROM (Read Only Memory). This is a memory whose contents have been defined permanently during manufacture and thereafter remain unalterable. The ROM installed in the microcontroller only has a restricted memory capacity, which means that an additional ROM is required in case of complicated applications.

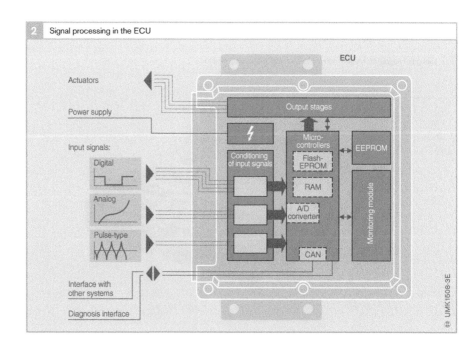

2   Signal processing in the ECU

EPROM

The data on an EPROM (Erasable Programmable **ROM**) can be erased by subjecting the device to UV light. Fresh data can then be entered using a programming unit.

The EPROM is usually in the form of a separate component, and is accessed by the CPU through the Address/Data-Bus.

Flash-EPROM (FEPROM)

The Flash-EPROM is electrically erasable so that it becomes possible to reprogram the ECU in the service workshops without having to open it. In the process, the ECU is connected to the reprogramming unit through a serial interface.

If the microcontroller is also equipped with a ROM, this contains the programming routines for the Flash programming. Flash-EPROMs are available which, together with the microcontroller, are integrated on a single microchip (as from EDC16).

Its decisive advantages have helped the Flash-EPROM to largely supersede the conventional EPROM.

Variable-data or main memory

Such a read/write memory is needed in order to store variable data (variables), such as, for example, arithmetic values and signal values.

RAM

Instantaneous values are stored in the RAM (Random Access Memory) read/write memory. If complex applications are involved, the memory capacity of the RAM incorporated in the microcontroller is insufficient so that an additional RAM module becomes necessary. It is connected to the ECU through the Address/Data-Bus.

When the ECU is switched off by turning the "ignition" key, the RAM loses its complete stock of data (volatile memory).

EEPROM (also known as the E²PROM)

As stated above, the RAM loses its information immediately its power supply is removed (e.g. when the "ignition switch" is turned to OFF). Data which must be retained, for instance the codes for the vehicle immobilizer and the fault-store data, must therefore be stored in a non-erasable (non-volatile) memory. The EEPROM is an electrically erasable EPROM in which (in contrast to the Flash-EPROM) every single memory location can be erased individually. It has been designed for a large number of writing cycles, which means that the EEPROM can be used as a non-volatile read/write memory.

ASIC

The ever-increasing complexity of ECU functions means that the computing powers of the standard microcontrollers available on the market no longer suffice. The solution here is to use so-called ASIC modules (Application Specific Integrated Circuit). These ICs (Integrated Circuits) are designed and produced in accordance with data from the ECU development departments and, as well as being equipped with an extra RAM and inputs and outputs, for instance, they can also generate and transmit PWM signals (see "PWM signals" below).

Monitoring module

The ECU is provided with a monitoring module. Using a "Question and Answer" cycle, the microcontroller and the monitoring module supervise each other, and as soon as a fault is detected one of them triggers appropriate back-up functions independent of the other.

Output signals

With its output signals, the microcontroller triggers driver stages which are usually powerful enough to operate the actuators directly. It is also possible for specific driver stage to trigger relays for particularly large power consumers (e.g., engine fans).

The driver stages are proof against shorts to ground or battery voltage, as well as against destruction due to electrical or thermal overload. Such malfunctions, together with open-circuit lines or sensor faults are identified by the driver-stage IC as an error and reported to the microcontroller.

Switching signals

These are used to switch the actuators on and off (for instance, for the engine fan).

PWM signals

Digital output signals can be in the form of PWM (Pulse-Width Modulated) signals. These are constant-frequency rectangular signals with variable on-times (Fig. 3), which can be used to move various actuators into any operating positions (e.g., exhaust-gas recirculation valve, fan, heater elements, charge-pressure actuator).

Communication within the ECU

In order to be able to support the microcontroller in its work, the peripheral components must communicate with it. This takes place using an address/data bus which, for instance, the microcomputer uses to issue the RAM address whose contents are to be accessed. The data bus is then used to transmit the relevant data. For former automotive applications, an 8-bit structure sufficed. This meant that the data bus comprised 8 lines which together could transmit 256 values simultaneously. The 16-bit address bus commonly used with such systems can access 65,536 addresses. Presently, more complex systems demand 16 bits, or even 32 bits, for the data bus. In order to save on pins at the components, the data and address buses can be combined in a multiplex system, i.e., addresses and data are dispatched through the same lines but offset from each other with respect to time.

Serial interfaces with only a single data line are used for data which need not be transmitted so quickly (e.g. data from the fault storage).

EoL programming

The extensive variety of vehicle variants with differing control programs and data records, makes it imperative to have a system which reduces the number of ECU types needed by a given manufacturer. To this end, the Flash EPROM's complete memory area can be programmed at the end of vehicle production with the program and the variant-specific data record (EoL, End-of-Line programming). A further possibility is to have a number of data variants available (e.g., gearbox variants), which can then be selected by special coding at the end of the line (EoL). This coding is stored in an EEPROM.

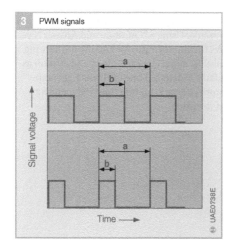

3 PWM signals

Signal voltage →

a

b

a

b

Time →

UAE0738E

Fig. 3
a   Fixed frequency
b   Variable on-time

Very severe demands are made on the ECU

Basically, the ECU in the vehicle functions the same as a conventional PC. Data is entered from which output signals are calculated. The heart of the ECU is the printed-circuit board (pcb) with microcontroller using high-precision microelectronic techniques. The automotive ECU though must fulfill a number of other requirements.

Real-time compatibility

Systems for the engine and for road/traffic safety demand very rapid response of the control, and the ECU must therefore be "real-time compatible". This means that the control's reaction must keep pace with the actual physical process being controlled. It must be certain that a real-time system responds within a fixed period of time to the demands made upon it. This necessitates appropriate computer architecture and very high computer power.

Integrated design and construction

The equipment's weight and the installation space it requires inside the vehicle are becoming increasingly decisive. The following technologies, and others, are used to make the ECU as small and light as possible:

- **Multilayer:** The printed-circuit conductors are between 0.035 and 0.07 mm thick and are "stacked" on top of each other in layers.
- **SMD components** are very small and flat and have no wire connections through holes in the pcb. They are soldered or glued to the pcb or hybrid substrate, hence SMD (**S**urface **M**ounted **D**evices).
- **ASIC:** Specifically designed integrated component (**A**pplication-**S**pecific **I**ntegrated **C**ircuit) which can combine a large number of different functions.

Operational reliability

Very high levels of resistance to failure are provided by integrated diagnosis and redundant mathematical processes (additional processes, usually running in parallel on other program paths).

Environmental influences

Notwithstanding the wide range of environmental influences to which it is subjected, the ECU must always operate reliably.

- **Temperature:** Depending upon the area of application, the ECUs installed in vehicles must perform faultlessly during continual operation at temperatures between −40°C and +60...125°C. In fact, due to the heat radiated from the components, the temperature at some areas of the substrate is considerably higher. The temperature change involved in starting at cold temperatures and then running up to hot operating temperatures is particularly severe.
- **EMC:** The vehicle's electronics have to go through severe electromagnetic compatibility testing. That is, the ECU must remain completely unaffected by electromagnetic disturbances emanating from such sources as the ignition, or radiated by radio transmitters and mobile telephones. Conversely, the ECU itself must not negatively affect other electronic equipment.
- **Resistance to vibration:** ECUs which are mounted on the engine must be able to withstand vibrations of up to 30g (that is, 30 times the acceleration due to gravity).
- **Sealing and resistance to operating mediums:** Depending upon installation position, the ECU must withstand damp, chemicals (e.g. oils), and salt fog.

The above factors and other requirements mean that the Bosch development engineers are continually faced by new challenges.

Sensors

Sensors register operating states (e.g. engine speed) and setpoint/desired values (e.g. accelerator-pedal position). They convert physical quantities (e.g. pressure) or chemical quantities (e.g. exhaust-gas concentration) into electric signals.

Automotive applications

Sensors and actuators represent the interfaces between the ECUs, as the processing units, and the vehicle with its complex drive, braking, chassis, and bodywork functions (for instance, the Engine Management, the Electronic Stability Program ESP, and the air conditioner). As a rule, a matching circuit in the sensor converts the signals so that they can be processed by the ECU.

The field of mechatronics, in which mechanical, electronic, and data-processing components are interlinked and cooperate closely with each other, is rapidly gaining in importance in the field of sensor engineering. These components are integrated in modules (e.g. in the crankshaft CSWS (Composite Seal with Sensor) module complete with rpm sensor).

Since their output signals directly affect not only the engine's power output, torque, and emissions, but also vehicle handling and safety, sensors, although they are becoming smaller and smaller, must also fulfill demands that they be faster and more precise. These stipulations can be complied with thanks to mechatronics.

Depending upon the level of integration, signal conditioning, analog/digital conversion, and self-calibration functions can all be integrated in the sensor (Fig. 1), and in future a small microcomputer for further signal processing will be added. The advantages are as follows:

- Lower levels of computing power are needed in the ECU
- A uniform, flexible, and bus-compatible interface becomes possible for all sensors
- Direct multiple use of a given sensor through the data bus
- Registration of even smaller measured quantities
- Simple sensor calibration

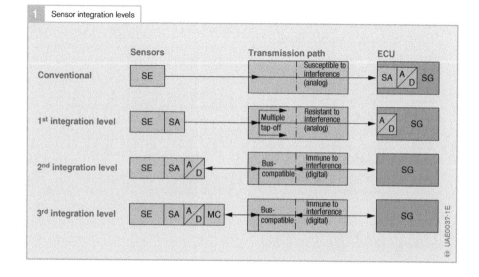

Fig. 1
SE Sensor(s)
SA Analog signal
    conditioning
A/D Analog-digital
    converter
SG Digital ECU
MC Microcomputer
    (evaluation
    electronics)

Temperature sensors

Applications

Engine-temperature sensor
This is installed in the coolant circuit (Fig. 1). The engine management uses its signal when calculating the engine temperature (measuring range −40…+130°C).

Air-temperature sensor
This sensor is installed in the air-intake tract. Together with the signal from the boost-pressure sensor, its signal is applied in calculating the intake-air mass. Apart from this, desired values for the various control loops (e.g. EGR, boost-pressure control) can be adapted to the air temperature (measuring range −40…+120°C).

Engine-oil temperature sensor
The signal from this sensor is used in calculating the service interval (measuring range −40…+170°C).

Fuel-temperature sensor
Is incorporated in the low-pressure stage of the diesel fuel circuit. The fuel temperature is used in calculating the precise injected fuel quantity (measuring range −40…+120°C).

Exhaust-gas temperature sensor
This sensor is mounted on the exhaust system at points which are particularly critical regarding temperature. It is applied in the closed-loop control of the systems used for exhaust-gas treatment. A platinum measuring resistor is usually used (measuring range −40…+1,000°C).

Design and operating concept

Depending upon the particular application, a wide variety of temperature sensor designs are available. A temperature-dependent semiconductor measuring resistor is fitted inside a housing. This resistor is usually of the NTC (Negative Temperature Coefficient, Fig. 2) type. Less often a PTC (Positive Temperature Coefficient) type is used. With NTC, there is a sharp drop in resistance when the temperature rises, and with PTC there is a sharp increase.

The measuring resistor is part of a voltage-divider circuit to which 5 V is applied. The voltage measured across the measuring resistor is therefore temperature-dependent. It is inputted through an analog to digital (A/D) converter and is a measure of the temperature at the sensor. A characteristic curve is stored in the engine-management ECU which allocates a specific temperature to every resistance or output-voltage.

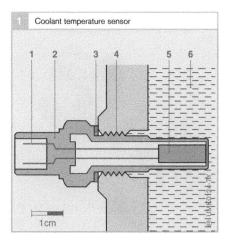

1   Coolant temperature sensor

1 cm

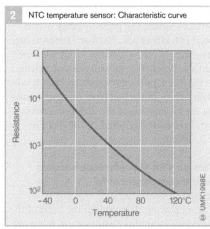

2   NTC temperature sensor: Characteristic curve

Resistance
Temperature

Fig. 1
1   Electrical connections
2   Housing
3   Gasket
4   Thread
5   Measuring resistor
6   Coolant

Micromechanical pressure sensors

Application

Manifold-pressure or boost-pressure sensor
This sensor measures the absolute pressure
in the intake manifold between the supercharger and the engine (typically 250 kPa
or 2.5 bar) and compares it with a reference
vacuum, not with the ambient pressure. This
enables the air mass to be precisely defined,
and the boost pressure exactly controlled in
accordance with engine requirements.

Atmospheric-pressure sensor
This sensor is also known as an ambient-
pressure sensor and is incorporated in the
ECU or fitted in the engine compartment.
Its signal is used for the altitude-dependent
correction of the setpoint values for the control loops. For instance, for the exhaust-gas
recirculation (EGR) and for the boost-pressure control. This enables the differing densities of the surrounding air to be taken into
account. The atmospheric-pressure sensor
measures absolute pressure (60...115 kPa or
0.6...1.15 bar).

Oil and fuel-pressure sensor
Oil-pressure sensors are installed in the oil
filter and measure the oil's absolute pressure.
This information is needed so that engine
loading can be determined as needed for the
Service Display. The pressure range here is
50...1,000 kPa or 0.5...10.0 bar. Due to its high
resistance to media, the measuring element
can also be used for pressure measurement
in the fuel supply's low-pressure stage. It is installed on or in the fuel filter. Its signal serves
for the monitoring of the fuel-filter contamination (measuring range: 20... 400 kPa or
0.2...4 bar).

Version with the reference vacuum on the component side

Design and construction
The measuring element is at the heart of the
micromechanical pressure sensor. It is com-

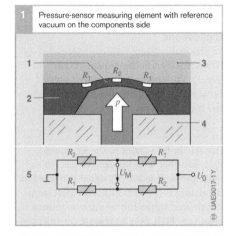

1 Pressure-sensor measuring element with reference vacuum on the components side

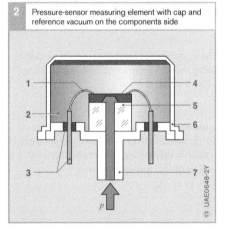

3 Pressure-sensor measuring element with cap and reference vacuum on the components side

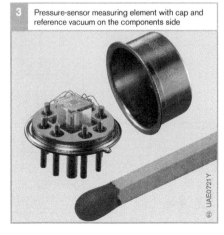

2 Pressure-sensor measuring element with cap and reference vacuum on the components side

prised of a silicon chip (Fig. 1, 2) in which a thin diaphragm has been etched micromechanically (1). Four deformation resistors (R_1, R_2) are diffused on the diaphram. Their electrical resistance changes when mechanical force is applied. The measuring element is surrounded on the component side by a cap which at the same time encloses the reference vacuum (Figs. 2 and 3). The pressure-sensor case can also incorporate an integral *temperature sensor* (Fig. 4, 1) whose signals can be evaluated independently. This means that at any point a single sensor case suffices to measure temperature and pressure.

Method of operation

The sensor's diaphragm deforms more or less (10...1,000 μm) according to the pressure being measured. The four deformation resistors on the diaphragm change their electrical resistances as a function of the mechanical stress resulting from the applied pressure (piezoresistive effect).

The four measuring resistors are arranged on the silicon chip so that when diaphragm deformation takes place, the resistance of two of them increases and that of the other two decreases. These deformation resistors form a Wheatstone bridge (Fig. 1, 5), and a change in their resistances leads to a change in the ratio of the voltages across them. This leads to a change in the measurement voltage U_M. This unamplified voltage is therefore a measure of the pressure applied to the diaphragm.

The measurement voltage is higher with a bridge circuit than would be the case when using an individual resistor. The Wheatstone bridge circuit thus permits a higher sensor sensitivity.

The component side of the sensor to which pressure is not supplied is subjected to a reference vacuum (Fig. 2, 2) so that it measures the absolute pressure.

The signal-conditioning electronics circuitry is integrated on the chip. Its assignment is to amplify the bridge voltage, compensate for

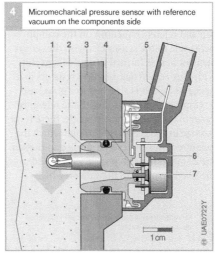

4 Micromechanical pressure sensor with reference vacuum on the components side

1 cm

Fig. 4
1 Temperature sensor (NTC)
2 Lower section of case
3 Manifold wall
4 Seal rings
5 Electrical terminal (plug)
6 Case cover
7 Measuring element

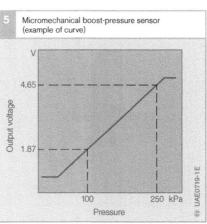

5 Micromechanical boost-pressure sensor (example of curve)

Output voltage

V

4.65

1.87

100   250 kPa

Pressure

temperature influences, and linearise the pressure curve. The output voltage is between 0...5 V and is connected through electrical terminals (Fig. 4, 5) to the engine-management ECU which uses this output voltage in calculating the pressure (Fig. 5).

Version with reference vacuum in special chamber

Design and construction

The *manifold or boost-pressure sensor* version with the reference vacuum in a special chamber (Figs. 6 and 7) is easier to install than the version with the reference vacuum on the components side of the sensor element. Similar to the pressure sensor with cap and reference vacuum on the components side of the sensor element, the sensor element here is formed from a silicon chip with four etched deformation resistors in a bridge circuit. It is attached to a glass base. In contrast to the sensor with the reference vacuum on the components side, there is no passage in the glass base through which the measured pressure can be applied to the sensor element. Instead, pressure is applied to the silicon chip from the side on which

the evaluation electronics is situated. This means that a special gel must be used at this side of the sensor to protect it against environmental influences (Fig. 8, 1). The reference vacuum is enclosed in the chamber between the silicon chip (6) and the glass base (3). The complete measuring element is mounted on a ceramic hybrid (4) which incorporates the soldering surfaces for electrical contacting inside the sensor.

A *temperature sensor* can also be incorporated in the pressure-sensor case. It protrudes into the air flow, and can therefore respond to temperature changes with a minimum of delay (Fig. 6, 4).

Operating concept
The operating concept, and with it the signal conditioning and signal amplification together with the characteristic curve, corresponds to that used in the pressure sensor with cap and reference vacuum on the sensor's structure side. The only difference is that the measuring element's diaphragm is deformed in the opposite direction and therefore the deformation resistors are "bent" in the other direction.

6 Micromechanical pressure sensor with reference vacuum in a chamber

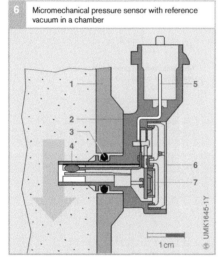

1 cm

Fig. 6
1 Manifold wall
2 Case
3 Seal ring
4 Temperature sensor (NTC)
5 Electrical connection (socket)
6 Case cover
7 Measuring element

7 Micromechanical pressure sensor with reference vacuum in a chamber and temperature sensor

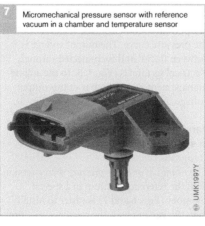

8 Measuring element of pressure sensor with reference vacuum in a chamber

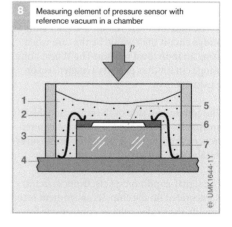

Fig. 8
1 Protective gel
2 Gel frame
3 Glass base
4 Ceramic hybrid
5 Chamber with reference volume
6 Measuring element (chip) with evaluation electronics
7 Bonded connection
p Measured pressure

High-pressure sensors

Application
In automotive applications, high-pressure sensors are used for measuring the pressures of fuels and brake fluids.

Diesel rail-pressure sensor
In the diesel engine, the rail-pressure sensor measures the pressure in the fuel rail of the Common Rail accumulator-type injection system. Maximum operating (nominal) pressure p_{max} is 160 MPa (1,600 bar). The fuel pressure is regulated in a control loop, and remains practically constant independent of load and engine speed. Any deviations from the setpoint pressure are compensated for by a pressure control valve.

Gasoline rail-pressure sensor
As its name implies, this sensor measures the pressure in the fuel rail of the DI Motronic with gasoline direct injection. Pressure is a function of load and engine speed and is 5...12 MPa (50...120 bar), and is used as an actual (measured) value in the closed-loop rail-pressure control. The rpm and load-dependent setpoint value is stored in a map and is adjusted at the rail by a pressure control valve.

Brake-fluid pressure sensor
Installed in the hydraulic modulator of such driving-safety systems as ESP, this high-pressure sensor is used to measure the brake-fluid pressure which is usually 25 MPa (250 bar). Maximum pressure p_{max} can climb to as much as 35 MPa (350 bar). Pressure measurement and monitoring is triggered by the ECU which also evaluates the return signals.

Design and operating concept
The heart of the sensor is a steel diaphragm onto which deformation resistors have been vapor-deposited in the form of a bridge circuit (Fig. 1, 3). The sensor's measuring range is a function of diaphragm thickness (thicker diaphragms for higher pressures,

thinner diaphragms for lower pressures). When the pressure is applied via the pressure connection (4) to one of the diaphragm faces, the resistances of the bridge resistors change due to diaphragm deformation (approx. 20 µm at 1,500 bar).

The 0...80 mV output voltage generated by the bridge is conducted to an evaluation circuit (2) which amplifies it to 0...5 V. This is used as the input to the ECU which refers to a stored characteristic curve in calculating the pressure (Fig. 2).

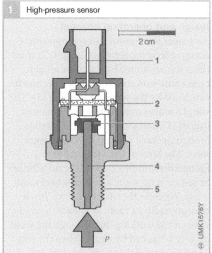

1   High-pressure sensor

2 cm

Fig. 1
1 Electrical connection (socket)
2 Evaluation circuit
3 Steel diaphragm with deformation resistors
4 Pressure connection
5 Mounting thread

2   High-pressure sensor (curve, example)

Output voltage

4.5

0.5

0   Pressure p_{max}

Inductive engine-speed sensors

Applications
Such engine-speed sensors are used for measuring:
- Engine rpm
- Crankshaft position (for information on the position of the engine pistons)

The rotational speed is calculated from the sensor's signal frequency. The output signal from the rotational-speed sensor is one of the most important quantities in electronic engine management.

Design and operating concept
The sensor is mounted directly opposite a ferromagnetic trigger wheel (Fig. 1, 7) from which it is separated by a narrow air gap. It has a soft-iron core (pole pin) (4), which is enclosed by the solenoid winding (5). The pole pin is also connected to a permanent magnet (1), and a magnetic field extends through the pole pin and into the trigger wheel. The level of the magnetic flux through the winding depends upon whether the sensor is opposite a trigger-wheel tooth or gap. Whereas the magnet's stray flux is concentrated by a tooth and leads to an increase in the working flux through the winding, it is weakened by a gap. When the trigger wheel rotates therefore, this causes a fluctuation of the flux which in turn generates a sinusoidal voltage in the solenoid

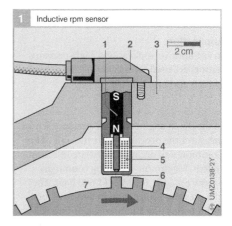

1   Inductive rpm sensor

2 cm

winding which is proportional to the rate of change of the flux (Fig. 2). The amplitude of the AC voltage increases strongly along with increasing trigger-wheel speed (several mV...>100 V). At least about 30 rpm are needed to generate an adequate signal level.

The number of teeth on the trigger wheel depends upon the particular application. On solenoid-valve-controlled engine-management systems for instance, a 60-pitch trigger wheel is normally used, although 2 teeth are omitted (7) so that the trigger wheel has 60 – 2 = 58 teeth. The very large tooth gap is allocated to a defined crankshaft position and serves as a reference mark for synchronizing the ECU.

There is another version of the trigger wheel which has one tooth per engine cylinder. In the case of a 4-cylinder engine, therefore, the trigger wheel has 4 teeth, and 4 pulses are generated per revolution.

The geometries of the trigger-wheel teeth and the pole pin must be matched to each other. The evaluation-electronics circuitry in the ECU converts the sinusoidal voltage, which is characterized by strongly varying amplitudes, into a constant-amplitude square-wave voltage for evaluation in the ECU microcontroller.

Fig. 1
1 Permanent magnet
2 Sensor housing
3 Engine block
4 Pole pin
5 Solenoid winding
6 Air gap
7 Trigger wheel with reference-mark gap

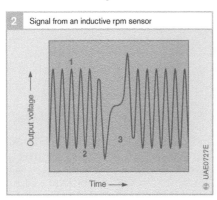

2   Signal from an inductive rpm sensor

Output voltage →

Time →

Fig. 2
1 Tooth
2 Tooth gap
3 Reference mark

Rotational-speed (rpm) sensors and incremental angle-of-rotation sensors

Application

The above sensors are installed in distributor-type diesel injection pumps with solenoid-valve control. Their signals are used for:

- The measurement of the injection pump's speed
- Determining the instantaneous angular position of pump and camshaft
- Measurement of the instantaneous setting of the timing device

The pump speed at a given instant is one of the input variables to the distributor pump's ECU which uses it to calculate the triggering time for the high-pressure solenoid valve, and, if necessary, for the timing-device solenoid valve.

The triggering time for the high-pressure solenoid valve must be calculated in order to inject the appropriate fuel quantity for the particular operating conditions. The cam plate's instantaneous angular setting defines the triggering point for the high-pressure solenoid valve. Only when triggering takes place at exactly the right cam-plate angle, can it be guaranteed that the opening and closing points for the high-pressure solenoid valve are correct for the particular cam lift. Precise triggering defines the correct start-of-injection point and the correct injected fuel quantity.

The correct timing-device setting as needed for timing-device control is ascertained by comparing the signals from the camshaft rpm sensor with those of the angle-of-rotation sensor.

Design and operating concept

The rpm sensor, or the angle-of-rotation sensor, scans a toothed pulse disc with 120 teeth which is attached to the distributor pump's driveshaft. There are tooth gaps, the number of which correspond to the number of engine cylinders, evenly spaced around the disc's circumference. A double differential magnetoresistive sensor is used.

Magnetoresistors are magnetically controllable semiconductor resistors, and similar in design to Hall-effect sensors. The double differential sensor has four resistors connected to form a full bridge circuit.

The sensor has a permanent magnet, and the magnet's pole face opposite the toothed pulse disc is homogenized by a thin ferromagnetic wafer on which are mounted the four magnetoresistors, separated from each other by half a tooth gap. This means that alternately there are two magnetoresistors opposite tooth gaps and two opposite teeth (Fig. 1). The magnetoresistors for automotive applications are designed for operation in temperatures of $\leq 170°C$ ($\leq 200°C$ briefly).

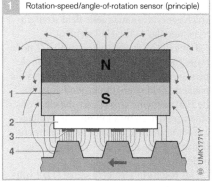

1 Rotation-speed/angle-of-rotation sensor (principle)

Fig. 1
1 Magnet
2 Homogenization wafer (Fe)
3 Magnetoresistor
4 Toothed pulse disc

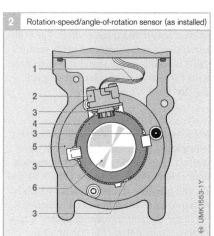

2 Rotation-speed/angle-of-rotation sensor (as installed)

Fig. 2
1 Flexible conductive foil
2 Rotation-speed (rpm)/angle-of-rotation sensor
3 Tooth gap
4 Toothed pulse wheel (trigger wheel)
5 Rotatable mounting
6 Driveshaft

Hall-effect phase sensors

Application

The engine's camshaft rotates at half the crankshaft speed. Taking a given piston on its way to TDC, the camshaft's rotational position is an indication as to whether the piston is in the compression or exhaust stroke. The phase sensor on the camshaft provides the ECU with this information.

Design and operating concept

Hall-effect rod sensors

As the name implies, such sensors (Fig. 2a) make use of the Hall effect. A ferromagnetic trigger wheel (with teeth, segments, or perforated rotor, 7) rotates with the camshaft. The Hall-effect IC is located between the trigger wheel and a permanent magnet (5) which generates a magnetic field strength perpendicular to the Hall element.

If one of the trigger-wheel teeth (Z) now passes the current-carrying rod-sensor element (semiconductor wafer), it changes the magnetic field strength perpendicular to the Hall element. This causes the electrons, which are driven by a longitudinal voltage across the element to be deflected perpendicularly to the direction of current (Fig. 1, angle α).

This results in a voltage signal (Hall voltage) which is in the millivolt range, and which is independent of the relative speed between sensor and trigger wheel. The evaluation electronics integrated in the sensor's Hall IC conditions the signal and outputs it in the form of a rectangular-pulse signal (Fig. 2b "High"/"Low").

Differential Hall-effect rod sensors

Rod sensors operating as per the differential principle are provided with two Hall elements. These elements are offset from each other either radially or axially (Fig. 3, S1 and S2), and generate an output signal which is proportional to the difference in magnetic flux at the element measuring points. A two-track perforated plate (Fig. 3a) or a two-track trigger wheel (Fig. 3b) are needed in order to generate the opposing signals in the Hall elements (Fig. 4) as needed for this measurement.

Such sensors are used when particularly severe demands are made on accuracy. Further advantages are their relatively wide air-gap range and good temperature-compensation characteristics.

Fig. 1
I Wafer current
I_H Hall current
I_V Supply current
U_H Hall voltage
U_R Longitudinal voltage
B Magnetic induction
α Deflection of the electrons by the magnetic field

Fig. 2
a   Positioning of sensor and single-track trigger wheel
b   Output signal characteristic U_A

1   Electrical connection (plug)
2   Sensor housing
3   Engine block
4   Seal ring
5   Permanent magnet
6   Hall-IC
7   Trigger wheel with tooth/segment (Z) and gap (L)

a Air gap
φ Angle of rotation

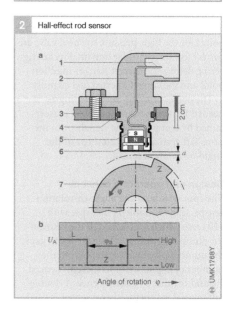

1   Hall element (Hall-effect vane switch)

2   Hall-effect rod sensor

Angle of rotation φ ⟶

3 Differential Hall-effect rod sensors

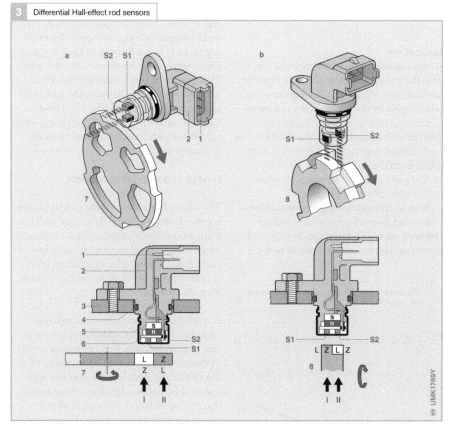

Fig. 3
a Axial tap-off
   (perforated plate)
b Radial tap-off
   (two-track trigger
   wheel)

1 Electrical connection
   (plug)
2 Sensor housing
3 Engine block
4 Seal ring
5 Permanent magnet
6 Differential Hall-IC
   with Hall elements S1
   and S2
7 Perforated plate
8 Two-track trigger
   wheel

I Track 1
II Track 2

4 Characteristic curve of the output signal U_A from a differential Hall-effect rod sensor

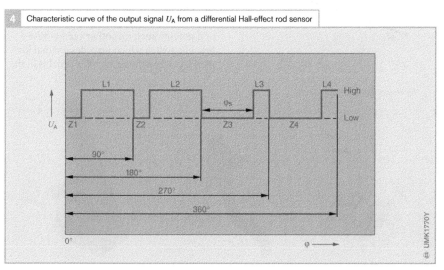

Fig. 4
Output signal "Low":
Material (Z) in front of
S1, gap (L) in front of S2

Output signal "High":
Gap (L) in front of S1,
material (Z) in front of S2

φ_S Signal width

Accelerator-pedal sensors

Application

In conventional engine-management systems, the driver transmits his/her wishes for acceleration, constant speed, or lower speed, to the engine by using the accelerator pedal to intervene mechanically at the throttle plate (gasoline engine) or at the injection pump (diesel engine). Intervention is transmitted from the accelerator pedal to the throttle plate or injection pump by means of a Bowden cable or linkage.

On today's electronic engine-management systems, the Bowden cable and/or linkage have been superseded, and the driver's accelerator-pedal inputs are transmitted to the ECU by an accelerator-pedal sensor which registers the accelerator-pedal travel, or the pedal's angular setting, and sends this to the engine ECU in the form of an electric signal. This system is also known as "drive-by-wire". The accelerator-pedal module (Figs. 2b, 2c) is available as an alternative to the individual accelerator-pedal sensor (Fig. 2a). These modules are ready-to-install units comprising accelerator pedal and sensor, and make adjustments on the vehicle a thing of the past.

Design and operating concept

Potentiometer-type accelerator-pedal sensor
The heart of this sensor is the potentiometer across which a voltage is developed which is a function of the accelerator-pedal setting. In the ECU, a programmed characteristic curve is applied in order to calculate the accelerator-pedal travel, or its angular setting, from this voltage.

A second (redundant) sensor is incorporated for diagnosis purposes and for use in case of malfunctions. It is a component part of the monitoring system. One version of the accelerator-pedal sensor operates with a second potentiometer. The voltage across this potentiometer is always half of that across the first potentiometer. This provides two independent signals which are used for trouble-shooting (Fig. 1). Instead of the second potentiometer, another version uses a low-idle switch which provides a signal for the ECU when the accelerator pedal is in the

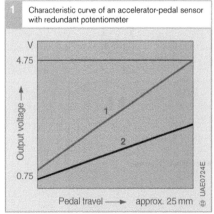

1 Characteristic curve of an accelerator-pedal sensor with redundant potentiometer

Output voltage →
V
4.75

1

2

0.75

Pedal travel ⟶ approx. 25 mm

UAE0724E

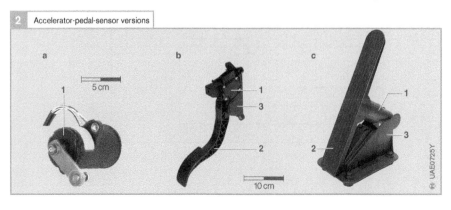

2 Accelerator-pedal-sensor versions

a   5 cm
1

b
1
3
2

c
1
3
2

10 cm

UAE0725Y

idle position. For automatic transmission vehicles, a further switch can be incorporated for a kick-down signal.

Hall-effect angle-of-rotation sensors

The ARS1 (**A**ngle of **R**otation **S**ensor) is based on the movable-magnet principle. It has a measuring range of approx. 90° (Figs. 3 and 4).

A semicircular permanent-magnet disc rotor (Fig. 4, 1) generates a magnetic flux which is returned back to the rotor via a pole shoe (2), magnetically soft conductive elements (3) and shaft (6). In the process, the amount of flux which is returned through the conductive elements is a function of the rotor's angle of rotation φ. There is a Hall-effect sensor (5) located in the magnetic path of each conductive element, so that it is possible to generate a practically linear characteristic curve throughout the measuring range.

The ARS2 is a simpler design without magnetically soft conductive elements. Here, a magnet rotates around the Hall-effect sensor. The path it takes describes a circular arc. Since only a small section of the resulting sinusoidal characteristic curve features good linearity, the Hall-effect sensor is located slightly outside the center of the arc. This causes the curve to deviate from its sinusoidal form so that the curve's linear section is increased to more than 180°.

Mechanically, this sensor is highly suitable for installation in an accelerator-pedal module (Fig. 5).

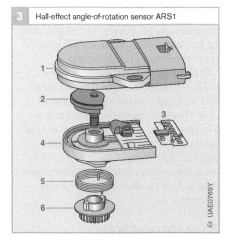

3 Hall-effect angle-of-rotation sensor ARS1

Fig. 3
1   Housing cover
2   Rotor (permanent magnet)
3   Evaluation electronics with Hall-effect sensor
4   Housing base
5   Return spring
6   Coupling element (e.g. gear)

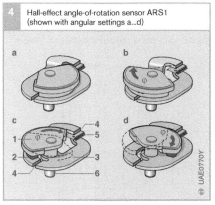

4 Hall-effect angle-of-rotation sensor ARS1 (shown with angular settings a...d)

Fig. 4
1   Rotor (permanent magnet)
2   Pole shoe
3   Conductive element
4   Air gap
5   Hall-effect sensor
6   Shaft (magnetically soft)

φ Angle of rotation

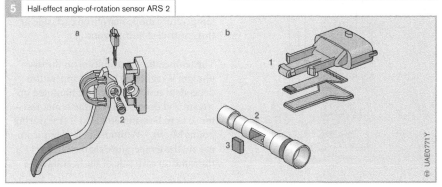

5 Hall-effect angle-of-rotation sensor ARS 2

Fig. 5
a   Installation in the accelerator-pedal module
b   Components

1   Hall-effect sensor
2   Pedal shaft
3   Magnet

Hot-film air-mass meter HFM5

Application

For optimal combustion as needed to comply with the emission regulations imposed by legislation, it is imperative that precisely the necessary air mass is inducted, irrespective of the engine's operating state.

To this end, part of the total air flow which is actually inducted through the air filter or the measuring tube is measured by a hot-film air-mass meter. Measurement is very precise and takes into account the pulsations and reverse flows caused by the opening and closing of the engine's intake and exhaust valves. Intake-air temperature changes have no effect upon measuring accuracy.

Design and construction

The housing of the HFM5 Hot-Film Air-Mass Meter (Fig. 1, 5) projects into a measuring tube (2) which, depending upon the engine's air-mass requirements, can have a

variety of diameters (for 370...970 kg/h). This tube is installed in the intake tract downstream from the air filter. Plug-in versions are also available which are installed inside the air filter.

The most important components in the sensor are the sensor element (4), in the air intake (8), and the integrated evaluation electronics (3). The partial air flow as required for measurement flows across this sensor element.

Vapor-deposition is used to apply the sensor-element components to a semiconductor substrate, and the evaluation-electronics (hybrid circuit) components to a ceramic substrate. This principle permits very compact design. The evaluation electronics are connected to the ECU through the plug-in connection (1). The partial-flow measuring tube (6) is shaped so that the air flows past the sensor element smoothly (without whirl effects) and back into the measuring tube via the air outlet (7). This method ensures efficient sensor operation even in case of extreme pulsation, and in addition to forward flow, reverse flows are also detected (Fig. 2).

Operating concept

The hot-film air-mass meter is a "thermal sensor" and operates according to the following principle:

A micromechanical sensor diaphragm (Fig. 3, 5) on the sensor element (3) is heated by a centrally mounted heater resistor and held at a constant temperature. The temperature drops sharply on each side of this controlled heating zone (4).

The temperature distribution on the diaphragm is registered by two temperature-dependent resistors which are mounted upstream and downstream of the heater resistor so as to be symmetrical to it (measuring points M_1, M_2). Without the flow of incoming air, the temperature characteristic (1) is the same on each side of the heating zone ($T_1 = T_2$).

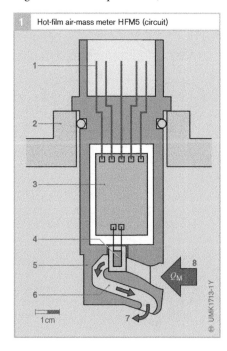

1   Hot-film air-mass meter HFM5 (circuit)

Fig. 1

1   Electrical plug-in connection
2   Measuring tube or air-filter housing wall
3   Evaluation electronics (hybrid circuit)
4   Sensor element
5   Sensor housing
6   Partial-flow measuring tube
7   Air outlet for the partial air flow Q_M
8   Intake for partial air flow Q_M

1 cm

As soon as air flows over the sensor element, the uniform temperature distribution at the diaphragm changes (2). On the intake side, the temperature characteristic is steeper since the incoming air flowing past this area cools it off. Initially, on the opposite side (the side nearest to the engine), the sensor element cools off. The air heated by the heater element then heats up the sensor element. The change in temperature distribution leads to a temperature differential (ΔT) between the measuring points M_1 und M_2.

The heat dissipated to the air, and therefore the temperature characteristic at the sensor element is a function of the air mass flow. Independent of the absolute temperature of the air flowing past, the temperature differential is a measure of the air mass flow. Apart from this, the temperature differential is directional, which means that the air-mass meter not only registers the mass of the incoming air but also its direction.

Due to its very thin micromechanical diaphragm, the sensor has a highly dynamic response (<15 ms), a point which is of particular importance when the incoming air is pulsating heavily.

The evaluation electronics (hybrid circuit) integrated in the sensor convert the resistance differential at the measuring points M_1 and M_2 into an analog signal of 0...5 V which is suitable for processing by the ECU. Using the sensor characteristic (Fig. 2) programmed into the ECU, the measured voltage is converted into a value representing the air mass flow [kg/h].

The shape of the characteristic curve is such that the diagnosis facility incorporated in the ECU can detect such malfunctions as an open-circuit line. A temperature sensor for auxiliary functions can also be integrated in the HFM5. It is located on the sensor element upstream of the heated zone.

It is not required for measuring the air mass. For applications on specific vehicles, supplementary functions such as improved separation of water and contamination are provided for (inner measuring tube and protective grid).

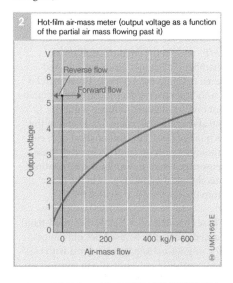

2 Hot-film air-mass meter (output voltage as a function of the partial air mass flowing past it)

3 Hot-film air-mass meter: Measuring principle

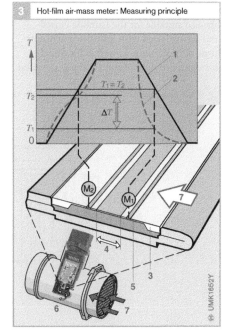

Fig. 3
1 Temperature profile without air flow across sensor element
2 Temperature profile with air flow across sensor element
3 Sensor element
4 Heated zone
5 Sensor diaphragm
6 Measuring tube with air-mass meter
7 Intake-air flow

M_1, M_2 Measuring points

T_1, T_2 Temperature values at the measuring points M_1 and M_2
ΔT Temperature differential

LSU4 planar broad-band Lambda oxygen sensor

Application

As its name implies, the broad-band Lambda oxygen sensor is used across a very extensive range to determine the oxygen concentration in the exhaust gas. The figures provided by the sensor are an indication of the air-fuel (A/F) ratio in the engine's combustion chamber. The excess-air factor λ is used when defining the A/F ratio. Broad-band Lambda sensors make precise measurements not only at the stoichiometic point $\lambda = 1$, but also in the lean range ($\lambda > 1$) and in the rich range ($\lambda < 1$). In combination with electronic closed-loop control circuitry, these sensors generate an unmistakable, continuous electrical signal (Fig. 2) in the range from $0.7 < \lambda < \infty$ (= air with 21% O_2).

These characteristics enable the broad-band Lambda sensor to be used not only in gasoline-engine-management systems with two-step control ($\lambda = 1$), but also in control concepts with rich and lean air-fuel (A/F) mixtures. This type of Lambda sensor is therefore also suitable for the Lambda closed-loop control used with lean-burn concepts on gasoline engines, as well as for diesel engines, gaseous-fuel engines and gas-powered

central heaters and water heaters (this wide range of applications led to the designation LSU: Lambda Sensor Universal (taken from the German), in other words Universal Lambda Sensor).

The sensor protrudes into the exhaust pipe and detects the flow of exhaust-gas flow from all cylinders.
In a number of systems, several Lambda sensors are installed for even greater accuracy. Here, for instance, they are fitted in the individual exhaust tracts of V-engines.

Design and construction

The LSU4 broad-band Lambda sensor (Fig. 3) is a planar dual-cell limit-current sensor. It features a zirconium-dioxide/ceramic (ZrO_2) measuring cell (Fig. 1), which is the combination of a Nernst concentration cell (sensor cell which functions in the same way as a two-step Lambda sensor) and an oxygen-pump cell for transporting the oxygen ions.

The oxygen pump cell (Fig. 1, 8) is so arranged with respect to the Nernst concentration cell (7) that there is a 10...50 μm diffusion gap (6) The gap is connected to the exhaust gas through a gas-access passage (10). The porous diffusion barrier (11) serves to limit the inflow of oxygen molecules from the exhaust gas.

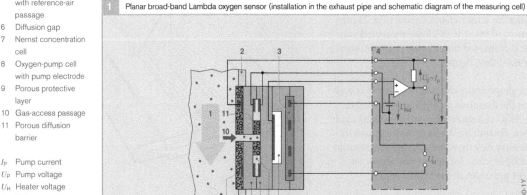

1   Planar broad-band Lambda oxygen sensor (installation in the exhaust pipe and schematic diagram of the measuring cell)

On the one side, the Nernst concentration cell is connected to the atmosphere by a reference-air passage (5), on the other it is connected to the exhaust gas in the diffusion gap.

The sensor must have heated up to at least 600...800°C before it generates a usable signal. It is provided with an integral heater (3), so that the required temperature is reached quickly.

Method of operation
The exhaust gas enters the actual measuring chamber (diffusion gap) of the Nernst concentration cell through the pump cell's gas-access passage. In order that the excess-air factor λ can be adjusted in the diffusion gap, the Nernst concentration cell compares the gas in the diffusion gap with that in the reference-air passage.

The complete process proceeds as follows:
By applying the pump voltage U_P across the pump cell's platinum electrodes, oxygen from the exhaust gas can be pumped into or out of the diffusion gap. With the help of the Nernst concentration cell, an electronic circuit in the ECU controls the voltage (U_P) across the pump cell in order that the composition of the gas in the diffusion gap

remains constant at $\lambda = 1$. If the exhaust gas is lean, the pump cell pumps the oxygen to the outside (positive pump current). On the other hand, if it is rich, due to the decomposition of CO_2 and H_2O at the exhaust-gas electrode the oxygen is pumped from the surrounding exhaust gas and into the diffusion gap (negative pump current). Oxygen transport is unnecessary at $\lambda = 1$ and pump current is zero. The pump current is proportional to the exhaust-gas oxygen concentration and is this a non-linear measure for the excess-air factor λ (Fig. 2).

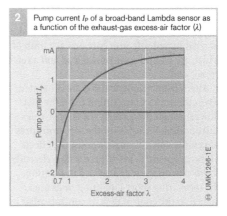

2 Pump current I_P of a broad-band Lambda sensor as a function of the exhaust-gas excess-air factor (λ)

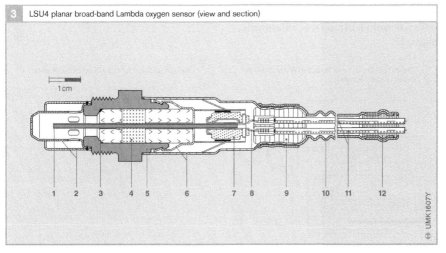

3 LSU4 planar broad-band Lambda oxygen sensor (view and section)

1 cm

1 2 3 4 5 6 7 8 9 10 11 12

© UMK1607Y

Fig. 3
1 Measuring cell (combination of Nernst concentration cell and oxygen-pump cell)
2 Double protective tube
3 Seal ring
4 Seal packing
5 Sensor housing
6 Protective sleeve
7 Contact holder
8 Contact clip
9 PTFE sleeve (Teflon)
10 PTFE shaped sleeve
11 Five connecting leads
12 Seal ring

Half-differential short-circuiting-ring sensors

Application

These sensors are also known as HDK (taken from the German) sensors, and are applied as position sensors for travel or angle, They are wear-free, as well as being very precise, and very robust, and are used as:

- Rack-travel sensors (RWG) for measuring the control-rack setting on in-line diesel injection pumps, and as
- Angle-of-rotation sensors in the injected-fuel-quantity actuators of diesel distributor pumps

Design and operating concept

These sensors (Figs. 1 and 2) are comprised of a laminated soft-iron core on each limb of which are wound a measuring coil and a reference coil.

Alternating magnetic fields are generated when the alternating current from the ECU flows through these coils. The copper rings surrounding the limbs of the soft-iron cores screen the cores, though, against the effects of the magnetic fields. Whereas the reference short-circuiting rings are fixed in position, the measuring short-circuiting rings are attached to the control rack or control-collar shaft (in-line pumps and distributor pumps respectively), with which they are free to move (control-rack travel s, or adjustment angle φ).

When the measuring short-circuiting ring moves along with the control rack or control-collar shaft, the magnetic flux changes and, since the ECU maintains the current constant (load-independent current), the voltage across the coil also changes.

The ratio of the output voltage U_A to the reference voltage U_{Ref} (Fig. 3) is calculated by an evaluation circuit. This ratio is proportional to the deflection of the measuring short-circuiting ring, and is processed by the ECU. Bending the reference short-circuiting ring adjusts the gradient of the characteristic curve, and the basic position of the measuring short-circuiting ring defines the zero position.

Fig. 1

1   Measuring coil
2   Measuring short-circuiting ring
3   Soft-iron core
4   Control-collar shaft
5   Reference coil
6   Reference short-circuiting ring

φ_{max} Adjustment-angle range for the control-collar shaft
φ Measured angle

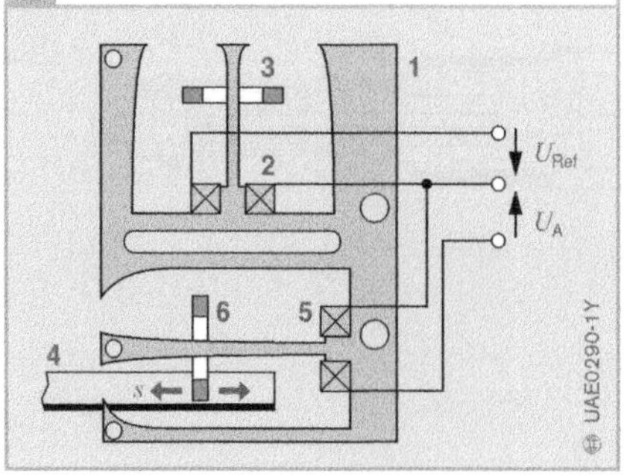

Fig. 2

1   Soft-iron core
2   Reference coil
3   Reference short-circuiting ring
4   Control rack
5   Measuring coil
6   Measuring short-circuiting ring

s Control-rack travel

Fig. 3

U_A Output voltage
U_{Ref} Reference voltage

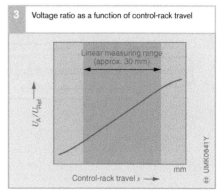

1 Design of the half-differential short-circuiting-ring sensor for diesel distributor pumps

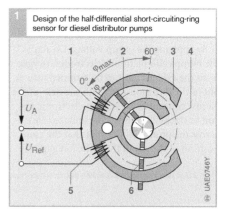

2 Design of the rack-travel sensor (RWG) for diesel in-line injection pumps

3 Voltage ratio as a function of control-rack travel

Fuel-level sensor

Application
It is the job of the fuel-level sensor to register the level of the fuel in the tank and send the appropriate signal to the ECU or to the display device in the vehicle's instrument panel. Together with the electric fuel pump and the fuel filter, it is part of the in-tank unit. These are installed in the fuel tank (gasoline or diesel fuel) and provide for an efficient supply of clean fuel to the engine (Fig. 1).

Design and construction
The fuel-level sensor (Fig. 2) is comprised of a potentiometer with wiper arm (wiper spring), printed conductors (twin-contact), resistor board (pcb), and electrical connections. The complete sensor unit is encapsulated and sealed against fuel. The float (fuel-resistant Nitrophyl) is attached to one end of the wiper lever, the other end of which is fixed to the rotatable potentiometer shaft (and therefore also to the wiper spring). Depending upon the particular version, the float can be either fixed in position on the lever, or it can be free to rotate). The layout of the resistor board (pcb) and the shape of the float lever and float are matched to the particular fuel-tank design.

Operating concept
The potentiometer's wiper spring is fixed to the float lever by a pin. Special wipers (contact rivets) provide the contact between the wiper spring and the potentiometer resistance tracks, and when the fuel level changes the wipers move along these tracks and generate a voltage ratio which is proportional to the float's angle of rotation. End stops limit the rotation range of 100° for maximum and minimum levels as well as preventing noise.

Operating voltage is 5...13 V.

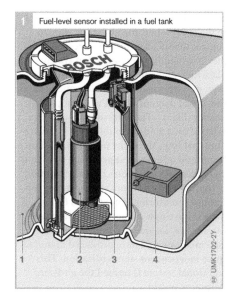

1 Fuel-level sensor installed in a fuel tank

Fig. 1
1 Fuel tank
2 Electric fuel pump
3 Fuel-level sensor
4 Float

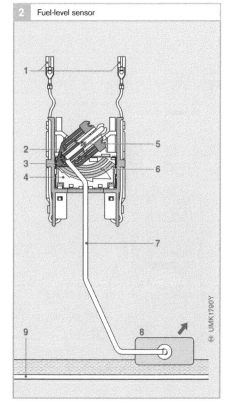

2 Fuel-level sensor

Fig. 2
1 Electrical connections
2 Wiper spring
3 Contact rivet
4 Resistor board
5 Bearing pin
6 Twin contact
7 Float lever
8 Float
9 Fuel-tank floor

Fault diagnostics

The rise in the sheer amount of electronics in the automobile, the use of software to control the vehicle, and the increased complexity of modern fuel-injection systems place high demands on the diagnostic concept, monitoring during vehicle operation (on-board diagnosis), and workshop diagnostics (Fig. 1). The workshop diagnostics is based on a guided troubleshooting procedure that links the many possibilities of on-board and offboard test procedures and test equipment. As emission-control legislation becomes more and more stringent and continuous monitoring is now called for, lawmakers have now acknowledged on-board diagnosis as an aid to monitoring exhaust-gas emissions, and have produced manufacturer-independent standardization. This additional system is termed the *on-board diagnostic system.*

Monitoring during vehicle operation (on-board diagnosis)

Overview

ECU-integrated diagnostics belong to the basic scope of electronic engine-management systems. Besides a self-test of the control unit, input and output signals, and control-unit intercommunication are monitored.

On-board diagnosis of an electronic system is the capability of a control unit to interpret and perform self-monitoring using "software intelligence", i.e. detect, store, and diagnostically interpret errors and faults. On-board diagnosis runs without the use of any additional equipment.

Monitoring algorithms check input and output signals during vehicle operation, and check the entire system and all its functions for malfunctions and disturbances. Any errors or faults detected are stored in the control-unit fault memory. Stored fault information can be read out via a serial interface.

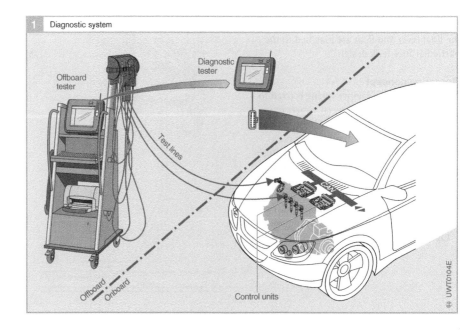

1   Diagnostic system

Diagnostic tester

Offboard tester

Test lines

CAN

Offboard

Onboard

Control units

UWT0104E

Input-signal monitoring

Sensors, plug connectors, and connecting lines (signal path) to the control unit (Fig. 2) are monitored by evaluating the input signal. This monitoring strategy is capable of detecting sensor errors, short-circuits in the battery-power circuit U_{Batt} and vehicle-ground circuit, and line breaks. The following methods are applied:

- Monitoring sensor supply voltage (if applicable).
- Monitoring detected values for permissible value ranges (e.g. 0.5....4.5 V).
- If additional information is available, a plausibility check is conducted using the detected value (e.g. comparison of crankshaft speed and camshaft speed).
- Critical sensors (e.g. pedal-travel sensor) are fitted in redundant configuration, which means that their signals can be directly compared with each other.

Output-signal monitoring

Actuators triggered by a control unit via output stages (Fig. 2) are monitored. The monitoring functions detects line breaks and short-circuits in addition to actuator faults. The following methods are applied:

- Monitoring an output signal by the output stage. The electric circuit is monitored for short-circuits to battery voltage U_{Batt}, to vehicle ground, and for open circuit.
- Impacts on the system by the actuator are detected directly or indirectly by a function or plausibility monitor. System actuators, e.g. exhaust-gas recirculation valves, throttle valves, or swirl flaps, are monitored indirectly via closed-control loops (e.g. continuous control variance), and also partly by means of position sensors (e.g. position of turbine geometry in the exhaust-gas turbocharger).

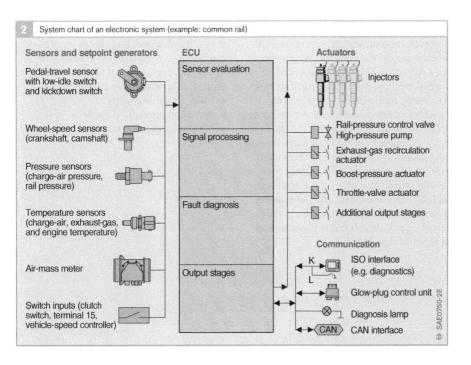

2 System chart of an electronic system (example: common rail)

Sensors and setpoint generators	ECU	Actuators
Pedal-travel sensor with low-idle switch and kickdown switch	Sensor evaluation	Injectors
Wheel-speed sensors (crankshaft, camshaft)	Signal processing	Rail-pressure control valve / High-pressure pump
Pressure sensors (charge-air pressure, rail pressure)		Exhaust-gas recirculation actuator
		Boost-pressure actuator
		Throttle-valve actuator
Temperature sensors (charge-air, exhaust-gas, and engine temperature)	Fault diagnosis	Additional output stages
		Communication
Air-mass meter	Output stages	ISO interface (e.g. diagnostics) / Glow-plug control unit
Switch inputs (clutch switch, terminal 15, vehicle-speed controller)		Diagnosis lamp / CAN interface

SAE0750-2E

Monitoring internal ECU functions

Monitoring functions are implemented in control-unit hardware (e.g. "intelligent" output-stage modules) and software to ensure that the control unit functions correctly at all times. The monitoring functions check each of the control-unit components (e.g. microcontroller, flash EPROM, RAM). Many tests are conducted immediately after startup. Other monitoring functions are performed during normal operation and repeated at regular intervals in order to detect component failure during operation. Test runs that require intensive CPU capacity, or that cannot be performed during vehicle operation for other reasons, are conducted in after-run more when the engine is switched off. This method ensures that the other functions are not interfered with. In the common-rail system for diesel engines, functions such as the injector switchoff paths are tested during engine runup or after-run. With a spark-ignition engine, functions such as the flash EPROM are tested in engine after-run.

Monitoring ECU communication

As a rule, communication with other ECUs takes place over the CAN bus (Controller Area Network). The CAN protocol contains control mechanisms to detect malfunctions. As a result, transmission errors are even detectable at CAN-module level. A number of other checks are also performed in the ECU. Since the majority of CAN messages are sent at regular intervals by the individual control units, the failure of a CAN controller in a control unit is detectable by testing at regular intervals. In addition, when redundant information is available in the ECU, the received signals are checked in the same way as all input signals.

Error handling

Error detection

A signal path is categorized at finally defective if an error occurs over a definite period of time. Until the defect is categorized, the system uses the last valid value detected. When the defect is categorized, a standby function is triggered (e.g. engine-temperature substitute value $T = 90°C$).

Most errors can be rectified or detected as intact during vehicle operation, provided the signal path remains intact for a definite period of time.

Fault storage

Each fault is stored as a fault code in the non-volatile area of the data memory. The fault code also describes the fault type (e.g. short-circuit, line break, plausibility, value range exceeded). Each fault-code input is accompanied by additional information, e.g. the operating and environmental conditions (freeze frame) at the time of fault occurrence (e.g. engine speed, engine temperature).

Limp-home function

If a fault is detected, limp-home strategies can be triggered in addition to substitute values (e.g. engine output power or speed limited). These strategies help to:
- Maintain driving safety
- Avoid consequential damage
- Minimize exhaust-gas emissions

On-board diagnosis system for passenger cars and light-duty trucks

The engine system and its components must be constantly monitored in order to comply with exhaust-gas emission limits specified by law in everyday driving situations. For this reason, regulations have come into force to monitor exhaust-gas systems and components, e.g. in California. This has standardized and expanded manufacturer-specific on-board diagnosis with respect to the monitoring of emission-related components and systems.

Legislation

OBD I (CARB)

1988 marked the coming into force of OBD I in California, that is, the first stage of CARB legislation (California Air Resources Board). The first OBD stage makes the following requirements:
- Monitoring emission-related electrical components (short-circuits, line breaks) and storage of faults in the control-unit fault memory.
- A Malfunction Indicator Lamp (MIL) that alerts the driver to the malfunction.
- The defective component must be displayed by means of on-board equipment (e.g. blink code using a diagnosis lamp).

OBD II (CARB)

The second stage of the diagnosis legislation (OBD II) came into force in California in 1994. OBD II became mandatory for diesel-engine cars with effect from 1996. In addition to the scope of OBD I, system functionality was now monitored (e.g. plausibility check of sensor signals).

OBD II stipulates that all emission-related systems and components must be monitored if they cause an increase in toxic exhaust-gas emissions in the event of a malfunction (by exceeding the OBD limits). Moreover, all components must be moni-

tored if they are used to monitor emission-related components or if they can affect the diagnosis results.

Normally, the diagnostic functions for all components and systems under surveillance must run at least once during the exhaust-gas test cycle (e.g. FTP 75). A further stipulation is that all diagnostic functions must run with sufficient frequency during daily driving mode. For many monitoring functions, the law defines a monitoring frequency (In Use Monitor Performance Ratio) in daily operation starting model year 2005.

Since the introduction of OBD II, the law has been revised in several stages (updates). The last update came into force in model year 2004. Further updates have been announced.

OBD (EPA)

Since 1994 the laws of the EPA (Environmental Protection Agency) have been in force in the remaining U.S. states. The scope of these diagnostics comply for the most part with the CARB legislation (OBD II).

The OBD regulations for CARB and EPA apply to all passenger cars with up to 12 seats and to light-duty trucks weighing up to 14,000 lbs (6.35 t).

EOBD (EU)

The OBD attuned to European conditions is termed EOBD and is based on the EPA-OBD.

EOBD has been valid for all passenger cars and light-duty trucks equipped with gasoline engines and weighing up to 3.5 t with up to 9 seats since January 2000. Since January 2003 the EOBD also applies to passenger cars and light-duty trucks with diesel engines.

Other countries

A number of other countries have already adopted or are planning to adopt EU or US-OBD legislation.

Requirements of the OBD system

The ECU must use suitable measures to monitor all on-board systems and components whose malfunction may cause a deterioration in exhaust-gas test specifications stipulated by law. The Malfunction Indicator Lamp (MIL) must alert the driver to a malfunction if the malfunction could cause an overshoot in OBD emission limits.

Emission limits

The U.S. OBD II (CARB and EPA) prescribes thresholds that are defined relative to emission limits. Accordingly, there are different permissible OBD emission limits for the various exhaust-gas categories that are applied during vehicle certification (e.g. TIER, LEV, ULEV). Absolute limits apply in Europe (Table 1).

Malfunction Indicator Lamp (MIL)

The Malfunction Indicator Lamp (MIL) alerts the driver that a component has malfunctioned. When a malfunction is detected, CARB and EPA stipulate that the MIL must light up no later than after two driving cycles of its occurrence. Within the scope of EOBD, the MIL must light up no later than in the third driving cycle with a detected malfunction.

If the malfunction disappears (e.g. loose contact), the malfunction remains entered in the fault memory for 40 trips (warmup cycles). The MIL goes out after three fault-free driving cycles.

Communication with scan tool

OBD legislation prescribes standardization of the fault-memory information and access to the information (connector, communication interface) compliant with ISO 15031 and the corresponding SAE standards (Society of Automotive Engineers). This permits the readout of the fault memory using standardized, commercially available testers (scan tools, see Fig. 1).

Depending on their scope of application, various communication protocols are used throughout the world: The most important are:

- ISO 9141-2 for European passenger cars.
- SAE J 1850 for U.S. passenger cars.
- ISO 14230-4 (KWP 2000) for European passenger cars and commercial vehicles.
- SAE J 1708 for U.S. commercial vehicles.

These serial interfaces operate at a bit rate (baud rate) of 5 to 10 k baud. They are designed as a single-wire interface with a common wire for both transmission and reception, or as a two-wire interface with a separate *data line* (K-line) and *initiate line* (L-line). Several electronic control units (such as Motronic, ESP, or EDC, and transmission-shift control, etc.) can be combined on one diagnosis connector.

Communication between the tester and ECU is set up in three phases:

- Initiate the ECU.
- Detect and generate baud rate.
- Read key bytes which identify the transmission protocol.

1 OBD limits for passenger cars and light-duty trucks				
	Otto passenger cars		**Diesel passenger cars**	

CARB	– Relative emission limits
	– Mostly 1.5 times the limit of
	  a specific exhaust-gas category
			– Relative emission limits
			– Mostly 1.5 times the limit of
			  a specific exhaust-gas category	

EPA	– Relative emission limits
	– Mostly 1.5 times the limit of
	  a specific exhaust-gas category
			– Relative emission limits (U.S. Federal)
			– Mostly 1.5 times the limit of
			  a specific exhaust-gas category	

EOBD	2000 CO: 3.2 g/km HC: 0.4 g/km NO$_X$: 0.6 g/km	2005 (draft) CO: 1.9 g/km HC: 0.3 g/km NO$_X$: 0.53 g/km	2003 CO: 3.2 g/km HC: 0.4 g/km NO$_X$: 1.2 g/km PM: 0.18 g/km	2005 (draft) CO: 3.2 g/km HC: 0.4 g/km NO$_X$: 1.2 g/km PM: 0.18 g/km

Table 1

Evaluation is performed subsequently. The following functions are possible:
- Identify the ECU
- Read the fault memory
- Erase the fault memory
- Read the actual values

In future, the CAN bus (ISO 15765-4) will be used increasingly to handle communication between ECUs and the tester. Starting 2008 diagnostics will only be permitted over this interface in the U.S.A.

To make it easier to read out the ECU fault-memory information, a standardized diagnosis socket will be fitted at an easily accessible place in every car (easy to reach from the driver's seat). The socket is used to connect the scan tool (Fig. 2).

Reading the fault information

Any workshop can use the scan tool to read out emission-relevant fault information from the ECU (Fig. 3). In this way, workshops not franchized to a particular manufacturer are also able to carry out repairs. Manufacturers are obliged to make the required tools and information available (on the internet) in return for a reasonable payment, to allow this.

Vehicle recall

If vehicles fail to comply with OBD requirements by law, the authorities may demand the vehicle manufacturer to start a recall at his own cost.

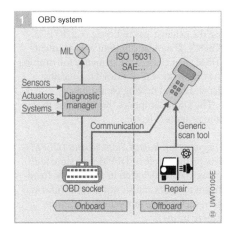

1   OBD system

Sensors
Actuators — Diagnostic manager
Systems

MIL ⊗   ISO 15031 SAE...

Communication   Generic scan tool

OBD socket   Repair

Onboard   Offboard

UWT0105E

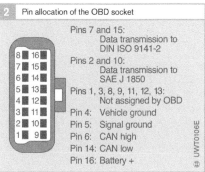

2   Pin allocation of the OBD socket

Pins 7 and 15:   Data transmission to DIN ISO 9141-2
Pins 2 and 10:   Data transmission to SAE J 1850
Pins 1, 3, 8, 9, 11, 12, 13:   Not assigned by OBD
Pin 4:   Vehicle ground
Pin 5:   Signal ground
Pin 6:   CAN high
Pin 14:   CAN low
Pin 16:   Battery +

UWT0106E

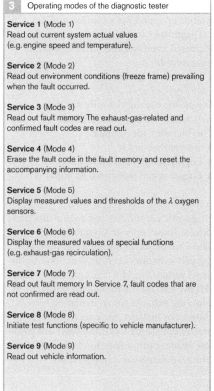

3   Operating modes of the diagnostic tester

Service 1 (Mode 1)
Read out current system actual values (e.g. engine speed and temperature).

Service 2 (Mode 2)
Read out environment conditions (freeze frame) prevailing when the fault occurred.

Service 3 (Mode 3)
Read out fault memory The exhaust-gas-related and confirmed fault codes are read out.

Service 4 (Mode 4)
Erase the fault code in the fault memory and reset the accompanying information.

Service 5 (Mode 5)
Display measured values and thresholds of the λ oxygen sensors.

Service 6 (Mode 6)
Display the measured values of special functions (e.g. exhaust-gas recirculation).

Service 7 (Mode 7)
Read out fault memory In Service 7, fault codes that are not confirmed are read out.

Service 8 (Mode 8)
Initiate test functions (specific to vehicle manufacturer).

Service 9 (Mode 9)
Read out vehicle information.

Functional requirements

Overview

Just as for on-board diagnosis, all ECU input and output signals, as well as the components themselves, must be monitored.

Legislation demands the monitoring of electrical functions (short-circuit, line breaks), a plausibility check for sensors, and a function monitoring for actuators.

The pollutant concentration expected as the result of a component failure (empirical values), and the monitoring mode partly required by law determine the type of diagnostics. A simple functional test (black/white test) only checks system or component operability (e.g. swirl flap opens and closes). The extensive functional test provides more detailed information about system operability. As a result, the limits of adaption must be monitored when monitoring adaptive fuel-injection functions (e.g. zero delivery calibration for a diesel engine, lambda adaption for a gasoline engine).

Diagnostic complexity has constantly increased as emission-control legislation has evolved.

Switchon conditions

Diagnostic capabilities only run if the switchon conditions are satisfied. This includes, for instance:

- Torque thresholds
- Engine-temperature thresholds
- Engine-speed thresholds or limits

Inhibit conditions

Diagnostic capabilities and engine functions cannot always operate simultaneously. There are inhibit conditions that prohibit the performance of certain functions. In the diesel system, the Hot-Film Air-Mass Meter (HFM) can only be monitored satisfactorily when the exhaust-gas recirculation valve is closed. For instance, tank ventilation (evaporative-emissions control system) in a gasoline system cannot function when catalytic-converter diagnosis is in operation.

Temporary disabling of diagnostic functions

Diagnostic capabilities may only be disabled under certain conditions in order to prevent false diagnosis. Examples include:

- Height too large.
- Low ambient temperature at engine switchon.
- Low battery voltage.

Readiness code

When the fault memory is checked, it is important to know that the diagnostic capabilities ran at least once. This can be checked by reading out the readiness code over the diagnostic interface. After erasing the fault memory in service, the readiness codes must be reset after checking the functions.

Diagnostic System Management (DSM)

The diagnostic capabilities for all components and systems checked must normally run in driving mode, but at least once during the exhaust-gas test cycle (e.g. FTP 75, NEDC). Diagnostic System Management (DSM) can change the sequence for running the diagnostic capabilities dynamically, depending on driving conditions.

The DSM comprises the following three components (Fig. 4):

Diagnosis Fault Path Management (DFPM)
The primary role of DFPM is to store fault states that are detected in the system. In addition to faults, it also stores other information, such as environmental conditions (freeze frame).

Diagnostic Function SCHEDuler (DSCHED)
DSCHED is responsible for coordinating assigned engine capabilities (MF) and diagnostic capabilities (DF). It obtains information from DVAL and DFPM to carry this out. In addition, it reports functions that require release by DSCHED to perform their readiness, after which the current system state is checked.

Diagnosis VALidator (DVAL)
The DVAL (only installed in gasoline systems to date) uses current fault-memory entries and additionally stored information to decide for each detected fault whether it is the actual cause, or a consequence of the fault. As a result, validation provides stored information to the diagnostic tester for use in reading out the fault memory.

In this way, diagnostic capabilities can be released in any sequence. All released diagnoses and their results are evaluated subsequently.

OBD functions

Overview

Whereas EOBD only contains detailed monitoring specifications for individual components, the requirements in CARB OBD II are much more detailed. The list below shows the current state of CARB requirements for gasoline-engined and diesel-engined passenger cars. Requirements that are also described in detail in the EOBD legislation are marked by (E):

- Catalytic converter (E), heated catalytic converter
- Combustion (ignition) misfire (E, for diesel system not for EOBD)
- Evaporation reduction system (fuel-tank leak diagnosis, only for gasoline system)
- Secondary-air injection
- Fuel system
- Lambda oxygen sensors (E)
- Exhaust-gas recirculation
- Crankcase ventilation
- Engine cooling system
- Cold-start emission reduction system (presently only for gasoline system)
- Air conditioner (components)
- Variable valve timing (presently only in use with gasoline systems)
- Direct ozone reduction system (presently only in use with gasoline systems)
- Particulate filter (particulate filter, only for diesel system) (E)
- Comprehensive components (E)
- Other emission-related components/systems (E)

"Other emission-related components/systems" refer to components and systems not mentioned in this last and that may exceed OBD emission limits, or block other diagnostic functions if they malfunction.

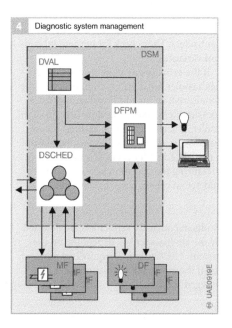

4 Diagnostic system management

DSM
DVAL
DFPM
DSCHED
MF   DF

UAE0919E

Catalytic converter diagnosis

In the diesel system, carbon monoxide (CO) and unburned hydrocarbons (HC) are oxidized in the oxidation-type catalytic converter. Work is ongoing on diagnostic capabilities for monitoring the operation of oxidation-type catalytic converters relating to temperature and differential pressure. One approach focuses on active secondary injection (intrusive operation). Here, heat is generated in the oxidation-type catalytic converter by an exothermic HC reaction. The temperature is measured and compared with calculated model values. These are used to derive the functionality of the catalytic converter.

Equally, work is ongoing on monitoring functions for the storage and regeneration capabilities of NO_x accumulator-type catalytic converters that will also be installed in the diesel system in future. The monitoring functions run based on loading and regeneration models, and the measured regeneration duration. This requires the use of lambda or NO_x sensors.

Combustion-miss detection

Incorrect fuel injection or loss of compression result in impaired combustion, and thus to changes in emission values. The misfire detector evaluates the time expired (segment time) from one combustion cycle to the next for each cylinder. This time is derived from the speed-sensor signal. A segment time that is longer than for the other cylinders indicates a misfire or loss of compression.

In the diesel system, diagnosis of combustion misses is only required and performed when the engine is at idle speed.

Fuel system diagnosis

In the common-rail system, a fuel-system diagnosis includes the electrical monitoring of injectors and rail-pressure control (high-pressure control). In the Unit Injector System, it also includes monitoring of the injector switching time. Special functions of the fuel-injection system that increase injected-fuel-quantity precision are also monitored. Examples of this include zero-fuel-quantity calibration, quantity-mean-value adaptation, and the AS MOD observer function (air-system model observer). The two last functions use information from the lambda oxygen sensor as input signals. From the models, they calculate fluctuations between setpoint and actual quantities.

Lambda-oxygen-sensor diagnosis

Modern diesel systems are fitted with broadband oxygen sensors. They require a different diagnostic procedure than two-stage sensors since their settings may deviate from $\lambda = 1$. They are monitored electrically (short-circuit, line interruption) and for plausibility. The heater element of the sensor heater is tested electrically and for permanent governor deviation.

Exhaust-gas recirculation system diagnosis

In the exhaust-gas recirculation system, the EGR valve and – if fitted – the exhaust-gas cooler are monitored.

The exhaust-gas recirculation valve is monitored for its electrical and functional operability. Functional monitoring is performed by air-mass regulators and position controllers. They check for permanent control variances.

If the exhaust-gas recirculation system has a cooler fitted, its function must also be monitored, i.e. an additional temperature measurement takes place downstream of the cooler. The temperature measured is compared with a setpoint value calculated from a model. If a fault occurs, it is detected by measuring the deviation between setpoint and actual values.

Crankcase ventilation diagnosis

Faults in crankcase ventilation are detected by the air-mass sensor, depending on the system. The legislation requires no monitoring if the crankcase ventilation has a "rugged" design.

Engine cooling system diagnosis
The cooling system comprises a thermostat and a coolant-termperature sensor. If the thermostat is defective, for instance, the engine temperature can only rise slowly and, consequently, the exhaust emission rates may increase. The diagnostic function for the thermostat uses the coolant-termperature sensor to check that a nominal temperature has been reached. A temperature model is also used for monitoring.

The coolant-termperature sensor is monitored to ensure that a minimum temperature has been reached in addition to monitoring for electrical faults by means of a dynamic plausibility function. Dynamic plausibility is performed as the engine cools down. These functions can monitor the sensor for "sticking" in both low and high temperature ranges.

Air-conditioner diagnosis
The engine can be operated at a different operating point in order to cover the air-conditioner's electrical load requirements. The required diagnosis must therefore monitor all electronic components in the air conditioner that may cause an increase in emissions if they malfunction.

Particulate-filter diagnosis
The particulate filter is currently monitored for filter breakage, removal, or blockage. A differential-pressure sensor is used to measure differential pressure (exhaust-gas backpressure downstream and upstream of the filter) at a specific volumetric flow. The measured value can be used to decide whether the filter is defective.

Comprehensive components
On-board diagnosis legislation requires that all sensors (e.g. air-mass meter, wheel-speed sensors, temperature sensors) and actuators (e.g. throttle valve, high-pressure pump, glow plugs) must be monitored if they either have an impact on emissions, or are used to monitor other components or systems (and consequently, may block other diagnoses).

Sensors monitor the following faults (Fig. 5):
- Electrical faults, i.e. short-circuits and line breaks *(signal-range check)*.
- Range faults *(out-of-range check)*, i.e. undercutting or exceeding voltage limits defined by the physical measurement range of a sensor.
- Plausibility faults *(rationality check);* these are faults that are inherent in the components themselves (e.g. drift), or which may be caused by shunts, for instance. Monitoring is carried out by a plausibility check on the sensor signals, either by using a model, or directly using other sensors.

Actuators must be monitored for electrical faults and – if technically possible – also for function. Functional monitoring means that, when a control command (setpoint value) is given, it is monitored by observing or measuring (e.g. by a position sensor) the system reaction (actual value) in a suitable way by using information from the system.

Besides all output stages, the following actuators are monitored:
- throttle valve
- exhaust-gas recirculation valve
- variable turbine geometry of the exhaust-gas turbocharger
- swirl flap
- glow plugs

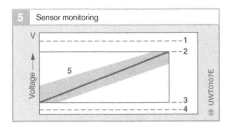

Fig. 5
1   Upper threshold for *signal-range check*
2   Upper threshold for *out-of-range check*
3   Lower threshold for *out-of-range check*
4   Lower threshold for *signal-range check*
5   Plausibility range *rationality check*

On-board diagnosis system for heavy-duty trucks

In Europe and the U.S.A., there exist draft laws that have not yet been adopted; they are based closely on legislation for passenger cars.

Legislation

In the EU, there are plans to introduce new type approvals in October 2005 (coinciding with Euro 4 emission-control legislation). With effect from October 2006 an OBD system will become obligatory for every commercial vehicle. In the U.S.A., the draft of the Californian CARB provides for the introduction of an OBD system for Model Year (MY) 2007. It is probable that EPA (U.S. Federal) will also follow with a draft in 2004 for subsequent introduction in MY 2007. Besides that, there are initiatives to promote worldwide harmonization (World Wide Harmonized (WWH) OBD). However, this is not expected until 2012. Japan is planning to introduce an OBD system in 2005.

EOBD for trucks and buses > 3.5 t

European OBD legislation provides for a two-stage introduction. Stage 1 (2005) requires monitoring:
- of the fuel-injection system for closed electrical circuit and total failure.
- of emission-related engine components or systems for compliance with OBD emission limits (Table 1).
- of the exhaust-gas treatment system for major functional faults (e.g. damaged catalytic converter, urea deficit in the SCR system).

Stage 2 (2008) requires:
- Monitoring of the exhaust-gas treatment system for emission limits.
- The OBD emission limits must be adapted to the prevailing state of the art (availability of exhaust-gas sensors).

Protocols for scan-tool communication over CAN have been approved using either ISO 15765 or SAE J1939.

CARB OBD for HD trucks > 14,000 lb. (6.35 t)

The present draft law is very close to passenger-car legislation in its function requirements, and also provides for a two-stage introduction:
- MY 2007: Monitoring for functional faults.
- MY 2010: Monitoring for OBD emission limits (Table 1).

The main changes compared with present passenger-car legislation are as follows:
- Erasing the OBD fault memory by scan tool is no longer possible. This is only possible by self-healing (e.g. after repair).
- SAE J1939 has also been approved as an alternative to CAN diagnostic communication to ISO 15765 (as for passenger cars).

1	OBD emission limits for heavy-duty trucks (draft)	
CARB	2007 – Functional check no limits	2010 – Relative limit – 1.5 times the value of each exhaust-gas category – Exception: catalytic converter, factor 1.75
EPA	– to be defined	– to be defined
EU	2005 – Absolute limit NO_x: 7.0 g/kWh PM: 0.1 g/kWh – Functional check for exhaust-gas treatment system	2008 – Absolute limit NO_x: 7.0 g/kWh PM: 0.1 g/kWh – Subject to review by EU Commission

Table 1

"Once you have driven an automobile, you will soon realize that there is something unbelievably tiresome about horses (…). But you do require a conscientious mechanic for the automobile (…)".

Robert Bosch wrote these words to his friend Paul Reusch in 1906. In those days, it was indeed the case that breakdowns could be repaired on the road or at home by an employed chauffeur or mechanic. However, with the growing number of motorists driving their own cars after the First World War, the need for workshops offering repair services in-creased rapidly. In the 1920s Robert Bosch started to systematically create a nationwide customer-service organization. In 1926 all the repair centers were uniformly named "Bosch Service" and the name was registered as a trademark.

Today's Bosch Service agencies retain the same name. They are equipped with the latest electronic equipment in order to meet the demands of 21st-century automotive technology and the quality expectations of the customers.

1 A repair shop in 1925 (photo: Bosch)

2 A Bosch car service workshop in the 21st century carried out with the very latest electronic testing equipment

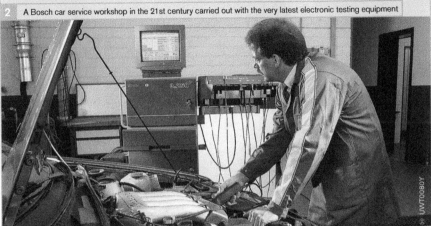

Service technology

More than 30,000 garages/workshops around the world are equipped with workshop technology, i.e., test technology and workshop software from Bosch. Workshop technology is becoming increasingly important as it provides guidance and assistance in all matters relating to diagnosis and troubleshooting.

Workshop business

Trends

Many factors influence workshop business. Current trends are, for example:

- The proportion of diesel passenger cars is rising
- Longer service intervals and longer service lives of automotive parts mean that vehicles are being checked into workshops less frequently
- Workshop capacity utilization in the overall market will continue to decline in the next few years

- The amount of electronic components in vehicles is increasing – vehicles are becoming "mobile computers"
- Internetworking of electronic systems is increasing, diagnostic and repair work covers systems which are installed and networked in the entire vehicle
- Only the use of the latest test technology, computers and diagnostic software will safeguard business in the future

Consequences
Requirements

Workshops must adapt to the trends in order to be able to offer their services successfully on the market in the future. The consequences can be derived directly from the trends:

- Professional fault diagnosis is the key to professional repairs
- Technical information is becoming the crucial requirement for vehicle repairs
- Rapid availability of comprehensive technical information safeguards profitability

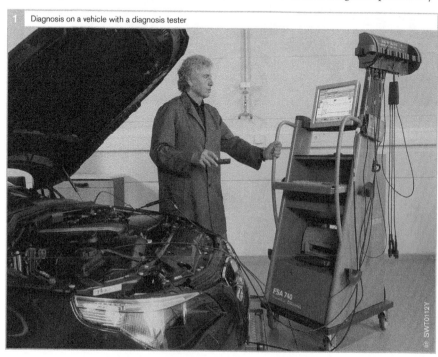

1   Diagnosis on a vehicle with a diagnosis tester

- The need for workshop personnel to be properly qualified is increasing dramatically
- Investment by workshops in diagnosis, technical information and training is essential

Measurement and test technology
The crucial step for workshops to take is to invest in the right test technology, diagnostic software, technical information, and technical training in order to receive the best possible support and assistance for all the jobs and tasks in the workshop process.

Workshop processes
The essential tasks which come up in the workshop can be portrayed in processes. Two distinct subprocesses are used for handling all tasks in the service and repair fields. The first subprocess covers the predominantly operations- and organizaton-based activity of *job-order acceptance*, while the second subprocess covers the predominantly technically based work steps of *service* and *repair implementation*.

Job-order acceptance
When a vehicle arrives in the workshop, the job-order acceptance system's database furnishes immediate access to all available information on the vehicle. The moment the vehicle enters the shop, the system offers access to the vehicle's entire service history, including all service and repairs that it has received in the past. Furthermore, this sequence involves the completion of all tasks relating to the customer's request, its basic feasibility, scheduling of completion dates, provision of resources, parts and working materials and equipment, and an initial examination of the task and extent of work involved. Depending on the process objective, all subfunctions of the ESI[tronic] product are used within the framework of the *service acceptance* process.

Service and repair implementation
Here, the jobs defined within the framework of the job-order acceptance are carried out. If it is not possible to complete the task in a single process cycle, appropriate repeat loops must be provided until the process result which is aimed for is achieved. Depending on the process objective, all subfunctions of the ESI[tronic] product are used within the framework of the service and repair implementation process.

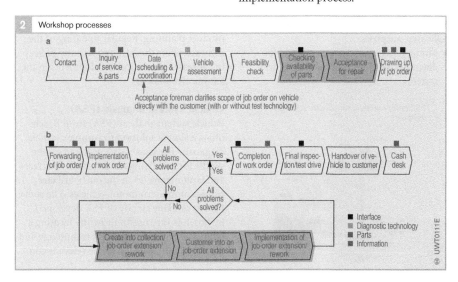

Fig. 2
a Job-order acceptance
b Service and repair implementation

Electronic Service Information (ESI[tronic])

System functions for supporting the workshop process

ESI[tronic] is a modular software product for the automotive-engineering trade. The individual modules contain the following information:

- Technical information on spare parts and automotive equipment
- Exploded views and parts lists for spare parts and assemblies
- Technical data and setting values
- Flat rate units and times for work on the vehicle
- Vehicle diagnosis and vehicle-system diagnosis
- Troubleshooting instructions for different vehicle systems
- Repair instructions for vehicle components, e.g., diesel power units
- Electronic circuit diagrams
- Maintenance schedules and diagrams
- Test and setting values for assemblies
- Data for costing maintenance, repair and service work

Application

The chief users of ESI[tronic] are motor garages/workshops, assembly repairers and the automotive-parts wholesale trade. They use the technical information for the following purposes:

- Motor garages/workshops: mainly for diagnosis, service and repair of vehicle systems

- Assembly repairers: mainly for testing, adjusting and repair of assemblies
- Automotive-parts wholesale trade: mainly for parts information

Garages/workshops and assembly repairers use this parts information in addition to diagnosis, repair and service information. Product interfaces enable ESI[tronic] to network with other (particularly commercial) software in the workshop environment and the automotive-parts wholesale trade in order, for instance, to exchange data with the accounting merchandise information system.

Benefit to the user of ESI[tronic]

The benefit of using ESI[tronic] lies in the fact that the system furnishes a large amount of information which is needed to conduct and safeguard the business of motor garages/workshops. This is made possible by the broadly conceived and modular ESI[tronic] product program. The information is offered on one interface with a standardized system for all vehicle marques.

Comprehensive vehicle coverage is important for workshop business in that the necessary information is always to hand. This is guaranteed by ESI[tronic] because country-specific vehicle databases and information on new vehicles are incorporated in the product planning. Regular updating of the software offers the best opportunity of keeping abreast of technical developments in the automotive industry.

Vehicle system analysis (FSA)

Vehicle system analysis (FSA) from Bosch offers a simple solution to complex vehicle diagnosis. The causes of a problem can be swiftly located thanks to diagnosis interfaces and fault memories in the on-board electronics of modern motor vehicles. The *component checking* facility of FSA developed by Bosch is very useful in swiftly locating a fault: The FSA measurement technology and display can be adjusted to the relevant com-

3   ESI[tronic] workshop software for all vehicle marques

SWE0020Y

ponent. This enables this component to be tested while it is still installed.

Measuring equipment

Workshop personnel can choose from various options for diagnosis and troubleshooting: the high-performance, portable KTS 650 system tester or the workshop-compatible KTS 520 and KTS 550 KTS modules in conjunction with a standard PC or laptop. The modules have an integrated multimeter, and KTS 550 and KTS 650 also have a 2-channel oscilloscope. For work applications on the vehicle, ESI[tronic] is installed in the KTS 650 or on a PC.

Example of the sequence in the workshop

The ESI[tronic] software package supports workshop personnel throughout the entire vehicle repair process A diagnosis interface allows ESI[tronic] to communicate with the electronic systems within the vehicle, such as the engine control unit. Working at the PC, the user starts by selecting the SIS (Service Information System) utility to initiate diagnosis of on-board control units and access the engine control unit's fault storage.

The diagnosis tester provides the data needed for direct comparisons of specified results and current readings, without the need for supplementary entries. ESI[tronic] uses the results of the diagnosis as the basis for generating specific repair instructions. The system also provides displays with other information, such as component locations, exploded views of assemblies, diagrams showing the layouts of electrical, pneumatic and hydraulic systems, etc. Working at the PC, users can then proceed directly from the exploded views to the parts lists with part numbers to order the required replacement components. All service procedures and replacement components are recorded to support the billing process. After the final road test, the bill is produced simply by pressing a few keys. The system also provides a clear and concise printout with the results of the vehicle diagnosis. This offers the customer a full report detailing all of the service operations and materials that went into the vehicle's repair.

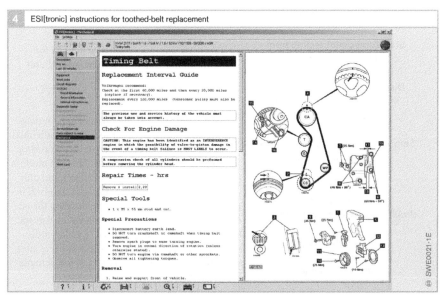

4 ESI[tronic] instructions for toothed-belt replacement

Diagnostics in the workshop

The function of this diagnosis is to identify the smallest, defective, replaceable unit quickly and reliably. The guided fault-finding procedure includes onboard information and offboard test procedures and testers. Support is provided by Electronic Service Information (ESI[tronic]). It contains instructions for further fault-finding for many possible problems (e.g. engine bucks) and faults (e.g. short-circuit in engine-temperature sensor).

Guided fault-finding

The main element is the guided fault-finding procedure. The workshop employee is guided by means of a symptom-dependent, event-controlled procedure – starting from the symptom (vehicle symptom or fault-memory entry). Onboard (fault-memory entry) and offboard facilities (actuator diagnosis and onboard testers) are used.

Guided fault-finding, fault-memory read-outs, workshop diagnostic functions, and electrical communication with offboard testers take place using PC-based diagnostic testers. This may be a specific workshop tester from the vehicle manufacturer or a universal tester (e.g. KTS 650 by Bosch).

Reading out fault-memory entries

Fault information (fault-memory entries) stored during vehicle operation are read out via a serial interface during vehicle service or repair in the customer-service workshop.

Fault entries are read out using a diagnostic tester. The workshop employee receives information about:

- Malfunctions (e.g. engine-temperature sensor)
- Fault codes (e.g. short-circuit to ground, implausible signal, static fault)
- Ambient conditions (measured values on fault storage, e.g. engine speed, engine temperature, etc.).

Once the fault information has been retrieved in the workshop and the fault corrected, the fault memory can be cleared again using the tester.

A suitable interface must be defined for communication between the control unit and the tester.

Actuator diagnostics

The control unit contains an actuator diagnostic routine in order to activate individual actuators at the customer-service workshop and test their functionality. This test mode is started using the diagnostic tester and only functions when the vehicle is at standstill below a specific engine speed, or when the engine is switched off. This allows an acoustic (e.g. valve clicking), visual (e.g. flap movement), or other type of inspection, e.g. measurement of electric signals, to test actuator function.

1 Flowchart of a guided fault-finding procedure with CAS[plus]

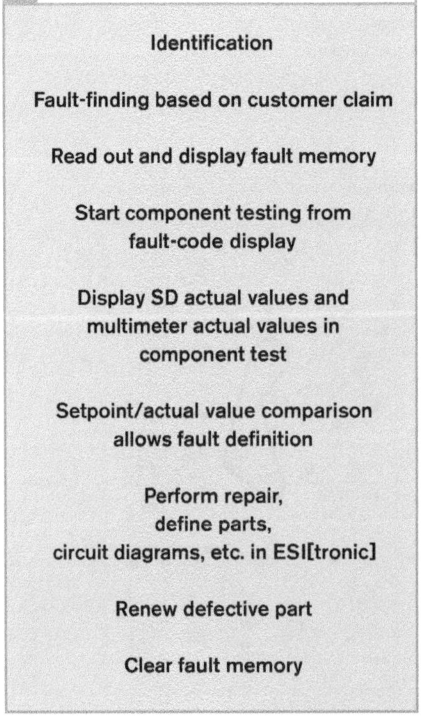

Identification

Fault-finding based on customer claim

Read out and display fault memory

Start component testing from fault-code display

Display SD actual values and multimeter actual values in component test

Setpoint/actual value comparison allows fault definition

Perform repair, define parts, circuit diagrams, etc. in ESI[tronic]

Renew defective part

Clear fault memory

Fig. 1
The CAS[plus] system (Computer Aided Service) combines control-unit diagnosis with SIS fault-finding instructions for even more efficient fault-finding. The decisive values for diagnostics and repair then appear immediately on screen.

Workshop diagnostic functions

Faults that the on-board diagnosis fails to detect can be localized using support functions. These diagnostic functions are implemented in the engine control unit and are controlled by the diagnostic tester.

Workshop diagnostic functions run automatically, either after they are started by the diagnostic tester, or they report back to the diagnostic tester at the end of the test, or the diagnostic tester assumes runtime control, measured data acquisition, and data evaluation. The control unit then implements individual commands only.

Example

During a compression test, the fuel-injection system is switched off while the engine is turned by the starter motor. The engine ECU records the crankshaft speed pattern. The compression in each of the cylinders can be deduced from speed fluctuations, i.e. the difference between the lowest and highest revolutions, thus giving an indication of engine condition

Offboard tester

The diagnostic capabilities are expanded by using additional sensors, test equipment, and external evaluators. In the event of a fault detected in the workshop, offboard testers are adapted to the vehicle.

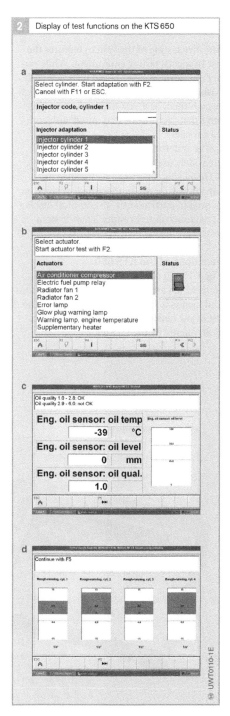

2 Display of test functions on the KTS 650

Fig. 2

a Adapting an injector
b Selecting an actuator test
c Reading out engine-specific data
d Evaluating smooth-running characteristics

Testing equipment

Effective testing of the system requires the use of special testing equipment. While earlier electronic systems could be tested with basic equipment such as a multimeter, ongoing advances have resulted in electronic systems that can only be diagnosed with complex testers.

The system testers in the KTS series are widely used by vehicle repairers. The KTS 650 (Fig. 1) offers a wide range of capabilities for use in the vehicle repairs, enhanced in particular by its graphical display of data such as test results. These system testers are also known as diagnosis testers.

Functions of the KTS 650

The KTS 650 offers a wide variety of functions which are selected by means of buttons and menus on the large display screen. The list below details the most important functions offered by the KTS 650.

Identification

The system automatically detects the connected ECU and reads actual values, fault memories and ECU-specific data.

Reading/erasing the fault memory

The fault information detected during vehicle operation by on-board diagnosis and stored in the fault memory can be read with the KTS 650 and displayed on screen in plain text.

Reading actual values

Current values which the engine control unit calculates can be read out as physical quantities (e.g, engine speed in rpm).

Actuator diagnostics

The electrical actuators (e.g., valves, relays) can be specifically triggered for function-testing purposes.

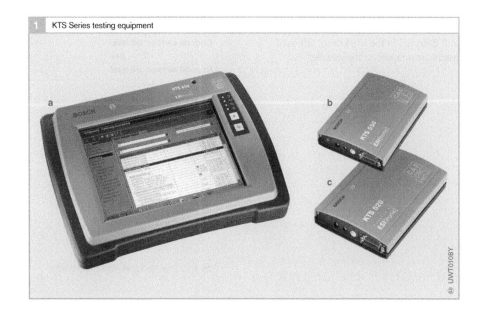

1   KTS Series testing equipment

⊕ UWT0108Y

Fig. 1
a   KTS 650
b   KTS 550 module
c   KTS 520 module

Engine test

The system tester initiates programmed test sequences in the engine control unit for checking the engine-management system or the engine (e.g., compression test).

Multimeter function

Electrical current, voltage and resistance can be tested in the same way as with a conventional multimeter.

Time graph display

The recorded measured values can be shown in graphic displays as signal curve similar to those available from an oscilloscope (e.g., Lambda-sensor voltage, signal voltage of hot-film air-mass sensor).

Additional information

Specific additional information relevant to the faults/components displayed can also be shown in conjunction with the Electronic Service Information (ESI[tronic]) (e.g., troubleshooting instructions, location of components in the engine compartment, test specifications, electrical circuit diagrams).

Printout

All data (e.g., list of actual values or document for the customer) can be printed out on standard PC printers.

Programming

The software of the engine control unit can be coded with the KTS 650 (e.g., automatic or manual transmission).

The extent to which the capabilities of the KTS 650 can be utilized in the workshop depends on the system to be tested. Not all ECUs support its full range of functions.

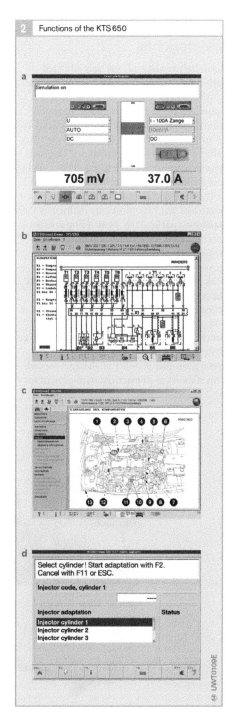

2 Functions of the KTS 650

Fig. 2
a   Multimeter function
b   Graphical display of a terminal diagram
c   Display of location of components in engine compartment
d   Function selection

Fuel-injection pump test benches

Accurately tested and precisely adjusted fuel-injection pumps and governor mechanisms are key components for obtaining optimized performance and fuel economy from diesel engines. They are also crucial in ensuring compliance with increasingly strict exhaust-gas emission regulations. The fuel-injection pump test bench (Fig. 1) is a vital tool for meeting these requirements.

The main specifications governing both test bench and test procedures are defined by ISO standards; particularly demanding are the specifications for rigidity and geometrical consistency in the drive unit (5).

As time progresses, so do the levels of peak pressure that fuel-injection pumps are expected to generate. This development is reflected in higher performance demands and power requirements for pump test benches. Powerful electric drive units, a large flyweight and precise control of rotational speed guarantee stability at all engine speeds. This stability is an essential requirement for repeatable, mutually comparable measurements and test results.

Flow measurement methods

An important test procedure is to measure the fuel pumped each time the plunger moves through its stroke. For this test, the fuel-injection pump is clamped on the test bench support (1), with its drive side connected to the test bench drive coupling. Testing proceeds with a standardized calibrating oil at a precisely monitored and controlled temperature. A special, precision-calibrated nozzle-and-holder assembly (3) is connected to each pump barrel. This strategy ensures mutually comparable measurements for each test. Two test methods are available.

Glass gauge method (MGT)

The test bench features an assembly with two glass gauges (Fig. 2, 5). A range of gages with various capacities are available for each cylinder. This layout can be used to test fuel-injection pumps for engines of up to 12 cylinders.

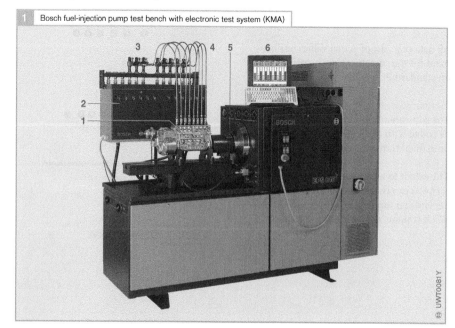

1   Bosch fuel-injection pump test bench with electronic test system (KMA)

Fig. 1

1   Fuel-injection pump on test bench

2   Quantity test system (KMW)

3   Test nozzle-and-holder assembly

4   High-pressure test line

5   Electric drive unit

6   Control, display and processing unit

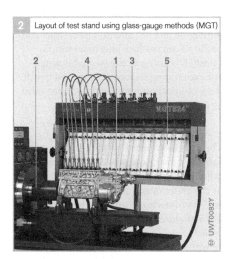

2   Layout of test stand using glass-gauge methods (MGT)

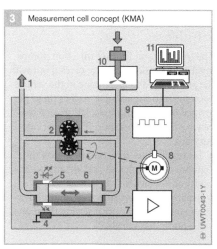

3   Measurement cell concept (KMA)

Fig. 2
1 Fuel-injection pump
2 Electric drive unit
3 Test nozzle-and-holder assembly
4 High-pressure test line
5 Glass gages

Fig. 3
1 Return line to calibrating oil tank
2 Gear pump
3 LED
4 Photocell
5 Window
6 Plunger
7 Amplifier with electronic control circuitry
8 Electric motor
9 Pulse counter
10 Test nozzle-and-holder assembly
11 Monitor (PC)

In the first stage, the discharged calibrating flows past the glass gages to return directly to the oil tank. As soon as the fuel-injection pump reaches the rotational speed indicated in the test specifications, a slide valve opens, allowing the calibrating oil from the fuel-injection pump to flow to the glass gages. Supply to the glass containers is then interrupted when the pump has executed the preset number of strokes.

The fuel quantity delivered to each cylinder in cm³ can now be read from each of the glass gages. The standard test period is 1,000 strokes, making it easy to interpret the numerical result in mm³ per stroke of delivered fuel. The test results are compared with the setpoint values and entered in the test record.

Electronic flow measurement system (KMA)
This system replaces the glass gauges with a control, display and processor unit (Fig. 1, 6). While this unit is usually mounted on the test bench, it can also be installed on a cart next to the test bench.

This test relies on continuous measuring the delivery capacity (Fig. 3). A control plunger (6) is installed in parallel with the input and output sides of a gear pump (2). When the pump's delivery quantity equals the quantity of calibrating oil emerging from the

test nozzle (10), the plunger remains in its center position. If the flow of calibrating oil is greater, the plunger moves to the left – if the flow of calibrating oil is lower, the plunger moves to the right. This plunger motion controls the amount of light traveling from an LED (3) to a photocell (4). The electronic control circuitry (7) records this deviation and responds by varying the pump's rotational speed until its delivery rate again corresponds to the quantity of fluid emerging from the test nozzle. The control plunger then returns to its center position. The pump speed can be varied to measure delivery quantity with extreme precision.

Two of these measurement cells are present on the test bench. The computer connects all of the test cylinders to the two measurement cells in groups of two, proceeding sequentially from one group to the next (multiplex operation). The main features of this test method are:
- Highly precise and reproducible test results
- Clear test results with digital display and graphic presentation in the form of bar graphs
- Test record for documentation
- Supports adjustments to compensate for variations in cooling and/or temperature

Testing in-line fuel-injection pumps

The test program for fuel-injection pumps involves operations that are carried out with the pump fitted to the engine in the vehicle (system fault diagnosis) as well as those performed on the pump in isolation on a test bench or in the workshop. This latter category involves
- Testing the fuel-injection pump on the pump test bench and making any necessary adjustments
- Repairing the fuel-injection pump/governor and subsequently resetting them on the pump test bench

In the case of in-line fuel-injection pumps, a distinction has to be made between those with mechanical governors and those which are electronically controlled. In either case, the pump and its governor/control system are tested in combination, as both components must be matched to each other.

The large number and variety of in-line fuel-injection pump designs necessitates variations in the procedures for testing and adjustment. The examples given below can, therefore, only provide an idea of the full extent of workshop technology.

Adjustments made on the test bench

The adjustments made on the test bench comprise
- Start of delivery and cam offset for each individual pump unit
- Delivery quantity setting and equalization between pump units
- Adjustment of the governor mounted on the pump
- Harmonization of pump and governor/control system (overall system adjustment)

For every different pump type and size, separate testing and repair instructions and specifications are provided which are specifically prepared for use with Bosch pump test benches.

The pump and governor are connected to the engine lube-oil circuit. The oil inlet connection is on the fuel-injection pump's camshaft housing or the pump housing. For each testing sequence on the test bench, the fuel-injection pump and governor must be topped up with lube oil.

Testing delivery quantity

The fuel-injection pump test bench can measure the delivery quantity for each individual cylinder (using a calibrated tube apparatus or computer operating and display terminal, see "Fuel-injection pump test benches"). The individual delivery quantity figures obtained over a range of different settings must be within defined tolerance limits. Excessive divergence of individual delivery quantity figures would result in uneven running of the engine. If any of the delivery quantity figures are outside the specified tolerances, the pump barrel(s) concerned must be readjusted. There are different procedures for this depending on the pump model.

Governor/control system adjustment

Governor

Testing of mechanical governors involves an extensive range of adjustments. A dial gauge is used to check the control-rack travel at defined speeds and control-lever positions on the fuel-injection pump test bench. The test results must match the specified figures. If there are excessive discrepancies, the governor characteristics must be reset. There are a number of ways of doing this, such as changing the spring characteristics by altering spring tension, or by fitting new springs.

Electronic control system

If the fuel-injection pump is electronically controlled, it has an electromechanical actuator that is operated by an electronic control unit instead of a directly mounted governor. That actuator moves the control rack and thus controls the injected fuel quantity. Otherwise, there is no difference in the mechanical operation of the fuel-injection pump.

During the tests, the control rack is held at a specific position. The control-rack travel must be calibrated to match the voltage signal of the rack-travel sensor. This done by adjusting the rack-travel sensor until its signal voltage matches the specified signal level for the set control-rack travel.

In the case of control-sleeve in-line fuel-injection pumps, the start-of-delivery solenoid is not connected for this test in order to be able to obtain a defined start of delivery.

Adjustments with the pump in situ
The pump's start of delivery setting has a major influence on the engine's performance and exhaust-gas emission characteristics. The start of delivery is set, firstly, by correct adjustment of the pump itself, and secondly, by correct synchronization of the pump's camshaft with the engine's timing system. For this reason, correct mounting of the injection pump on the engine is extremely important. The start of delivery must therefore be tested with the pump mounted on the engine in order to ensure that it is correctly fitted.

There are a number of different ways in which this can be done depending on the pump model. The description that follows is for a Type RSF governor.

On the governor's flyweight mount, there is a tooth-shaped timing mark (Fig. 1). In the governor housing, there is a threaded socket which is normally closed off by a screw cap. When the piston that is used for calibration (usually no. 1 cylinder) is in the start-of-delivery position, the timing mark is exactly in line with the center of the threaded socket. This "spy hole" in the governor housing is part of a sliding flange.

Fitting the fuel-injection pump
Locking the camshaft
The fuel-injection pump leaves the factory with its camshaft locked (Fig. 1a) and is mounted on the engine when the engine's crankshaft is set at a defined position. The pump lock is then removed. This tried and tested method is economical and is adopted increasingly widely.

Start-of-delivery timing mark
Synchronizing the fuel-injection pump with the engine is performed with the aid of the start-of-delivery timing marks, which have to be brought into alignment. Those marks are to be found on the engine as well as on the fuel-injection pump (Fig. 2 overleaf). There are several methods of determining the start of delivery depending on the pump type.

Normally, the adjustments are based on the engine's compression stroke for cylinder no. 1 but other methods may be adopted for reasons related to specific engine designs. The engine manufacturer's instructions must therefore always be observed. On most diesel engines, the start-of-delivery timing mark is on the flywheel, the crankshaft pulley or the vibration damper. The vibration damper is generally mounted on the crankshaft in the position normally occupied by the V-belt pulley, and the pulley then bolted to the vibration damper. The complete assembly then looks rather like a thick V-belt pulley with a small flywheel.

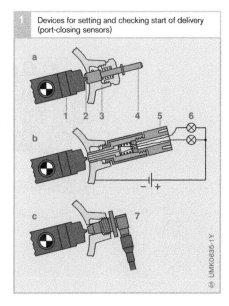

1 Devices for setting and checking start of delivery (port-closing sensors)

Fig. 1
Illustration shows Type RSF governor; other types have a sliding flange

a Locked in position by locking pin
b Testing with an optical sensor (indicator-lamp sensor)
c Testing with an inductive sensor (governor signal method)

1 Governor flyweight mount
2 Timing mark
3 Governor housing
4 Locking pin
5 Optical sensor
6 Indicator lamp
7 Inductive speed sensor

Checking static start of delivery
Checking with indicator-lamp sensor
The tooth-shaped timing mark can be located with the aid of an optical sensor, the indicator-lamp sensor (Fig. 1b), which is screwed into the socket in governor housing. When it is opposite the sensor, the two indicator lamps on the sensor light up. The start of delivery in degrees of crankshaft rotation can then be read off from the flywheel timing marks, for example.

High-pressure overflow method
The start-of-delivery tester is connected to the pressure outlet of the relevant pump barrel (Fig. 3). The other pressure outlets are closed off. The pressurized fuel flows through the open inlet passage of the pump barrel and exits, initially as a jet, into the observation vessel (3). As the engine crankshaft rotates, the pump plunger moves towards its top dead center position. When it reaches the start-of-delivery position, the pump plunger closes off the barrel's inlet passage. The injection jet entering the observation vessel thus dwindles and the fuel flow is reduced to a drip. The start of delivery in degrees of crank shaft rotation is read off from the timing marks.

Checking dynamic start of delivery
Checking with inductive sensor
An inductive sensor that is screwed into the socket in the governor housing (Fig. 1c) supplies an electrical signal every time the governor timing mark passes when the engine is running. A second inductive sensor supplies a signal when the engine is at top dead center (Fig. 4). The engine analyzer, to which the two inductive sensors are connected, uses those signals to calculate the start of delivery and the engine speed.

Checking with a piezoelectric sensor and a stroboscopic timing light
A piezoelectric sensor is fixed to the high-pressure delivery line for the cylinder on which adjustment is to be based. As soon as the fuel-injection pump delivers fuel to that cylinder, the high-pressure delivery line expands slightly and the piezoelectric sensor transmits an electrical signal. This signal is received by an engine analyzer which uses it to control the flashing of a stroboscopic timing light. The timing light is pointed at the timing marks on the engine. When illuminated by the flashing timing light, the flywheel timing marks appear to be stationary. The angular value in degrees of crankshaft rotation can then be read off for start of delivery.

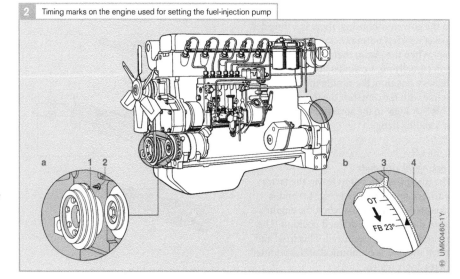

2   Timing marks on the engine used for setting the fuel-injection pump

Fig. 2
a   V-belt pulley timing marks
b   Flywheel timing marks

1   Notch in V-belt pulley
2   Marker point on cylinder block
3   Graduated scale on flywheel
4   Timing mark on crankcase

Venting

Air bubbles in the fuel impair the proper operation of the fuel-injection pump or disable it entirely. Therefore, if the system has been temporarily out of use it should be carefully vented before being operated again. There is generally a vent screw on the fuel-injection pump overflow or the fuel filter for this purpose.

Lubrication

Fuel-injection pumps and governors are normally connected to the engine lube-oil circuit as the fuel-injection pump then requires no maintenance.

Before being used for the first time, the fuel-injection pump and the governor must be filled with the same type of oil that is used in the engine. In the case of fuel-injection pumps that are not directly connected to the engine lube-oil circuit, the pump is filled through the filler cap after removing the vent flap or filter. The oil level check takes place at the same time as the regular engine oil changes and is performed by removing the oil check plug on the governor. Excess oil (from leak fuel) is then drained off or the level topped up if required. Whenever the fuel-injection pump is removed or the engine overhauled,

the oil must be changed. Fuel-injection pumps and governors with separate oil systems have their own dipsticks for checking the oil level.

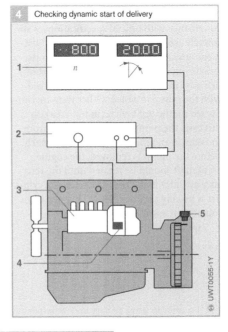

4 Checking dynamic start of delivery

Fig. 4

Schematic diagram of in-line fuel-injection pump and governor using port-closing sensor system

1 Engine analyzer
2 Adaptor
3 In-line fuel-injection pump and governor
4 Inductive speed sensor (port-closing sensor)
5 Inductive speed sensor (TDC sensor)

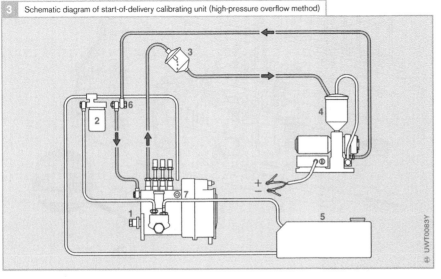

3 Schematic diagram of start-of-delivery calibrating unit (high-pressure overflow method)

Fig. 3

1 Fuel-injection pump
2 Fuel filter
3 Observation vessel
4 Start-of-delivery calibrating unit
5 Fuel tank
6 Oversize banjo bolt and nut
7 Screw cap

Testing helix and port-controlled distributor injection pumps

Good engine performance, high fuel economy and low emissions depend on correct adjustment of the helix and port-controlled distributor injection pump. This is why compliance with official specifications is absolutely essential during testing and adjustment operations on fuel-injection pumps.

One important parameter is the start of delivery (in service bay), which is checked with the pump installed. Other tests are conducted on the test bench (in test area). In this case, the pump must be removed from the vehicle and mounted on the test bench. Before the pump is removed, the engine crankshaft should be rotated until the reference cylinder is at TDC. The reference cylinder is usually cylinder No. 1. This step eases subsequent assembly procedures.

Test bench measurements

The test procedures described here are suitable for use on helix and port-controlled axial-piston distributor pumps with electronic and mechanical control, but not with solenoid-controlled distributor injection pumps.

Test bench operations fall into two categories:
- Basic adjustment and
- Testing

The results obtained from the pump test are entered in the test record, which also lists all the individual test procedures. This document also lists all specified minimum and maximum results. The test readings must lie within the range defined by these two extremes.

A number of supplementary, special-purpose test steps are needed to assess all the different helix and port-controlled axial-piston distributor pumps; detailed descriptions of every contingency, however, extend beyond the bounds of this chapter.

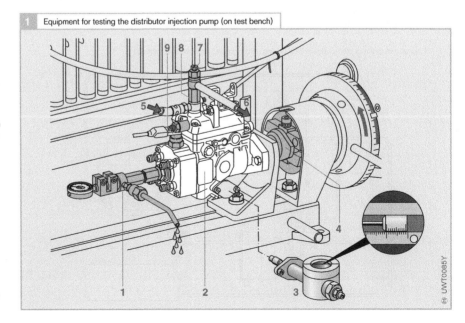

1   Equipment for testing the distributor injection pump (on test bench)

Fig. 1

1 Test layout with drain hose and dial gauge
2 Distributor injection pump
3 Timing device travel tester with vernier scale
4 Pump drive
5 Calibrating oil inlet
6 Return line
7 Overflow restrictor
8 Adapter with connection for pressure gauge
9 Electric shutoff valve (ELAB) (energized)

UWT0085Y

Basic adjustments

The first step is to adjust the distributor injection pump to the correct basic settings. This entails measuring the following parameters under defined operating conditions.

LPC adjustment

This procedure assesses the distributor plunger lift between Bottom Dead Center (BDC) and the start of delivery. The pump must be connected to the test-bench fuel supply line for this test. The technician unscrews the 6-point bolt from the central plug fitting and then installs a test assembly with drain tube and gauge in its place (Fig. 1, 1).

The gauge probe rests against the distributor plunger, allowing it to measure lift. Now the technician turns the pump's input shaft (4) by hand until the needle on the gauge stops moving. The control plunger is now at Top Dead Center (TDC).

A supply pressure of roughly 0.5 bar propels the calibrating oil into the plunger chamber behind the distributor plunger (5). For this test, the solenoid-operated shutoff valve (ELAB) (9) is kept energized to maintain it in its open position. The calibrating oil thus flows from the plunger chamber to the test assembly before emerging from the drain hose.

Now the technician manually rotates the input shaft in its normal direction of rotation. The calibrating oil ceases to flow into the plunger chamber once its inlet passage closes. The oil remaining in the chamber continues to emerge from the drain hose. This point in the distributor plunger's travel marks the start of delivery.

The lift travel between Bottom Dead Center (BDC) and the start of delivery indicated by the gauge can now be compared with the setpoint value. If the reading is outside the tolerance range, it will be necessary to dismantle the pump and replace the cam mechanism between cam disk and plunger.

Supply-pump pressure

As it affects the timing device, the pressure of the supply pump (internal pressure) must also be tested. For this procedure, the overflow restrictor (7) is unscrewed and an adapter with a connection to the pressure gauge (8) is installed. Now the overflow restrictor is installed in an adapter provided in the test assembly. This makes it possible to test the pump's internal chamber pressure upstream of the restrictor.

A plug pressed into the pressure-control valve controls the tension on its spring to determine the pump's internal pressure. Now the technician continues pressing the plug into the valve until the pressure reading corresponds to the setpoint value.

Timing device travel

The technician removes the cover from the timing device to gain access for installing a travel tester with a vernier scale (3). This scale makes it possible to record travel in the timing device as a function of rotational speed; the results can then be compared with the setpoint values. If the measured timing device travel does not correspond with the setpoint values, shims must be installed under the timing spring to correct its initial spring tension.

Adjusting the basic delivery quantity

During this procedure the fuel-injection pump's delivery quantity is adjusted at a constant rotational speed for each of the following four conditions:
- Idle (no-load)
- Full-load
- Full-load governor regulation and
- Starting

Delivery quantities are monitored using the MGT or KMA attachment on the fuel-injection pump test bench (refer to section on "Fuel-injection pump test benches").

First, with the control lever's full-load stop adjusted to the correct position, the full-load governor screw in the pump cover is adjusted to obtain the correct full-load delivery quantity at a defined engine speed. Here, the governor adjusting screw must be turned back to prevent the full-load stop from reducing delivery quantity.

The next step is to measure the delivery quantity with the control lever against the idle-speed stop screw. The idle-speed stop screw must be adjusted to ensure that the monitored delivery quantity is as specified.

The governor screw is adjusted at high rotational speed. The measured delivery quantity must correspond to the specified full-load delivery quantity.

The governor test also allows verification of the governor's intervention speed. The governor should respond to the specified rpm threshold by first reducing and then finally interrupting the fuel flow. The breakaway speed is set using the governor speed screw.

There are no simple ways to adjust the delivery quantity for starting. The test conditions are a rotational speed of 100 rpm and the control lever against its full-load shutoff stop. If the measured delivery quantity is below a specified level, reliable starting cannot be guaranteed.

Testing

Once the basic adjustment settings have been completed, the technician can proceed to assess the pump's operation under various conditions. As during the basic adjustment procedure, testing focuses on
- Supply-pump pressure
- Timing device travel
- Delivery quantity curve

The pump operates under various specific conditions for this test series, which also includes a supplementary procedure.

Overflow quantity

The vane-type supply pump delivers more fuel than the nozzles can inject. The excess calibrating oil must flow through the overflow restriction valve and back to the oil tank. It is the volume of this return flow that is measured in this procedure. A hose is connected to the overflow restriction valve; de-

pending on the selected test procedure. The other end is then placed in a glass gauge in the MGT assembly, or installed on a special connection on the KMA unit. The overflow quantity from a 10-second test period is then converted to a delivery quantity in liters per hour.

If the test results fail to reach the setpoint values, this indicates wear in the vane-type supply pump, an incorrect overflow valve or internal leakage.

Dynamic testing of start of delivery

A diesel engine tester (such as the Bosch ETD 019.00) allows precise adjustment of the distributor injection pump's delivery timing on the engine. This unit registers the start of delivery along with the timing adjustments that occur at various engine speeds with no need to disconnect any high-pressure delivery lines.

Testing with piezo-electric sensor and stroboscopic timing light

The piezo-electric sensor (Fig. 2, 4) is clamped onto the high-pressure delivery line leading to the reference cylinder. Here, it is important to ensure that the sensor is mounted on a straight and clean section of tubing with no bends; the sensor should also be positioned as close as possible to the fuel-injection pump.

The start of delivery triggers pulses in the fuel-injection line. These generate an electric signal in the piezo-electric sensor. The signal controls the light pulses generated by the timing light (5). The timing light is now aimed at the engine's flywheel. Each time the pump starts delivery to the reference cylinder, the timing light flashes, lighting up the TDC mark on the flywheel. This allows correlation of timing to flywheel position. The flashes occur only when delivery to the reference cylinder starts, producing a static image. The degree markings (6) on the crankshaft or flywheel show the crankshaft position relative to the start of delivery.

Engine speed is also indicated on the diesel engine tester.

Setting start of delivery

If the results of this start-of-delivery test deviate from the test specifications, it will be necessary to change the fuel-injection pump's angle relative to the engine.

The first step is to switch off the engine. Then the technician rotates the crankshaft until the reference cylinder's piston is at the point at which delivery should start. The crankshaft features a reference mark for this operation; the mark should be aligned with the corresponding mark on the bellhousing. The technician now unscrews the 6-point screw from the central plug screw. As for basic adjustment process on the test bench, the technician now installs a dial-gauge assembly in the opening. This is used to observe distributor plunger travel while the crankshaft is being turned. As the crankshaft is turned counter to its normal direction of rotation (or in the normal direction on some engines), the plunger retracts in the pump. The technician should stop turning the crankshaft once the needle on the gauge stops moving. The plunger is now at bottom dead center. Now the dial gauge is reset to zero. The crankshaft is then rotated in its normal direction of rotation as far as the TDC mark. The dial gauge now indicates the travel executed by the distributor plunger on

its way from its bottom dead center position to the TDC mark on the reference cylinder. It is vital to comply with the precise specification figure for this travel contained in the fuel-injection pump's datasheet. If the dial gauge reading is not within the specification, it will be necessary to loosen the attachment bolt on the pump flange, turn the pump housing and repeat the test. It is important to ensure that the cold-start accelerator is not active during this procedure.

Measuring the idle speed

The idle speed is monitored with the engine heated to its normal operating temperature, and in a no-load state, using the engine tester. The idle speed can be adjusted using the idle-speed stop screw.

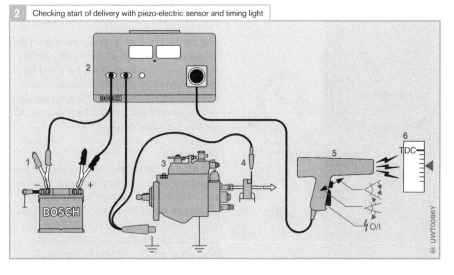

2   Checking start of delivery with piezo-electric sensor and timing light

Fig. 2
1   Battery
2   Diesel tester
3   Distributor injection pump
4   Piezo-electric sensor
5   Stroboscopic timing light
6   Angle and TDC marks

Nozzle tests

The nozzle-and-holder assembly consists of the nozzle and the holder. The holder includes all of the required filters, springs and connections.

The nozzle affects the diesel engine's output, fuel economy, exhaust-gas composition and operating refinement. This is why the nozzle test is so important.

An important tool for assessing nozzle performance is the nozzle tester.

Nozzle tester

The nozzle tester is basically a manually operated fuel-injection pump (Fig. 1). For testing, a high-pressure delivery line (4) is used to connect the nozzle-and-holder assembly (3) to the tester. The calibrating oil is contained in a tank (5). The required pressure is generated using the hand lever (8). The pressure gage (6) indicates the pressure of the calibrating oil; a valve (7) can be used to disconnect it from the high-pressure circuit for specific test procedures.

1   Nozzle tester with nozzle-and-holder assembly

1   2   3   4   5   6   7   8

UWT0078Y

Fig. 1
1   Suction equipment
2   Injection jet
3   Nozzle-and-holder assembly
4   High-pressure test line
5   Calibrating oil tank with filter
6   Pressure gage
7   Valve
8   Hand lever

The EPS100 (0684200704) nozzle tester is specified for testing nozzles of Sizes P, R, S and T. It conforms to the standards defined in ISO 8984. The prescribed calibrating oil is defined in ISO standard 4113. A calibration case containing all the components is required to calibrate inspect the nozzle tester.

This equipment provides the basic conditions for reproducible, mutually compatible test results.

Test methods

Ultrasonic cleaning is recommended for the complete nozzle-and-holder assemblies once they have been removed from the engine. Cleaning is mandatory on nozzles when they are submitted for warranty claims.

Important: Nozzles are high-precision components. Careful attention to cleanliness is vital for ensuring correct operation.

The next step is to inspect the assembly to determine whether any parts of the nozzle or holder show signs of mechanical or thermal wear. If signs or wear are present, it will be necessary to replace the nozzle or nozzle-and-holder assembly.

The assessment of the nozzle's condition proceeds in four test steps, with some variation depending on whether the nozzles are pintle or hole-type units.

Chatter test

The chatter test provides information on the smoothness of action of the needle. During injection, the needle oscillates back and forth to generate a typical chatter. This motion ensures efficient dispersion of the fuel particles.

The pressure gage should be disconnected for this test (close valve).

Pintle nozzle

The lever on the nozzle tester is operated at a rate of one to two strokes per second. The pressure of the calibrating oil rises, ultimately climbing beyond the nozzle's opening pressure. During the subsequent discharge, the nozzle should produce an audible chatter; if it fails to do so, it should be replaced.

When installing a new nozzle in its holder, always observe the official torque specifications, even on hole-type nozzles.

Hole-type nozzle
The hand lever is pumped at high speed. This produces a hum or whistling sound, depending on the nozzle type. No chatter will be present in some ranges. Evaluation of chatter is difficult with hole-type nozzles. This is why the chatter test is no longer assigned any particular significance as an assessment tool for hole-type nozzles.

Spray pattern test
High pressures are generated during this test. Always wear safety goggles.

The hand lever is subjected to slow and even pressure to produce a consistent discharge plume. The spray pattern can now be evaluated. It provides information on the condition of the injection orifices. The prescribed response to an unsatisfactory spray pattern is to replace the nozzle or nozzle-and-holder assembly.

The pressure gage should also be switched off for this test.

Pintle nozzle
The spray should emerge from the entire periphery of the injection orifice as even tapered plume. There should be no concentration on one side (except with flatted pintle nozzles).

Hole-type nozzle
An even tapered plume should emerge from each injection orifice. The number of individual plumes should correspond to the number of orifices in the nozzle.

Checking the opening pressure
Once the line pressure rises above the opening pressure, the valve needle lifts from its seat to expose the injection orifice(s). The specified opening pressure is vital for correct operation of the overall fuel-injection system.

The pressure gage must be switched back on for this test (valve open).

Pintle nozzle and hole-type nozzle with single-spring nozzle holder
The operator slowly presses the lever downward, continuing until the gage needle indicates the highest available pressure. At this point, the valve opens and the nozzle starts to discharge fuel. Pressure specifications can be found in the "nozzles and nozzle-holder components" catalog.

Opening pressures can be corrected by replacing the adjustment shim installed against the compression spring in the nozzle holder. This entails extracting the nozzle from the nozzle holder. If the opening pressure is too low, a thicker shim should be installed; the response to excessive opening pressures is to install a thinner shim.

Hole-type nozzle with two-spring nozzle holder
This test method can only be used to determine the initial opening pressure on two-spring nozzle-and-holder assemblies.

The is no provision for shim replacement on some nozzle-and-holder assemblies. The only available response with these units is to replace the entire assembly.

Leak test
The pressure is set to 20 bar above the opening pressure. After 10 seconds, formation of a droplet at the injection orifice is acceptable, provided that the droplet does not fall.

The prescribed response to an unsuccessful leak test is to replace the nozzle or nozzle-and-holder assembly.

Exhaust-gas emissions

Increasing energy consumption, which is mainly covered by the energy contained in fossil fuels, have made air quality a critical issue. The quality of the air we breathe depends on a wide range of factors. In addition to emissions from industry, homes, and power plants, the exhaust gas generated by motor vehicles also plays a significant role. In developed countries, this accounts for about 20% of total emissions.

Overview

The statutory limits restricting pollutant emissions from motor vehicles have been progressively tightened in recent years. In order to achieve compliance with these limits, vehicles have been equipped with supplementary emissions-control systems.

Combustion of the air/fuel mixture

A basic rule that applies to all internal-combustion engines is that absolutely complete combustion does not occur inside the engine's cylinders. This rule remains valid even when the combustion mixture contains excess air. Less efficient combustion leads to an increase in levels of toxic components containing carbon in the exhaust gas. In addition to a high percentage of nontoxic elements, the internal-combustion engine's gas also contains byproducts which – at least when present in high concentrations – represent potential sources of environmental damage. These are classified as pollutants.

Positive crankcase ventilation

Additional emissions stem from the engine's crankcase ventilation system. Combustion gases travel along the cylinder walls and into the crankcase. From there, they are returned to the intake manifold for renewed combustion within the engine.

Since nothing more than pure air is compressed in the diesel's compression stroke, diesel engines generate only negligible amounts of these bypass emissions. The gases that make their way into the crankcase contain only about 10% as much pollution as the bypass gases in a gasoline engine. Despite this fact, closed crankcase-ventilation systems are now mandatory on diesel engines.

Evaporative emissions

Additional emissions can escape from vehicles powered by gasoline engines when volatile components in the fuel evaporate and emerge from the fuel tank, regardless of whether the vehicle is moving or parked. These emissions consist primarily of hydrocarbons. To prevent these gases from evaporating directly to the atmosphere, vehicles must be equipped with an evaporative-emissions control system designed to store them for subsequent combustion in the engine.

Evaporative emissions from diesels are not a major concern, as diesel fuel possesses virtually no high-volatility components.

Major components

Assuming the presence of sufficient oxygen, ideal, complete combustion of pure fuel would be possible in the following chemical reaction:

$$n_1 \, C_xH_y + m_1 \, O_2 \rightarrow n_2 \, H_2O + m_2 \, CO_2$$

The absence of ideal conditions for combustion combines with the composition of the fuel itself to produce a certain number of toxic components in addition to the primary combustion products water (H_2O) and carbon dioxide (CO_2) (Fig. 1).

Water (H_2O)

During combustion, the water chemically bound within the fuel is transformed into water vapor, most of which condenses when its cools. This is the source of the exhaust plume visible on cold days. The proportion of water in the exhaust gas is dependent on the operating point in diesel engines.

Carbon dioxide (CO_2)

In complete combustion, the hydrocarbons in the fuel's chemical bonds are transformed to carbon dioxide (CO_2). Its proportion is also dependent on the operating point. Here again, the proportion depends on the engine operating conditions. The amount of converted carbon dioxide in the exhaust gas is directly proportional to fuel consumption. Thus the only way to reduce carbon-dioxide emissions when using standard fuels is to reduce fuel consumption.

Carbon dioxide is a natural component of atmospheric air, and the CO_2 contained in automotive exhaust gases is not classified as a pollutant. However, it is one of the sub-stances responsible for the greenhouse effect and the global climate change that this causes. In the period since 1920, atmospheric CO_2 has risen continuously, from roughly 300 ppm to approx. 450 ppm in the year 2001. This renders efforts to reduce carbon-dioxide emissions and fuel consumption more important than ever.

Nitrogen (N_2)

Nitrogen is the primary constituent (78%) of the air drawn in by the engine. Although it is not directly involved in the combustion process, it is the largest single component within the exhaust gas, at approximately 69...75%.

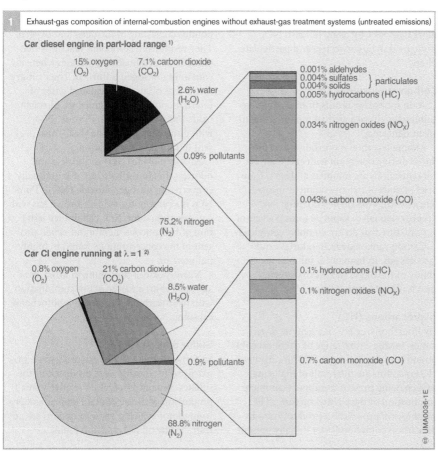

1 Exhaust-gas composition of internal-combustion engines without exhaust-gas treatment systems (untreated emissions)

Car diesel engine in part-load range [1]

15% oxygen (O_2)
7.1% carbon dioxide (CO_2)
2.6% water (H_2O)
0.09% pollutants
75.2% nitrogen (N_2)

0.001% aldehydes
0.004% sulfates } particulates
0.004% solids
0.005% hydrocarbons (HC)
0.034% nitrogen oxides (NO_x)
0.043% carbon monoxide (CO)

Car CI engine running at $\lambda = 1$ [2]

0.8% oxygen (O_2)
21% carbon dioxide (CO_2)
8.5% water (H_2O)
0.9% pollutants
68.8% nitrogen (N_2)

0.1% hydrocarbons (HC)
0.1% nitrogen oxides (NO_x)
0.7% carbon monoxide (CO)

Fig. 1
Figures in percent by weight

The concentrations of exhaust-gas components, especially pollutants, may vary. Among other factors, they depend on the engine operating conditions and the ambient conditions (e.g. air humidity).

[1] NO_x and particulate emissions can be reduced by more than 90% by the use of NO_x storage catalytic converters and particulate filters.

[2] Catalytic converters, which are standard equipment today, can reduce pollutant emissions by up to 99%.

UMA0036-1E

Combustion byproducts

During combustion, the air-fuel mixture generates a number of byproducts. The most significant of these combustion byproducts are:
- Carbon monoxide (CO)
- Hydrocarbons (HC), and
- Nitrogen oxides (NO_X)

Engine modifications and exhaust-gas treatment can reduce the amount of pollutants produced.

Since they always operate with excess air, diesel engines inherently produce much smaller amounts of CO and HC than gasoline engines. The main emphasis is therefore on NO_X and particulate emissions. Both of these types of emissions can be reduced by more than 90% by using modern NO_X storage catalytic converters and particulate filters.

Carbon monoxide (CO)
Carbon monoxide results from incomplete combustion in rich air/fuel mixtures due to an air deficiency.

Although carbon monoxide is also produced during operation with excess air, the concentrations are minimal, and stem from brief periods of rich operation or inconsistencies within the air/fuel mixture. Fuel droplets that fail to vaporize form pockets of rich mixture that do not combust completely.

Carbon monoxide is an odorless and tasteless gas. In humans, it inhibits the ability of the blood to absorb oxygen, thus leading to asphyxiation.

Hydrocarbons (HC)
Hydrocarbons, or HC, is a generic designation for the entire range of chemical compounds uniting hydrogen H with carbons C. HC emissions are the result of inadequate oxygen being present to support complete combustion of the air/fuel mixture. The combustion process also produces new hydrocarbon compounds not initially present in the original fuel (by separating extended molecular chains, etc.).

Aliphatic hydrocarbons (alkanes, alkenes, alkines and their cyclical derivatives) are virtually odorless. Cyclic aromatic hydrocarbons (such as benzol, toluol and polycyclic hydrocarbons) emit a discernible odor.

Some hydrocarbons are considered to be carcinogenic in long-term exposure. Partially oxidized hydrocarbons (aldehydes, ketones, etc.) emit an unpleasant odor. The chemical products that result when these substances are exposed to sunlight are also considered to act as carcinogens under extended exposure to specified concentrations.

Nitrogen oxides (NO_X)
Nitrogen oxides, or oxides of nitrogen, is the generic term embracing chemical compounds consisting of nitrogen and oxygen. They result from secondary reactions that occur in all combustion processes where air containing nitrogen is burned. The primary forms that occur in the exhaust gases of internal-combustion engines are nitrogen oxide (NO) and nitrogen dioxide (NO_2), with dinitrogen monoxide (N_2O) also present in minute concentrations.

Nitrogen oxide (NO) is colorless and odorless. In atmospheric air, it is gradually converted to nitrogen dioxide (NO_2). Pure NO_2 is a poisonous, reddish-brown gas with a penetrating odor. NO_2 can induce irritation of the mucous membranes when present in the concentrations found in highly-polluted air.

Nitrous oxides contribute to forest damage (acid rain) and also act in combination with hydrocarbons to generate photochemical smog.

Sulfur dioxide (SO_2)
Sulfurous compounds in exhaust gases – primarily sulfur dioxide – are produced by the sulfates contained in fuels. A relatively small proportion of these pollutant emissions stem from motor vehicles. These emissions are not restricted by official emission limits.

It is not possible to use a catalytic converter to convert sulfur dioxide. Sulfur forms deposits within catalytic converters, reacting with the active chemical layer and inhibiting the catalytic converter's ability to remove other pollutants from the exhaust gases. While sulfur contamination can be reversed in the NO_X storage catalytic converters used for emissions control with direct-injection gasoline engines, this process requires a considerable amount of energy, and consequently negate the fuel-economy benefits achieved by direct injection.

The earlier limits on sulfur concentrations within fuel of 500 ppm (parts per million, 1,000 ppm = 0.1%), valid until the end of 1999, have now been tightened by EU legislation. The new limits, valid from 2000 onward, are 150 ppm for gasoline and 350 ppm for diesel fuels. A further reduction to 50 ppm for both types of fuel is slated for 2005. In practice, however, sulfur-free fuel will be introduced sooner. Gasoline and diesel fuel with a sulfur content of ≤ 10 ppm will already be available throughout Germany in 2003 (throughout the EU by 2005).

Particulates

The problem of particulate emissions is primarily associated with diesel engines. Levels of particulate emissions from gasoline engines with multipoint injection systems are negligible.

Particulates result from incomplete combustion. While exhaust-gas composition varies as a function of the combustion process and engine operating condition, these particulates basically consist of hydrocarbon chains (soot) with an extremely extended specific surface ratio. Uncombusted and partly combusted hydrocarbons form deposits on the soot, where they are joined by aldehydes, with their penetrating odor. Aerosol components (minutely dispersed solids or fluids in gases) and sulfates bond to the soot. The sulfates result from the sulfur content in the fuel. Consequently, these pollutants do not occur if sulfur-free fuel is used.

▶ Greenhouse effect

Shortwave solar radiation penetrates the earth's atmosphere and continues to the ground, where it is absorbed. This process promotes warming in the ground, which then radiates long-wave heat, or infrared energy. A portion of this radiation is reflected by the atmosphere, causing the earth to warm.

Without this natural greenhouse effect the earth would be an inhospitable planet with an average temperature of −18°C. Greenhouse gases within the atmosphere (water vapor, carbon dioxide, methane, ozone, dinitrogen oxide, aerosols and particulate mist) raise average temperatures to approximately +15°C. Water vapor, in particular, retains substantial amounts of heat.

Carbon dioxide has risen substantially since the dawn of the industrial age more than 100 years ago. The primary source of this increase has been the combustion of coal and petroleum products. In this process, the carbon bound in the fuels is released in the form of carbon dioxide.

The processes that influence the greenhouse effect within the earth's atmosphere are extremely complex. While some scientists maintain that anthropogenic (of human origin) emissions are the primary source of climate change, this theory is challenged by other experts, who believe that the warming of the earth's atmosphere is being caused by increased solar activity.

There is, however, a large degree of unanimity in calling for reductions in energy use to lower carbon-dioxide emissions and combat the greenhouse effect.

Emission-control legislation

The state of California assumed a pioneering role in efforts to restrict toxic emissions emanating from motor vehicles. This development arises from the fact that the geography of large cities like Los Angeles prevents wind from dispersing exhaust gases, causing a blanket of fog that smothers the city. The resulting smog not only has damaging impacts on the health of city dwellers, but also impairs visibility severely.

Overview

California introduced the first emission-control legislation for gasoline in the 1960s. These regulations became progressively more stringent in the ensuing years. Meanwhile all industrialized countries have introduced emission-control laws which define limits for gasoline and diesel engines, as well as the test procedures employed to confirm compliance.

The most important legal restrictions on exhaust-gas emissions are (Fig. 1):
- CARB legislation (California Air Resources Board)
- EPA legislation (Environmental Protection Agency), U.S.A.
- EU legislation (European Union)
- Japanese legislation

Classifications

Countries with legal limits on motor-vehicle emissions divide vehicles into various classes:
- Passenger cars: Emission testing is conducted on a chassis dynamometer.
- Light-duty trucks: Depending on national legislation, the top limit for gross weight rating is 3.5...6.35 t. Testing is conducted on a chassis dynamometer (as for passenger cars).
- Heavy-duty trucks: Gross weight rating over 3.5...6.35 t. The test is conducted on an engine test bench. No provision is made for vehicle testing.
- Off-highway (e.g. construction, agricultural, and forestry vehicles): Tested on an engine test bench, as for heavy-duty trucks.

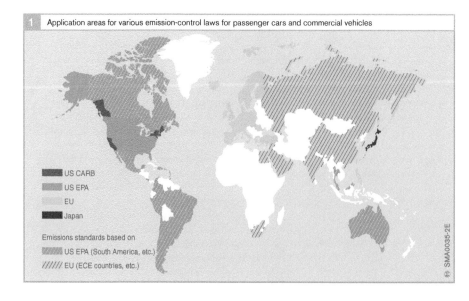

1   Application areas for various emission-control laws for passenger cars and commercial vehicles

US CARB
US EPA
EU
Japan

Emissions standards based on
US EPA (South America, etc.)
////// EU (ECE countries, etc.)

SMA0035-2E

Test procedures

Japan and the European Union have followed the lead of the United States by defining test procedures for certifying compliance with emission limits. These procedures have been adopted in modified or unrevised form by other countries.

Legal requirements prescribe any of three different tests, depending on vehicle class and test objective:
- Type Approval (TA) to obtain General Certification
- Random testing of vehicles from serial production conducted by the approval authorities (Conformity of Production)
- In-field monitoring for testing certain exhaust-gas components in vehicle operation

Type approval

Type approvals are a precondition for granting General Certification for an engine or vehicle type. This process entails proving compliance with stipulated emission limits in defined test cycles. Different countries have defined individual test cycles and emission limits.

Test cycles

Dynamic test cycles are specified for passenger cars and light-duty trucks. The country-specific differences between the two procedures are rooted in their respective origins:
- Test cycles designed to mirror conditions recorded in actual highway operation, e.g. Federal Test Procedure (FTP) test cycle in the U.S.A.
- Synthetically generated test cycles consisting of phases at constant cruising speed and acceleration rates, e.g. Modified New European Driving Cycle (MNEDC) in Europe.

The mass of toxic emissions from each vehicle is determined by operating it in conformity with speed cycles precisely defined for the test cycle. During the test cycle, the exhaust gases are collected for subsequent analysis to determine the pollutant mass emitted during the driving cycle.

For heavy-duty trucks, steady-state exhaust-gas tests (e.g. 13-stage test in the EU), or dynamic tests (e.g. Transient Cycle in the U.S.A.) are carried out on the engine test bench.

All the test cycles are depicted at the end of this section.

Testing serial-production vehicles

Testing serial-production vehicles is usually conducted by the vehicle manufacturer as part of quality control that accompanies the production process. The same test procedures and limits are generally applied as for type approval. The registration authorities may demand confirmation testing as often as necessary. EU and ECE directives (Economic Commission for Europe) take account of production tolerances by carrying out random testing on minimum 3 to maximum 32 vehicles. The most stringent requirements are encountered in the U.S.A., and particularly in California, where the authorities require what is essentially comprehensive and total quality monitoring.

In-field monitoring

Random emission-control tests are conducted in driving mode on vehicles whose running performance and age are within specific limits. The emission-control test procedure is simplified compared with type-approving testing.

CARB legislation (passenger cars/LDT)

CARB, or California Air Resources Board emission limits for passenger cars and Light-Duty Trucks (LDT) are defined in standards specifying exhaust-gas emissions:
- LEV I
- LEV II (Low Emission Vehicle)

The LEV I standard applies to passenger cars and light-duty trucks with a gross weight rating up to 6,000 lb. manufactured in model years 1994 through 2003. Starting model year 2004 the LEV II standard has applied to all new vehicles with a gross weight rating up to 14,000 lb.

Phase-in

Following introduction of the LEV II standard, at least 25% of new vehicle registrations must be certified to this standard. The phase-in rule stipulates that an additional 25% of vehicles must then conform to the LEV II standard in each consecutive year. As of 2007 all new vehicle registrations must then be certified according to the LEV II standard.

Emission limits

The CARB regulations define limits on:
- Carbon monoxide (CO)
- Nitrogen oxides (NO_x)
- Non-Methane Organic Gases (NMOG)
- Formaldehyde (LEV II only)
- Particulate emissions (diesel engines: LEV I and LEV II; gasoline-engines: LEV II only)

Actual emission levels are determined using the FTP 75 (Federal Test Procedure) driving cycle. Limits are defined in relation to distance and specified in grams per mile.

Within the period 2001 through 2004 the SFTP (Supplement Federal Test Procedure) standard was introduced together with other test cycles. There are also further limits that require compliance in addition to FTP emission limits.

Fig. 1

[1] For Tier 1, the NMHC limit applies instead of the NMOG limit (NMHC: Non-Methane Hydrocarbons).

[2] Limit in each case for "full useful life" (10 years/ 100,000 miles for LEV I or 120,000 miles for LEV II).

[3] Limit value in each case for "intermediate useful life" (5 years/ 50,000 miles).

[4] Only limits for "full useful life" (see the section on "Long-term compliance")

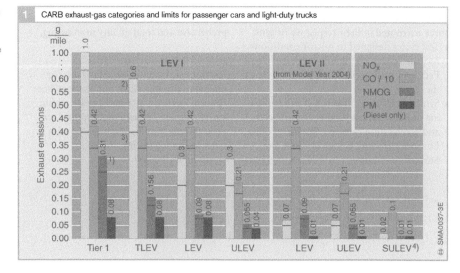

CARB exhaust-gas categories and limits for passenger cars and light-duty trucks

Exhaust-gas categories

Automotive manufacturers are at liberty to deploy a variety of vehicle concepts within the permitted limits, providing they maintain a fleet average (see the section on "Fleet averages"). The concepts are allocated to the following exhaust-gas categories, depending on their emission values for NMOG, CO, NO_X, and particulate emissions (Fig. 1):
- Tier 1 (LEV I only)
- TLEV (Transitional Low-Emission Vehicle; LEV I only)
- LEV (Low-Emission Vehicle), applicable to both exhaust and evaporative emissions
- ULEV (Ultra-Low-Emission Vehicle)
- SULEV (Super Ultra-Low-Emission Vehicle)

In addition to the categories of LEV I and LEV II, two categories define zero-emission and partial zero-emission vehicles:
- ZEV (Zero-Emission Vehicle), vehicles without exhaust-gas or evaporative emissions

- PZEV (Partial ZEV), which is basically SULEV, but with more stringent limits on evaporative emissions and stricter long-term performance criteria

Since 2004 the LEV II exhaust-gas emission standard has been in force. The categories comprising Tier 1 and TLEV are phased out, and in their place, SULEV is added with much lower emission limits. The LEV and ULEV categories remain in place. The CO and NMOG limits from LEV I remain unchanged, but the NO_x limit is substantially lower for LEV II. The LEV II standard also includes new, supplementary limits governing formaldehyde.

Long-term compliance

To obtain approval for each vehicle model (type approval), the manufacturer must prove compliance with the official emission limits for pollutants. Compliance means that the limits may not be exceeded for the following mileages or periods of useful life:
- 50,000 miles or 5 years ("intermediate useful life")
- 100,000 miles (LEV I)/120,000 miles (LEV II) or 10 years ("full useful life")

Manufacturers also have the option of certifying vehicles for 150,000 miles using the same limits that apply to 120,000 miles. The manufacturer then receives a bonus when the NMOG fleet average is defined (see the section entitled "Fleet averages").

The relevant figures for the PZEV emission-limit category are 150,000 miles or 15 years ("full useful life").

For this type of approval test, the manufacturer must furnish two vehicle fleets from serial production:
- One fleet in which each vehicle must cover 4,000 miles before testing.
- One fleet for endurance testing, in which deterioration factors for individual components are defined.

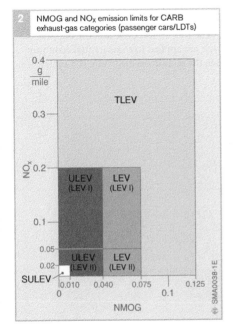

2 NMOG and NO_X emission limits for CARB exhaust-gas categories (passenger cars/LDTs)

Endurance testing entails subjecting the vehicles to specific driving cycles over distances of 50,000 and 100,000/120,000 miles. Exhaust-gas emissions are tested at intervals of 5,000 miles. Service inspections and maintenance are restricted to the standard prescribed distances.

Countries that base their regulations on the U.S. test cycles allow application of defined deterioration factors to simplify the certification process.

Fleet averages (NMOG)

Each vehicle manufacturer must ensure that exhaust-gas emissions for its total vehicle fleet do not exceed a specified average. NMOG emissions serve as the reference category for assessing compliance with these averages. The fleet average is calculated from the average figures produced by all of the manufacturer's vehicles that comply with NMOG limits and are sold within one year. Different fleet averages apply to passenger cars and light-duty trucks.

The compliance limits for the NMOG fleet average are lowered in each subsequent year (Fig. 3). To meet the lower limits, manufacturers must produce progressively cleaner vehicles in the more stringent emissions categories in each consecutive year.

The fleet averages apply irrespective of LEV I or LEV II standards.

Fleet consumption

U.S. legislation specifies the average amount of fuel an automotive manufacturer's vehicle fleet may consume per mile. The prescribed CAFE value (Corporate Average Fuel Economy) currently (2004) stands at 27.5 mpg. This corresponds to 8.55 liters per 100 kilometers in metric terms. At the present time it is not planned to reduce this limit.

The value for light-duty trucks is 20.7 miles per gallon or 11.4 liters per 100 kilometers. From 2005 to 2007 fuel economy will be raised each year by 0.6 mpg. There are no regulations for heavy-duty trucks.

At the end of each year the average fuel economy for each manufacturer is calculated based on the numbers of vehicles sold. The manufacturer must remit a penalty fee of $5.50 per vehicle for each 0.1 mpg its fleet exceeds the target. Buyers will also have to pay a gas-guzzler tax on vehicles with especially high fuel consumption. Here, the limit is 22.5 miles per gallon (corresponding to 10.45 liters per 100 kilometers in metric terms).

These penalties are intended to spur development of vehicles offering greater fuel economy.

The FTP 75 test cycle and the highway cycle are applied to measure fuel economy (see the section entitled "U.S. test cycles").

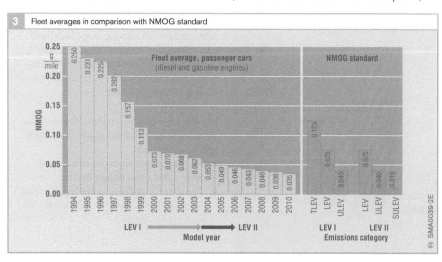

3   Fleet averages in comparison with NMOG standard

Zero-emission vehicles

Starting 2003 10% of new vehicle registrations will have to meet the requirements of the ZEV (Zero-Emission Vehicle) exhaust-gas category. These vehicles may emit no exhaust gas or evaporative emissions in operation. This category mainly refers to electric cars.

A 10% percentage may partly be covered by vehicles of the PZEV (Partial Zero-Emission Vehicles) exhaust-gas category. These vehicles are not zero-emission, but they emit very few pollutants. They are weighted using a factor of 0.2 to 1, depending on the emission-limit standard. The minimum weighting factor of 0.2 is granted when the following demands are met:
- SULEV certification indicating long-term compliance extending over 150,000 miles or 15 years.
- Warranty coverage extending over 150,000 miles or 15 years on all emission-related components.
- No evaporative emissions from the fuel system (0 EVAP (Zero Evaporation)), achieved by extensive encapsulation of the tank and fuel system.

Special regulations apply to hybrid vehicles combining diesel engines and electric motors. These vehicles can also contribute to achieving compliance with the 10% limit.

In-field monitoring

Unscheduled testing
Random emission testing is conducted on in-use vehicles using the FTP 75 test cycle and an evaporation test. Only vehicles with mileages of less than 50,000 or 75,000 miles (varies according to the certification status of the individual vehicle model) are selected for testing.

Vehicle monitoring by the manufacturer
Official reporting of claims and damage to specific emissions-related components and systems has been mandatory for vehicle manufacturers since model year 1990. The reporting obligation remains in force for a period of 5 or 10 years, or 50,000 or 100,000 miles, depending on the warranty period applying to the component or assembly.

Reporting is divided into three stages with an incremental amount of detail:
- Emissions Warranty Information Report (EWIR)
- Field Information Report (FIR)
- Emission Information Report (EIR)

Information concerning:
- problem reports
- malfunction statistics
- defect analysis
- impacts on emissions

is then forwarded to the environment-protection authorities. The authorities use the FIR as the basis for issuing mandatory recall orders to the manufacturer.

EPA legislation (passenger cars/LDT)

EPA (Environment Protection Agency) legislation applies to all of the Federal states where the more stringent CARB stipulations from California are not in force. CARB regulations have already been adopted by some states, such as Maine, Massachusetts, and New York.

EPA regulations in force since 2004 conform to the Tier 2 standard.

Emission limits

EPA legislation define emission limits for the following pollutants:
- Carbon monoxide (CO)
- Nitrogen oxides (NO_X)
- Non-Methane Organic Gases (NMOG)
- Formaldehyde (HCHO)
- Particulates

Pollutant emissions are determined using the FTP 75 driving cycle. Limits are defined in relation to distance and specified in grams per mile.

The SFTP (Supplemental Federal Test Procedure) standard, comprising further test cycles, has been in force since 2002. The applicable limits require compliance in addition to FTP emission limits.

Since the introduction of Tier 2 standards, vehicles with diesel and spark-ignition engines have been subject to an identical exhaust-gas emission standards.

Exhaust-gas categories

Tier 2 (Fig. 1) classifies limits for passenger cars in 10 bins (emission standards) and 11 bins for heavy-duty trucks. After 2007 bins 9 through 11 will be phased out.

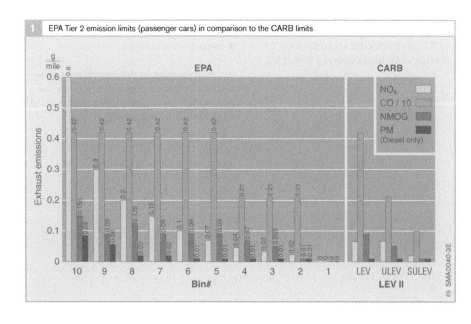

1   EPA Tier 2 emission limits (passenger cars) in comparison to the CARB limits

The transition to Tier 2 have produced the following changes:

- Introduction of fleet averages for NO_x.
- Formaldehydes (HCHO) are subject to a separate pollutant category.
- Passenger cars and light-duty trucks with a GWR up to 6,000 lb. (2.72 metric tons) will be combined in a single vehicle class.
- MDPV (Medium-Duty Passenger Vehicle) forms a separate vehicle category; previously assigned to HDV (Heavy-Duty Vehicles).
- "Full useful life" is extended to 120,000 miles (192,000 kilometers).

Phase-in

At least 25% of all new passenger-car and LLDT (Light Light-Duty Trucks) registrations are required to conform to Tier 2 standards when they take effect in 2004. The phase-in rule stipulates that an additional 25% of vehicles are then required to conform to Tier 2 standards in every following year. All vehicles are required to conform to Tier 2 standards by 2007. For the HLDT/MDPV category, the phase-in period ends 2009.

Fleet averages

NO_x emissions are used to determine fleet averages for individual manufacturers under EPA legislation. CARB regulations, however, are based on NMOG emissions.

Fleet fuel economy

The regulations defining average fleet fuel consumption in the 49 states are the same as those applied in California. Here again, the limit applicable to passenger cars is 27.5 miles per gallon (8.55 liters per 100 kilometers). Beyond this figure, manufacturers are required to pay a penalty tax. The purchaser also pays a penalty tax on vehicles that consume more than 22.5 miles per gallon.

In-field monitoring

Unscheduled testing

In analogy to CARB legislation, the EPA regulations require random exhaust-gas emission testing based on the FTP 75 test cycles for in-use vehicles. Testing is conducted on low-mileage (10,000 miles, roughly one year old), and higher mileage vehicles (50,000 miles, and at least one vehicle per test group with 75,000/90,000 miles, approx. 4 years old). The number of vehicles tested varies according to the number sold.

Vehicle monitoring by the manufacturer

For vehicles after model year 1972, the manufacturer is obliged to make an official report concerning damage to specific emission-related components or systems if at least 25 identical emission-related parts in a model year are defective. The reporting obligation ends five years after the end of the model year. The report comprises a description of damage to the defective component, presentation of the impacts on exhaust-gas emissions, and suitable corrective action by the manufacturer. The environment-protection authorities use this report as the basis for deciding whether to issue recall orders to the manufacturer.

EU legislation (passenger cars/LDT)

The regulations contained in European Union directives are defined by the EU Commission. Emission-control legislation for passenger cars/light-duty trucks is Directive 70/220/EEC from 1970. For the first time it defined exhaust-gas emission limits, and the provisions have been updated ever since.

The emission limits for passenger cars and Light-Duty Trucks (LDT) are contained in the following exhaust-gas emission standards:
- Euro 1 (as from 1 July 1992)
- Euro 2 (as from 1 January 1996)
- Euro 3 (as from 1 January 2000)
- Euro 4 (as from 1 January 2005)

A new exhaust-gas emission standard is generally introduced in two stages. In the first stage, compliance with the newly defined emission limits is required for vehicle models subject to new Type Approvals (TA). In the second stage – usually one year later –

every new registration must comply with the new limits (First Registration (FR)). The authorities can also inspect serial-production vehicles to verify compliance with emission limits (Conformity of Production (COP)).

EU directives allow tax incentives for vehicles that comply with upcoming exhaust-gas emission standards before they actually become law. Depending on a vehicle's emission standard, there are a number of different motor-vehicle tax rates in Germany.

Emission limits

The EU standards define limits for the following pollutants:
- Carbon monoxide (CO)
- Hydrocarbons (HC)
- Nitrogen oxides (NO_X)
- Particulates, although these limits are initially restricted to diesel vehicles

The limits for hydrocarbons and nitrogen oxides for the Euro 1 and Euro 2 stages are combined into an aggregate value ($HC+NO_X$). Since Euro 3, there has been a special restriction for NO_X emissions in addition to the aggregate value.

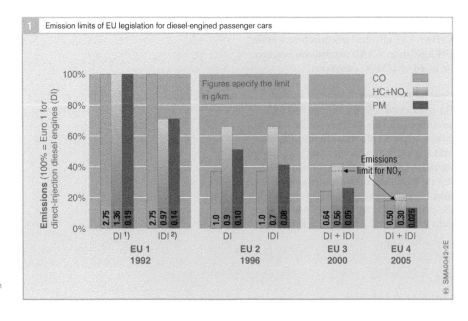

1   Emission limits of EU legislation for diesel-engined passenger cars

Figures specify the limit in g/km.

CO
HC+NO_X
PM

Emissions limit for NO_X

Emissions (100% = Euro 1 for direct-injection diesel engines (DI))

	DI[1]	IDI[2]	DI	IDI	DI + IDI	DI + IDI
CO	2.75	2.75	1.0	1.0	0.64	0.50
HC+NO_X	1.36	0.97	0.9	0.7	0.56	0.30
PM	0.19	0.14	0.10	0.08	0.05	0.025

EU 1 — 1992   EU 2 — 1996   EU 3 — 2000   EU 4 — 2005

SMA0042-2E

Fig. 1
[1] For engines with direct injection
[2] For indirect-injection engines

The limits are defined based on mileage and are specified in grams per kilometer (g/km) (Fig. 1). Since EU 3, emissions are measured on a chassis dynamometer using the MNEDC (Modified New European Driving Cycle).

The limits are different for vehicles with diesel or gasoline engines. In future, they will merge (probably by Euro 5).

The limits for the LDT category are not uniform. There are three classes (1 to 3) into which LDTs are subdivided, depending on the vehicle reference weight (unladen weight + 100 kg). The limits for Class 1 are the same as for cars.

Type approval

While type-approval testing basically corresponds to U.S. procedures, deviations are found in the following areas: Measurements of the pollutants HC, CO, NO_x are supplemented by particulate and exhaust-gas opacity measurements on diesel vehicles. Test vehicles absolve an initial run-in period of 3,000 kilometers before testing. Deterioration factors used to assess test results are defined in the legislation for every pollutant component; manufacturers are also allowed to present documentation confirming lower factors obtained during specified endurance testing over 80,000 km (100,000 km starting with EU 4).

Compliance with the specified limits must be maintained over a distance of 80,000 km (Euro 3), or 100,000 km (Euro 4), or after 5 years.

Type tests

There are six different type tests for type approval. Of those, the Type I test and the Type V test apply to diesel-engined vehicles.

The Type I test evaluates exhaust-gas emissions after cold-starting. Exhaust-gas opacity is also assessed on vehicles with diesel engines.

The Type V test assesses the long-term durability of the emission-reducing equipment. It may involve a specific test sequence, or it may be subjected to deterioration factors specified by the legislation.

CO_2 emissions

There are no legal emission limits for CO_2. However, the vehicle manufacturers (Association des Constructeurs Européen d'Automobiles (ACEA)(Association of European Automobile Manufacturers) have pledged to uphold a voluntary program. The objective is to achieve CO_2 emissions of max. 140 g/km for passenger cars by 2008 – this is equivalent to a fuel consumption of 5.8 l/100 km.

In Germany, vehicles with specially low CO_2 emissions (so-called 5-liter and 3-liter cars) will be tax-exempt until the end of 2005.

In-field monitoring

EU legislation also calls for conformity-verification testing on in-use vehicles as part of the Type I test cycle. The minimum number of vehicles of a vehicle type under test is three, while the maximum number varies according to the test procedure.

Vehicles under test must meet the following criteria:
- Mileages vary from 15,000 km to 80,000 km, and vehicle age from 6 months to 5 years (Euro 3). In Euro 4, the mileage specified ranges from 15,000 km to 100,000 km.
- Regular service inspections were carried out as specified by the manufacturer.
- The vehicle must show no indications of non-standard use (e.g. manipulation, major repairs, etc.).

If emissions from an individual vehicle fail substantially to comply with the standards, the source of the high emissions must be determined. If more than one vehicle displays excessive emissions in random testing, no matter what the reason, the results of the random test must be classified as negative.

If there are various reasons, the test schedule may be extended, providing the maximum sample size is not reached.

If the type-approval authorities detect that a vehicle type fails to meet the criteria, the vehicle manufacturer must devise suitable action to eliminate the defect. The action catalog must be applicable to all vehicles with the same defect. If necessary, a recall action must be started.

Periodic emissions inspections (AU)

In Germany, all passenger cars and light-duty trucks are required to undergo emissions inspection (AU) three years after their first registration, and then every two years. For gasoline-engined vehicles, the main focus is on CO levels, while for diesel vehicles, the opacity test is the main criterion.

Since the introduction of on-board diagnosis, the exhaust-gas test also tests whether the OBD system is functioning correctly to ensure monitoring of exhaust-gas-related components and systems in use.

Japanese legislation (passenger cars/LDTs)

The permitted emission values are also subject to gradual stages of severity in Japan. For 2005 it was decided to introduce a further stage of severity for emission limits.

Vehicles with a gross weight rating up to 2.5 t (starting 2005: 3.5 t) are divided into three categories: passenger cars (up to 10 seats), LDV (Light-Duty Vehicle) up to 1.7 t, and MDV (Medium-Duty Vehicle) up to 2.5 t (starting 2005: 3.5 t). Slightly higher limits for NO_X and particulates apply to the MDV category compared with the other two vehicle classes.

Emission limits

Japanese legislation specifies limits for the following emissions (Fig. 1):
● Carbon monoxide (CO)
● Nitrogen oxides (NO_X)
● Hydrocarbons (HC)
● Particulates (diesel vehicles only)
● Smoke (diesel vehicles only)

Pollutant emissions are measured in the $10 \cdot 15$-Mode Test (see the section entitled "Japanese test cycle for passenger cars and LDTs"). A modified $10 \cdot 15$ Mode Test, including a cold start to be introduced in 2005, is under discussion.

Fleet fuel economy

In Japan, measures for reducing the CO_2 emissions of cars are planned. One proposal plans to fix the average fuel economy (CAFE value) for the entire passenger-car fleet. The proposal foresees a gradual phase-in of emission limits by vehicle weight.

Fig. 1
1) For vehicles with an unladen weight up to 1,265 kg
2) For vehicles with an unladen weight over 1,265 kg
3) Limit for vehicles up to 1,265 kg
4) Limit for vehicles over 1,265 kg

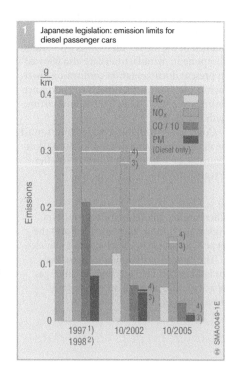

1   Japanese legislation: emission limits for diesel passenger cars

SMA0049-1E

U.S. legislation (heavy-duty trucks)

Heavy-duty trucks are defined in EPA legislation as vehicles with a gross weight rating over 8,500 or 10,000 lb. (equivalent to 3.9 and 4.6 t), depending on vehicle type.

In California, all vehicles over 14,000 lb. (equivalent to 6.4 t) are classified as heavy-duty trucks. To a great extent, Californian legislation is identical to parts of EPA legislation. However, there is an additional program for city buses.

Emission limits

The U.S. standards for diesel engines define limits for:
- Hydrocarbons (HC)
- NMHCs in some cases
- Carbon monoxide (CO)
- Nitrogen oxides (NO_X)
- Particulate
- Exhaust-gas opacity

The permissible limits are related to engine power output and specified in g/kW. The emissions are measured on the engine test bench during the dynamic test cycle with cold starting sequence (HDTC, Heavy-Duty Transient Cycle); the exhaust-gas opacity is measured using the Federal Smoke Test.

New, more stringent regulations apply to vehicles starting model year 2004, with significantly reduced NO_X emission limits. Non-methane hydrocarbons and nitrogen oxides are grouped together in one aggregate (NMHC + NO_X). CO and particulate emission limits remain at the same level as model year 1998.

Another very drastic tightening of emission restrictions comes into force in model year 2007. The NO_X and particulate emission limits will then be 10 times lower than previous levels. This is probably not achievable without the use of emission-control systems (e.g. NO_X catalytic converters and particulate filters).

A gradual phase-in will take place for NO_X and NMHC emission limits between model years 2007 and 2010.

To help compliance with severe particulate limits, the maximum permitted sulfur content in diesel fuel will be reduced from 500 ppm at present to 15 ppm from mid-2006.

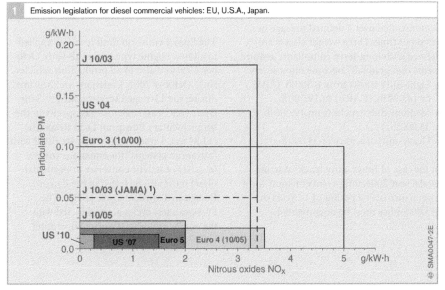

1   Emission legislation for diesel commercial vehicles: EU, U.S.A., Japan.

Fig. 1
[1]) Voluntary obligation of Japanese Association of Automotive Manufacturers (JAMA): one engine type per manufacturer

For heavy-duty trucks – in contrast with cars and LDTs – there are no limits specified for average fleet emissions and fleet consumption.

Consent Decree

In 1998 a legal agreement was reached between EPA, CARB, and a number of engine manufacturers. It provides for sanctions against manufacturers if they make illegal modifications to engines to achieve optimized consumption in the highway cycle, resulting in higher NO_X emissions. The "Consent Decree" specifies that the applicable emission limits must also undercut the steady-state European 13-stage test in addition to the dynamic test cycle. Furthermore, emissions are not allowed to exceed the limits for model year 2004 by more than 25%, regardless of driving mode within a specified engine-speed/torque range ("not-to-exceed" zone).

These additional tests are applicable to all diesel commercial vehicles starting model year 2007. However, emissions in the not-to-exceed zone may be up to 50% above the emission limits.

Long-term compliance

Compliance with emission limits must be demonstrated over a defined mileage or period of time. Three weight classes are defined with long-term compliance requirements that gradually become more severe:
- Light-duty trucks from 8,500 lb. (EPA) or 14,000 lb. (CARB) to 19,500 lb.
- Medium-duty trucks from 19,500 lb. to 33,000 lb.
- Heavy-duty trucks over 33,000 lb.

In the case of heavy-duty trucks starting model year 2004, long-term emission-limit compliance over a period of 13 years or 435,000 miles must be documented.

EU legislation (heavy-duty trucks)

In Europe, all vehicles with a permissible gross weight of over 3.5 t, or capable of transporting more than 9 persons, are classified as heavy-duty trucks. The emission-limit regulations are set down in Directive 88/77/EEC, which is subject to continuous updating.

As for cars and light-duty trucks, new emission limits for heavy-duty trucks are introduced in two stages. New engine designs must meet the new emission limits during type approval. One year later, compliance with the new limits will be a condition for awarding a general vehicle approval. The legislator can inspect Conformity of Production (COP) by taking engines out of serial production and testing them for compliance with the new emission limits.

Emission limits

For commercial-vehicle diesel engines, the Euro standards define emission limits for hydrocarbons (HC and NMHCs), carbon monoxide (CO), nitrogen oxides (NO_X), particulates, and exhaust-gas opacity. The permissible limits are related to engine power output and specified in g/kW.

The Euro 3 emission limit level has applied to all new engine type approvals since October 2000 and also to all production vehicles since October 2001. Emissions are measured during the 13-stage European Steady-State Cycle (ESC), and exhaust-gas opacity in the supplementary European Load Response (ELR) test. Diesel engines that are fitted with "advanced systems" for emissions control (e.g. NO_X catalytic converter or particulate filter) must also be tested in the dynamic European Transient Cycle (ETC). The European test cycles are conducted with the engine running at normal operating temperature.

In the case of small engines, i.e. engines with a capacity of less than 0.75 liter per cylinder and a rated speed of over 3000 rpm, slightly higher particulate emission levels are permitted than for large engines. There are separate emission limits for the ETC – for example, particulate limits are approximately 50% higher than specified for the ESC because of the soot-emission peaks expected under dynamic operating conditions.

In October 2005 the emission limit level Euro 4 enters into force initially for new type approvals, and for serial production one year later. All emission limits are significantly lower than specified by Euro 3, but the biggest increase in severity applies to particulates, for which the limits have been reduced by approximately 80%. The following changes will also apply after introduction of Euro 4:
- The dynamic exhaust-gas emission test (ETC) will be obligatory – in addition to ESC and ELR – for all diesel engines.
- The continued functioning of emissions-related components must be documented for the entire service life of the vehicle.

The EU 5 level of emission limits will be introduced in October 2008 for all new engine approvals, and one year later for all new serial-production vehicles. Only the NO_X emission limits will be more severe compared to Euro 4.

Very low-emission vehicles
The EU Directives allow for tax incentives for early compliance with the limits specified by a particular phase of the EU standards, and for EEVs (Enhanced Environmentally Friendly Vehicles).

Voluntary emission limits are defined for the EEV category for the ESC, ETC, and ELR emission-limit tests. The NO_X and particulate emission limits comply with the Euro 5 ESC emission limits. The standards for HC, NMHC, CO, and exhaust gas opacity are stricter than for Euro 5.

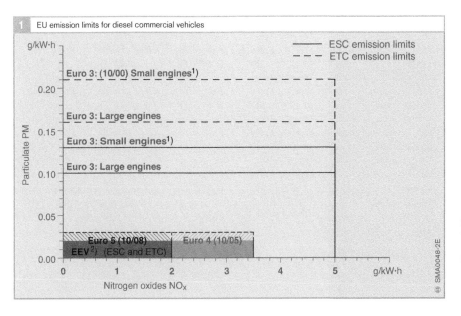

1 EU emission limits for diesel commercial vehicles

Fig. 1
[1] $V_{cyl} \leq 0.75$ l, $n_{rated} \geq 3,000$ rpm
[2] Enhanced Environmentally Friendly Vehicle (voluntary limits)

Japanese legislation (heavy-duty trucks)

In Japan, vehicles with a permissible gross weight of over 2.5 t (from 2005: 3.5 t), or capable of transporting more than 10 persons are classified as heavy-duty trucks.

Emission limits

The "New Short-Term Regulation" has been in force since October 2003. It specifies limits for HC, NO_X, CO, particulates, and exhaust-gas opacity. Emission levels are measured by the fixed Japanese 13-stage test (hot test), and exhaust-gas opacity by the Japanese smoke test. Long-term compliance for emission limits must be documented over a distance of 80,000 to 650,000 km (depending on the permitted vehicle weight).

The "New Long-Term Regulation" comes into force in October 2005. Compared with 2003, the emission limits are halved, and particulate limits are even cut by 75%.

A dynamic Japanese test cycle will also be introduced for this limit stage.

Regional programs

In addition to the nationwide regulations for new vehicles, there are also regional requirements for the overall vehicle population aimed at reducing existing emission levels by replacing or upgrading old diesel vehicles.

Since 2003 the "Vehicle NO_x Law" has been in force for vehicles with a permitted vehicle weight over 3,500 kg, for example in the greater Tokyo region. The regulation stipulates that 8 to 12 years after first vehicle registration, the NO_x and particulate limits must comply with the prevailing emission limit levels (e.g. the 1998 limits starting 2003). The same principle also applies to particulate emissions. Here, the regulation will already apply 7 years after first vehicle registration.

▶ Ozone and smog

Exposure to the sun's radiation splits nitrogen-dioxide molecules (NO_2). The products are nitrogen oxide (NO) and atomic oxygen (O), which combine with the ambient air's atomic oxygen (O_2) to form ozone (O_3). Ozone formation is also promoted by volatile organic compounds. This is why higher ozone levels must be anticipated on hot, windless summer days when high levels of air pollution are present.

In normal concentrations, ozone is essential for human life. However, in higher concentrations, it leads to coughing, irritation of the throat and sinuses, and burning eyes. It adversely affects lung function, reducing performance potential.

There is no direct contact or mutual movement between the ozone formed in this way at ground level, and the stratospheric ozone that reduces the amount of ultraviolet radiation penetrating the earth's atmosphere.

Smog is not limited to the summer. It can also occur in winter in response to atmospheric layer inversions and low wind speeds. The temperature inversion in the air layers prevents the heavier, colder air containing the higher pollutant concentrations from rising and dispersing.

Smog leads to irritation of the mucous membranes, eyes, and respiratory system. It can also impair visibility. This last factor explains the origin of the term smog, which is a contraction of "smoke" and "fog".

U.S. test cycles for passenger cars and LDTs

FTP 75 test cycle

The FTP 75 test cycle (Federal Test Procedure) consists of speed cycles that were actually recorded in commuter traffic in Los Angeles (Fig. 1a).

This test is also in force in some countries of South America besides the U.S.A. (including California).

Preconditioning

The vehicle is subjected to an ambient temperature of 20...30°C in a climatic chamber for a period of 12 hours.

Collecting pollutants

The vehicle is started and driven on the specific speed cycle. The emitted pollutants are collected in separate bags during various phases.

Phase ct (cold transient):
Collection of exhaust gas during the cold test phase.

Phase s (stabilized):
Start of stabilized phase 505 seconds after start. The exhaust gas is collected without interrupting the driving cycle. At the end of Phase s, after a total of 1,365 seconds, the engine is switched off for a period of 600 seconds.

Phase ht (hot transient):
The engine is restarted for the hot test. The speed cycle is identical to the cold transient phase (Phase ct).

Analysis

The bag samples from the first two phases are analyzed during the pause before the hot test. This is because samples may not remain in the bags for longer than 20 minutes. The sample exhaust gases contained in the third bag are also analyzed on completion of the driving cycle. The total result includes emissions from the three phases rated at different weightings.

The pollutant masses of Phases ct and s are aggregated and assigned to the total distance of these two phases. The result is then weighted at a factor of 0.43.

The same process is applied to the aggregated pollutant masses from Phases ht and s, related to the total distance of these two phases, and weighted at a factor of 0.57. The test result for the individual pollutants (HC, CO, and NO_x) is obtained from the sum of the two previous results.

The emissions are specified as the pollutant emission per mile.

SFTP schedules

Test according to the SFTP standard (Supplemental Federal Test Procedure) were phased in from 2001 to 2004. These are composed of the following driving cycles:
- FTP 75 cycle
- SC03 cycle (Fig. 1b)
- US06 cycle (Fig. 1c)

The extended tests are intended to examine the following additional driving conditions:
- Aggressive driving
- Radical changes in vehicle speed
- Engine start and acceleration from standing start
- Operation with frequent minor variations in speed
- Periods with vehicle parked
- Operation with air conditioner on

After preconditioning, the SC03 and US06 cycles proceed through the FTP 75 ct phase without exhaust-gas collection. Other preconditioning procedures may also be used.

The SC03 cycle is carried out at a temperature of 35°C and 40% relative humidity (vehicles with air conditioning only). The individual driving cycles are weighted as follows:

- Vehicles with air conditioning:
  35% FTP 75 + 37% SC03 + 28% US06
- Vehicles without air conditioning:
  72% FTP 75 + 28% US06

The SFTP and FTP 75 test cycles must be successfully completed on an individual basis.

Test cycles for determining fleet averages

Each vehicle manufacturer is required to provide data on fleet averages. Manufacturers that fail to comply with the emission limits are required to pay penalties.

Fuel consumption is determined from the exhaust-gas emissions produced during two test cycles: the FTP 75 test cycle (weighted at 55%) and the highway test cycle (weighted at 45%). An unmeasured highway test cycle (Fig. 1b) is conducted once after preconditioning (vehicle allowed to stand with engine off for 12 hours at 20...30°C). The exhaust-gas emissions from a second test run are then collected. The CO_2 emissions are used to calculate fuel consumption.

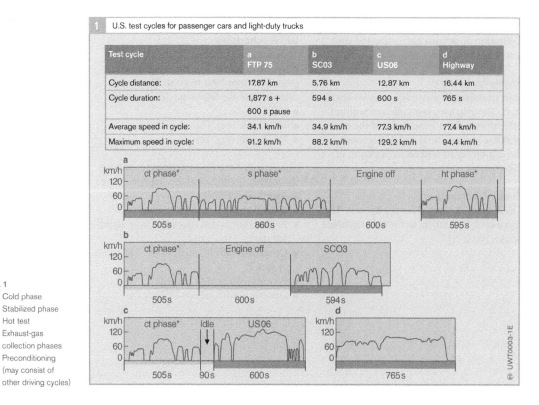

1   U.S. test cycles for passenger cars and light-duty trucks

Test cycle	a FTP 75	b SC03	c US06	d Highway
Cycle distance:	17.87 km	5.76 km	12.87 km	16.44 km
Cycle duration:	1,877 s + 600 s pause	594 s	600 s	765 s
Average speed in cycle:	34.1 km/h	34.9 km/h	77.3 km/h	77.4 km/h
Maximum speed in cycle:	91.2 km/h	88.2 km/h	129.2 km/h	94.4 km/h

Fig. 1
* ct Cold phase
  s Stabilized phase
  ht Hot test
  ▪ Exhaust-gas collection phases
  ▫ Preconditioning (may consist of other driving cycles)

European test cycle for passenger cars and LDTs

MNEDC
The Modified New European Driving Cycle (MNEDC) has been in force since Euro 3. Contrary to the New European Driving Cycle (Euro 2), that only starts 40 seconds after the vehicle has started, the MNEDC also includes a cold-start phase.

Preconditioning
The vehicle is allowed to start at an ambient temperature of 20...30°C for a minimum period of 6 hours.

Collecting pollutants
The exhaust gas is collected in bags during two phases:
● Urban Driving Cycle (UDC) at a maximum of 50 km/h
● Extra-urban cycle at speeds up to 120 km/h

Analysis
The pollutant mass measured by analyzing the bag contents is referred to the distance covered.

Japanese test cycle for passenger cars and LDTs

The 10 · 15-mode test cycle (Fig. 2) is conducted once with a hot start. This test cycle simulates characteristic driving behavior in Tokyo. Top speed is lower than in the European test cycle due to the higher traffic density in Japan. This generally results in lower driving speeds.

The preconditioning procedure for the hot test includes the prescribed exhaust-gas test at idle. The procedure is as follows: After the vehicle is allowed to warm up for approximately 15 minutes at 60 km/h, the concentrations of HC, CO, and CO_2 are measured in the exhaust pipe at idle. The 10 · 15-mode hot test commences after a second warm-up phase, consisting of 5 minutes at 60 km/h.

The pollutants are defined relative to distance, and are indicated in grams per kilometer.

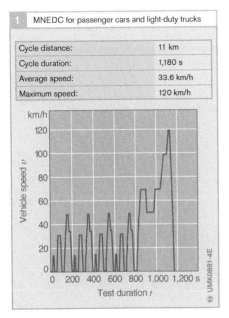

1   MNEDC for passenger cars and light-duty trucks

Cycle distance:	11 km
Cycle duration:	1,180 s
Average speed:	33.6 km/h
Maximum speed:	120 km/h

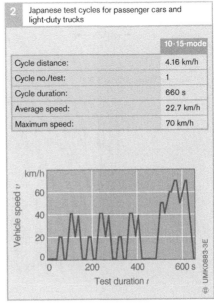

2   Japanese test cycles for passenger cars and light-duty trucks

	10 · 15-mode
Cycle distance:	4.16 km/h
Cycle no./test:	1
Cycle duration:	660 s
Average speed:	22.7 km/h
Maximum speed:	70 km/h

Test cycles for heavy-duty trucks

For heavy-duty trucks, all test cycles are run on the engine test bench. For nonstationary test cycles, the exhaust gases are collected and analyzed using the CVS method (cf. the section entitled "Measuring procedures"), while untreated exhaust gases are measured in the steady-state test cycles. Emissions are specified in g/kWh.

Europe

For vehicles over 3.5 t permitted vehicle weight and more than 9 seats, the new 13-stage test, the European Steady-State Cycle (ESC), has been in force in Europe since the introduction of Euro 3 (October 2000).

The test procedure specifies measurements in 13 steady-state operating states calculated from the engine full-load curve. The emissions measured at each operating point are weighted according to certain factors. This also apples to power output (Fig. 1). The test results are obtained for each pollutant by calculating the total of the weighted emissions divided by the total of the weighted power output.

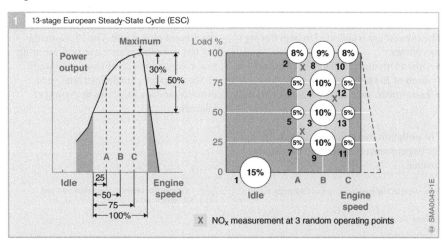

1   13-stage European Steady-State Cycle (ESC)

X   NOₓ measurement at 3 random operating points

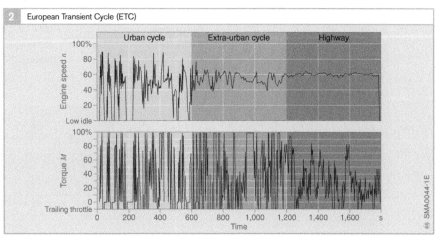

2   European Transient Cycle (ETC)

For certification, an additional three NO_x measurements can be taken over the tested range. The NO_x emissions may not vary by a significant degree from the levels measured at the adjacent operating points. The aim of the additional measurements is to prevent any modification of engines specifically for testing purposes.

As well as Euro 3, the ETC (European Transient Cycle, Fig. 2) was also introduced to determine gaseous emissions and particulate, and the ELR (European Load Response) test to measure exhaust-gas opacity. Under the Euro 3 standards, the ETC applies only to commercial vehicles with "advanced" emission-control equipment (particulate filters, NO_x catalytic converters); starting with Euro 4 (October 2005), it will be obligatory for all vehicles.

The test cycle is derived from realistic road-driving patterns and is subdivided into three sections: an urban section, an extra-urban section, and a highway section. The length of the test is 30 minutes, and the periods of time for which engine speeds and torque levels must be maintained are specified in seconds.

All European test cycles are performed with the engine running at normal operating temperature.

Japan

Pollutant emissions are measured using the Japanese 13-stage steady-state test (hot test). The engine operating points, their sequence, and weighting are different from those defined by the European 13-stage test, however. Compared with the ESC; the test focuses on lower engines speeds and loads. A dynamic Japanese test cycle that comes into force in 2005 will also be introduced for this limit stage.

U.S.A.

Since 1987 engines for heavy-duty trucks have been tested on an engine test bench according to a steady-state test cycle (Transient Cycle), including a cold-start sequence (Fig. 3). The test cycle is basically equivalent to operating an engine under realistic road-traffic conditions. It includes significantly more idle sections than the European ETC.

An additional test, the Federal Smoke Cycle, tests exhaust-gas opacity under dynamic and quasi steady-state conditions.

Starting with model year 2007, U.S. emission limits must also comply with the European 13-stage test (ESC). Furthermore, emissions in the not-to-exceed zone (i.e. with any driving mode within a specified engine speed/torque range) may be max. 50% above the emission limits.

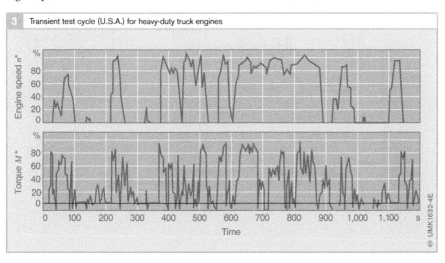

3   Transient test cycle (U.S.A.) for heavy-duty truck engines

Fig. 3
The standardized engine speed n^* and the standardized torque M^* are specified by law.

Exhaust-gas measuring techniques

Exhaust-gas test for type approval

During type-approval testing to obtain General Certification for passenger cars and light-duty trucks, the exhaust-gas test is conducted with the vehicle mounted on a chassis dynamometer. The test differs from exhaust-gas tests that are conducted using workshop measuring devices for in-field monitoring.

For the type approval of heavy-duty trucks, exhaust-gas tests are carried out on engine test benches.

The prescribed test cycles on the chassis dynamometer stipulate that practical on-road driving mode must be simulated as closely as possible. Testing on a chassis dynamometer offers many advantages compared with on-road testing:

- The results are easy to reproduce since the ambient conditions are constant.
- The tests are comparable since a specified speed/time profile is driven irrespective of traffic flow.
- The required measuring instruments are set up in a stationary environment.

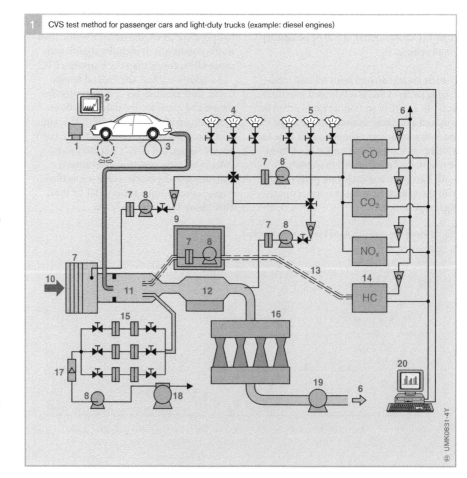

1   CVS test method for passenger cars and light-duty trucks (example: diesel engines)

Fig. 1

1   Cooling fan
2   Driver display monitor
3   "Rolling road" with dynamometer
4   Air-sample bag
5   Exhaust-gas bag
6   Extraction
7   Filter
8   Pump
9   Heated prefilter and pump
10   Dilution air
11   Dilution tunnel
12   Heat exchanger
13   Heated pipe
14   Gas analyzers
15   Measuring filter
16   Quadruple venturi tubes
17   Flow meter
18   Gas meter
19   CVS blower
20   PC with monitor

Besides type-approval testing, exhaust-gas measurements on the chassis dynamometer are conducted during the development of engine components.

Test setup

The test vehicle is placed on a chassis dynamometer with its drive wheels on the rollers (Fig. 1, 3). The test cycle is repeated by a driver. During this cycle, the required and current vehicle speeds are displayed on a driver monitor. In some cases, an automated driving system replaces the driver to increase the reproducibility of test results by driving the test cycle with extreme precision.

This means that the forces acting on the vehicle, i.e. the vehicle's moments of inertia, rolling resistance, and aerodynamic drag, must be simulated so that the test cycle on the test bench reproduces emissions comparable to those obtained during an on-road test. For this purpose, asynchronous machines, direct-current machines, or even electrodynamic retarders on older test benches, generate a suitable speed-dependent load that acts on the rollers for the vehicle to overcome. More modern machines use electric flywheel simulation to reproduce this inertia. Older test benches use real flywheels of different sizes attached by rapid couplings to the rollers to simulate vehicle mass. A blower mounted a short distance in front of the vehicle provides the necessary engine cooling.

The test-vehicle exhaust pipe is generally a gas-tight attachment to the exhaust-gas collection system – the dilution system is described below. A proportion of the diluted exhaust gas is collected there. At the end of the test cycle, the gas is analyzed for pollutants limited by law (hydrocarbons, nitrogen oxides, and carbon monoxide), and carbon dioxide (to determine fuel consumption).

In addition, and for development purposes, part of the exhaust gas flow can be extracted continuously from sampling points along the vehicle's exhaust-gas system or dilution system to analyze pollutant concentrations.

The complete sampling system, including the exhaust-gas measuring instrument for hydrocarbons, is heated to 190°C to avoid any condensation of hydrocarbons that boil at high temperatures.

There is also a dilution tunnel with high internal flow turbulence, and a particulate filter whose loading is analyzed to determine particulate emissions.

CVS dilution procedure

The most commonly used method of collecting exhaust gases emitted from an engine is the Constant Volume Sampling (CVS) method. It was introduced for the first time in the U.S.A. in 1972 for passenger cars and light-duty trucks. In the meantime it has been updated in several stages. The CVS method is used in other countries, such as Japan. It has also been in use in Europe since 1982.

During the dilution process, exhaust gas is mixed with air, then part of this mixture is collected in bags. The exhaust gas is only analyzed at the end of the test. Dilution avoids condensation of water vapor contained in the exhaust gases and also prevents the loss of gas components that are dissolvable in water. Dilution also avoid secondary reactions in the collected exhaust gas, and simulates actual dilution conditions in the atmosphere.

Principle of the CVS method

Exhaust gas emitted by the test vehicle is diluted with ambient air (10) at an average ratio of 1:5...1:10, and extracted using a special pump setup (7, 8). This ensures that the total volumetric flow, comprising exhaust gas and dilution air, remains constant. The admixture of dilution air is therefore dependent on the momentary exhaust-gas volumetric flow. A sample is continuously extracted from the diluted exhaust-gas flow and is collected in one or more (5) exhaust-gas sample bags. Filling the sample bags generally corresponds to the phases in which the test cycles are divided (e.g. the ht phase in the FTP 75-test cycle).

As the exhaust-gas sample bags are filled, a sample of dilution air is taken and collected in one or more (4) air-sample bags in order to measure the pollutant concentration in the dilution air.

The sampling volumetric flow is constant during the bag-filling phase. The pollutant concentration in the exhaust-gas sample bags at the end of the test cycle corresponds to the average value of the concentrations in the diluted exhaust gas for the sample-bag filling period. The pollutant masses emitted during the test are calculated from these concentrations and from the total air/exhaust gas mixture conveyed from the volume – taking into account the pollutants contained in the dilution air.

Dilution systems

There are two alternative methods to obtain a constant volumetric flow of diluted exhaust gas:

- Positive Displacement Pump (PDP) method: A rotary-piston blower (Roots blower) is used.
- Critical Flow Venturi (CFV) method: A venturi tube and a standard blower are used in the critical state.

Advances in the CVS method

Diluting the exhaust gas causes a reduction in pollutant concentrations as a factor of dilution. As pollutant emissions have dropped significantly in the past few years due to the growing severity of emission limits, the concentrations of some pollutants (in particular hydrogen compounds) in the diluted exhaust gas are equivalent to concentrations in diluted air in certain test phases (or are even lower). This poses a problem from the measuring-process aspect, as the difference between the two values is crucial for measuring exhaust-gas emissions. A further challenge is presented by the precision of analyzers used to measure small concentrations of pollutants.

To confront these problems, more recent CVS dilution systems apply the following measures:

- Reduce dilution: This requires precautions to avoid the condensation of water, e.g. by heating parts of the dilution system.
- Reduce and stabilize pollutant concentrations in the dilution air, e.g. by using activated charcoal filters.
- Optimize the measuring instruments (including dilution systems), e.g. by selecting or preconditioning the materials used and system setups; by using modified electronic components.
- Optimize processes, e.g. by applying special purge procedures.

Bag Mini Diluter

As an alternative to advances in CVS technology described above, a new type of dilution system was developed in the U.S.A.: the Bag Mini Diluter (BMD). Here, part of the exhaust-gas flow is diluted at a constant ratio with dried, heated zero gas (e.g. cleaned air). During the test, part of the diluted exhaust-gas flow that is proportional to the exhaust-gas volumetric flow is filled in (exhaust-gas) sample bags and analyzed at the end of the driving test. Diluting the exhaust gas with a pollution-free zero gas dispenses with air-sample bag analysis followed by taking the difference between the exhaust-gas and air-sample bag concentrations. However, a more complex procedure is required than for the CVS method, e.g. one requirement is to determine the (undiluted) exhaust-gas volumetric flow and the proportional sample-bag filling.

Testing commercial vehicles

The transient test method for testing emissions from diesel engines in heavy-duty trucks over 8,500 lb. (U.S.) or 3.5 t (Europe) is performed on dynamic engine test benches and also uses the CVS test method. This test came into force in the U.S.A. starting model year 1986 and is slated for 2005 in Europe. However, the size of the engines demands a test setup with a substantially higher throughput in order to keep to the same dilution ratios as for cars and light-duty trucks. Double dilution (by means of a secondary tunnel) permitted by law helps to minimize equipment costs.

Under critical conditions, the volumetric flow rate of diluted exhaust gas is controllable, either using a calibrated Roots blower, or venturi tubes.

Exhaust-gas measuring devices

Emission-control legislation in the EU, the U.S.A., and Japan defines standard test procedures for emission-limit pollutants in order to measure the pollutant concentrations in exhaust-gas and air-sample bags:
- Measurement of CO and CO_2 concentrations with Non-Dispersive InfraRed (NDIR) analyzers.
- Measurement of NO_X concentrations (aggregate of NO and NO_2) using ChemiLuminescence Detectors (CLD).
- Measurement of total hydrocarbon concentrations (THC) using Flame Ionization Detectors (FID).
- Gravimetric measurement of particulate emissions.

NDIR analyzer

The NDIR (Non-Dispersive InfraRed) analyzer uses the property of certain gases to absorb infrared radiation within a narrow characteristic wavelength band. The absorbed radiation is converted into vibrational and rotational energy of the absorbing molecules.

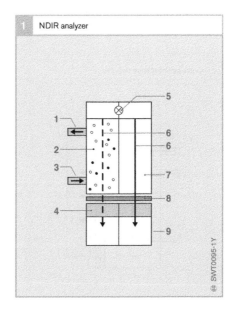

1 NDIR analyzer

Fig. 1
1   Gas outlet
2   Absorption cell
3   Test-gas inlet
4   Optical filter
5   Infrared light source
6   Infrared radiation
7   Reference cell
8   Rotating chopper
9   Detector

In the NDIR analyzer, the analysis gas flows through the absorption cell (vessel) (Fig. 1, 2) where it is exposed to infrared radiation. The gas absorbs radiation energy within the characteristic wavelength band of the pollutant, whereby the radiation energy is proportional to the concentration of the pollutant under analysis. A reference cell (7) arranged in parallel to the absorption cell is filled with an inert gas (e.g. nitrogen (N_2)).

The detector (9) is located at the opposite end of the cell to the infrared light source and measures the residual energy from infrared radiation in the measurement and reference cells. The detector comprises two chambers connected by a membrane and containing samples of the gas components under analysis. The reference-cell radiation characteristic for this component is absorbed in one chamber. The other absorbs radiation from the test-gas vessel. The difference between the radiation received and the radiation absorbed in the two detector chambers results in a pressure difference, and thus a deflection in the membrane between the measuring and reference detectors. This deflection is a measure of the pollutant concentration in the test-gas vessel.

[1] The absorption of infrared radiation within a particular wavelength band is possible not only with the gas component measured, but also with water vapor.

A rotating chopper (8) interrupts infrared radiation cyclically, causing an alternating deflection of the membrane, and thus a modulation of the sensor signal.

NDIR analyzers have a strong cross-sensitivity[1] to water vapor in the test gas since H_2O molecules absorb infrared radiation across a broad wavelength band. This is the reason why NDIR analyzers are positioned downstream of a test-gas treatment system (e.g. a gas cooler) to dry the exhaust gas when they are used to make measurements on undiluted exhaust gas.

ChemiLuminescence Detector (CLD)

Due to its measuring principle, the CLD is limited to determining NO concentrations. Before measuring the aggregate of NO_2 and NO concentrations, the test gas is first routed to a converter that reduces NO_2 into NO.

The test gas is mixed with ozone in a reaction chamber (Fig. 2) to determine the nitrogen monoxide concentration (NO). The nitrogen monoxide contained in the test gas oxidizes in this environment to form NO_2; some of the molecules produced are in a state of excitation. When these molecules return to their basic state, energy is released in the form of light (chemiluminescence).

Fig. 2
1   Reaction chamber
2   Ozone inlet
3   Test-gas inlet
4   Gas discharge
5   Filter
6   Detector

Fig. 3
1   Gas discharge
2   Collector electrode
3   Amplifier
4   Combustion air
5   Test-gas inlet
6   Combustion gas
    (H_2/He)
7   Burner

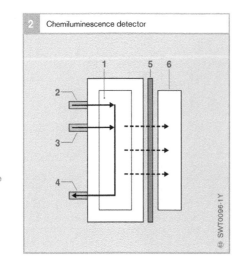

2   Chemiluminescence detector

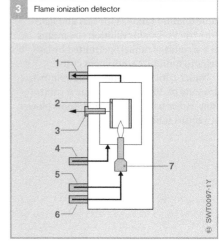

3   Flame ionization detector

A detector, e.g. a photomultiplier, measures the emitted luminous energy; under specific conditions, it is proportional to the nitrogen-monoxide concentration in the test gas.

Flame Ionization Detector (FID)

The hydrocarbons present in the test gas are burned off in a hydrogen flame (Fig. 3). This forms carbon radicals; some of the radicals are ionized temporarily. The radicals are discharged at a collector electrode; the current produced is measured and is proportional to the number of carbon atoms in the test gas.

Measuring particulate emission

A gravimetric process is a process specified by law to measure particulate emissions during type-approval testing.

Gravimetric process (particulate filter process)

Part of the diluted exhaust gas is sampled from the dilution tunnel during the driving test and then channelled through particulate filters. The quantity of particulate emissions is calculated from the increase in weight of the (conditioned) particulate filter, taking into account volumetric flow. The gravimetric process has the following disadvantages:

- Relatively high detection limit, only reducible to a limited extent by using intensive instrument resources (e.g. to optimize tunnel geometry).
- It is not possible to measure particulate emissions continuously.
- The process requires numerous resources since the particulate filter requires conditioning to minimize environmental influences.
- Only particulate mass is measured; however, it is not possible to determine the chemical composition of the particulate or particle size.

Due to the disadvantages discussed above, as well as the drastic reduction in limits for particulate emissions anticipated in future, the lawmakers are considering replacing the gravimetric process, or supplementing it in order to determine particle-size distribution or particle quantity. However, an alternative process has not yet been found.

The measuring instruments that show particulate-size distribution in exhaust gas include the following:

- Scanning Mobility Particle Sizer (SMPS)
- Electrical Low Pressure Impactor (ELPI)
- Photo-Acoustic Soot Sensor (PASS)

Exhaust-gas measurement in engine development

For development purposes, many test benches also include the continuous measurement of pollutant concentrations in the vehicle exhaust-gas system or dilution system. The reason is to capture data on emission-limit components, as well as other components not subject to legislation. Other test procedures than those mentioned are required for this, e.g.:

- GC FID and Cutter FID to measure methane concentrations (CH_4).
- Paramagnetic method to measure oxygen concentrations (O_2).
- Opacity measurement to determine particulate emissions.

Other analyses can be conducted using multicomponent analyzers:

- Mass spectroscopy
- FTIR (Fourier Transform InfraRed) spectroscopy
- IR laser spectroscopy

GC FID and Cutter FID

There are two equally common methods to measure methane concentration in the test gas. Each method consists of the combination of a CH_4-separating element and a flame ionization detector. Either a gas-chromotography column (GC FID), or a heated catalytic converter, oxidizes the non-CH_4 hydrocarbons (cutter FID) in order to separate methane. Unlike the cutter FID, the GC FID can only determine CH_4 concentration discontinuously (typical interval between two measurements: 30...45 seconds).

ParaMagnetic Detector (PMD)

There are different designs of paramagnetic detectors (dependent on the manufacturer). The measuring principle is based on inhomogenous magnetic fields that exert forces on molecules with paramagnetic properties (such as oxygen), causing the molecules to move. The movement is proportional to the concentration of molecules in the test gas and is sensed by a special detector.

Opacity measurement

An opacity meter (opacimeter) is used in development and in diesel smoke-emission testing in the workshop during exhaust-gas testing (see the section entitled "Emissions testing (opacity measurement)").

The smoke-emission test equipment (Fig. 1) used in development extracts a specified quantity of diesel exhaust gas (e.g. 0.1 or 1 l) through a strip of filter paper. As a requirement for the high-precision reproducibility of results, the volume extracted is recorded for every test sequence and converted to the standardized quantity. The system also takes account of pressure and temperature impacts, as well as dead volume between the exhaust-gas sample probe and the filter paper.

The blackened filter paper is analyzed optoelectronically using a reflective photometer. The results are generally indicated as the Bosch smoke number or mass concentration (mg/m³).

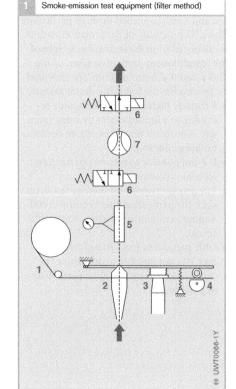

1   Smoke-emission test equipment (filter method)

Fig. 1
1   Filter paper
2   Gas penetration
3   Reflective photometer
4   Paper transport
5   Volume measuring device
6   Purge-air switchover valves
7   Pump

Emissions testing (opacity measurement)

The procedure for emissions testing in the workshop comprises the following steps for a diesel-engined vehicle:

- Identifying the vehicle.
- Visually inspecting the exhaust-gas system.
- Testing engine speed and temperature.
- Detecting the average idle speed.
- Detecting the average breakaway speed.
- Opacity measurement: Initiating at least three accelerator bursts (unrestricted acceleration) to determine exhaust-gas opacity. If opacity figures are below the limit, and all three measured values are within a bandwidth of $< 0.5\,\mathrm{m^{-1}}$, the vehicle passes the emissions test.

With effect from 2005 Germany also stipulates an on-board diagnosis as part of the emissions test.

Opacity meter (absorption method)

During unrestricted acceleration, a certain amount of exhaust gas is taken from the vehicle's exhaust pipe (without vacuum assistance), using an exhaust-gas sampling probe and a hose leading to the measuring chamber. This method avoids impacts arising from exhaust-gas backpressure and its fluctuations on test results, since pressure and temperature are controlled (Hartridge tester).

In the measuring chamber, a light beam passes through the diesel exhaust gas. Attenuation of the light is measured photoelectrically and displayed as a percentage opacity T or absorption coefficient k. High precision and good reproducibility of test results are dependent on a specific measuring-chamber length and keeping the inspection windows free from soot.

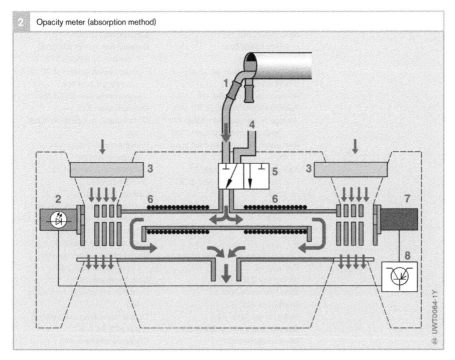

2   Opacity meter (absorption method)

Fig. 2
1   Exhaust-gas sample
    probe
2   LED
3   Fan
4   Purge air
5   Calibrating valve
6   Heater
7   Receiver
8   Evaluation
    electronics
    and display

Index of technical terms

Abbreviations

A

ABS: Antilock Braking System

AC: Air Conditioner

ACC: Adaptive Cruise Control

ACEA: Association des Constructeurs Européens d'Automobiles (Association of european automobile constructors)

ACK: Acknowledgement

A/D: Analog/Digital converter

ADA: Atmospheric-pressure sensitive full-load stop (German: Atmosphärendruckabhängiger Volllastanschlag)

ADM: Application Data Manager

ALFB: Load-dependent start of delivery with deactivation feature (German: Abschaltbarer lastabhängiger Förderbeginn)

ARD: Active surge dumping (German: Aktive Ruckeldämpfung)

ARS: Angle of Rotation Sensor

ASIC: Application Specific Integrated Circuit

ASTM: American Society for Testing and Materials

ATDC: After Top Dead Center (piston/crankshaft)

ATL: Exhaust-gas turbocharger (German: Abgasturbolader)

AU: German emissions inspection (German: Abgasuntersuchung)

AWN: Bosch workshop network

AZG: Adaptive cylinder equalization (German: Adaptive Zylindergleichstellung)

B

BDC: Bottom Dead Center (piston/crankshaft)

bhp: Brake horse power (1 bhp = 0.7355 kW)

BIP: Begin of Injection Period

BMD: Bag Mini Diluter

BTDC: Before Top Dead Center (piston/crankshaft)

C

CA: Camshaft

CA: Camshaft Angle

CAD: Computer-Aided Design

CAFE: Corporate Average Fuel Economy

CAN: Controller Area Network

CARB: California Air Resources Board

CAS[plus]: Computer Aided Service

CCRS: Current Controlled Rate Shaping

CDM: Calibration Data Manager

CDPF: Catalyzed Diesel Particulate Filter

CFPP: Cold Filter Plugging Point

CFR: Cooperative Fuel Research

CFV: Critical Flow Venturi

cks: Crankshaft

CLD: ChemiLuminescence Detector

COP: Conformity of Production

CPU: Central Processing Unit

CR: Common Rail

CRC: Cyclic Redundancy Check

CRS: Common Rail System

CRT: Continuously Regenerating Trap

CSF: Catalyzed Soot Filter

CT phase: Cold Transient Phase

CVS: Constant Volume Sampling

CN: Cetane Number

D

DCU: DENOXTRONIC Control Unit

DDS: Diesel-engine immobilizers (German: Diesel-Diebstahl-Schutz)

DFPM: Diagnosis Fault Path Management

DHK: Nozzle-and-holder assembly (German: Düsenhalterkombination)

DI: Diesel engine

DI: Direct Injection

DIN: German Institute for Standardization (German: Deutsches Institut für Normung)

DME: Dimethylether

DMV: Diesel solenoid valve (German: Diesel-Magnetventil)

DOC: Diesel Oxidation Catalyst

DPF: Diesel Particulate Filter

DSCHED: Diagnostic Function Scheduler

DSM: Diagnostic System Management

DVAL: Diagnosis Validator

DWS: Angle-of-rotation sensor (German: Drehwinkelsensor)

DZG: Speed sensor (German: Drehzahlgeber (Drehzahlsensor))

E

EC: End of Combustion

EC: Exhaust Closes

ECE: Economic Commission for Europe

ECM: Electrochemical Machining

ECU: Electronic Control Unit

EDC: Electronic Diesel Control

EDR: Maximum-speed governor (German: Enddrehzahlregelung)

EDP: Electronic Data Processing

EEPROM: Electrically Erasable Programmable Read Only Memory

EEV: Enhanced Environmentally-Friendly Vehicle

EGR: Exhaust-Gas Recirculation

EGS: Electronic transmission-shift control (German: Elektronische Getriebesteuerung)

EHAB: Electrohydraulic shutoff device (German: Elektro-Hydraulische Abstellvorrichtung)

EI: End of Ignition

EIR: Emission Information Report

EKP: Electric fuel pump (German: Elektrokraftstoffpumpe)

ELAB: Solenoid-operated shutoff valve (German: Elektrisches Abstellventil)

ELPI: Electrical Low Pressure Impactor

ELR: Electronic idle-speed control system (German: Elektronische Leerlaufregelung)

ELR: European Load Response

EMI: Injected-fuel-quantity indicator (German: Einspritzmengenindikator)

EMC: Electromagnetic Compatibility

EO: Exhaust Opens

EOBD: European OBD

EOL: End of Line

EPA: Environmental Protection Agency

EPROM: Erasable Programmable Read Only Memory

ESC: European Steady-State Cycle

ESI[tronic]: Electronic Service Information

ESP: Electronic Stability Program

ETC: European Transient Cycle

EU: European Union

EUDC: Extra Urban Driving Cycle

EWIR: Emissions Warranty Information Report

F

FAME: Fatty Acid Methyl Ester

FGB: Maximum-speed limiter (German: Fahrgeschwindigkeitsbegrenzer (Limiter))

FGR: Vehicle-speed controler (Cruise Control) (German: Fahrgeschwindigkeitsregler (Tempomat))

FID: Flame Ionisation Detector

FIR: Field Information Report

FR: First Registration

FSA: Vehicle system analysis (German: Fahrzeug-System-Analyse)

FSS: Delivery-signal sensor
(German: Fördersignalsensor)
FTIR: Fourier Transform Infrared
(spectroscopy)
FTP: Federal Test Procedure

G

GC: Gas Chromatography
GDI: Gasoline Direct Injection
GDV: Constant-pressure valve
(German: Gleichdruckventil)
GPS: Global Positioning System
GRV: Constant-volume valve
(German: Gleichraumventil)
GSK: Glow plug
(German: Glühstiftkerze)
GST: Graduated (or adjustable) start
quantity (German: Gestufte Start-
menge)
GZS: Glow plug control unit
(German: Glühzeitsteuergerät)

H

HBA: Hydraulically controlled torque
control device (German: Hydrau-
lisch betätigte Angleichung)
HCCI: Homogeneous Compressed
Combustion Ignition
HD: High-pressure
(German: Hochdruck)
HDK: Sensor with semidifferential
short-circuiting ring (German: Halb-
Differenzial-Kurzschlussringsensor)
HDTC: Heavy-Duty Transient Cycle
HDV: Heavy-Duty Vehicle
HFM: Hot-Film Air-Mass Meter
HFRR: High Frequency Reciprocating
Rig
HGB: Maximum-speed limiter
(German: Höchstgeschwindigkeits-
begrenzung)
HGV: Heavy Goods Vehicles
HLDT: Heavy Light-Duty Truck
HSV: Hydraulic start-quantity locking
device (German: Hydraulische
Startmengenverriegelung)
HWL: Urea/water solution (German:
Harnstoff-Wasser-Lösung)

I

IC: Integrated Circuit
IDE-Bit: Identifier Extension Bit
IDI: Indirect Injection
IGL: Ignition Lag
IL: Injection Lag
IMA: Injector delivery compensation
(German: Injektormengenabgleich)

INCA: Integrated Calibration and
Acquisition System
IO: Inlet Opens
IP: Injection Point
ISO: International Organization for
Standardization
IWZ: Incremental angle-time signal
(German: Inkremental-Winkel-Zeit-
Signal)

J

JAMA: Japanese Automobile Manufacturers Association

K

KMA: Electronic flow measurement
system (German: Kontinuierliche
Mengenanalyse)
KSB: Cold-start accelerator
(German: Kaltstartbeschleuniger)

L

LDA: Manifold-pressure compensator
(German: Ladedruckabhängiger
Volllastanschlag)
LDT: Light-Duty Truck
LED: Light-Emitting Diode
LEV: Low-Emission Vehicle
LFG: Idle-speed spring attached to
governor housing (German:
Leerlauffeder – gehäusefest)
LLDT: Light Light-Duty Truck
LSF: Two-point oxygen sensor Type
LSF (German: (Zweipunkt-)Finger-
Lambda-Sonde)
LSU: Broadband oxygen sensor Type
LSU (German: (Breitband-)Lambda-
Sonde-Universal)
LTCC: Low Temperature Cofired
Ceramic

M

MAB: Fuel cutout
(German: Mengenabstellung)
MAR: Control of injected-fuel-quantity
compensation (German: Mengen-
ausgleichsregelung)
MBEG: Fuel limitation
(German: Mengenbegrenzung)
MC: Microcomputer
MDA: Measure Data Analyzer
MDPV: Medium Duty Passenger Vehicle
MGT: Glass gauge method
(German: Messglas-Technik)
MI: Main Injection
MIL: Malfunction Indicator Lamp

MMA: Fuel-quantity mean-value
adaptation (German: Mengen-
mittelwertadaption)
MNEDC: Modified New European
Driving Cycle
MOSFET: Metal Oxide Semiconductor
Field Effect Transistor
MSG: Engine ECU
(German: Motorsteuergerät)
MV: Solenoid valve
(German: Magnetventil)

N

NBS: Needle-motion sensor
(German: Nadelbewegungssensor)
ND: Low pressure
(German: Niederdruck)
NDIR Analyser: Non-Dispersive
Infrared Analyser
NEDC: New European Driving Cycle
NLK: Hydraulically assisted timing
device (German: Nachlaufkolben
(-Spritzversteller))
NMHC: Non-Methane Hydro Carbons
NMOG: Non-Methane Organic Gases
NSC: NO_x Storage Catalyst
NTC: Negative Temperature Coefficient
NYCC: New York City Cycle

O

OBD: On-Board-Diagnosis
OHW: Off-Highway

P

PASS: Photo-Acoustic Soot Sensor
PCB: Printed-Circuit Board
PDP: Positive Displacement Pump
PF: Particulate Filter
pHCCI: partly Homogeneous
Compressed Combustion Ignition
PI: Pre Injection (Pilot Injection)
PLA: Pneumatic idle-speed increase
(German: Pneumatische Leerlauf-
anhebung)
PM: Particle Mass
PMD: Paramagnetic Detector
PNAB: Pneumatic shutoff device
(German: Pneumatische Abstell-
vorrichtung)
POI: Post Injection
PROF: Programming of Flash-EPROM
PSG: Pump ECU
(German: Pumpensteuergerät)
PTC: Positive Temperature Coefficient
PWM: Pulse-Width Modulation
PZEV: Partial Zero-Emission Vehicle

R

RAM: Random Access Memory

RDV: Orifice check valve
(German: Rückstromdrosselventil)

RME: Rape Seed Oil Methyl Ester

ROM: Read Only Memory

RSD: Orifice check valve
(German: Rückströmdrosselventil)

RTR: Remote Transmission Request

RWG: Control-rack travel sensor
(German: Regelweggeber)

RCP: Roller-Cell Pump

S

S phase: Stabilization Phase
(Test phase during emissions test)

SAE: Society of Automotive Engineers

SC: Start of Combustion

SCR: Selective Catalytic Reduction

SD: Start of Delivery

SE: Secondary Electron

SEM: Secondary Electron Microscope

SFTP: Supplemental Federal Test
Procedure

SI: Start of Ignition

SIS: Service Information System

SMD: Surface Mounted Devices

SME: Soya Methyl Ester

SMPS: Scanning Mobility Particle
Sizer

SRC: Smooth Running Control

SULEV: Super Ultra-Low-Emission
Vehicle

T

TA: Type Approval

TAS: Temperature-compensating
start-quantity stop (German: Tem-
peraturabhängiger Startanschlag)

TCS: Traction Control System

TDC: Top Dead Center
(piston/crankshaft)

TLA: Temperature-controlled idle-
speed increase (German: Tempera-
turabhängige Leerlaufanhebung)

TLEV: Transitional Low-Emission
Vehicle

TME: Tallow Methyl Ester

TQ signal: Fuel-consumption signal

U

UDC: Urban Driving Cycle

UDDS: Urban Dynamometer Driving
Schedule

UFOME: Used Frying Oil Methyl Ester

UI: Unit Injector

UIS: Unit Injector System

ULEV: Ultra-Low-Emission Vehicle

UP: Unit Pump

UPS: Unit Pump System

V

V_h: Swept volume of an engine
cylinder

V_H: Overall cylinder capacity of an
engine

VST: Variable Sleeve Turbine

VTG: Variable Turbine Geometry

W

WOT: Wide Open Throttle

WSD: Wear Scar Diameter

WWH-OBD: World Wide Harmonized
On Board Diagnostics

Z

ZDR: Intermediate-speed control
(German: Zwischendrehzahl-
regelung)

ZEV: Zero-Emission Vehicle

Bosch reference books – First-hand technical knowledge

Gasoline-Engine Management

Starting with a brief review of the beginnings of automotive history, this book discusses the basics relating to the method of operation of gasoline-engine control systems. The descriptions of cylinder-charge control systems, fuel-injection systems (intake manifold and gasoline direct injection), and ignition systems provide a comprehensive, firsthand overview of the control mechanisms indispensable for operating a modern gasoline engine. The practical implementation of engine management and control is described by the examples of various Motronic variants, and of the control and regulation functions integrated in this particular management system. The book concludes with a chapter describing how a Motronic system is developed.

Hardcover,
17 x 24 cm format,
2nd Edition, completely revised and extended,
418 pages,
hardback,
with numerous illustrations.

ISBN 1-86058-434-9

Contents
- Basics of the gasoline engine
- Gasoline-engine control system
- History of gasoline-engine management system development
- Cylinder-charge control systems
- Fuel supply and delivery
- Intake-manifold fuel injection
- Gasoline direct injection
- Inductive ignition system
- Ignition coils
- Spark plugs
- Motronic engine management
- Sensors
- Electronic control unit
- Electronic control and regulation
- Electronic diagnostics (on-board diagnosis, OBD)
- Data transmission between electronic systems
- Pollutant reduction
- Catalytic emission-control systems
- Emission-control legislation
- Service technology
- Electronic control unit development

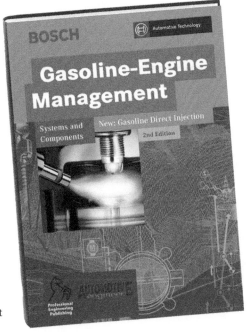

Bosch reference books –
First-hand technical knowledge

Automotive Electrics/
Automotive Electronics

The rapid pace of development in automotive electrics and electronics has had a major impact on the equipment fitted to motor vehicles. This simple fact necessitated a complete revision and amendment of this authoritative technical reference work.
The 4th Edition goes into greater detail on electronics and their application in the motor vehicle. The book was amended by adding sections on "Microelectronics" and "Sensors". As a result, the basics and the components used in electronics and microelectronics are now part of this book. It also includes a review of the measured quantities, measuring principles, a presentation of the typical sensors, and finally a description of sensor-signal processing.

Hardcover,
17 x 24 cm format,
4th Edition, completely revised and extended,
503 pages,
hardback,
with numerous illustrations.

ISBN 1-86058-436-5

Contents
- Automotive electrical systems, including calculation of wire dimensions, plug-in connections, circuit diagrams and symbols
- Electromagnetic compatibility and interference suppression
- Batteries
- Alternators
- Starters
- Lighting technology
- Windshield and rear-window cleaning
- Microelectronics
- Sensors
- Data processing and transmission in motor vehicles.

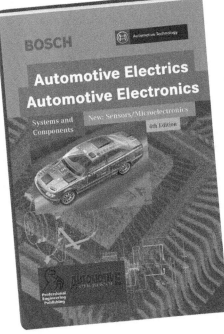

Automotive Handbook

The Bosch Automotive Handbook is a comprehensive reference work. Many of the topics featured in the 6th English edition have again been newly written, revised or updated by experts from Bosch, from the automotive industry and from universities and technical colleges.

The present state of automotive engineering is comprehensively presented in an easy-to-understand format. Many system representations, illustrations and tables provide further insight into a fascinating branch of engineering.

Topics new to the 6th Edition
Hydrostatics ● Fluid mechanics ● Mechatronics ● Coating systems ● Frictional joints ● Positive or form-closed joints ● Engine lubrication ● Emission reduction systems ● Diagnosis ● Truck brake management as a platform for truck driver assistance systems ● Analog and digital signal transmission ● Mobile information services ● Fleet management ● Multimedia systems ● Developments methods and application tools for electronic systems ● Sound design ● Vehicle wind tunnels ● Environmental management ● Service technology.

Completely updated, revised or extended topics
Basic equations used in mechanics ● Threaded fasteners ● Springs ● Air filtration ● Engine lubrications ● Engine cooling ● Turbochargers and superchargers ● Exhaust-gas system ● Exhaust-gas measuring techniques ● Engine management for spark-ignition (SI) engines (including Motronic engine management) ● Alternative spark-ignition engine operation ● Diesel-engine management ● Driving-stabilization systems ● Car radio with auxiliary equipment ● Vehicle antennas ● Mobile and data radio ● Passenger-car driver-assistance systems.

The "Blue Book" from Bosch has become a bestseller in the field of technical literature. All over the world it is t h e reference work for exact and concise information on the subject of automotive engineering.

This edition is also available in German, French and Spanish; previous editions have also been published in Dutch, Russian, Finnish, Italian, Hungarian, Chinese and Japanese.

Format 12 x 18 cm,
6th Edition, completely revised and extended
1,232 pages,
bound,
with numerous illustrations.

ISBN 1-86058-474-8

Printed and bound in the UK by
CPI Antony Rowe, Eastbourne

Printed and bound by CPI Group (UK) Ltd, Croydon, CR0 4YY

27/10/2024

14580149-0001